First Latin-American Conference on

HIGH TEMPERATURE SUPERCONDUCTIVITY

SERIES ON PROGRESS IN HIGH TEMPERATURE SUPERCONDUCTIVITY

Published

Vol. 1 — Proceedings of the Adriatico Research Conference on High Temperature Superconductors
(*eds. S. Lundqvist, E. Tosatti, M. Tosi and Yu Lu*)

Vol. 2 — Proceedings of the Beijing International Workshop on High Temperature Superconductivity
(*eds. Z. Z. Gan, G. J. Cui, G. Z. Yang and Q. S. Yang*)

Vol. 3 — Proceedings of the Drexel International Conference on High Temperature Superconductivity (*eds. S. M. Bose and S. D. Tyagi*)

Vol. 4 — Proceedings of the 2nd Soviet-Italian Symposium on Weak Superconductivity (*eds. A. Barone and A. Larkin*)

Vol. 5 — Proceedings of the IX Winter Meeting on Low Temperature Physics — High Temperature Superconductors
(*eds. J. Heiras, R. A. Barrio, T. Akachi and J. Taguena*)

Vol. 7 — Chemical and Structural Aspects of High Temperature Superconductors
(*ed. C. N. R. Rao*)

Vol. 8 — World Congress on Superconductivity
(*eds. C. G. Burnham and R. D. Kane*)

Vol. 14 — Towards the Theoretical Understanding of High T_c Superconductors
(*eds. S. Lundqvist et al.*)

Progress in High Temperature Superconductivity — Vol. 9

Proceedings of the I LACHTS

First Latin-American Conference on

HIGH TEMPERATURE SUPERCONDUCTIVITY

4 — 6 May 1988, Rio de Janeiro, Brazil

Editors

Roberto Nicolsky
Instituto de Fisica
Universidade Federal do Rio de Janeiro
Rio de Janeiro, Brazil

Rafael A. Barrio
Instituto Investigaciones en Materiales
Universidade Nacional Autonoma de Mexico
Mexico D.F., Mexico

Oscar Ferreira de Lima
Instituto de Fisica Gleb Wataghin
Universidade Estadual de Campinas
Campinas, Brazil

Roberto Escudero
Instituto de Investigaciones en Materiales
Universidade Nacional Autonoma de Mexico
Mexico D.F., Mexico

World Scientific
Singapore • New Jersey • London • Hong Kong

Published by

World Scientific Publishing Co. Pte. Ltd.
P O Box 128, Farrer Road, Singapore 9128

USA office: World Scientific Publishing Co., Inc.
687 Hartwell Street, Teaneck, NJ 07666, USA

UK office: World Scientific Publishing Co. Pte. Ltd.
P O Box 379, London N12 7JS, England

FIRST LATIN-AMERICAN CONFERENCE ON HIGH TEMPERATURE SUPERCONDUCTIVITY

ISBN 9971-50-645-9
ISBN 9971-50-646-7 (pbk)

Printed in Singapore by General Printing Services Pte. Ltd.

FOREWORD

The I LACHTS — Latin-American Conference on High Temperature Superconductivity started a new series of high level conferences concerned with new superconductors to be held in Latin-American countries.

The main aim of the initiative is to provide a periodic update on the high-T_c superconductors for the Latin-American researchers and, in addition, to enable them to present their work on the subject. The Proceedings of the I LACHTS contains 28 invited papers by leading scientists covering the essential topics on the new superconductors, and 47 contributed papers generated in our region showing a wide range of theoretical and experimental investigations on these materials.

We hope this volume may contribute to a better understanding of the exciting field of high temperature superconductivity.

The Editors

I LACHTS - First Latin-American Conference on
High Temperature Superconductivity

4 - 6 May, 1988, Rio de Janeiro, Brazil

NATIONAL ORGANIZING COMMITTEE

Armando Paduan, USP, São Paulo
Elisa Baggio Saitovitch, CBPF, Rio de Janeiro
Erich Meyer, UFRJ, Rio de Janeiro
Heitor C. Basso, USP, São Carlos
Oscar F. de Lima, UNICAMP, Campinas
Paulo Pureur, UFRGS, Porto Alegre
Ramayana Gazzinelli, UFMG, Belo Horizonte
Sergio M. Rezende, UFPe, Recife
Roberto Nicolsky, UFRJ, Rio de Janeiro - Chairman

LATIN AMERICAN ADVISORY COMMITTEE

Francisco de la Cruz, CAB, Bariloche, Argentina
Miguel Kiwi, PUC, Santiago, Chile
Miguel Octavio, IVIC, Caracas, Venezuela
Roberto Escudero, UNAM, México D.F., Mexico
Virgilio Niño, UN, Bogotá, Colombia

SPONSORS

SBF - Sociedade Brasileira de Física
CLAF - Centro Latino Americano de Física

CNPq - Conselho Nacional de Desenvolvimento Científico e Tecnológico
FINEP - Financiadora de Estudos e Projetos
FAPERJ - Fundação de Amparo a Pesquisa do Estado do Rio de Janeiro
FAPESP - Fundação de Amparo a Pesquisa do Estado de São Paulo

IBM Brasil
VARIG Brazilian Airline - Official Carrier

CONTENTS

CONTRIBUTED PAPERS

INVITED PAPERS

PROGRESS IN UNDERSTANDING HIGH-TEMPERATURE SUPERCONDUCTIVITY

K. Alex Müller

IBM Research Division, Zurich Research Laboratory, 8803 Rüschlikon, Switzerland

The discovery of high-T_c superconductivity[1)] initiated three main thrusts of activity: the search for new compounds, the experimental analysis of those found, and the proposal of a relatively large number of theoretical models.

In the first part of the Lecture, the structure of the newest layered compounds was presented. They contain either Bi- or Tl-oxide layers sandwiched between the copper-oxide layers. The latter consist of either corner-connected octahedra, pyramids or planar squares: the transition temperature T_c increases with the density of copper-oxide layers per unit cell and is accounted for by single-electron interlayer tunneling according to Schneider and Baeriswyl[2)].

The existence of Cooper pairs was discussed next. Their short coherence lengths ξ, especially along the axis perpendicular to the copper-oxide sheets, was emphasized. An immediate consequence is the very high critical fields H_{c_2} found and the critical phenomena present near T_c. A further result of the extremely small Ginzburg-Landau parameter $\kappa = \xi / R_L$, where R_L is the London penetration depth, is the observation of many very sharp microwave-absorption spikes in single crystals of $BaY_2Cu_3O_7$. As a function of an external magnetic field, they are about two milliGauss wide and equally spaced by some 250 milliGauss. This spacing is narrowest along [110] directions and proportional to $1/\cos\theta$, where θ is the angle between the magnetic field and the [110] direction. They result from microwave excitation of fluxons trapped but mobile in [110] twin boundaries[3)]. The existence of a superconducting glass state and both its static and dynamic properties were also discussed.

The main part of the review closely followed the Nobel Lecture[4)]. An important part of the efforts is aimed at determining the superconducting gap Δ and its anisotropy. Some measurements of infrared reflectivity on single crystals and nuclear magnetic resonance in powders indicate that the ratio $2\Delta / kT_c \gtrsim 7$, whereas others yield values near the BCS weak coupling limit of 3.5. Determination of this ratio is important, because it indicates whether the coupling is weak or strong. It was also pointed out that the tunneling experiments yield gaps which are somewhat too low because Δ is depressed at the surface owing to the intrinsic short coherence lengths[5)].

The electronic properties of the new copper-oxide superconductors are of relevance to infer which types of mechanisms are operative. Clear evidence for the presence of holes in oxygen 2p orbits was reviewed especially from X-ray absorption. The anisotropy of resistivity and its temperature dependence along the c-axis are a quite crucial test for the proposed resonant valence bond theory. Important insight stems from the magnetic properties. Both neutron scattering and NMR prove the existence an antiferromagnetic phase for the insulating La_2CuO_4 and $BaY_2CuO_{6.4}$ compounds, their magnetic correlations being nearly two-dimensional. Upon introduction of holes, by doping with Sr^{2+} in the former and by oxidizing in the latter compounds, the AFM Néel temperature T_N is strongly depressed, giving way to a spin-glass state followed by a state showing superconductivity. The phase diagram described can be accounted for by *magnetic* coupling of the O^- spin with the nearest antiferromagnetically correlated Cu^{2+} spins[6].

REFERENCES

1. Müller, K. A. and Bednorz, J. G., Science 237, 1133 (1987).

2. Schneider, T. and Baeriswyl, Z. Phys. B (1988) submitted.

3. Blazey, K. W., Portis, A. M., Müller, K. A. and Holtzberg, F. H., Europhys. Lett. 6, 457 (1988).

4. Bednorz, J. G. and Müller, K. A., "Perovskite-type oxides – the new appproach to high-T_c superconductivity," in: *Les Prix Nobel en 1987* (Nobel Foundation, Stockholm, 1988) in press.

5. Deutscher, G. and Müller, K. A., Phys. Rev. Lett. 59, 1745 (1987).

6. Aharony, A., Birgenau, R. J., Coniglio, A., Kastner, M. A. and Stanley, H. E., Phys. Rev. Lett. 60, 1330 (1988).

CONSEQUENCES OF A VERY SHORT COHERENCE LENGTH IN HIGH-T_c SUPERCONDUCTORS

F. de la Cruz, L. Civale and H. Safar
Centro Atómico Bariloche-Instituto Balseiro
8400 Bariloche, Argentina

The short coherence length of high T_c superconductors allows for a low energy cost of the spatial modulation of the superconducting order parameter at zero applied field. We believe that the order parameter can reach its maximum value at regions of the sample where the stoichiometry and lattice structure are ideal. Experiments on deoxygenated samples clearly show that superconductivity is strongly reduced in regions encircling superconducting islands. Superconductivity between islands is induced at temperatures well below the critical one. Remanent magnetic moment experiments indicate the existence of two types of superconducting materials. It is shown that the reversible flux magnetization near T_c is a consequence of the flux expulsion from the non interacting superconducting islands. The reduction of oxygen content in less than one percent reduces the Hc1 of the superconducting islands in more than a factor of 30.

The small coherence length, $\xi(0)$, of the new high T_c superconductors (1) can be taken (2) as an intrinsic property. In these superconductors the distance between pairs is estimated (2) to be of the order of $\xi(0)$. As a consequence, the inhomogeneous character (2,3) of their superconducting properties can be understood considering that its superconducting order parameter is fully developed in regions of the sample where the stoichiometry as well as the lattice structure are ideal and strongly depressed in regions where those conditions are not fulfilled.

In traditional superconductors, on the contrary, the distance between pairs is smaller than the coherence length. In this case, non ideal material conditions confined to regions smaller than the coherence length, do not change the homogeneous characteristics of the superconducting state at zero field. The critical temperature will be eventually depressed but the superconducting transition will remain sharp.

It was shown (3) that the flux expulsion of oxygenated $La_{1.8}Sr_{.2}CuO_{4-y}$ is characteristic of two different superconducting shielding currents flowing in the material. At low temperatures and very low external fields, Ha, the induction field, B, is a parabolic function of Ha (the induction is defined as $B = \phi/A$, where ϕ is the magnetic flux through the sample geometrical cross section, A). At higher fields, B changes linearly with Ha. The slope dB/dH indicates (3) that a minimum of 50% of the sample remains in the Meissner state and the deviation of linearity at higher fields is assumed to be due to vortices penetration in that portion of the sample. Defining Hc1 as the field where deviations from linearity are evident we have obtained, for a fully oxygenated sample (2,3) the results shown in fig. 1.

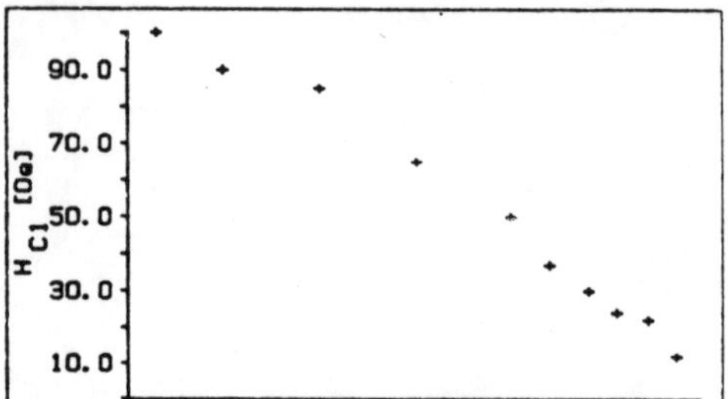

Fig.1.- Lower critical field, H_{C1}, of an oxygenated sample, obtained as explained in the text.

The parabolic dependence in the low field region is (3) typical of a critical state, well described by the Bean model (4). Supporting this picture it was found (2,3) that the flux penetration into the sample is not reversible with temperature, even at fields as low as 1 mOe. There is no linear B dependence at the lowest accessible field region, indicating the lack of a Meissner state in the whole sample. Therefore, there are portions of the sample that have a zero Hc1 or, if finite, smaller than 1 mOe. This statement points out the difference between the two approaches used to model the superconducting behavior. One model assumes (5) that superconductivity is nucleated in islands, at a temperature Tco. At lower temperatures the superconductivity is induced in the rest of the sample, by some proximity effect. In this case the minimum Gibbs free energy, at low enough fields, corresponds to a Meissner state. Within the model the order parameter results to be inhomogeneous but finite everywhere in the sample. The other model, suggested by Bednorz and Müller (6), corresponds to a spin glass like picture (7). The superconductivity is also nucleated in islands at Tco, although the order parameter remains identically zero in regions encircled by superconducting paths. The ground state in the presence of a field is degenerate due to the disorder induced by the field, in the multiple connected superconducting network.

From our point of view there is not enough experimental evidence to support one or the other picture. The time dependence of the magnetization (8) can be explained either by flux transfer within the superconducting network in the glass model or, alternatively by flux creep of superconducting vortices, once some extremal low Hc1 is exceeded. Up to now the discussion on the difference between one or the other point of view is based on quantitative arguments rather than in qualitative ones. The experimental phase diagram (6) separating the reversible (ergodic) from the non-reversible region has been used (6) as an argument in favor of the glass model (7). Nevertheless, this result has recently also been found (9) compatible with the flux creep model.

The specific heat, the magnetization and the electrical resistivity measurements, performed on the oxygenated material, have shown (10) that the specific heat anomaly occurs somewhere above Tco. The superconducting magnetic shielding is significant (10) only below the temperature where the resistivity becomes zero, that is, at the temperature where the superconducting percolation through the sample occurs. The analysis of the specific heat, magnetization and resistivity data clearly shows that superconductivity in the oxygenated (2,3) material is nucleated in islands, occupying (3) more than 50% of the sample. Since the resistivity is finite at Tco, the superconducting islands have to be surrounded by regions where the order parameter is zero. The percolation at lower temperatures is induced by superconducting tunneling among islands.

In a previous work (2) it has been shown that deoxygenation induces strong modifications in the electrical transport properties and in the width of the superconducting transition. In fig. 2 we reproduce those results, showing the resistivity as a function of temperature, for different oxygen concentrations. The whole resistive increase is due to a decrease in oxygen concentration of less than 1%.

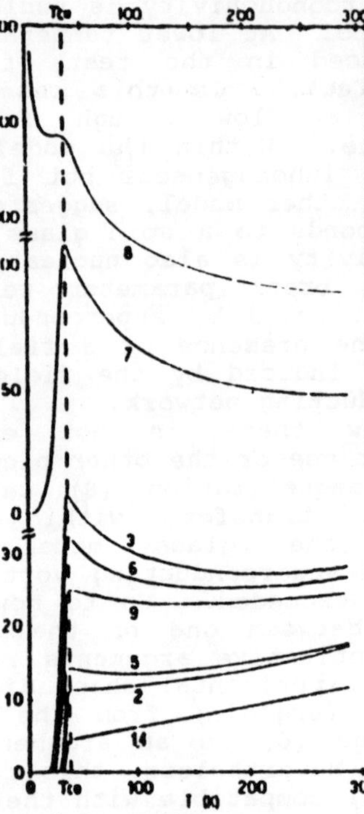

Fig.2.- Resistivity as a function of temperature for successive heat treatments at 770 K. The deoxygenated sample studied in this work has a resistivity indicated by curve A in the figure. For more details see ref.2

The change in the resistivity is reversible with oxygen concentration. As a consequence, oxygen depletion turns out to be a convenient tool to study the transport properties in the normal state as well as the percolating superconducting behavior, below T_c. The results are consistent with the presence of superconducting islands, nucleated at Tco, inducing superconductivity in links interconnecting the islands. Nevertheless, the electrical resistivity by itself cannot provide information neither about the superconducting connectivity below the percolation threshold nor about the identification of the sample regions affected by the oxygen depletion. Within the link bond picture, the increase of the transition width could be due either to a deeper depression of the order parameter in the links, leaving the superconducting islands without modification, or due to a link dilution or link lengthening caused by a decrease of the superconducting sample fraction, or both.

In this work we do not provide evidence to decide if the glass like model is appropiate to describe the superconducting behavior of ceramics but we provide new experimental data, imposing conditions that realistic models should take into account.

The results discussed here have been obtained in the sample of $La_{1.8}Sr_{0.2}CuO_{4-y}$ used in previous experiments (2,3). Its preparation technique and experimental details have been already described (2). The magnetization measurements were performed using a SQUID and the ambient field was reduced to 20 μOe.

One of the points we study in this work is the effect induced by oxygen depletion in the magnetization. In principle, the magnetization behavior as a function of field and temperature, should be able to indicate if the oxygen is leaving only from localized regions or from everywhere in the sample. Taking this goal into account we decided to study the magnetization behavior of a deoxygenated sample, in field cooling (FC) and zero field cooling (ZFC) experiments. In the FC measurements the sample is cooled in a presence of Ha, while the flux change is measured. In the ZFC case the sample is cooled in the lowest remanent field (20 μOe in our case) to the desired temperature, the field is then applied, the temperature is raised and the change of flux is recorded. The results obtained provide unexpected insight on the macroscopic superconducting properties of these materials.

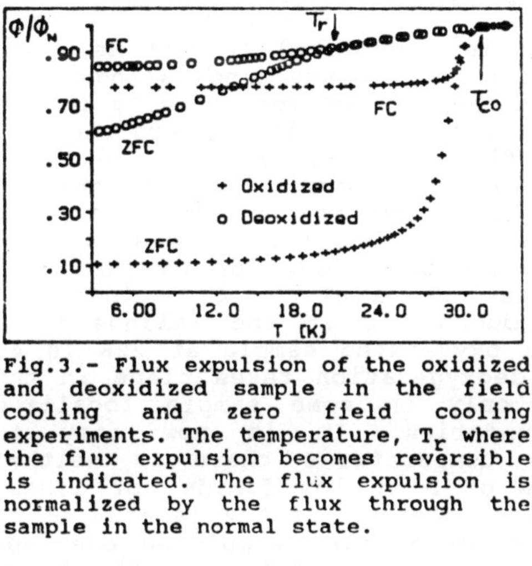

Fig.3.- Flux expulsion of the oxidized and deoxidized sample in the field cooling and zero field cooling experiments. The temperature, T_r where the flux expulsion becomes reversible is indicated. The flux expulsion is normalized by the flux through the sample in the normal state.

Figure 3 shows the flux expulsion as a function of temperature for the oxidized sample and for the same sample after a deoxygenation corresponding to curve 9 in fig.2. It is evident that the reversible temperature region (extended from Tr to Tco) has been noticeably enlarged in the deoxygenated material. It is also obvious that in the deoxygenated sample the shielding capability and flux expulsion, in the whole temperature range, has been reduced.

Contrary to what is found (3) at low temperatures in the oxygenated materials, the flux expulsion is neither parabolic, at low fields, nor linear at higher ones.

It is interesting to mention that the flux expulsion in the reversible temperature region scales with field up to 1.5 Oe. Although we defer to a future publication the elaboration of these data, we consider that it is a relevant result, since temperature reversibility and field scaling are necessary conditions to define the Meissner state in some isolated regions of the sample volume.

The resistive transition was measured in the same apparatus and in the same experiment where the flux expulsion data was obtained. Figure 4 shows the resistive transition together with the flux expulsion. It is seen that the superconducting percolation occur at almost the same temperature Tr, above which flux scaling and reversi-

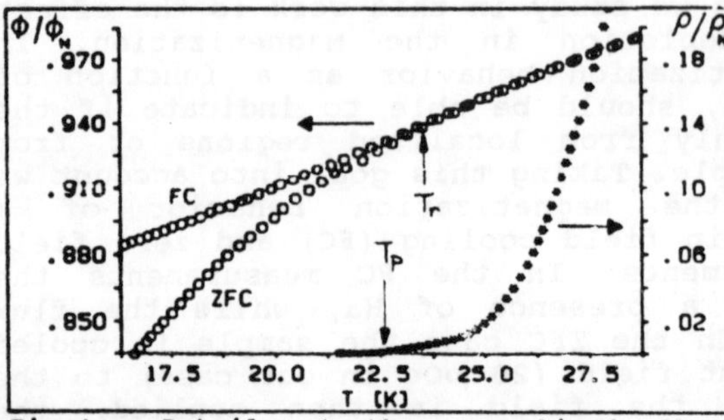

Fig.4.- Detail of the normalized flux expulsion, see fig.3, showing the reversible region together with the normalized resistivity near the superconducting percolation temperature.

bility is observed. This is a noticiable result providing useful information to understand the magnetic behavior of the oxide superconductors. The lack of superconducting percolation above Tr indicates that the superconducting islands remain essentially isolated. As a result the reversible flux expulsion is associated to the Meissner state of practically isolated superconducting islands. The reversibility results from the lack of connectivity in the superconducting network and not from its reversible (6) intrinsic properties. At fields higher than 1.5 Oe there is neither flux scaling, nor reversible temperature behavior. Since the reversibility is associated to the Meissner state of the superconducting islands the irreversibility indicates the break down of the Meissner state at fields as low as 1.5 Oe at Tr. This indicates that oxygen depletion reduces Hc1 in the islands by a factor of 30 (Hc1 of the oxygenated sample at 22K is 40 Oe). It also shows that deoxygenation takes place in the whole material and not only in some sample localized region, i.e. grain boundaries. It is now easy to understand why at low temperatures there is neither parabolic flux dependence of B at low fields nor linear dependence at higher fields. The Hc1 of the islands is so much depressed that its Meissner flux expulsion overlaps with the flux variation due to the critical state of the links.

It is possible to make other experiments to cross check the validity of our assumptions concerning the topological distribution of links and islands. Following our thoughts and the experimental results shown in fig.3 and 4, we should expect trapped flux into the sample at low temperatures, once the external field is switched off. This flux will appear as a temperature dependent remanent moment. Since we assume that there are superconducting islands and links, the moment will be associated to trapped flux due to current induced in the links and, eventually, to vortices pinned into the islands. This suggests different ways to induce the magnetic moment. If the sample is cooled in zero field and a field, lower than Hc1, is later switched on and off the remanent moment will be due to flux trapped in the links, while the islands will remain in the Meissner state. If the field exceeds Hc1 the remanent moment will be increased by the vortices contribution. The moment induced after zero field cooling will be called ZFCM. Another way to induce a remanent moment is to switch off the field, after cooling the sample in the presence of Ha. The moment obtained using this procedure will be called FCM. The FCM will be due to currents in the links plus the contribution of the vortices, nucleated and pinned into the islands, after crossing the phase diagram when cooling. Following these arguments, the magnetic moment should keep memory of the procedure followed when induced. As a consequence the temperature dependence of the remanent moment will depend on the history of the experiment. The contribution of the superconducting links to the magnetization has to be essentially zero at the superconducting percolation temperature, 22K, see figure 4. Consequently, when Hc1 has not been reached, the ZFCM should disappear at that temperature. On the other hand, the remanent moment associated to vortices pinned into the grains should go to zero at Tco.

The ZFCM induced by fields lower than Hc1 can loose their memory, when the currents in the links induce a critical behavior (3) in the whole sample. If the field is low enough the flux will penetrate only partially into the sample, due to the critical currents of the links near its surface. If the field is increased, without reaching the Hc1 of the islands, more flux will penetrate and the remanent moment will increase. This behavior will proceed until the applied field is big enough to surpass the links critical current in the whole sample. Then the remanent moment will be quenched in a value determined by the sample critical state. The remanent moment will result field independent up to the Hc1 of the islands. At this and higher fields, the moment should increase again with the external field, due to vortices induced in the superconducting islands. Figure 5 shows ZFCM and the FCM induced by a field of 1OE, as a function of temperature. It is evident that the ZFCM is much smaller than the FCM, showing clear evidence of the remanent moment memory.

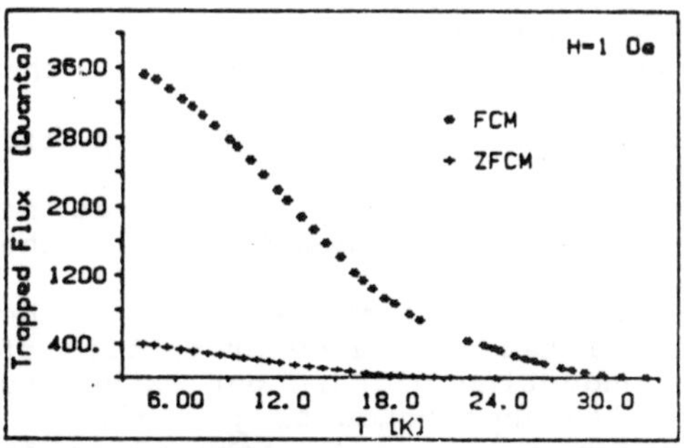

Fig.5.- Trapped flux at zero applied field as a function of temperature in the deoxidized sample. The magnetic moments (ZFCM and FCM) as defined in the text were induced by a field of 1 Oe.

In fig. 6 it is plotted, in an expanded scale, the ZFCM as a function of temperature, for different moment inducing

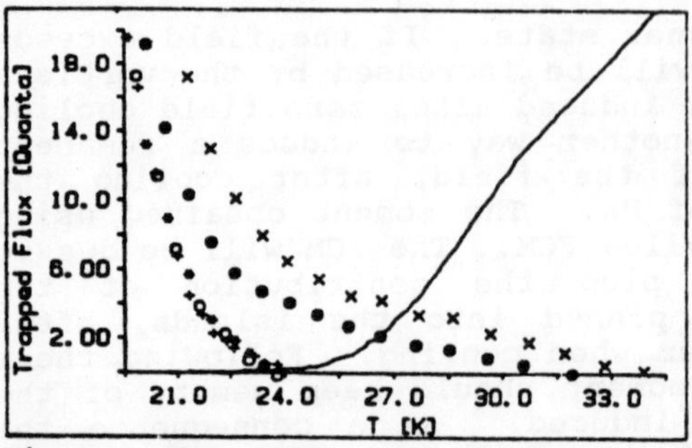

Fig.6.- The ZFCM (see text) as a function of temperature in an expanded scale for different inducing fields. +, 0.2 Oe; *, 0.4 Oe; o, 1 Oe; ●, 2 Oe; x, 5 Oe. The solid line shows the resistive transition (arbitrary units) near the superconducting percolation.

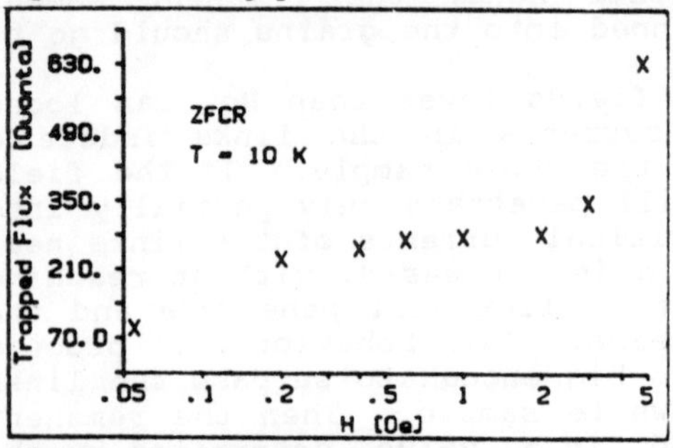

Fig.7.- ZFCM (see text) at 10K as a function of inducing field.

fields. In the same figure we plot the resistive transition. When the flux trapped is induced by fields lower than the lower critical field of the islands the ZFCM is seen to go to zero at Tr, as expected. When the field is switched off from a value higher than Hc1 the ZFCM tends to zero at much higher temperatures, very close to Tco. The influence of the field in the ZFCM is shown in fig. 7: At low fields the remanent moment increases with Ha, then it saturates, as expected from the previous discussion. When Ha becomes larger than Hc1 the vortices penetrate into the islands, and the ZFCM increases again with field.

In conclusion we have shown that:

- Sample deoxygenation modifies the topology of the superconducting network
- The grain-bond model correctly describes the magnetic flux response of the material.
- The grains or superconducting islands should not be associated only to ceramic grains, as demonstrated by the magnetic response to sample deoxygenation.
- The reversible flux expulsion temperature range in the deoxidized material is not related to glassy properties of a multiple connected superconducting picture but rather to the Meissner state of decoupled superconducting islands.
- Since the oxygen depletion is shown to take place in the whole sample, the suggestion (2) that the electrical resistivity behavior is characteristic of the bulk of the material is reinforced.
- The field Hc1 of the superconducting islands is decreased by a large factor when reducing oxygen concentration. Preliminary results in a more deoxygenated sample indicate no Meissner grain expulsion in the whole

temperature range, for fields as low as 0.50 Oe. The origin of this interesting behavior is not known.

The results we have discussed strongly suggest the convenience of extending similar measurements to samples partially oxidized. This will allow to follow the correlation between the high temperature reversible behavior and the percolation region. If the glassy model is correct the percolation temperature should increase, with oxygen, faster than the one determining the reversible region. It will also allow to study the increase of the lower critical field with oxygen. A careful comparison of the flux expulsion behavior with resistivity measurements, might relate this unexpected behavior with the peculiar transport properties showing (2) metallic behavior at high temperatures, localization when the temperature is decreased and superconductivity at the lowest ones.

References

1) J.G. Bednorz and K.A. Müller, Z. Phys. B 64, 189 (1986); C.W. Chu, P.H. Hor, R.L. Meng, L. Gao, Z.J. Haung and Y.O. Wang, Phys. Rev. Lett. 58, 405 (1987).

2) E. Osquiguil, L. Civale, R. Decca, H. Safar, E. Martinez, G. Nieva and F. de la Cruz, Materials Research Society 1987, Fall Meeting, Boston, MA, to be published; E. Osquiguil, R. Decca, G. Nieva, L. Civale and F. de la Cruz, Solid State Commun. 65, 491 (1987).

3) L. Civale, H. Safar, F. de la Cruz, D.A. Esparza and C.A. D'Ovidio, Solid State Commun. 65, 129 (1987); F. de la Cruz, L. Civale, E. Osquiguil, H. Safar and R. Decca "IX Winter Meeting on Low Temperature Physics High-Tc Superconductors", Tequesquitengo, Mexico.

4) C.P. Bean, Phys. Rev. Letters 8, 250 (1962).

5) John R. Clem, International Conference on High Temperature Superconductivity, Interlaken, Switzerland, Feb. 20-March 4, 1988) and references therein.

6) K.A. Müller, M. Takashige and J.G. Bednorz, Phys. Rev. Lett. 58, 1143 (1987).

7) C. Ebner and D. Stroud, Phys. Rev. B 31, 165 (1985); I. Morgenstern, K.A. Müller and J.G. Bednorz, Z. Phys. B 69, 33 (1987).

8) A.C. Mota, A. Pollini, K.A. Müller and J.G. Bednorz, Phys. Rev. B 36, 401 (1987); M. Tuominen, A.M. Goldman and M.L. Mecartney, Phys. Rev. B 37, 548 (1988).

9) Y. Yeshurun and A.P. Malozemoff, Phys. Rev. Lett, to be published.

10) G. Nieva, E. Martinez, F. de la Cruz, D.A. Esparza and C. D'Ovidio, Phys. Rev. B 36, 8780 (1987).

A REVIEW OF CURRENT THEORIES OF HIGH-T_c SUPERCONDUCTORS

T. M. RICE
Theoretische Physik, ETH-Hönggerberg
8093 Zürich, Switzerland.

ABSTRACT

At present there is no consensus on the key question whether high-T_c superconductivity is due to weak attraction in a Fermi liquid or whether it arises from a new type of quantum liquid. This difference appears also in different views of the underlying electronic structure of the Cu-oxides. The arguments in favor of superconductivity as a property of a Resonant Valence Bond liquid obtained by doping a Mott insulator will be emphasized.

1. INTRODUCTION

The discovery of high temperature superconductivity in several Cu-oxide systems has caused great excitement and has led to a great many proposals and models to explain these remarkable superconductors. In a single talk it is absolutely impossible to attempt a survey of even the most popular models that are currently being actively pursued - there have been so many publications in the past one and one half years. Further there is still no consensus among the experts as to which model represents the correct starting point. This was very clear at the meeting this past spring in Interlaken[1)] where a wide range of theoretical models were presented. So this talk will be a brief and very incomplete survey and from the standpoint of a theorist who favors the strongly correlated models. This survey will be based on this viewpoint and will do no more than indicate how and where the other models depart from this viewpoint.

Although the microscopic theory remains controversial there is a general consensus that the macroscopic theory is not too different if at all, from conventional superconductors. The observation of flux quantization with charge 2e[2)] and of Andreev scattering[3)] leads naturally to the view that the superconducting state must have a valid description in terms of Cooper electron pairs. The observation of an Abrikosov flux lattice[4)] which is a consequence of the Ginzburg-Landau equations also points to a strong similarity to conventional superconductors. So one concludes that a total break with the past is not called for. That is not to say that the order parameter is simply that of a conventional s-wave singlet superconductor. The symmetry of the superconducting order parameter is notoriously hard to determine because of the lack of experimental probes which couple directly to it, witness the continuing controversy over the symmetry of the superconducting state of the heavy electron metals.

In Fig. 1 a plot of the maximum value of T_c vs. time is made. A quick glance at this curve is sufficient to show that a radical break with the past has been made. Indeed if one remembers that underneath the curve are countless other compounds and alloys which had smaller values of T_c, it is even more apparent that T_c is determined by a new set of rules and hence by a new mechanism. Further if one accepts that the Cu-oxides, as members of the more general class of materials, namely oxides and in particular transition metal oxides with partly filled 3d-shells, belong to a class of materials characterized by very strong interactions,both electron-electron and electron-phonon then it is clear that this breakthrough in T_c has been achieved not by making more metallic or more itinerant metals but by finding a strong coupling system with a superconducting groundstate. I mean strong coupling here,not in the usual sense in the theory of superconductivity,but rather in the sense that the carrier bandwidth is not dominant with respect to the interactions. Rather the very many different types of ground states (almost all insulating) that 3d-transition metal oxides display

Fig. 1: The development of the maximum value of the superconducting transition temperature, T_c with time up to present. [Courtesy of O. Fischer]

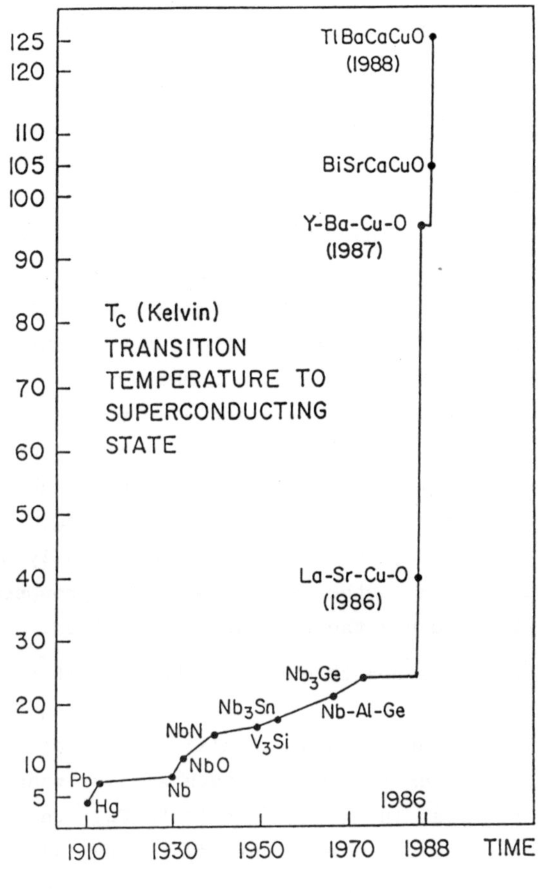

point to interacting strengths exceeding the carrier bandwidths. This point was implicit in Anderson's paper[5)] which was written very soon after the first system was discovered.

2. EFFECTIVE HAMILTONIAN FOR CuO_2-PLANES

There is naturally a clear connection between the overall view of the electronic structure of the CuO_2-planes and the type of effective Hamiltonian that one obtains to describe the low energy states. We know now that stoichiometric CuO_2-planes (formal valence Cu^{2+}) form a nonconducting antiferromagnetically (AF) ordered structure[6]. This can be interpreted in two ways. Either one says that this AF-structure is a consequence of the special bandstructure of a half-filled tight-binding model and that doping the planes (i.e. making $Cu^{(2+\delta)+}$, with $\delta > 0$) leads to a destruction of this special nesting structure and therefore to a transition to a paramagnetic Fermi liquid state. From this starting point one looks for an unusually strong attractive interaction in this Fermi liquid and then uses a BCS-theory to obtain superconductivity. There are a number of proposals for the source of the attraction, for example in the papers by Friedel and by Fulde at Interlaken[1] the electron-phonon interaction is stressed while also at the conference Abrahams (see also Varma's paper here), Ginzburg and Weber proposed an excitonic mechanism involving the interchange of a low lying electronic excitation. On the other hand Schrieffer proposed that AF spin fluctuations cause the necessary attraction.

However there is reason to question the assumption that an itinerant approach applies, if one takes as a working definition of a Mott insulator that it is an insulator in which the energy scale of the magnetic excitations is much smaller than that of the charge excitations. In stoichiometric La_2CuO_4 the magnetic energy scale[7] is $\sim$ 0.1eV whereas the lowest optical absorption[8] is around 2eV with a strong additional absorption edge around 4eV. Therefore the conditions for a Mott insulator would seem to be fulfilled. The effect of doping then should not be to destroy the local moments but rather to disorder the local moments. There are three sets of experiments which seem to confirm this point of view, namely the Raman scattering experiments of Lyons et al[9], the neutron scattering experiments of Birgeneau et al[10] and the infrared optical experiments of Thomas, Orenstein et al[11]. These experiments reduced to their simplest interpretations, show that there are low energy spin excitations in the doped samples which are superconducting, and that there is strong coupling between the holes which carry the charge and the spin excitations. It seems logical then that is these strong interactions which cause the superconductivity.

However there are two models which are consistent with the viewpoint. They are distinguished by the type of O-orbital the holes go into. The choice is whether the holes go into the 2p-orbitals oriented perpendicular to the Cu-O-Cu directions, or into the antibonding 2p-orbitals oriented parallel to the Cu-O-Cu directions. The former are non-bonding orbitals with respect to the $d_{x^2-y^2}$ orbital of the Cu^{2+} ion. Birgeneau, Kastner and Aharony[12] have proposed this choice using a ionic argument namely that these orbitals are oriented away from the positive charges and towards a region of highly negative charge. A series of cluster calculations by Guo,Langlois and Goddard[13] favor this choice. In this case the holes would move simply in a Bloch band formed from these perpendicular orbitals and the interaction with the Cu^{2+} spin-1/2 moments comes through potential exchange and is ferromagnetic. This leads to an effective Hamiltonian

$$H_{BKA} = \sum_{\substack{<\ell\ell'> \\ \sigma}} t_{O-O}\, p^{+}_{\ell\perp\sigma}\, p_{\ell'\perp\sigma} + \sum_{<i,\ell>} J_{Cu-O}\, \vec{S}_i \cdot \vec{S} + \sum_{<ij>} J\, \vec{S}_i \cdot \vec{S}_j \qquad (1)$$

where the labels i, j run over Cu-sites, ℓ, ℓ' over O-orbitals $\perp$ to the Cu-O-Cu directions and the second term denotes a strong ferromagnetic ($J_{Cu-O} < 0$) interaction between hole spins on O and neighboring Cu-orbitals.

Using this model Aharony et al[14] obtained a good description of the magnetic phase diagram for $La_{2-x}Sr_xCuO_4$ - the rapid suppression of the AF-phase and the occurrence of a spin-glass phase are consequences of the magnetic frustration introduced by the ferromagnetic interactions. Birgeneau et al[12)] then went on to calculate an attractive interaction between holes in Eq. (1) due to spin mediated interactions in the AF background. Their approach is based on fixing the holes in space and calculating the energy of the relaxed spin configurations. Therefore it is appropriate for the limit $t_{O-O} \ll J, J_{Cu-O}$ but the a priori estimates place us in the opposite limit. At present it is not clear how much this affects their attractive interaction. It is perhaps worth noting that Eq. (1), is similar to the Hamiltonian for a rare earth metal say - local spins coupled to itinerant electrons and there one finds a magnetic ordering of the local spins and a Fermi liquid for the itinerant electrons as the ground state.

Lastly we come to the case when the holes go into the antibonding states. This is the case which was initially proposed by Anderson[5,15)] who argued for a strong hybridization between the $d_{x^2-y^2}$ Cu-orbital and the combination of the parallel p O-orbitals with the same symmetry. This problem was investigated in more detail by Zhang and Rice[16)] who showed that, at least in the case of not too large hybridization, the model reduces to the same effective Hamiltonian as one derives in the strong coupling limit of the one-band Hubbard model. The result is a form composed of two terms, namely a hopping term for the holes and a Heisenberg term

$$H_{eff} = H_t + H_s$$

with

$$H_t = (-t) \sum_{\substack{<ij> \\ \sigma}} (1-n_{i-\sigma})\, d^+_{i,\sigma}\, d_{j,\sigma}\, (1-n_{j-\sigma}) + h.c.$$

$$H_s = J \sum_{<ij>} \vec{S}_i \cdot \vec{S}_j \qquad (2)$$

where $d^+_{i,\sigma}$, $\vec{S}_i$ and $n_{i,\sigma}$ are the creation, spin and number operators on site, i. This effective Hamiltonian is restricted to the Hilbert subspace with one or less electrons on each site.

This choice is favored by the hybridization terms and at least, in fits to the energy levels determined from density functional calculations, no large energy shift is observed between parallel and perpendicular p O-orbitals. Note since the Hartree term is accurately included in a density functional calculation, ionic energy level shifts should be included. These calculations (e.g. see Schlüter et al[17)] and Mila[18)]) favor this latter picture and by combining estimates one obtains a value of $t \approx 0.5eV$ in (2) which in turn is substantially larger than the experimental value of $J \approx 0.1eV$.

Therefore we shall concentrate on the Hamiltonian (2) which seems to include the two most relevant processes namely hopping of holes and spin-spin interactions. These two processes are strongly coupled to each other so that even this simplified effective Hamiltonian is difficult to solve in a reliable way. Until this is done it would seem hardly worthwhile to explore yet more complex Hamiltonians. Anderson[15)] recently referred to (2) as the last hard unsolved problem in manybody physics.

3. RESONANT-VALENCE-BOND (RVB) THEORY

In the first theoretical paper on the new superconductors which appeard very shortly after the discovery of the first class of $Cu\ O_2$ oxides, Anderson[5)] considered the Hamiltonian (2) and suggested that an appropriate wavefunction would be a coherent superposition of singlet bonds between spins described by the form

$$|\Psi_{RVB}\rangle = P_{D=0}\left(\sum_{ij} a_{ij}\, b^{+}_{ij}\right)^{N/2}|vac\rangle \tag{3}$$

where $P_{D=0}$ is a Gutzwiller projection operator

$$P_{D=0} = \prod_{i} (1-n_{i\uparrow}\, n_{i\downarrow}) \tag{4}$$

which enforces the condition for the Hilbert subspace. The operator $b^{+}_{ij} = d^{+}_{i\uparrow}\, d^{+}_{j\downarrow} - d^{+}_{i\downarrow}\, d^{+}_{j\uparrow}$ creates a singlet bond state on the sites (ij) and a_{ij} is a form factor for these bonds. This wavefunction looks like a Bose condensation of singlet bonds and away from half-filling when the number of electrons $N < N_s$ (number of sites), it goes over to BCS-wavefunction when written in $\vec{k}$-space i.e.

$$|\Psi_{RVB}\rangle = P_{D=0}(\sum_{\vec{k}} a(\vec{k})\, d^+_{\vec{k}\uparrow}\, d^+_{-\vec{k}\downarrow})^{N/2}|vac\rangle \quad (5)$$

looks a BCS-wavefunction projected on to a fixed number of particles, N. Anderson's reasoning was that the large quantum fluctuation in spin-1/2 systems which lower the singlet energy so much, would favor this RVB state at and near 1/2-filling and that away from 1/2-filling because of its BCS form it would describe a superconductor.

This RVB-form can be regarded as part of a more general wavefunction form

$$|\Psi\rangle = P_{D=0}|\Psi_o\rangle \quad (6)$$

where $|\Psi_o\rangle$ is any Hartree-Fock-Boguliubov wavefunction. Formulated in this way it is natural to try a mean field theory. This has been done now by many workers whose work we cannot cover in detail here. It turns out that the correct formulation of the meanfield theory in terms of the fermion operators ($d^+_{i\sigma}$, $d_{i\sigma}$), because of the large redundacy of the latter at half-filling, is a subtle business and it was only after a series of papers by several groups that the proper mean field formulation was achieved (see the recent papers of Zhang et al[19] and Kotliar and Liu[20]). The difficulty stems from the Gutzwiller projection operator which is very difficult to handle analytically.

There are however numerical techniques to handle this projection operator exactly namely a Monte-Carlo summation over spatial configurations which has been used by Gros[21] and by Yokoyama and Shiba[22] to obtain accurate values for the expectation value of H_{eff} in wavefunctions of the form (6) where the unprojected wavefunction $|\Psi_o\rangle$ is of the form of a Fermi liquid, a BCS-form or a spin density wave (SDW)-form. The latter then has long range AF order.

To make a direct comparison between the two methods it is necessary to introduce approximate renormalization factors similar to the original approximations of Gutzwiller for the effects of correlation. This was done by Zhang et al[19] and then one can directly compare the renormalized mean field (RMF) results with the variational Monte-Carlo (VMC) results. In both cases one finds[19,20] a large class of paired states with a BCS-form for $|\Psi_o\rangle$ is degenerate at half filling including a state with d-wave symmetry (but not one with s-wave).

In Table 1, the results are summarized. In the RMF-approximation the d-wave paired state is clearly favored over the AF or SDW state whereas in the exact VMC calculations the AF state is slightly favored over the d-wave state.

Table 1; Energies at 1/2-filling

		FL	d-wave RVB	AF
$2\langle \underline{S}_i \underline{S}_j \rangle$	-RMF-	-0.49	-0.69	-0.60
	-VMC-	-0.534 ±0.006	-0.636 ±0.004	-0.642 ±0.002
	'Exact'			-0.668 ±0.002

Table 1; The energies at $\delta = 0$, extrapolated to infinite systems. The variational Monte-Carlo (VMC) results for FL (Fermi liquid) and d-wave RVB states are taken from Gros[21], for AF (projected SDW) from Yokoyama and Shiba[22]. The results labelled exact are the values obtained by Oitmaa and Betts[23] refined by Huse[24] by extrapolation from small system calculations. Note the exact results predict AF order in the ground state.

Turning to the case, $\delta > 0$, away from half-filling, the kinetic energy (expectation value of H_t) per hole can also be calculated. This is most favorable for the d-wave state where values of -2.71t (RMF)[19] and -2.44t (VMC)[21] are obtained as compared with -2.16t (RMF)[19] and -2.54t (VMC)[21] which are found for the AF state.

The conclusions are, first, both states give favorable values for both the spin-spin interaction and kinetic energy terms. However that is not to say that it might not be possible to construct even better wavefunctions which contain some additional effects not included here. There is a sufficient separation between the exact values at half-filling ($\sim$ 4%) for example, to make for caution. Further the recent success of Liang, Doucot and Anderson[25] in finding variational RVB wavefunctions, expressed however in spin operators not fermion operators,with an energy separation < 1% points in this direction also. Secondly the difference between AF and RVB states is sufficiently small that it is difficult to say that one is definitely favored especially at finite values of δ or that smaller additional terms in the Hamiltonian might not be important.Thirdly it can be shown either in the RMF[19] or,by Monte Carlo calculations[21] that $|\Psi_{RVB}\rangle$ has a finite off diagonal long range order with a value $\propto \delta$.

The transport properties of the normal state are very anomalous particularily the positive sign of the Hall coefficient and the thermo power point towards positively charged holes as the carriers, not electrons as expected in a Fermi liquid description, and a highly unusual resistivity linear in the temperature is ubiquitous in the Cu-oxides. These properties require a physical picture of the excitations out of RVB states. Kivelson et al[26] and Anderson[15] have proposed that the charge and spin excitations are separate. The former since it has no spin would behave as $+|e|$ boson called a holon by Anderson. The latter would behave as a neutral spin-1/2 fermion called a spinon by

Anderson. There is a nice physical picture of these excitations. Consider removing an electron from the RVB state. Then one must break a singlet bond to give a hole or empty site and an unbonded spin-1/2. But these may migrate away from each other under the influence of the H_t and H_s operators to create a separate holon and spinon. Anderson and Zou[27)] proposed to explain the linear temperature dependence of the resistivity using a model of holons being scattered off thermally excited spinons whose density varied linearly with the temperature. The latter would require a constant density of states for the spinons in the limit of zero excitation energy. However Kallin and Berlinsky[28)] pointed out that the holons cannot be taken simply as free bosons - a point accepted by Anderson and Zou[27)]. Clearly the last word has not been said on this subject and further work is required to understand this very strong coupled system.

4. CONCLUSIONS

A brief review like this can only skim the surface of a field as active as high-T_c superconductivity. There are many interesting other ideas that are being pursued. For example what is the role of interlayer coupling? We have heard elsewhere in this conference about the empirical trend that having closely spaced CuO_2 plane enhances T_c. This trend fits nicely to the suggestion of Wheatley, Hsu and Anderson[29)] that T_c is determined by this interaction through the exchanging of holon pairs. The basic idea is that individual holons are confined to a single layer but a pair is physically like a pair of real holes in an overall singlet state and so can tunnel coherently. But this assumes that a single layer by itself cannot have a superconducting ground state, an issue which to me still is not clear. Another very interesting set of ideas concerns the coupling of holes to possible defects in spin order to stabilize the spin liquid as Wiegmann[30)] proposes.

There are many interesting ideas and open questions at present but to finish on an optimistic note, I believe we are near to the final answers and am optimistic that the 50 years span between the discovery of superconductivity in Pb and the BCS theory will be considerably shorter for the high-T_c materials.

ACKNOWLEDGMENTS: I am grateful for the close cooperation of my colleagues at ETH-Zürich on the subjects discussed here, particularily F.C. Zhang, C. Gros, F. Mila, D. Poilblanc and M. Sigrist and to many other colleagues for stimulating discussions on these topics, especially G. Aeppli, A. Aharony, O.K. Andersen, B. Batlogg, R.J. Birgeneau, W.F. Brinkman, P.A. Lee, M. Schlüter and H. Shiba.

REFERENCES

1. Proceedings of the International Conference on High-T_c Superconductors and Materials and Mechanisms of Superconductivity, Interlaken 1988 edited by J. Müller and J.L. Olsen, Physica C, in press.
2. Gough C.E., et al Nature 326, 855 (1987) and in Ref. 1
3. van Bentum P.J.M. et al in Ref. 1
4. Gammel P.L. and Bishop D.J. and coworkers Phys. Rev. Lett. 59 (1987) 2592
5. Anderson P.W., Science, 235, 1196 (1987)
6. See Sinha S.K., this conference
7. Shirane G. et al, Phys. Rev. Lett. 59, 1613 (1987)
8. Orenstein J. et al, Phys. Rev. B36, 8892 (1987)
9. Lyons K.B., Fleury P.A., Schneemeyer L.F. and Wuszcak J.V., Phys. Rev. Lett. 60, 732 (1988)
10. Birgeneau R.J. et al, preprint (1988)
11. Thomas G.A. et al, preprint (1988), Orenstein J. et al, in Ref. 1
12. Birgeneau R.J., Kastner M.A. and Aharony A., Zeit. f. Physik (in press)
13. Guo Y., Langlois J.-M. and Goddard W.A., Science, 239, 896 (1988)
14. Aharony A., Birgeneau R.J., Coniglio A., Kastner M.A. and Stanley H.E., preprint (1988)
15. Anderson P.W., Proc. of the International School of Physics 'Enrico Fermi', Varenna, July 1987 ("Frontiers and Borderlines in Many-Particle Physics") North Holland, Amsterdam, in press.

16. Zhang, F.C. and Rice T.M., Phys. Rev. B37, (1988) 3759

17. Schlüter, M., Hybertsen M.S. and Christensen N.E., in Ref. 1 and this conference

18. Mila F., preprint (1988)

19. Zhang F.C., Gros C., Rice T.M. and Shiba H., Superconductor Science and Tech. 1, 36 (1988)

20. Kotliar G. and Liu J., Phys. Rev. B (in press)

21. Gros C., Phys. Rev. B (in press)

22. Yokoyama H. and Shiba H., J. Phys. Soc. Japan 56, 3570 (1987) and in press

23. Oitmaa J. and Betts D.D., Can. J. Phys. 56, 897 (1978)

24. Huse D.A., Phys. Rev. B37, 3280 (1988)

25. Liang S., Doucot B. and Anderson P.W., preprint (1988)

26. Kivelson S.A., Rohhsar D.S. and Sethna J.P., Phys. Rev. B35, 8865 (1987)

27. Anderson P.W. and Zou Z., Phys. Rev. Lett. 60, 132 (1988); ibid 2557 (1988)

28. Kallin C. and Berlinsky A.J., Phys. Rev. Lett. 60, 2556 (1988)

29. Wheatley J.M., Hsu T.C. and Anderson P.W., preprint

30. Wiegmann P.B., Phys. Rev. Lett. 60, 821 (1988) and preprints.

THE PHENOMENOLOGY OF HIGH TEMPERATURE CERAMIC SUPERCONDUCTORS: STRUCTURE, OXIDATION STATES AND SUPERCONDUCTIVITY

IVAN K. SCHULLER
PHYSICS DEPARTMENT B-019
UNIVERSITY OF CALIFORNIA, SAN DIEGO
LA JOLLA, CALIFORNIA 92093
U.S.A.

ABSTRACT

We present a brief overview of the phenomenology of high temperature oxide superconductors particularly aimed towards understanding which general structural and/or electronic features determine the superconductivity. A systematic study of the phenomenology in the $YBa_2Cu_3O_{7-\delta}$ system shows that structural features as well as oxidation states affect strongly the superconductivity and that in certain regions of the phase diagram universal relations exist which relate certain structural features to the superconducting transition temperatures.

Although the study of high temperature oxide superconductors has proceeded at a feverish level for the last two years, to date no clear-cut consensus has emerged regarding the main phenomenological features which determine the superconductivity.[1-3] In general, two approaches are being used; in one correlation are searched between the different systems, in the other a variety of systematic changes are induced in one particular system and their effect on the superconducting properties are studied.

The general structural feature which is present in all high temperature oxides is the presence of Cu-O and/or Bi-O planes with a distinct asymmetry in the coordination of Cu or Bi and the O (Ba-K-Bi-O[4,5], La-Cu-O[6-8], Y-Ba-Cu-O[9-22], Bi-Ca-Sr-Cu-O[23-32], Tl-Ca-Ba-Cu-O[33-39]). In addition, all systems contain some sort of dopping which changes the formal oxidation state of Cu from 2+ to 3+ or that of O from 2- to 1-. For instance, in the case of the La_2CuO_4 family, the dopant can be a metal ion such as Sr or in the case of $YBa_2Cu_3O_y$ family the role of the dopant is taken by the oxygen deficiency. It may be argued, by comparing different systems, that the transition temperature increases with increasing number of closely packed Cu-O planes;[40-42] for instance, the La_2CuO_4 type superconductors have one closed packed plane and a T_c of ~ 40 K, $YBa_2Cu_3O_y$ has two closed packed planes and a T_c of ~ 90 K. Similar effects have been claimed earlier in multilayered superconductors of Pb/Ge, Sn/Ge or Al/Ge where the T_c was found to increase with the number of layers.[43] However, controlled studies are difficult to perform in order to vary the number of closely spaced planes and only three closely spaced planes in a variety of inter-grown configurations have been realized to date.[41] The main reason for this, is that the number of closely spaced planes are determined by the metallurgy, stoichiometry and thermodynamics of a particular compound and therefore are not controlable at will. Moreover, to date the detailed structure (particularly the oxygen ordering) of some of the compounds (BiCaSrCuO and TlCaSrCuO) is not uniquely determined and therefore the conclusions should be taken with some caution.

A possible alternate approach is to perform well-controlled changes in a particular system which contains all the hopefully relevant features.[44] This approach has been quite successfully used in the YBaCuO system. The reasons this sytem is particularly attractive are the following: a) the transition temperature is quite high, b) a large number of substitutions which have varied degrees of influence on the superconductivity can be performed, c) changes in the oxygen stoichiometry can easily be accomplished by a variety of means and perhaps most importantly d) almost single phase compounds can be prepared which allows a very precise and detailed determination of the structure.

As is well-known by now, the superconductivity in this system occurs in the single phase compound $YBa_2Cu_3O_{7-\delta}$ with $\delta < 0.5$. The structure consists of two dimpled Cu-O planes surrounding an Y plane and these units weakly linked together by "fence" like Cu-O chains as shown in Fig. 1.[9] This structure is commonly known in the literature as the "1-2-3 phase". Based on the valency of the constituents, an oxygen stoichiometry of 6.5 implies that the formal oxidation state of copper is 2+ and that of oxygen, 2-. An increase in the oxygen stoichiometry above this value, naively implies the existence of some Cu^{3+} or some O^{1-} (in practice, it is hard to distinguish these two cases experimentally). The fundamental question therefore is which of the structural features (if any) are "responsible" (i.e. crucial) for the superconductivity and whether the existence of trivalent Cu or monovalent O is crucial.

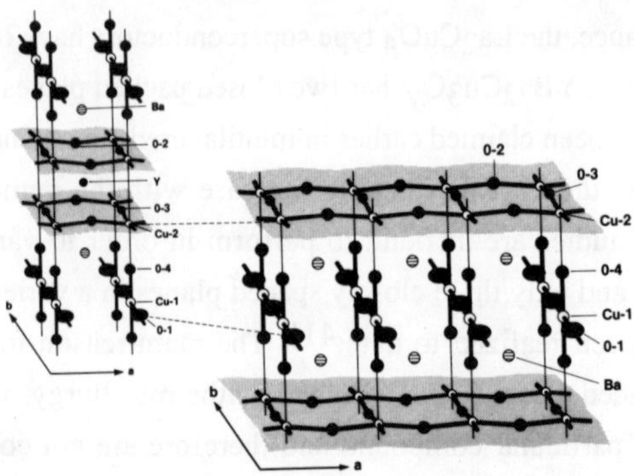

Figure 1 Orthorhombic structure of $YBa_2Cu_3O_{7-\delta}$ showing the full structure (above) and the coupling by the Cu-O "fence" of the Cu-O planes. The occupied oxygen site on the chain is labeled 01, the unoccupied 05, the bridging oxygen 04 and the oxygens in the plane are 02 and 03. Cu in the chains are labeled Cu1 and in the plane Cu2.

The answer to many of these type of questions lie in a series of well-controlled experiments in which the structure and superconductivity for various oxygen difficiencies and metal ion substitutions have been performed.

Figure 2 shows the resistivity of two $YBa_2Cu_3O_{7-\delta}$ samples, one slowly cooled and another quenched from a high temperature.[45] The slow cooled sample shows a low oxygen deficiency and normal state resistivity and a high T_c, whereas the quenched sample has a higher oxygen deficiency and normal state resistivity, and a depressed T_c. Detailed studies show that the transition temperature decreases to zero close to an oxygen stoichiometry of 6.5 (i.e. only Cu^{2+}or O^{1-} present) and close to the occurrance of the orthorhombic to tetragonal phase transition. Recent gas evolution experiments[46] have shown that in fact, for low enough vacancy concentration in the Cu-O planes which do not disrupt conduction, the T_c and the orthorhombic-tetragonal phase transitions are determined mostly by the concentration of oxygens in the Cu-O chains.

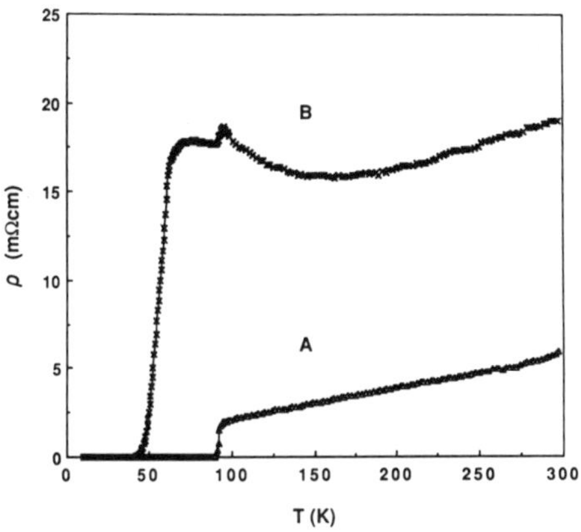

Figure 2 Resistivity versus temperature for a slowly cooled sample (A) and the same sample after quenching from a high temperature (B).

A variety of other experiments, especially in the $La(Ba_{2-x}La_x)Cu_3O_y$[47] and the Nd $(Ba_{2-x}Nd_x)$ Cu_3O_y[48] systems show that again the superconducting transition temperatures are correlated with the concentration of oxygen in the Cu-O chains.

Moreover, a plot of two types of quenching experiments (fixed times different temperature, or fixed temperature different times), the $La(Ba_{2-x}La_x)Cu_3O_y$ and Nd $(Ba_{2-x} Nd_x)$ Cu_3O_y systems are found to scale universally above a $T_c \sim 50$ K, with the chain occupancy (i.e. number of 0 in the 01 chain site - number of 0 in the empty 05 sites).[49] Below this T_c, the different types of experiments diverge from each other. Whether this universal behavior above 50 K is of fundamental importance or accidental requires further studies. At this point it should be stressed that the concentration of oxygens in the chains is simultaneously related to the overall oxygen stoichiometry, and structural parameters such cell and bond lengths, etc. Because of this, it is difficult to ascertain unequivocally which of these parameters is key in determining the superconductivity. Therefore, it is necessary to search for universality between the different types of experiments; i.e. the various kinds of quenching and substitutions. Moreover, if a definite statement is to be made detailed structural models, *experimentally determined* are necessary.

The substitutions of a number of rare earths in the Y site show very little effect on the superconducting properties. However, three elements (Ce, Pr, Tb) have not been successfully incorporated into the superconducting 1-2-3 phase. These are also the elements that form mixed valent compounds and are occasionally found in the tetravalent state, unlike Y which is always trivalent. Fig. 3 shows the temperature dependence of the resistivity for a series of samples in which Pr is substituted in the $Y_{1-x}Pr_xBa_2Cu_3O_y$ system.[50] Again, as a function of increasing x the normal state resistivity increases and the T_c decreases. These changes are quite similar to changes which occur with increasing oxygen deficiency δ in the $YBa_2Cu_3O_{7-\delta}$ system as described above. It is tempting to assign all these changes to a change in the amount of Cu^{3+} or equivalently O^{1-} present in the sample. As the Pr concentration increases the amount of Cu^{3+} (or O^{1-}) decreases and therefore the resistivity increases and the T_c decreases. Unfortunately, no *detailed* neutron refinements are presently available in this system to make a definite statement as far as the role of Pr valency as opposed to the oxygen ordering is concerned. It should be stressed again that in all cases a variety of phenomena occur simultaneously and therefore assigning changes in the superconductivity to one particular feature may be misleading.

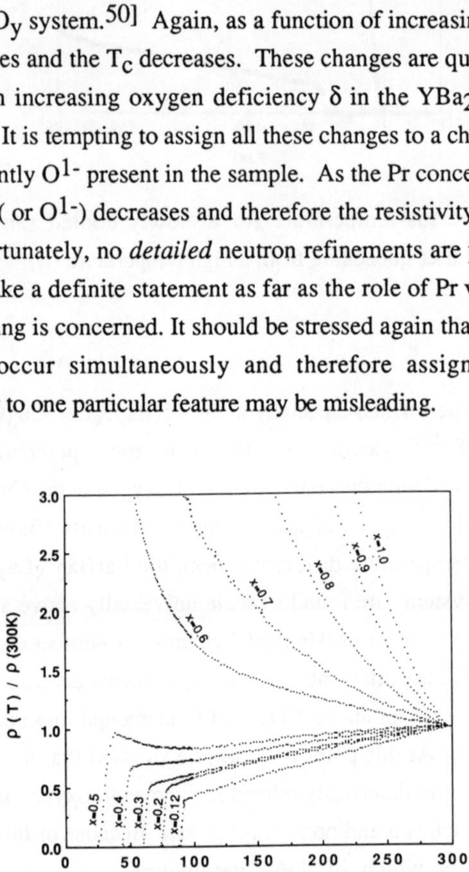

Figure 3 Relative resistivity (normalized to room temperature) as a function of temperature for selected $Y_{1-x}Pr_xBa_2Cu_3O_{7-\delta}$ samples. Notice the systematic decrease of T_c and increase of normal state resistivity with increasing x.

This brief review has outlined some of the main issues in the field of ceramic superconductors. Namely, which is the main feature which determines the superconductivity; a definite structural feature (bond length, cell edge, oxygen ordering, etc.) or electronic feature, particularly the presence of Cu^{3+} or O^{1-}. In order to ascertain which of these features is critical it is argued that universality in behavior among a variety of different well-controlled experiments should be searched. Experiments to date imply that both the formal oxidation state of Cu or O as well as the oxygen ordering are of importance. Above a T_c of ~ 50 K, the superconducting transition temperatures obtained in quenching and substitution (Nd, La) experiments seem to scale universally with the oxygen chain order parameter. Below this T_c no universal behavior is found.

I would like to thank by collaborators in this field: M. Beno, Y. Bruynseraede, D.W. Capone, J.-P. Locquet, D. Hinks, J. Jorgensen, C. Segre, L. Soderholm and K. Zhang, with whom I had an extremely active and pleasureable interaction which will be remembered fondly for a long time. Many thanks to Professor R. Nicolsky for organization of the Latin-American Conference on High Temperature Superconductivity which stimulated an interesting and fruitful interaction with Latin-American colleagues.

Work supported by the National Science Foundation Grant DMR-88/03185. International Travel was provided by NATO.

REFERENCES

1. See for instance, "Novel Superconductivity", S.A. Wolf and V.Z. Kresin, eds., Plenum Press, New York (1987).
2. See for instance, "High Temperature Superconductors", S. Lundquist, E. Tosatti, M.P. Tosi and Y. Lu, eds., World Scientific, Singapore (1987).
3. See for instance, "High Temperature Superconductors", M.B. Brodsky, R.C. Dynes, K. Kitazawa and H.L. Tuller, eds., Material Research Society, Pittsburgh (1988).
4. R.J. Cava, B. Batlogg, J.J. Krajevski, R. Farrow, L.W. Rupp, Jr., A.E. White, K. Short, W.F. Peck and T. Kometani, *Nature* **332**, 814 (1988).
5. D.G. Hinks, B. Dabrowski, J.D. Jorgensen, A.W. Mitchell, D.R. Richards, S. Pei and D. Shi, *Nature* (In Press).
6. C. Michel and B. Raveau, *Rev. Chim. Miner.* **21**, 407 (1984).

7. J.G. Bednorz, M. Takashige and K.A. Müller, *Europhys. Lett.* **3**, 379 (1987).
8. D.E. Cox, S.C. Moss, R.L. Meng, P.H. Hor and C.W. Chu, *J. Mater. Res.* (In Press).
9. M.A. Beno, L. Soderholm, D.W. Capone II, D.G. Hinks, J.D. Jorgensen, J.D. Grace, I. K. Schuller, C.U. Segre and K. Zhang, *Appl. Phys. Lett.* **51**, 57 (1987).
10. J.E. Greedan, A. O'Reilly and C.V. Stager, *Phys. Rev. B.* **35**, 8770 (1987).
11. J.J. Capponi, C. Chaillout, A.W. Hewat, P. LeJay, M. Marezio, N. Nguyen, B. Raveau, J.L. Soubeyroux, J.L. Tholence and R. Tournier, *Europhys. Lett.* **3**, 1301 (1987).
12. F. Izumi, H. Asano, T. Ishigaki, E.T. Muromachi, Y. Uchida, N. Watanabe and T. Nishikawa, *Jpn. J. Appl. Phys.* **26**, L649 (1987).
13. S. Sueno, I. Nakai, F. Okamura and A. Ono, *Jpn. J. Appl. Phys.* **26**, 309 (1987).
14. R.M. Hazen, L.W. Finger, R.J. Angel, C.T. Prewitt, N.L. Ross, H.K.Mao, C.G. Hadidiacos, P.H. Hor, R.L. Meng and C.W. Chu, *Phys. Rev. B.* **35**, 7238 (1987).
15. M.A. Alario-Franco, J.J. Capponi, C. Chaillout, J. Chenavas and M. Marezio, pg. 41, ref. 3, ibid.
16. D.E. Cox, A.R. Moodenbaugh, J.J. Hurst and R.H. Jones, *J. Phys. Chem. Sol.* **49**, 47 (1988).
17. H. You, R.K. McMullan, J.D. Axe, D.E. Cox, J.Z. Liu, G.W. Crabtree and D.J. Lam, *Sol. St. Comm.* **64**, 739 (1987).
18. R. Beyers, G. Lim, E.M. Engler, R.J. Savoy, T.M. Shaw, T.R. Dinger, W.J. Gallagher and R.L. Sandstrom, *Appl. Phys. Lett.* **50**, 1918 (1987).
19. Y. Syono, M. Kikuchi, K. Oh-ishi, K. Hiraga, H. Arai, Y. Matsui, N.Koboyashi, T. Sasoka and Y. Muto, *Jpn. Jour. Appl. Phys.* **26**, L498 (1987).
20. P.M. Grant, R.B. Beyers, E.M. Engler, G. Lim, S.S.P. Parkin, M.L. Ramirez, V.Y. Lee, A. Nazzal, J.E. Vasquez and R.J. Savoy, *Phys. Rev. B.* **35**, 7242 (1987).

21. S. Katano, S. Funahashi, T. Hatano, A. Matsushita, K. Nakamura, T. Matsumoto and K. Ogawa, *Jpn. Jour. Appl. Phys.* **26**, L1046 (1987).
22. W.I.F. David, W.T.A. Harrison, J.M.F. Gunn, O. Moze, A.K. Soper, P. Day, J.D. Jorgensen, M.A. Beno, D.W. Capone II, D.G. Hinks, I. K. Schuller, L. Soderholm, C.U. Segre, K. Zhang and J.D. Grace, *Nature* **328**, 604 (1987).
23. J.B. Torrance, Y. Tokura, S.J. LaPlaca, T.C. Huang, R.J. Savoy and A.T. Nazzal, *Sol. St. Comm.* **66**, 103 (1988).
24. M.A. Subramanian, C.C. Torardi, J.C. Calabrese, J. Gopalkrishnan, K.J. Morrisey, T.R. Askew, R.B. Flippen, U. Chowdry and A.W. Sleight, *Science* **239**, 1015 (1988).
25. R.M. Hazen, C.T. Prewitt, R.J. Angel, N.L. Ross, L.W. Finger, C.G. Hadidiacos, D.R. Veblen, P.J. Heaney, P.H. Hor, R.L. Meng, Y.Y. Sun, Y.G. Wang, Y.Y. Xue, Z.J. Huang, L. Gao, J. Bechtold and C.W. Chu, *Phys. Rev. Lett.* **60**, 1174 (1988).
26. D.R. Veblen, P.J. Heaney, R.J. Angel, L.W. Finger, R.M. Hazen, C.T. Prewitt, N.L.Ross, C.W. Chu, P.H. Hor and R.L. Meng, *Nature* **332**, 334 (1988).
27. J.M. Tarascon, Y. LePage, P. Barboux, B.G. Bagley, L.H. Greene, W.R. McKinnon, G.W. Hull, M. Giroud and D.M. Hwang, *Phys. Rev. B.* (In Press).
28. K. Kuwahara, S. Yaegashi, K. Kishio, T. Hasegawa and K. Kitazawa, this conference.
29. R.J. Cava, B. Batlogg, S.A. Sunshine, T. Siegriest, R.M. Fleming, K. Rabe, L.F. Schneemeyer, D.W. Murphy, R.V. van Dover, P.K. Gallagher, S.H. Glarum, S. Nakahara, R.C. Farrow, J.J. Krajewsky, S.M. Zahurak, J.V. Waszczak, J.H. Marshall, P. Marsh, L.W. Rupp. W.F. Peck and E.A. Rietman, *Physica B&C* (In Press).
30. G. van Tendeloo, H.W. Zandbergen, J. Van Landuyt and S. Amelinckx, *Physics C* (In Press).
31. H.W. Zandbergen, Y.K. Huang, M.J.V. Menken, J.N. Li, K. Kadowaki, A.A. Menovsky, G. van Tendeloo and S. Amelinckx, *Nature* (In Press).
32. P. Bordet, J.J. Capponi, C. Chaillout, J. Chenavas, A.W. Hewat, E.A. Hewat, J.L. Hodeau, M. Marezio, J.L. Tholence and D. Tranqui, *Physics C* (Submitted).

33. R.M. Hazen, L.W. Finger, R.J. Angel, C.T. Prewitt, N.L. Ross, C.G. Hadidiacos, P.J. Heaney, D.R. Veblen, Z.Z. Sheng, A. El Ali and A.M. Herman, *Phys. Rev. Lett.* **60**, 1657 (1988).
34. M.A. Subramanian, J.C. Calabrese, C.C. Torardi, J. Gopalkrishnan, T.R. Askew, R.B. Flippen, K.J. Morrisey, U. Chowdry and A.W. Sleight, *Nature* **332**, 420 (1988).
35. C.C. Torardi, M.A. Subramanian, J.C. Calabrese, J. Gopalkrishnan, K.J. Morrisey, T.R. Askew, R.B. Flippen, U. Chowdry and A.W. Sleight, *Nature* **240**, 631 (1988).
36. S.S. P. Parkin, V.Y. Lee, E.M. Engler, A.I. Nazzal, T.C. Huang, G. Gorman, R. Savoy and R. Beyers, *Phys. Rev. Lett.* (Submitted).
37. S.S.P. Parkin, V.Y. Lee, A.I. Nazzal, R. Savoy, R. Beyers and S.J. LaPlaca, *Phys. Rev. Lett.* (Submitted).
38. M. Kikuchi, N. Kobayashi, H. Iwasaki, D. Shindo, T. Oku, A. Tokiwa, T. Kajitani, K. Hiraga, Y. Syono and Y. Muto, *Jpn. J. Appl. Phys.* (In Press).
39. H.W. Zandbergen, G. van Tendeloo, J. Van Landuyt and S. Amelinckx, *Appl. Phys. C.* (In Press).
40. To my knowledge, this idea was first advanced by P. Chu in a number of invited talks.
41. S.S.P. Parkin, V.Y. Lee, A.I. Nazzal, R. Savoy, T.C. Huang, G. Gorman and R. Beyers, *Phys. Rev. B.* (Submitted).
42. P.M. Grant, this conference.
43. M. Strogin et al, *Phys. Rev. Lett.* **21**, 1320 (1968).
44. I. K. Schuller, D.G. Hinks, J.D. Jorgensen, L. Soderholm, M. Beno, K. Zhang, C.U. Segre, Y. Bruynseraede and J.-P. Locquet, in ref. 1, ibid.
45. I. K. Schuller, D.G. Hinks, M.A. Beno, D.W. Capone II, L. Soderholm, J.-P. Locquet, Y. Bruynseraede, C.U. Segre and K. Zhang, *Sol. St. Comm.* **63**, 385 (1987).
46. J.-P. Locquet, J. Vanacken, B. Wuyts, Y. Bruynseraede and I. K. Schuller (Submitted).
47. C.U. Segre, B. Dabrowski, D.G. Hinks, K. Zhang, J.D. Jorgensen, M.A. Beno and I. K. Schuller, *Nature* **329**, 227 (1987).
48. D.G. Hinks, B. Dabrowski, K. Zhang, C.U. Segre, J.D. Jorgensen, L. Soderholm and M.A. Beno, pg. 9, ref. 3, ibid.

49. I.K. Schuller, unpublished.

50. L. Soderholm, K. Zhang, D.G. Hinks, M.A. Beno, J.D. Jorgensen, C.U. Segre and I.K. Schuller, *Nature* **328**, 604 (1987).

MICROSTRUCTURAL PHASE ANALYSIS OF Bi-Sr-Ca-Cu-O SYSTEM WITH $T_c(\rho=0)=108K$ PREPARED BY LOW TEMPERATURE HEAT TREATMENT

K. Kuwahara, S. Yaegashi*, K. Kishio, T. Hasegawa and K. Kitazawa

Department of Industrial Chemistry, Faculty of Engineering, University of Tokyo, Hongo, Bunkyo-ku, Tokyo 113, Japan.
*Central Research Laboratories, Nippon Mining and Co. Ltd., Niizo-Minami, Todashi 335, Japan.

ABSTRACT

The microstructural distribution of phases with different compositions was determined by EPMA for the nominal bulk composition $BiSrCaCu_2O_y$ sintered at 865°C for 70 h and annealed for 20 h at much lower temperature, 400°C. The results were compared for the as-sintered and annealed specimens, referring to the electrical and magnetic properties and to X-ray diffraction patterns. The involvement of a disproportionation reaction in the solid state of the 80K phase into the 110K and low T_c (<20K) phases was observed during the annealing process. The reason why it is difficult to obtain the 110K phase in a pure state is explained in terms of the large scale phase separation mainly of the 80K phase and the phase composed of (Sr,Ca)O and CuO occurring in the initial sintering process.

INTRODUCTION

Recently a multi-phase Bi-Sr-Ca-Cu-O system was found to exhibit the superconductivity transition at 80K and 115K by Maeda et al.[1)], following a much lower T_c(7-22K) observation in Bi-Sr-Cu-O system by Michel et al.[2)] Similarly, after the finding of Tl-Ba-Cu-O to exhibit T_c near 20K by Sato et al.[3)] the highest ever T_c superconducting system Tl-Ba-Ca-Cu-O (100 and 130K) was discovered by Sheng and Hermann[4)]. These two series of materials are believed to possess the corresponding crystal structures to each other[5-8)].

One of the largest problems with these material systems is the difficulty to obtain the highest T_c phase in a pure state especially for the Bi system. Because of this, the properties of the highest T_c have not been reported.

In order to find the way to obtain the pure phase of the highest T_c phase, we have performed a microstructural composition analysis on the specimens processed under various conditions.

EXPERIMENTAL

The starting powders Bi_2O_3, $SrCO_3$, $CaCO_3$, and CuO were mixed in the ratio to give $BiSrCaCu_2O_y$ in agate mortar with ethanol, calcined at 830°C for 10 h, reground, pelletized and sintered in air at 865°C for 70 h.

Parts of the as-sintered specimens were measured for resistivity-temperature relationships by the usual four probe dc method with silver paste electrodes, X-ray diffraction by Cu-K_α radiation and dc susceptibility by a SQUID susceptometer, HOXAN HSM-2000 in the field cooling mode (Meissner effect). The microstructural analysis was performed by an electron probe micro-analyzer JEOL JXA-8600 for SEM image, reflected electron image and the fluorescent X-ray image. The quantitative analysis of the component elements was made by fixing the electron beam of energy 15keV at the desired positions and by measuring the intensity of the characteristic fluorescent X-ray by the wavelength dispersion analyzers against the standard samples of CaF_2, $BaBiO_3$, La_2CuO_4 and $SrTiO_3$.

The rests of the sintered specimens were subjected to low temperature annealing in oxygen atmosphere; 300-600°C for 20 h. In a previous report[9)], we found the low temperature annealing near 400°C was effective to increase the volume fraction of the 110K phase and to raise the zero resistivity critical temperature, $T_c(\rho=0)$ up to 108K. Hence the annealing effect was mainly examined at 400°C. After this heat treatment the specimens were subjected to the same measurements and compared with the results for the un-annealed ones.

RESULTS

The resistivity-temperature(ρ–T) relationships of the typical as-sintered specimens are given in Fig. 1. Although the sharp resistivity drops were observed near 115K in most of the specimens, $T_c(\rho=0)$ was attained only below 75K. The X-ray diffraction patterns are shown in Fig. 2a, indicating that the main phase has the c lattice spacing of 30.7 Å, which is the one reported for the 80K phase. The ac susceptibility showed two step changes starting at 115K and also at 80K as shown in Fig. 3. The extent of the stepwise change was smaller for the former than for the latter.

When the specimen was annealed at 400°C in 1 atm O_2 for 20 h, the ρ-T curve changed as shown in Fig. 4. The value $T_c(\rho=0)$ was raised significantly up to 108K. This change was sensitively dependent on the characteristics of the as-sintered starting specimens. The requirement to attain the highest $T_c(\rho=0)$ was to use the specimens which did not show the distinct stepwise drop in the ρ-T curve at 80K. The initial $T_c(\rho=0)$ was not necessarily high but could even be lower than 4.2K provided that the ρ-T curve started dropping at 110K without showing any large distinct drop at 80K. The X-ray diffractogram in Fig. 2b showed the weakening of the peaks for the phase c=30.7 Å in association with the appearance of the new peaks for c=37.05 Å and c=24.5 Å. The susceptibility change was still two-step like but the change near 110K was greater than the

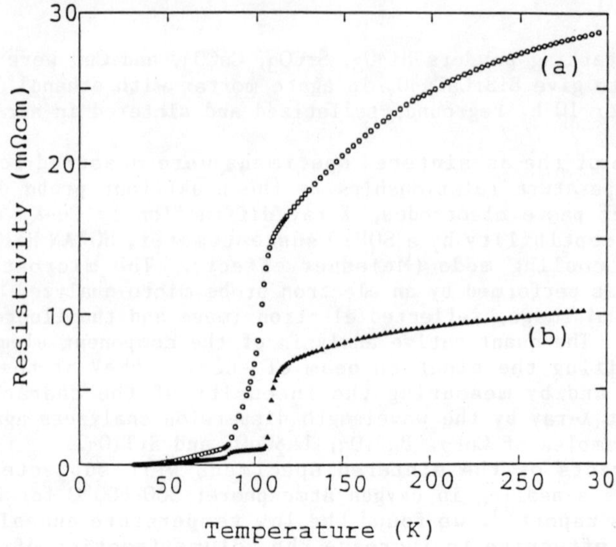

Fig. 1. The resistivity-temperature relationships of two batches, (a) and (b), of as-sintered specimens with nominal composition of $BiSrCaCu_2O_y$. Sintered in air at 865°C for 70 h and rapidly cooled.

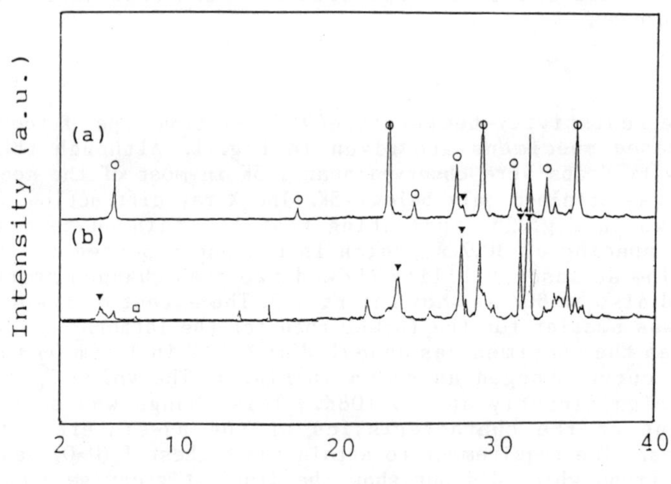

Fig. 2. Comparison of the X-ray diffractograms between the as-sintered $BiSrCaCu_2O_y$ (a) and subsequently low temperature annealed (400°C for 20 h) specimens (b). The characteristic peaks are indicated with marks for the phase with
○ c_0=30.7 Å, □ for c_0=24.6 Å and △ for c_0=37.0 Å.

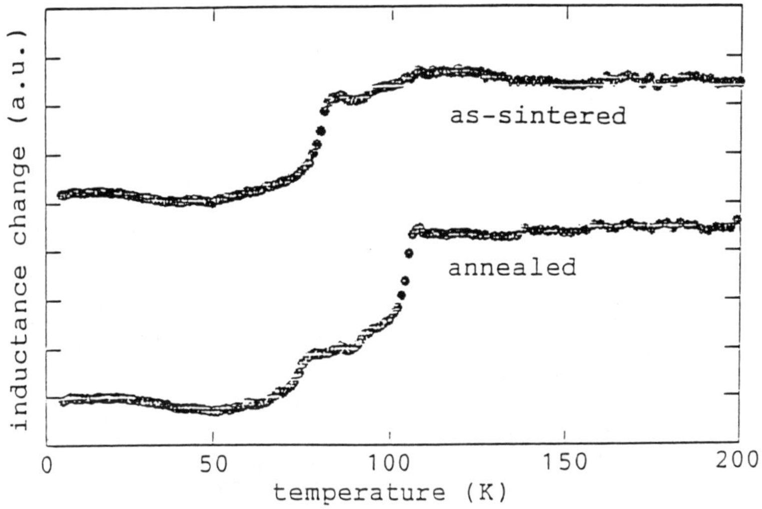

Fig. 3. Comparison of the ac susceptibility (1kHz) between the as-sintered $BiSrCaCu_2O_y$ (a) and annealed specimens (b).

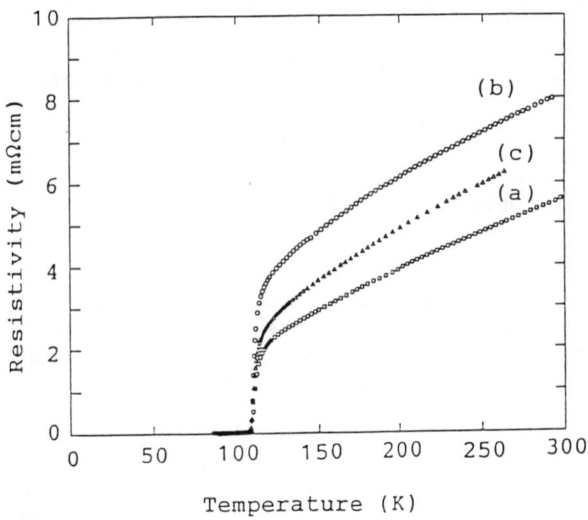

Fig. 4. The resistivity-temperature relationships of the annealed specimens at 400°C for 20 h in oxygen atmosphere at various pressures; (a) 35 atm; (b) 10 atm; (c) 1 atm.

one near 80K, as shown in Fig. 3(b).

The reflected electron image, of which the brightness represents the average atomic number of the composition, obtained by EPMA is given in Fig. 5. There were four phases present with differing compositions. The darkest areas were the voids. The second darkest areas, about 15% of the total area, showed the composition $Sr_{1.0}Ca_{1.2}Cu_{3.6}O_{5.8}$, which approximately corresponds to the compound $(Sr,Ca)_3Cu_5O_8$. Another remarkable feature in Fig. 5 was the presence of an elongated phase penetrating through the other areas. The curved and branched needles in the picture had the typical length as long as several hundred micrometers. This phase gave the averaged composition of $Bi_{1.0}Sr_{0.88}Ca_{0.91}Cu_{1.4}O_{4.2}$, closely corresponding to the ideal composition $Bi_2Sr_2Ca_2Cu_3O_y$. This phase seems to have the lattice parameter c_0=37.0 with the triple CuO_2 layered structure between the two Bi_2O_2 double layers.

The remaining matrix areas were composed of two fine grained phases. One of them raised the brighter reflected electron image and had an elongated structure with typical length of a few micrometers and was dispersed in the other phase. The composition of this phase was determined to be $Bi_{1.0}Sr_{0.4}Ca_{0.4}Cu_{0.48}O_{2.5}$, corresponding to the ideal composition $Bi_2(Sr,Ca)_2CuO_y$ which has a single CuO_2 layer between the two Bi_2O_2 double layers and may have much lower T_c of about 7-22K[2]. The plausibleness of this phase to be the low T_c superconductor was shown in the susceptibility-T curve as given in Fig. 6 in which a clear decrease in the susceptibility was seen below 8K. The other phase had the composition $Bi_{1.0}Sr_{0.8}Ca_{0.76}Cu_{1.2}O_{4.5}$, approximating the ideal composition $Bi_2(Sr,Ca)_3Cu_2O_y$, 80K phase, which has the double CuO_2 layers between the two Bi_2O_2 layers. A summary of these observations is given in Table 1.

Table 1. The microscopical phase distribution in the low temperature (400°C) annealed specimens. The nominal composition was $Bi_1Sr_1Ca_1Cu_2O_y$.

# of CuO_2 layers	ideal composition (Bi)	(Sr,Ca)	(Cu)	composition analyzed by EPMA (Bi)	(Sr)	(Ca)	(Cu)	morphology observed	T_c
-	-	3	5	0.0	1.00	1.21	3.68	round island	-
1	2	2	1	1.0	0.40	0.40	0.48	tiny platelet	(7-22K)
2	2	3	2	1.0	0.80	0.76	1.17	residual matrix	80K
3	2	4	3	1.0	0.88	0.91	1.40	penetrating giant platelet	110K

(a)

(b)

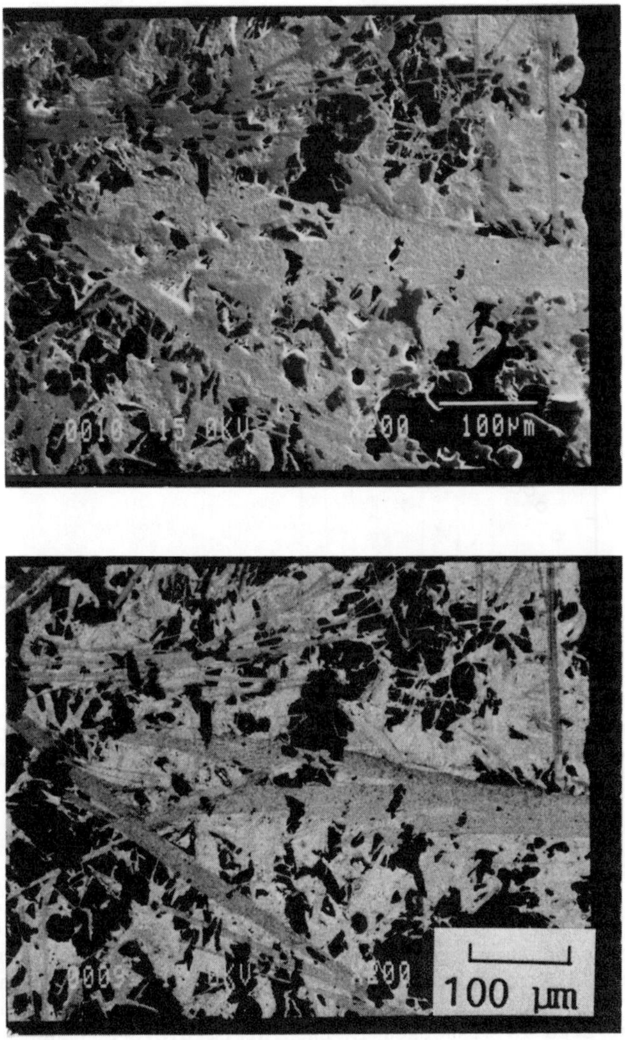

Fig. 5. Secondary (a) and reflected (b) electron images by EPMA of the annealed specimen of $BiSrCaCu_2O_y$. The contrast is enhanced in (b) than in (a) both of which are taken in the same area. The second darkest areas in (a) is completely dark in (b) and hence cannot be distinguished from the void areas. Beam energy used was 15 keV.

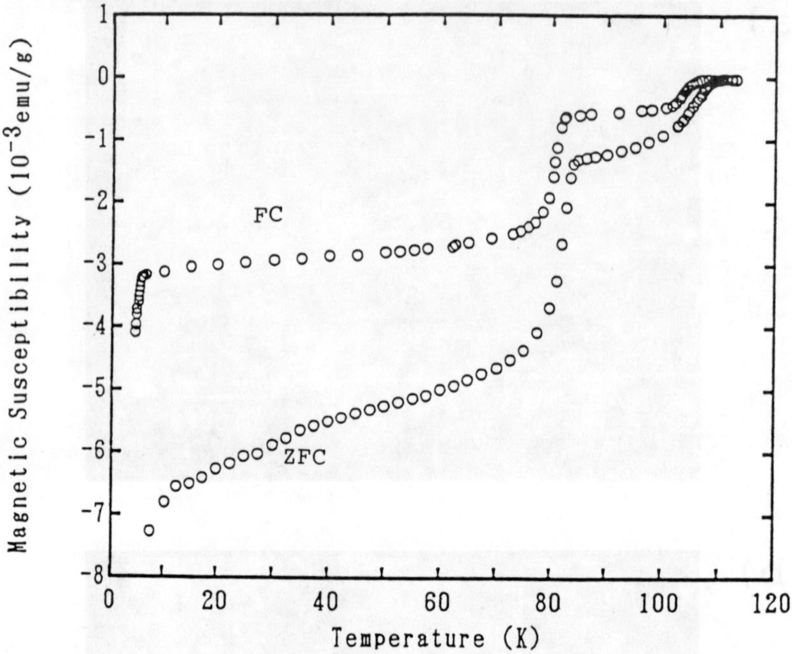

Fig. 6. The dc susceptibility signal of an annealed specimen when the annealing was done at 600°C. The dip below 8K was not clearly observed for the specimens annealed at 400°C. Field cooled at H=3.4Oe.

The microstructure of the as-sintered specimens is shown in Fig. 7. The major difference between Figs. 5 and 7 was the significant growth of the 110K phase in the former. The characteristic morphology of this phase, the intergranular penetration, was essentially the same in the two figures. Fig. 8 is a magnified images of Fig. 7 in order to show the fine structures. The brightest area is dispersed in the intermediate contrast region and has elongated structure with length about several micrometers. The composition obtained was $Bi_{1.0}Sr_{0.49}Ca_{0.32}Cu_{0.48}O_{2.62}$ and hence corresponded to $Bi_2(Sr,Ca)_2CuO_6$, the low T_c phase. The amount of this low T_c phase was larger in Fig. 5 than in Fig. 8. The intermediate contrast region appeared to be composed of very fine grains with submicron size and gave the EPMA composition of $Bi_{1.0}Sr_{0.80}Ca_{0.48}Cu_{0.96}O_{3.67}$. This seemed to be the phase of $Bi_2(Sr,Ca)_3Cu_2O_8$ with T_c=80 K. The composition of the 110 K phase was determined to be $Bi_{1.0}Sr_{0.83}Ca_{0.70}Cu_{1.19}O_{4.24}$, which was slightly depleted of copper in comparison with that in the annealed specimens.

(a)

(b)

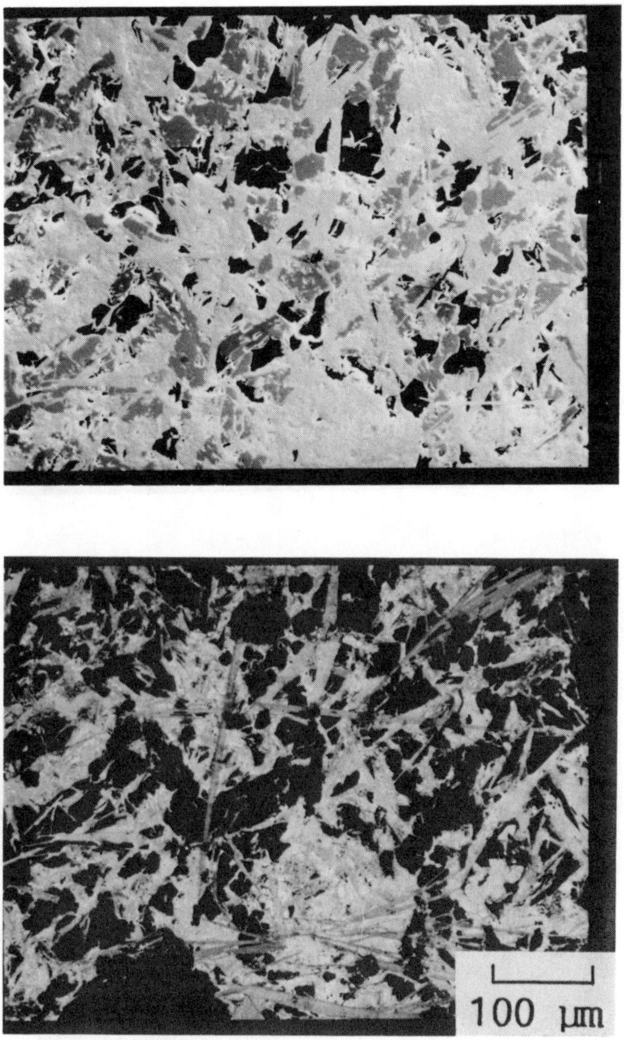

Fig. 7. Secondary (a) and reflected (b) electron images by EPMA of the as-sintered specimen of $BiSrCaCu_2O_y$. The contrast is enhanced in (b) than in (a) similarly to Fig. 5. Beam energy used was 15 keV.

(a)

(b)

Fig. 8. Magnified secondary (a) and reflected (b) electoron images by EPMA of the lower central region of Fig. 7

DISCUSSION

The results stated above as a whole indicate that the giant curving platelets of the 110K phase grow during the low T annealing process at the sacrifice of the 80 K phase. Simultaneously, the low T_c(7-22K) phase grows in order to compensate the composition[9], as

$$Bi_2(Sr,Ca)_3Cu_2O_y \longrightarrow Bi_2(Sr,Ca)_4Cu_3O_y + Bi_2(Sr,Ca)_2CuO_y.$$

The occurrence of this disproportionation reaction means that the 80K phase is thermodynamically unstable at temperatures near 400°C. The reason that some 80 K phase still remained in the annealed specimens should then be explained by the slow kinetics.

Zandbergen et al.[10] have suggested a possibility that the 110K phase might be composed of quadruple layer structure based on the TEM-EDX chemical analysis. However, since there are no appropriate standard specimens available for TEM chemical analysis, we believe that the composition determined in the present experiments, which utilized the standard samples for the analysis, is more reliable. Hence we may conclude that the 110K phase has the triple CuO_2 layer structure and has the ideal composition $Bi_2(Sr,Ca)_4Cu_3O_y$. The actual composition could be slightly deficient in the content of Sr and Ca as seen in Table 1.

Thermodynamically speaking, one should be able to obtain the 110K single phase if started from the right composition. In the actual process, however, it is understood from Figs. 5 and 7 that the phase $(Sr,Ca)Cu_2O_3$ has been separated spatially too far from the 80K phase, by a few tens of micrometers already in the initial sintering process at 865°C. Hence the further annealing at the lower temperature is thought to contribute to the only incomplete reaction due to the slow material transport process. In order for the 80K phase to change into the 110K phase, it must react with CaO and CuO. Since the latter components are not sufficiently provided, it is thought that the disproportionation reaction takes place instead of the desired reaction

$$Bi_2Sr_2CaCu_2O_y + (Sr,Ca)O + CuO \longrightarrow Bi_2Sr_2Ca_2Cu_3O_y.$$

In conclusion, four distinct phases were observed as summarized in Table 1 by EPMA analysis. The changes in the microscopical morphology and in the relative abundance of the phases indicated that the disproportionation reaction from the 80K phase into 110K phase and the low T_c (8K) phases took place. The occurrence of this reaction was concluded to be due to the insufficient supply of CaO and CuO to the reaction site because of the large spatial separation of these substances from the site. The large separation was found to occur already in the initial high temperature process. Preparation of the 110 K phase, therefore, should be designed so that to avoid the large scale phase separation during the process.

REFERENCES

1) H. Maeda, Y. Tanaka, M. Fukutomi and T. Asano, Jpn. J. Appl. Phys., 27, L209 (1988).
2) C. Michel, M. Hervieu, M. M. Borel, A. Grandin, F. Deslandes, J. Provost and B. Raveau, Z. Phys. B68, 421 (1987).
3) S. Kondoh, Y. Ando, M. Onoda, M. Sato and J. Akimitsu, Solid State Commun., 65, 1329 (1988).
4) Z. Z. Sheng and A. M. Hermann, Nature, 332, 55 (1988).
5) M. A. Subramanian, C. C. Torardi, J. C. Calabrese, J. Gopalakrishnan, K. J. Morrissey, T. R. Askew, R. B. Flippen, U. Chowdry and A. W. Sleight, Science, 239, 1015 (1988).
6) R. M. Hazen, C. T. Prewitt, R. J. Angel, N. L. Ross, L. W. Finger, C. G. Hadidiacos, D. R. Veblen, P. J. Heaney, P. H. Hor, R. L. Meng, Y. Y. Sun, Y. Q. Wang, Y. Y. Xue, Z. J. Huang, L. Gao, J. Bechtold and C. W. Chu, Phys. Rev. Lett., 60, 1174 (1988).
7) N. Kijima, H. Endo, J. Tsuchiya, A. Sumiyama, M. Mizuno and Y. Oguri, to be published in Jpn. J. Appl. Phys., 27(No.5).
8) E. Takayama-Muromachi, Y. Uchida, Y. Matsui, M. Onoda and K. Kato, Jpn. J. Appl. Phys., 27, L556 (1988).
9) K. Kitazawa, S. Yaegashi, K. Kishio, T. Hasegawa, N. Kanazawa, K. Park and K. Fueki, presented at Amer. Cer. Soc. 90th Annual Meeting, Cincinatti, Ohio, May 1988, to be published in Adv. Cer. Mat.
10) H. W. Zandbergen, Y. K. Huang, M. J. V. Menken, J. N. Li, K. Kadowaki, A. A. Menovsky, G. van Tendeloo and S. Amelinckx, Nature, 332, 620 (1988).

High Temperature Superconductivity at 30K in (Ba, K)BiO_3

B. Batlogg, R. J. Cava, L. W. Rupp, Jr., J. J. Krajewski and W. F. Peck, Jr.

AT&T Bell Laboratories
Murray Hill, NJ 07974

ABSTRACT

Several basic physical properties of the 30K superconductor $Ba_{0.6}K_{0.4}BiO_3$ have been measured. From the upper and lower critical field slopes near T_c, the density of states of E_F was estimated and found to be 2.2 ± 0.4 times higher than in $BaPb_{0.75}Bi_{0.25}O_3$, similar to results of band structure calculations. The normal state susceptibility is essentially temperature independent and consists of a dominant diamagnetic core contribution and a smaller paramagnetic contribution, which is proportional to the superconducting volume.

Introduction

With critical temperatures T_c near 30K,[1)] the Bismuth-oxide based superconductors[2,3)] are speculated to be "promising runners-up"[4)] to the high temperature superconductors containing Copper-oxide. In addition to the T_c well in excess of all traditional intermetallic compounds, the two families of compounds share in common the metal-oxygen network which is essential for superconductivity. The central question is whether the same pairing mechanism leads to the high T_c. Here we present the results of a study in which we measured several basic physical properties to characterize the new superconductor and to shed some light on this question. First we find that (Ba, K)BiO_3 clearly belongs to the group of high T_c superconductors. From analysis of the critical fields we find that the T_c of ~30K is 5-10 times higher than for traditional compounds with the same low electronic density of states, joining $Ba(Pb, Bi)O_3$ and the cuprate superconductors.

Magnetism is invoked by several models to explain superconductivity in the cuprates. So far Bismuth is not known to be supporting magnetism. We carefully studied the normal state susceptibility of the bismuthates and found them to be diamagnetic. This diamagnetism is essentially temperature independent to 400K and is dominated by the electronic cores. The small variations of a small paramagnetic contribution among the samples are correlated with the superconducting volume fraction. Therefore, we conclude that high T_c superconductivity in the bismuthates is not associated with magnetism.

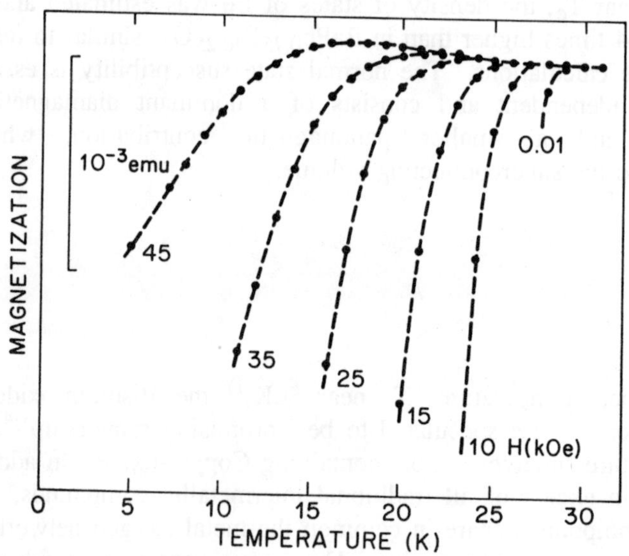

Fig. 1 Temperature dependence of the magnetization in magnetic fields up to 50 kOe, for determination of the upper critical field $H_{c2}(T)$.

Experimental Aspects

Employing a sealed silver tube reaction technique,[1] we have identified the perovskite (Ba, K)$BiO_{3-\delta}$ as the bulk superconductor in the Ba/K/Bi/O chemical system[2] and found optimal properties for an A-site composition close to $Ba_{0.6}K_{0.4}$. The perovskite unit cell (a_o =4.29Å) is cubic. More recent studies[6] found no symmetry change on cooling to 14K. (For additional information on preparation, see Ref. 1). The samples are in the form of pressed powders. All measurements reported here are done in SQUID magnetometers. To determine the normal state susceptibility, the magnetization up to H=5T was measured at various temperatures up to 400K. Magnetic measurements are also well suited to determine superconducting parameters because of the isotropic electronic structure of the cubic perovskite. In addition, the superconducting parameters can be derived from intensive quantities, such as the lower and upper critical fields, and the results are independent of the superconducting volume fraction of the sample.

Upper Critical Field H_{c2}

The upper critical field $H_{c2}(T)$ was measured magnetically. Fig. 1 shows magnetization curves as functions of temperature in constant fields. For clarity, all curves have been shifted vertically such as to coincide in the normal state. (Normal state susceptibility and sample holder contributions are thus eliminated.) The magnetization scale is greatly expanded to facilitate determination of $T_c(H)$, which is taken as the temperature where M(T) starts deviating from the high temperature curve. All M(H) curves were measured during warm-up, and no time dependence of M was observed within 10-15 minutes after the 1K temperature steps. The resulting $H_{c2}(T)$ is given in the upper part of Fig. 2, and the slope near T_c is $-5\pm$ 0.5 kOe/K.

Lower Critical Field H_{c1}

The lower critical field was deduced from magnetization measurements after cooling in zero field. At lowest temperatures, H_{c1} is 100-110 Oe, and dH_{c1}/dT in the linear range near T_c is $-4.5=\pm -0.5$ Oe/K. The values of the critical field slopes are similar to the ones measured in $BaPb_{.75}Bi_{.25}O_3$ crystals (7kOe/K, 2 Oe/K).[7]

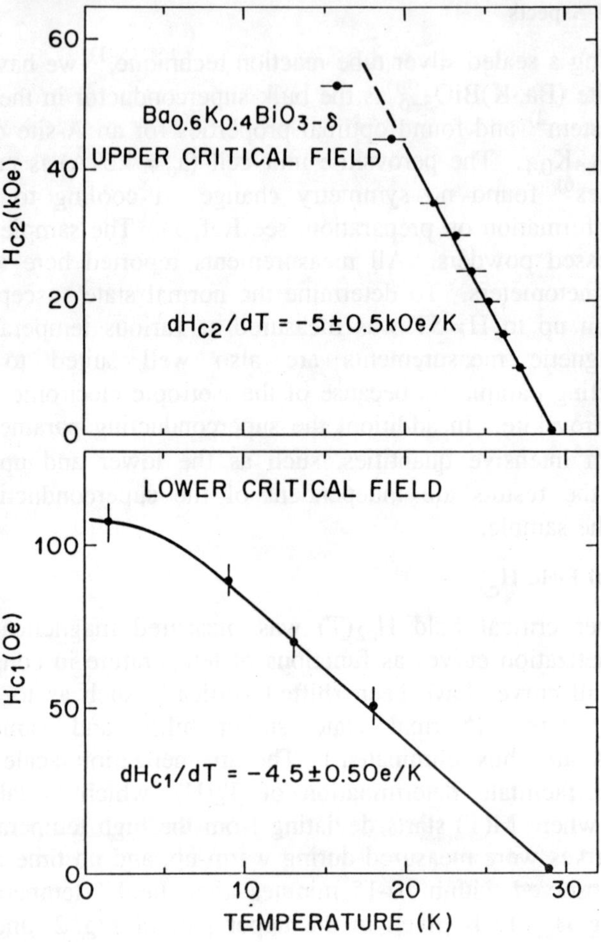

Fig. 2 Upper and lower critical fields $H_{c1}(T)$ and $H_{c2}(T)$ for $Ba_{0.6}K_{0.4}BiO_3$. The resulting slope of the thermodynamic critical field dH_c/dT is 75 ± 10 Oe/K.

Normal State Susceptibility

The normal state susceptibility up to 400K was derived from magnetization curves up to 50kOe. Over the entire temperature range, the susceptibility is negative and only weakly temperature dependent. (As we had suspected, the earlier data[1)] were affected by parasitic paramagnetic impurities). The measured value of χ is dominated by the core

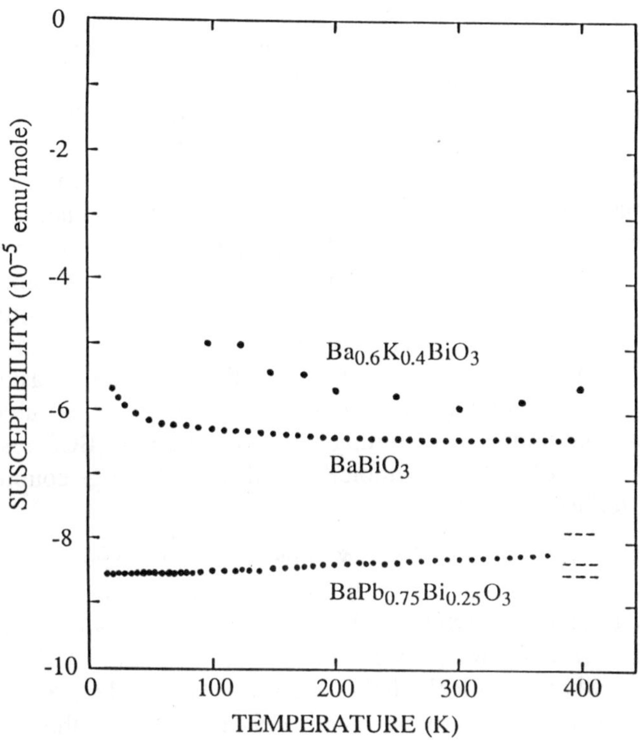

Fig. 3 Normal state susceptibility of $Ba_{0.6}K_{0.4}BiO_3$, $BaBiO_3$ and $BaPb_{0.75}Bi_{0.25}O_3$. The broken lines, associated with each curve, indicate the susceptibility of the core electrons only. The paramagnetic contribution in $Ba_{0.6}K_{0.4}BiO_3$ is proportional to the superconducting volume fraction.

diamagnetism. From systematic studies of reference compounds and from published data, we estimate χ_{core} to be $-7.8\,10^{-5}$ emu/mole for $Ba_{0.6}K_{0.4}BiO_3$. The result shown in Fig. 3 thus implies a paramagnetic contribution of $\sim 2\ 10^{-5}$ emu/mole. We observed that this paramagnetism is sample dependent. The larger the Meissner signal (= superconducting volume fraction), the larger the paramagnetic susceptibility. A detailed discussion of magnetic properties will be given elsewhere.

Numerical Analysis

Here we combine the above experimental results to extract some of the basic parameters of this superconductor. Provided the crystallographic unit cell volume is known, it is sufficient to know the slopes of the critical fields H'_{c1} and H'_{c2} to calculate the specific heat anomaly at T_c. This procedure relies only on thermodynamics and does not require the entire sample to be superconducting. (H_{c1} and H_{c2} are intensive quantities.) First, we combine $H'_{c1} = 4.5 \pm 0.5$ Oe/K and $H'_{c2} = 5000 \pm 500$ Oe/K to get the slope of the thermodynamic critical field $H'_c = 75 \pm 10$ Oe/K and a Ginzbury-Landau parameter $\chi \cong 47$. Using Rutgers' relationship, with a molar volume of 47.55 cm^3, the specific heat anomaly at T_c is calculated as 64 mJ/mole K^2. While the analysis so far relied only on measured quantities, the next step, extracting the Sommerfeld parameter γ, is subject to the well known uncertainties. Assuming the weak-coupling BCS result of $\Delta C = 1.43\gamma T_c$, a γ of 1.5 in J/moleK^2 follows. Strong coupling corrections would reduce γ.

The alternative way to estimate γ involves the measured (or extrapolated) values of $H_{c1}(0)$ and $H_{c2}(0)$, but uncertainties of a different kind enter into the analysis, which will be discussed elsewhere. For comparison with $(Ba, K)BiO_3$, we re-analyze earlier data on $BaPb_{0.75}Bi_{0.25}O_3$, where critical field slopes $-H'_{c2} = 7$ kOe/K and $-H'_{c1} = 2$ Oe/K had been measured on single crystals. Within the same framework a γ of 0.68 ± 0.14 mJ/mole K^2 is found, which is somewhat smaller than earlier values for γ, demonstrating the limitations in the various approaches to calculate γ.

Instead of discussing the absolute values of γ, which are subject to systematic uncertainties, we notice that ΔC (and the derived value for γ) in $Ba_{0.6}K_{0.4}BiO_3$ is 2.2 ± 0.4 times larger than in $BaPb_{.75}Bi_{0.25}O_3$. This ratio compares well with the results of band structure calculations,[9)] which give a ratio of ~1.9. The T_c enhancement from 12K to 30K is thus of least partially due to the higher density of states at E_F.

In Fig. 4, we have added $Ba_{0.6}K_{0.4}BiO_3$ to the well-known "γ–T_c" plot. Due to its position, far separated from conventional superconductors, the title compound could be argued to have the relatively highest transition temperature of all superconductors. Its T_c is ~7.7 times the T_c of Hg, whereas T_c of $Ba_2YCu_3O_7$ is 4 times higher than the T_c of Nb_3Ge.

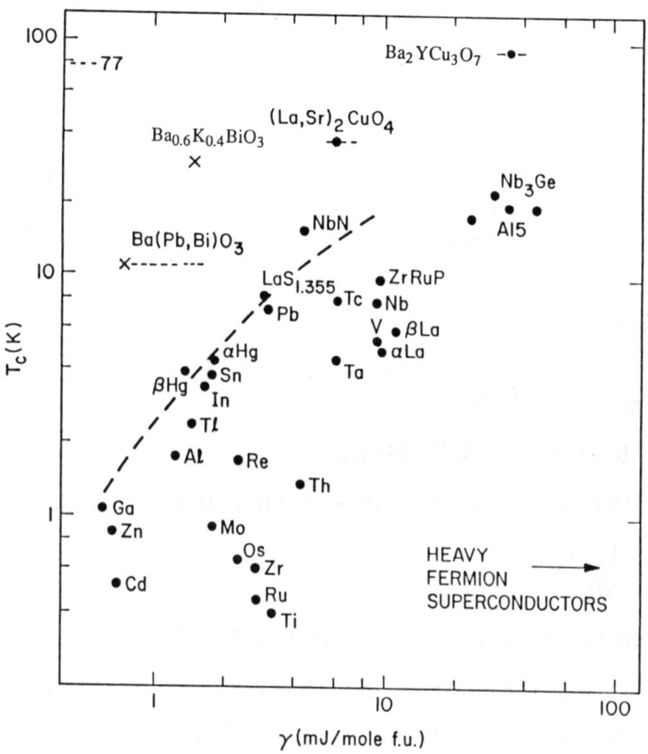

Fig. 4 The "γ–T_c" plot, including an estimate of γ for the Bismuth-oxide and Copper-oxide based superconductors. Note that these estimates assume a BCS value for the specific heat anomaly ΔC at T_c.

Summary

By analyzing the measured upper and lower critical fields of $Ba_{0.6}K_{0.4}BiO_3$, we have found that the density of states at the Fermi level for this 30K superconductor is 2.2 ± 0.4 times higher than in the closely related $BaPb_{0.75}Bi_{0.25}O_3$ with T_c of ~12K. The normal state susceptibility in both compounds is dominated by the diamagnetism of the core electrons, with the band electricity contributions being small and consistent with expectations.

REFERENCES

1) R. J. Cava, B. Batlogg, J. J. Krajewski, R. C. Farrow, L. W. Rupp, Jr., A. E. White, K. T. Short, W. F. Peck, Jr., and T. Y. Kometani, Nature **332,** 814 (1988).

2) L. F. Mattheiss, E. M. Gyorgy, and D. W. Johnson, Jr., Phys. Rev. B **37,** 3745 (1988).

3) A. W. Sleight, J. L. Gillson and P. E. Bierstedt, Solid State Commun. **17,** 27 (1975).

4) T. M. Rice, Nature **332,** (1988).

5) J. G. Bednorz and K. A. Müller, Z. Phys. B **64,** 189 (1986).

6) R. M. Fleming, P. Marsh, R. J. Cava and J. J. Krajewski (to be published).

7) B. Batlogg, Physica **126** B+C, 275 (1984).

8) K. Kitazawa, S. Uchida and S. Tanaka, Physica **135** B, 505 (1985).

9) L. F. Mattheiss and D. R. Hamann, Phys. Rev. Lett. **60,** 2681 (1988).

CHEMICAL STOICHIOMETRY AND STRUCTURE OF HIGH TEMPERATURE SUPERCONDUCTORS

P. Day
Oxford University
Inorganic Chemistry Laboratory
South Parks Road
Oxford OX1 3QR
U.K.

ABSTRACT
Each of the groups of high temperature superconductors discovered in the last 18 months belongs to a different, but related, structure type long familiar in solid state inorganic chemistry. These are respectively the K_2NiF_4 (40K), defect perovskite (90K) and Aurivillius structure (110K). Using neutron diffraction and electron microscopy data from the Oxford group we survey the chemical stoichiometry and structural parameters for examples of all three classes. In particular we show how the lattice distortion in $La_{2-x}Sr_xCuO_4$ varies with x, and how structural superlattices develop in two members of the Bi-Ca-Sr-Cu-O series. We present a microscopic model for the modulated structure of $Bi_2Sr_{1.67}CaCu_2O_8$.

1. INORGANIC CHEMISTRY

Confronted with the extraordinary phenomenon of high temperature superconductivity the first question that an inorganic chemist asks is what part of the Periodic Table produces the compounds that show this property. The answer, of course, is among the oxides of Group 11, specifically of Cu. Given that (at least at the time of writing) four distinct categories of Cu oxide superconductors are known, next follows the question 'what features of crystal and electronic structure do they share?' Furthermore, a full perspective from the chemical viewpoint demands that the compounds are placed in the context of their congenors, from an electronic and structural standpoint. This means that if we are to ask ' why Cu oxides?' we should also ask 'why not Ag oxides or Cu sulphides?' Finally, it must be emphasised that superconductivity in oxides is not new, so we should seek the common factor between the older and newer phases - what lattice types?

what valence electron configurations? The aim of this paper is to examine such questions, using the language of inorganic chemistry rather than solid state physics. The structural answers and generalizations will be based largely on the work of our own group.

2. CHEMICAL STOICHIOMETRY AND Cu OXIDATION STATE

In each of the separate Cu oxide series it has been important to establish the precise chemical stoichiometry of the superconducting phases, and the compositional range over which superconductivity occurs. Genuine non-stoichiometry, and hence a non-integral average oxidation state of Cu, appears to be present in every case, though it is brought about by different chemical means. A word about the chemists' meaning of the phrase 'oxidation state' is in order. Conventionally anions, and cations outside the d- and p-blocks, are assumed to have closed shell electron configurations, e.g. O^{2-} ($2s^22p^6$) and Sr^{2+}($4s^24p^6$)[1]. Lanthanum is an f-block element, but nearly always occurs as La^{3+}, though metallic compounds like LaI_2 are known, in which the La configuration is $5d^1$[1]. It is overlap between the latter orbitals that leads to the formation of the conduction band. Following this prescription, the oxidation state of Cu in La_2CuO_4 is Cu^{2+}($3d^9$). Given the symmetry of the ligand field about the Cu (see below) the unique half-occupied orbital would be x^2-y^2, oriented in the basal plane with its lobes directed towards the four O^{2-} that form a square around it.

Definitions of single atom oxidation states become blurred when, as in the present case, there is strong cation-anion covalency. The Cu(x^2-y^2) overlaps O(2pσ) so the partly filled orbital is not purely Cu but an admixture of Cu and O. This is the origin of the suggestion from photoemission work that "O^-" exists in the superconductors[2]. In fact, there is just a conduction band of local x^2-y^2 symmetry. Nevertheless, it is convenient for book-keeping purposes to count the extent to which it is filled by *arbitrarily* assigning O as $2p^6$ and varying Cu$(x^2$-$y^2)^p$ from p=0-2 as the band goes from empty to full: p=0 then corresponds formally to Cu(III), p=1 to Cu(II) and p=2 to Cu(I). Note the use of Roman numerals in place of Cu^+, Cu^{2+}, Cu^{3+} to denote that full ionicity is not assumed. In some papers[3] the notation $(Cu$-$O)^+$ has been coined for the same reason. In the conventional chemists' notation that we shall use, a mixed oxidation state or 'mixed valency' implies a mean value of p that is not an integer.

2.1 $La_{2-x}M_xCuO_{4-y}$(M=Ca,Sr,Ba)

The origin of the mixed valency or partial x^{2}-y^{2} band occupancy in the 40K series $La_{2-x}M_xCuO_{4-y}$ (M=Cu,Sr,Ba) is primarily in the A-group cation lattice, where a divalent replaces a trivalent ion. For x<0.20, it has been shown that the conduction band filling as exemplified by the normal state electrical properties scales with x [3], i.e. that y is unimportant. However, for x>0.20 such scaling no longer holds, indicating that further increases in x beyond 0.20 are largely compensated by increases in y. This implies that there is an effective maximum in the Cu oxidation state that can be sustained within the K_2NiF_4 lattice type. The mode of oxygen elimination is not known, but most likely occurs through intergrowth of double, and higher multiplicity, perovskite block structures into the single layer K_2NiF_4 structure. Such intergrowths have been observed by high resolution electron microscopy in La_2CuO_4 itself, when the La:Cu ratio is made less than two[5)]. Thus the K_2NiF_4 structure of La_2CuO_4 consists of single (001) layers of the cubic perovskite lattice, separated by La^{3+}. Similar (001) layers consisting of n blocks of perovskite joined through vertices along (001) and separated by La^{3+} would have the overall stoichiometry $La_{n+1}Cu_nO_{3n+1}$ so that when n is infinite the three dimensional cubic perovskite structure of $LaCuO_3$ is achieved. Assuming for the moment that the anions have closed electronic shells, the Cu oxidation state is progressively turned from (II) to (III). Examples are shown in Fig.1.

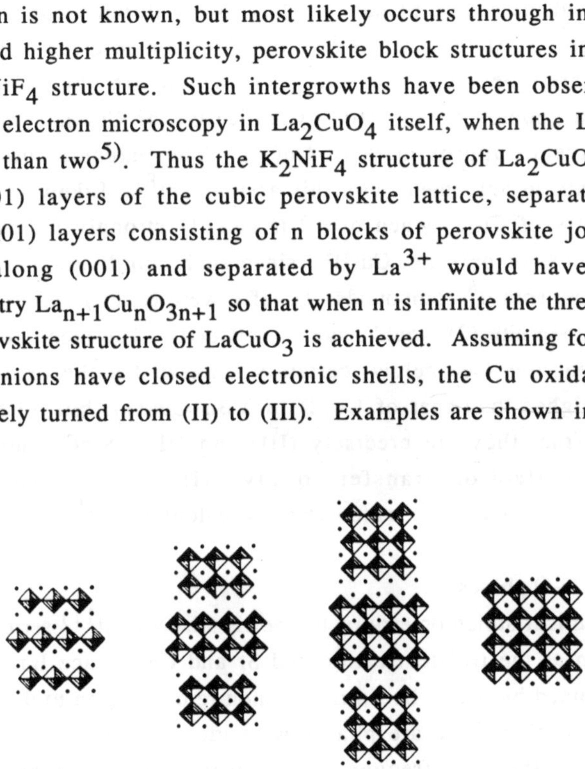

Figure 1. Evolution of perovskite blocks from K_2NiF_4 to cubic perovskite structure (Ref. 5)

2.2 $LaBa_2Cu_3O_{7-\delta}$

In the $LnBa_2Cu_3O_{7-\delta}$ (Ln=lanthanide) series, the origin of Cu mixed valency is two-fold. Many investigators[6)] have shown how δ can be tuned from 0 to 1 by controlled atmosphere annealling, corresponding to Cu mean oxidation states of 2.33 and 1.67 (we have to use arabic instead of Roman numerals here since the Romans did not invent decimals!). Unfortunately the non-integral mean oxidation state cannot be interpreted straightforwardly because in the $LnBa_2Cu_3O_{7-\delta}$ series, when δ is small, there are two crystallographically distinguishable Cu sites, called Cu(1) and Cu(2), with the square pyramidal $Cu(2)O_5$ units joined through equatorial vertices to form layers, and the square planar $Cu(1)O_4$ joined through opposite vertices to form chains[7)]. At least when δ=1 the situation is clear because there are no O shared between the sites (1), which are therefore linearly coordinated, a geometry very strongly indicative of Cu(I). Then the site (2) would have to approximate to Cu(II). The magnetic properties of $YBa_2Cu_3O_6$ are in agreement with this assignment[8)]. Likewise the square planar geometry of Cu(1) when δ = 0 is equally strongly indicative to an inorganic chemist of p=0, i.e. Cu(III), since such coordination is extremely common for low spin d^8 compounds (4). If δ were exactly zero, and site (1) were Cu(III), then site (2) would have to be Cu(II). The mean bond lengths around the two sites (see below) certainly suggest that the oxidation state of Cu(1) is higher than that of Cu(2) in $YBa_2Cu_3O_7$, but do not tell us unequivically that they are precisely (III) and (II). Since they share an anion, partial electron transfer to give (III -q), (II +q) is quite possible, and is indicated by band structure calculations[10)].

2.3 Bi-Sr-Ca-Cu-O phases

The chemical stoichiometry of the newly discovered[11)] series in which Bi replaces Ln as the trivalent cation and Sr and Ca replace Ba, appears to have been assumed by many, but in my opinion remains open to debate. What seems certain is that there are phases with ideal Bi:Cu ratios of 2:1, 2:2 and 2:3, though they are frequently intergrown on a unit cell scale, as shown later. The first phase reported in this category had a T_c of only 20K, and contained only Sr[12)]. Phases containing Ca as well as Sr have T_c up to 110K. From the chemical similarity of Ca^{2+} and Sr^{2+} one might expect that they could replace one another: their ionic radii are respectively 0.94

and 1.10 A, less different than those of Y^{3+} and Ba^{2+} (0.90 and 1.29 A). If that were the case the Sr:Cu and Ca:Cu ratios would vary from sample to sample, and even from crystallite to crystallite in the sintered preparation.

The method of choice to determine chemical stoichiometry on a microscopic scale is analytical electron microscopy, in the form of energy dispersive X-ray fluorescence (EDX). This method has indeed been applied to the Bi,Sr,Ca,Cu oxides by several groups[13-15], but the results do not appear to have been subjected to statistical analysis. Furthermore, in a number of cases a stoichiometry has been assumed, for example in crystal structure analyses by X-ray[14] and neutron diffraction[16], that does not correspond precisely to what is observed by EDX. Thus for example Sunshine et al[17] report $Bi_{2.2}Sr_{1.7}Ca_2Cu_2O_{8.2}$ by EDX (averaged over an unrecorded number of crystallites) and H_2 reduction, but formulate the phase as $Bi_2Sr_2CaCu_2O_8$.

We have applied EDX to 35 crystallites of the phase approximating to Bi:Cu of 1:1 and 21 of the phase having Bi:Cu near 2:1[18]. The experiments were done in a JEOL 2000FX TEMSCAN instrument. Each sample was finely ground, dispersed in $CHCl_3$ and deposited on a nylon grid coated with a holey-carbon film. Data were analysed using the ratio method, which is applicable in the thin crystal limit[19]. As calibration standards we used Bi_2CuO_4, $SrCuO_2$ and Ca_2CuO_3. Each of the preparations examined in this way contained a single major phase, accounting for at least 80% of the sample. A.c. susceptibility also showed that each sample had at least 60% ideal diamagnetism at 77K so in each case the major phase present was the superconducting one. Minor (<5%) amounts of identifiable impurities such as Ca_2CuO_3 and ones containing only Ca or Sr were also found. It is important to notice that no crystallites of the 1:1 phase were detected in the 2:1 sample or vice versa.

Results of our EDX are displayed in Fig.2 as histograms of the various stoichiometric ratios, with the mean values of the ratios and estimated error bars. The latter are somewhat larger for the 2:1 than the 1:1 phase, but are quite sufficient to indicate approximate formulae. The 2:1 phase can be approximated as $Bi_2Sr_{1.25}Ca_{0.25}CuO_x$ and, bearing in mind the structure determinations (next section), the 1:1 phase is $Bi_2Sr_{1.67}CaCu_2O_x$. Note that the Sr:Cu ratio of 0.84(3) and the (Sr+Ca):Cu ratio of 1.32(3) are

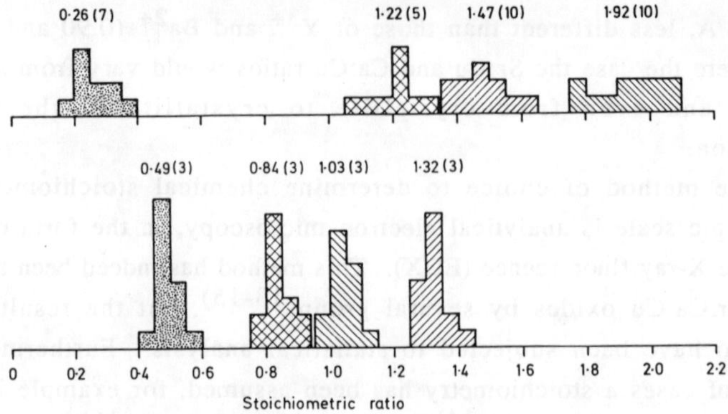

Figure 2. Stoichiometric ratios of Bi,Sr,Ca,Cu,O crystallites determined by EDX (Ref.18). Dotted Ca/Cu; cross hatched Sr/Cu; righthand hatched Bi/Cu; lefthand hatched (Sr + Ca)/Cu.

far from those suggested by the "ideal" formula $Bi_2Sr_2CaCu_2O_x$ that has been widely quoted. A similar situation occurs with the 2:1 compound, for which a formula $Bi_2Sr_{2-x}Ca_xCuO_6$ has been suggested[20]. Our interpretation would be $Bi_2Sr_{1.5-x}Ca_xCuO_6$.

The implications of these variations in formula for the microstructures of the 1:1 and 2:1 phases are considered in the next section. Nevertheless it is becoming clear that in contrast to the two earlier series (La,Sr) and (Ln,Ba), each of which represented a single structure type, the Bi, Sr, Ca, Cu compounds form a family of closely related structure types, most probably constituting a homologous series, of stoichiometry Bi_2Cu_n. We ourselves have observed intergrowths of an n=3 phase in n=2 samples by high resolution electron microscopy, and Zandbergen et al[21] recently reported n=4 intergrowths in a sample consisting predominantly of n=3.

3. STRUCTURAL COMPARISON BETWEEN SUPERCONDUCTING Cu PHASES

In this summary we draw mainly on neutron diffraction data obtained by our group, working on the High Resolution Powder Diffractometers (HRPD) at the ISIS pulsed neutron source of the Rutherford Appleton Laboratory, UK and D2B at the High Flux Reactor of the Institut Laue-Langevin, Grenoble,

France. Neutron diffraction is a particularly effective tool for locating O atoms since the scattering cross-section of this element is comparable to those of the higher atomic number elements La, Y, Cu etc. It is also valuable to use powder samples to avoid problems due to micro-twinning.

3.1 CuO_2 Layers

The principal feature of all the Cu high T_c phases is the presence in the crystal of square, or almost square, CuO_4 groups sharing corners to form infinite layers of CuO_2 stoichiometry. The Cu-O equatorial bond lengths in these layers are collected in Table 1. They are all close to 1.9 A, though one should note the significant distortion of the $Cu(2)O_4$ unit in YBa_2Cu_3-$O_{7-\delta}$. The environment of the Cu within these layers is not exclusively square coplanar, however, since further O lie above and below the Cu, at distances which vary quite considerably from one compound to another. Also, whereas there are two such apical oxygens in the La series, producing an elongated octahedral coordination about the Cu, in $LnBa_2Cu_3O_{7-\delta}$ and $Bi_2Sr_{1.67}CaCu_2O_8$ there is only one, so the Cu coordination is square pyramidal. On the other hand the Cu in "$Bi_2Sr_2CuO_6$", the pure Sr phase with T_c = 6K studied by Torrance et al[20] is elongated octahedral. One must presume that the structure of our $Bi_2Sr_{1.25}Ca_{0.25}CuO_x$ phase with T_c = 93K[18] is very similar, since the unit cell constants are almost identical. Clearly T_c is extraordinarily sensitive to the Sr:Ca ratio in the Bi_2Cu_1 series.

If the perovskite based phases are assumed to be tetragonal the equatorial Cu-O bond lengths can be estimated with precision from the unit cell *a* parameter since *a* equals the Cu-O-Cu distance. Thus these estimates are quite precise, even for those phases such as the Bi ones, for which only limited X-ray data are available. On the other hand the axially coordinated O is not on a special position, so the axial Cu-O bond length estimates rely for their precision on the quality and quantity of intensity data. Note, in particular, that the values of this bond length given for $Bi_2Sr_{1.67}CaCu_2O_8$ in Table 1 differ widely. One would anticipate that the neutron diffraction estimate will be the most precise.

3.2 Orthorhombic Distortion

Searching for further systematic features in the structures of Cu oxide superconductors one should also point out that on close examination,

Table 1. Structural comparison between Cu oxide superconductors

		$La_{1.85}Sr_{0.15}CuO_4$ (ref 22)	$YBa_2Cu_3O_{7-\delta}$ (ref 7a)	$Bi_2Sr_{1.67}$ (ref 16)	$CaCu_2O_8$ (ref 20)	$Bi_2Sr_2CuO_6$ (ref 20)
Cu-O bond lengths in CuO_2 layers (A):						
equatorial	(300K)	1.887	1.929, 1.958	1.914	1.906	1.905
	(20K)	1.882				
axial	(300K)	2.399	2.306	2.645	2.30	2.264
	(20K)	2.395				
Mean Cu-O in CuO_2 layers (A):	(300K)	2.058	2.016 Cu(2) 1.892 Cu(1)	2.060	1.985	2.025
Cu-O-Cu angle in CuO_2 layers	(300K)	180°	165°	?		
	(20K)	176.6°	165°	?		
Orthorhombic distortion $(a-b)/(a+b)\times10^3$		2.3(max) 2.0(min)	8.4	2.7 (ref 13) 1.4 (ref 15) 0.6 (ref 17)		

none of the CuO_2 layers has strictly tetragonal symmetry. This is shown in Table 1, where we have indicated the magnitude of orthorhombic distortion for the main classes in dimensionless units of the basal plane unit cell parameters (a-b)/(a+b). Thugh small, the distortions are easily detectable but, especially in the new Bi phases, are still subject to some experimental uncertainty, as indicated by the divergent estimates in Table 1. The crystallographic origin of the lowering in symmetry, like the origin of the non-integral Cu oxidation state, is different in each case.

In the 40K (La,M^{II}) series it comes from the tilting of the elongated CuO_6 octahedra about the c-axis, alternately along the (110) direction of the primitive perovskite unit cell. Thus rows of O atoms normal to this direction lie alternately above and below the plane of the Cu. The extent of tilting, and hence the Cu-O-Cu angle, varies with tmeprature since there is a second order transition to a high temperature tetragonal phase[22], at a temperature that varies with x in $La_{2-x}Sr_xCuO_{4-y}$. The limiting low temperature value of the Cu-O-Cu angle also varies smoothly with x, increasing from 173.5^o in undoped La_2CuO_4 to 180° when x=0.2. The latter composition is also the one beyond which superconductivity is no longer observed, though the samples remain metallic. On the other hand, as noted in the previous section, it may be that the loss of superconductivity with increasing x is better correlated with increasing y. In Table 2 we show the quantitative variation of Cu-O-Cu angle with x, measured at 300K and 5K by powder neutron diffraction[23]. The important point is that at the optimum x for superconductivity, about 0.15, the orthorhombic distortion is established well above T_c, so that at T_c itself the Cu-O-Cu angle is 3.0(3)^o [24].

Table 2. Variation of Cu-O-Cu angle in $La_{2-x}Sr_xCuO_{4-y}$ with x (ref 23)

x:		0	0.01	0.03	0.06	0.15	0.20
Cu-O-Cu angle:	5K	173.46	173.51	173.73	174.40	176.55	180
	300K	174.04	-	-	-	180	180

The orthorhombic distortion in the $LnBa_2Cu_3O_{7-\delta}$ compounds arises, not from tilting of the $Cu(2)O_5$ square pyramids, but from the chains of corner-linked square planar $Cu(1)O_4$, which are aligned exclusively along he b-axis, and share the apical O of $Cu(1)O_5$. That provides the reason for the two different equatorial Cu-O bond lengths (Table 1). There is no evidence that these bond lengths, or the Cu-O-Cu angles, are temperature dependent. The transition to a tetragonal phase at high temperature is an order-disorder transition of the O in the basal plane about Cu(1)[25]. A further tetragonal phase arises from removal, rather than disordering, to give phases with $\delta > 0.5$ that are not superconducting (or, when $\delta \sim 1$, even metallic) at low temperature[26].

In the new Bi phases, there was some doubt at first whether the true lattice symmetry was tetragonal or orthorhombic. Most early structure determinations were based on the assumption of an average structure that was tetragonal, but there is increasing evidence (especially from electron microscopy) that the microstructures are more complex. Major electron diffraction reflections index to a tetragonal cell based on the primitive perovskite lattice repeat *a* of Cu-O-Cu in the basal plane (about 3.8 A). However, closer examination shows further weak satellite reflections corresponding to a superlattice along the b-axis of almost, but not quite 5 $\sqrt{(2)}a$, i.e. the structure is incommensurate[15].

3.3 Modulated Structure of the Bi_2Cu_n Phases

High resolution lattice images show a similar type of incommensurate modulation in all the Bi_2Cu_n phases, from the low T_c Bi_2Cu_1 phase of Michel et al.[12] to the most recent Bi_2Cu_3 samples[21]. In projection along (001) a clear modulated contrast variation is observed, exclusively parallel to (010). Because all these crystals have a strong plate-like habit, it is not easy to observe lattice images in other orientations, especially perpendicular to the Cu and Bi layers. Fortunately, using ion milling, Dr. P.L. Gai in the Metallurgy Department at Oxford, has succeeded in obtaining samples that give high quality lattice images in this orientation in the 400kV microscope[27]. The latter has a point to point resolution of about 1.4 A, so individual heavy Bi atoms are resolved, while the lighter Cu, Sr and Ca show as columns normal to the Bi layers.

The most striking result of these measurements is that the nearest neighbour Bi-Bi separation along the *b* axis is not constant, but exhibits a sinusoidal modulation. Figure 3 shows the result of directly measuring this separation in two adjacent layers of $Bi_2Sr_{1.67}CaCu_2O_8$ from the lattice image: the maximum and minimum Bi-O-Bi distances of 6.6 and 5.2 A occur approximately five unit cells apart, so the variation in spacing from one unit cell to the next is in the region of 4%. Furthermore, the modulations in neighbouring layers are out of phase. Why are they formed?

Two different types of explanation could be envisaged for the modulations. First, there might exist a basic incompatibility between the Bi-O distances, which have a strong preference for 2.7 A in a wide range of Bi oxide layer compounds of the Aurivillius type[28)], and the preferred Cu-O distance within the CuO_2 layers, which, as we have seen, is equally strongly fixed at 1.9 A. In the ideal tetragonal structures of the Bi_2Cu_n phases, $d(\text{Bi-O}) = \sqrt{2}d(\text{Cu-O})$. However, if the energy-distance relationships for the two types of bonds taken separately were such that their equilibrium bond lengths did not quite satisfy the requirement for the tetragonal structure, competition would take place between them. Competing interactions are a classic cause of incommensurability.

The situation would be analogous to that found in discotic mesophases, which are composed of stacks of disc-like aromatic molecules with long chain aliphatic hydrocarbon groups attached to their peripheries. The equilibrium distance between the aromatic discs is different from that which would minimize the interaction energy between the aliphatic side-chains, yet the

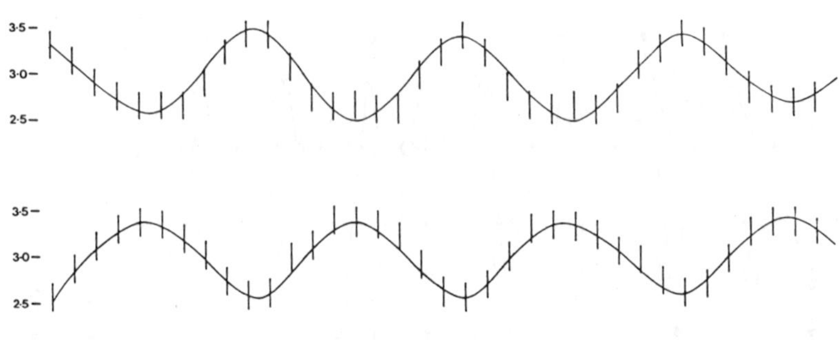

Figure 3. Sinusoidal modulation of Bi-Bi spacing in the basal plane of $Bi_2Sr_{1.67}CaCu_2O_8$ (Ref.27).

discs and their sidechains are connected by covalent bonds. The total energy of the combined system is minimized by alternating regions of the stack where the intermolecular spacing is determined primarily by the disc-disc interaction with regions where the chain-chain interaction 'wins'. The result is a long wavelength sinusoidal modulation in molecular spacing, that has been called 'pincement de Skoulios'[29)]. In the present case the 'connections between the CuO_2 and Bi_2O_2 layers are made through the apical O of the CuO_5 square pyramids, which are shared with the Bi, so there is a Cu-O-Bi 'hinge'.

A second type of explanation relies on a consideration of the chemical stoichiometry determined by EDX which, as we pointed out in the previous section, is crucially different from what has been assumed in the ideal formulae of the Bi_2Cu_n phases. Since the Bi_2O_2 and CuO_2 layers share common O vertices, the modulation in Bi-Bi separation must be accompanied by a corresponding modulation in the tilt angle of the CuO_5 to either side of the c-axis. Such a tilt will also cause a displacement of all the cations between the layers, and explain why the Bi-Bi modulations are out of phase in adjacent layers. The resulting structure of one 15.5 A thick layer in $Bi_2Sr_{1.67}BaCu_2O_8$ is shown schematically, and with exaggerated tilting angle, in Figure 4. Now we can see that there must also be a modulation in the sizes of the O cavities containing the Sr and Ca. If the modulation along b were precisely five-fold, instead of 4.8, smaller cavities would appear in the Sr layer every five repeat units. Suppose we assume that these are too small to accommodate an Sr, and are vacant, as shown in Figure 4. Then the

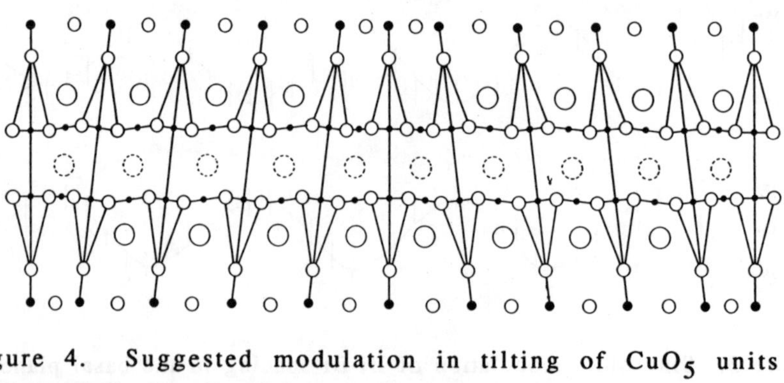

Figure 4. Suggested modulation in tilting of CuO_5 units in $Bi_2Sr_{1.67}CaCu_2O_8$. Small dots are Cu, large dots Bi, small open circles O, large open circles Sr, large dotted circles Ca.

chemical formula becomes $Bi_2Sr_{1.6}CaCu_2O_8$, very close to what is found[18]. Alternatively, because the structure is actually incommensurate, one could envisage that in the smaller cavities of the Sr layer, Sr is replaced by Ca, so that an equivalent number of vacancies appear in the Ca layer. The latter alternative provides a readier explanation for the fractional number of group 2 cations, since the Ca can be displaced more easily in the b-direction than the Sr.

Now perhaps all of this has nothing to do with the superconductivity of the Bi_2Cu_n phases. However, I think it does, because the modulated cation vacancies provide a simple structural mechanism for achieving an average Cu oxidation state greater than 2, while retaining a completely ordered lattice of O vacancies. Alternative formulae, like those of refs.(13-17) postulated random addition of extra O in the Ca layer, not, in my view, a very plausible hypothesis, in view of the fact that the chemical stoichiometry is fixed. It is unlikely that a structural model as complex as Figure 4 will be verified by powder neutron diffraction and single crystal work is needed. Modelling of the lattice images is also in progress.

4. CONCLUSION

The existence of several families of ternary and quaternary Cu oxide superconductors enables us to identify the structural and electronic constraints that govern the appearance of what, to a chemist, appears a bizarre phenomenon. Structurally, one needs an infinite layer formed by corner-sharing planar or nearly planar CuO_4 units, producing a stoichiometry CuO_2. Chains are not required. The Cu-O bond lengths within the layer must be within 2% of 1.9 A, but the Cu-O-Cu angle need not be 180°. Deviations up to $6\text{-}7^o$ seem to be tolerable. The symmetry of the layer should be close to, but not quite, tetragonal though the microstructural nature of the distortion lowering the symmetry to orthorhombic may differ from one structure to another. Electronically, the requirement is for a mean Cu oxidation state (defined as in section 2) greater than 2, i.e. an $(x^2\text{-}y^2)$ band that is less than half filled. The ideal stoichiometry of the Bi_2Cu_2 phase indicated by our EDX studies yields a mean Cu oxidation state of 2.33, comparable to that in $YBa_2Cu_3O_{7-\delta}$. Enough electronic and structural evidence is now available to be quite sure that any new Cu oxide superconductors will have to fulfil the above criteria.

ACKNOWLEDGEMENT

Thanks are due to all my collaborators whose names appear in the references, to the ILL and ISIS neutron sources for beam time and SERC for financial support.

REFERENCES

(1) Jorgensen, C.K., Oxidation Numbers and Oxidation States, Springer, 1971, passim.

(2) Sarma, D.D., Sreedan, K., Ganguly, P. and Rao, C.N.R., Phys.Rev.*B* 36, 371 (1987).

(3) Shafer, M.W., Penney, T. and Olson, B.L. Phys. Rev. B 36, 4047 (1987).

(4) Robin, M.B. and Day, P. Adv. Inorg. Chem. Radiochem. 247, (1967).

(5) Tilley, R.J.D. and Davies, A.H. Nature 326, 859 (1987).

(6) e.g. Steinfink, H., Swinnear, J.S., Sui, Z.T., Hsu, H.M., Goodenough, J.B., J. Am. Chem. Soc. 109, 3348 (1987).

Cava, R.J., Batlogg, B., van Dover, R.B., Murphy, D.W., Sunshine, S.A., Siegrist, T., Remeika, Rietman, E.A., Zahurak, S., and Espinosa, G.P. Phys. Rev. Lett. 58, 1676 (1987).

Manthiram, A., Swinnea, J.S., Sui, Z.T., Steinfink, H., and Goodenough, J.B., J. Am. Chem. Soc. 109, 6667 (1987).

Gallagher, P.K., O'Bryan, H.M., Sunshine, S.A., and Murphy, D.W., Mat. Res. Bull. 22, 995 (1987).

(7) David, W.I.F., Harrison, W.T.A., Gunn, J.M.F., Moze, O., Soper, A., Day, P., Jorgensen, J.D., Beno, M.A., Capone, D.W., Hinke, D.C., Schuttler, I.K., Soderholm, L., Segre, C.V., Zhang, K. and Grace, J.D., Nature 327, 310 (1987); Capponi, J.J., Chaillot, C., Hewat, A.W., Lejay, P., Marezio, M., Nguyen, N., Raveau, B., Soubeyroux, J.L., Tholence, J.L., and Tournier, R., Europhys. Lett. 3, 301 (1987).

(8) Tranquada, J.M., Cox, D.E., Kunnmann, W., Moudden, H., Shirane, G., Suerage, M., Zollyker, P., Phys. Rev.B, in press.

(9) Cotton, F.A., and Wilkinson, G., Advanced Inorganic Chemistry, J. Wiley, 1984.

(10) Whangbo, M-H., Evain, M., Beno, M.A. and Williams, J.M., ch. in 'High Temperature Superconductivity: Synthesis, Properties and Processing' ed. W.E. Hatfield, Marcel Dekker, 1988.

(11) Maeda, A.H., Tanaka, Y., Fukutomi, N., and Asano, T., Jap. J. Appl. Phys. 27, 2 (1988).

(12) Michel, C., Hervieu, M., Bond, M.M., Grandin, A., Deslandes, F., Provost, J. and Raveau, B., Z. Phys. B68, 421 (1987).

(13) Hazen, R.M., Prewitt, C.T., Angel, R.J., Ross, N.L., Finger, L.W., Hadidacos, C.G., Veblen, D.R., Heaney, P.J., Hor, P.H., Meng, R.L., Sun, Y.Y., Wang, Y.Q., Xue, Y.Y., Huang, Z., Gao, L., Bechtold, J., and Chou, C.W., Phys,. Rev. Lett. 60, 1015 (1988).

(14) Subramanian, M.A., Torardi, C.C., Calabrese, J.C., Gopalkrishnanm, J., Morrissey, K., Askew, T.R., Fippen, R.B., Chowdhry, U., and Sleight, A.W., Science, 239, 1015 (1988).

(15) Shaw, T.M., Shivshankar, C.A., LaPlaca, S.J., Cuomo, J.J., Mcguire, T.R., Roy, R.A., Kellher, K.H., and Yee, D.S., submitted to Phys. Rev. Lett.

(16) Bordet, P., Capponi, J.J., Chaillout, C., Chenevas, J., Hewat, A.W., Hewat, E.A., Hodeau, J.L., Marezio, M., Tholence, J.L., and Tranqui, D., to appear in Proceedings of the Interlaken Conf. on High Tc Superconductors, 29 Feb-4 Mar. 1988, Editors K.A. Muller and P. Olsen, Physica C (Superconductors).

(17) Sunshine, S.A., Siegrist, T., Schneemeyer, L.F., Murphy, D.W., Cava R.J., Batlogg, B., van Dover, R.B., Fleming, R.M., Glarum, S.H., Nakahara, S., Farrow, R., Krajewski, J.J., Zahurak, S.M., Waszczak, J.V., Marshall, J.H., Marsh, P., Rupp, L.W.Jr., and Peck, W.F. (1988) preprint.

(18) Chippendale, A.M., Hibble, S.J., Hriljac, J.A., Cowey, L., Bagguley, D.M.S., Day, P., and Cheetham, A.K., Physica-C, 152, 154 (1988).

(19) Cheetham, A.K., and Skarnulis, A.J., Anal. Chem., 53, 1060 (1987).

(20) Torrance, J.B., Tokura, Y., LaPlaca, S.J., Huang, T.C., Savoy, R.J., and Nazzal, A.I., Sol. St. Commun., in press.

(21) Zandbergen, H.W., Huang, Y.K., Menken, M.J.V., Li, J.N., Kadowski, K., Manovsky, A.A., Van Tendeloo, G., and Amelinkx, S., Nature, 332, 620 (1988).

(22) Day, P., Rosseinsky, M.J., Prassides, K., David, W.I.F., Moze, O., and Soper, A., J. Phys. C: Sol. St. Phys., 20, L429 (1987).

(23) Rosseinsky, M.J., Prassides, K. and Day, P., unpublished.

(24) Day, P., ch. in 'High Temperature Superconductivity: Preparations, Properties, Processing', W.E. Hatfield, ed., New York, Marcel Dekker, 1988.

(25) Manthiram, A. and Goodenough, J.B., Nature, 329, 701 (1987).

(26) Swinnea, J.S. and Steinfink, H., J.Mat.Res. 2, 424 (1987).

(27) Gai, P.L., and Day, P., Physica C, in press.

(28) Newnham, R.E., Wolf, R.W. and Dorrian, J.F., Mat. Rev. Bull., 6, 1029 (1971).

(29) de Gennes, P.G., J. Physique (Lettres) 44, L657 (1983).

Energy Gap Measurements of High T_c Superconductors with Tunneling Methods

K. W. Ng, S. Pan and A. L. de Lozanne

Physics Department, University of Texas, Austin, TX 78705, U.S.A.

Abstract

We measured the energy gap of high T_c superconductors with a low temperature STM. High quality stable curves can be obtained by a proper choice of tip material. Gap values for YBaCuO samples occurred more frequently in the ranges of 40 to 45 mV and 80 to 90 mV. The smaller range is nearly 3 times as large as the BCS value, 13.6 mV.

The observation of Shapiro steps in Josephson tunneling is strong evidence that electrons are paired in the superconducting state of the high T_c materials [1]. It is natural to ask how large is the energy gap and how does the gap behave in comparison with the conventional theories of superconductivity. The gap has been measured extensively by the tunneling technique, but no consistent gap value has been obtained yet. This is probably due to the poor surface condition of this type of high T_c materials. In the future, improved sample quality will certainly improve the measurements. Besides the energy gap, tunneling spectroscopy can also provide the density of electron states, which may include crucial information on the pairing mechanisms.

Tunneling experiments can be done by either conventional sandwich type junctions or mechanical junctions, like breakable junctions, point contacts or low temperature STM. All these methods have been tried by different groups already. For YBaCuO, Moreland et al. [2] reported a gap of 20 to 40 mV with the breakable junction method. Iguchi et al. [3] reported a gap size of 20 to 25 mV by evaporating a lead electrode on bulk YBaCuO. The energy gap also shows a BCS-like temperature dependence, but it vanished at a temperature of about 48K. So far no good temperature dependence results have been obtained from mechanical junctions. This is due to the instability of the mechanical device against temperature change (which produces thermal drift). With point-contact junctions (of typical resistance 10 Kohm), Gray et al. [4] obtained gap values between 25 to 35 mV. With a low temperature STM, Kirtley et al. [5] reported values that are closest to the BCS value (13.6 mV), between 13 and 20 mV. On the other hand, with the same technique, we rarely observe energy gap smaller than 40mV [6,7]. From all the values above, we can see that although the gap values vary, they are always larger than or equal to the BCS value.

Another interesting feature from tunneling data is the occurrence of multiple gaps. This was first observed by Naito et al. [8] in the case of the LaSrCuO system. This feature raised skepticism about the validity of the tunneling results. Some authors [9] have interpret the multiple gaps as a result of charging effects for small particle tunneling. No conclusion can be drawn until we can prove that the gaps are related to superconductivity by changing temperature or applying an external field. Unfortunately, as mentioned before, mechanical junctions are not very suitable for this purpose. In the Iguchi et al. [3] report, they did observe three features with a BCS-like temperature dependence that vanished together at 48K.

All of our tunneling work was performed with a low temperature STM with an operating temperature of about 10K. A load lock and manipulator system allow us to break the sample (bulk) in situ at low temperature before measurement. Two types of sample have been measured: sintered bulk samples and thin films. They behaved differently only in that thin films gave a higher probability of obtaining a good junction and a better quality of the tunneling curve, there was no significant difference in gap values.

We are currently using Al and In tips, after some experience with W and PtRh tips [6]. For the best data, we found that tip material had no effect on the magnitude of gap values. Some of the tunneling curves are severely smeared out, especially for tungsten tips. This can be explained by current depairing due to small tunneling area. Al or In tips can lightly touch the sample with their oxide layer as a barrier, making the junction more stable. At the same time, the tip can also be deformed to match the surface, hence increasing the effective tunneling area and the barrier thickness. The gap smearing and curvature at high bias are much reduced in this case. (Fig. 1)

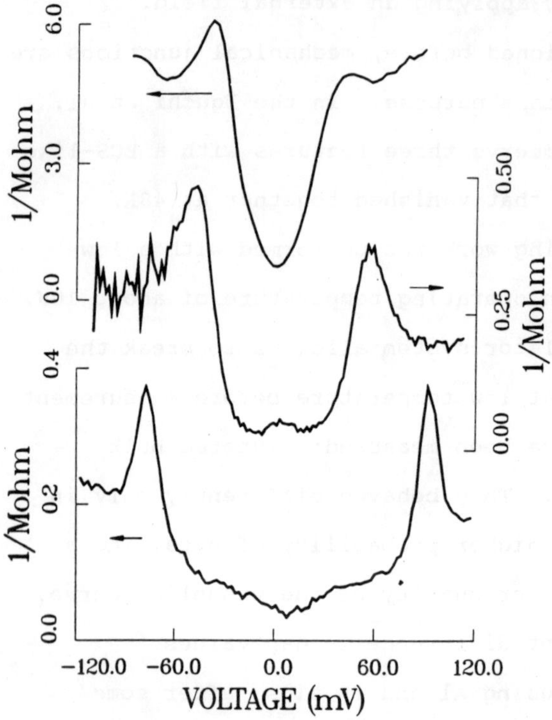

Fig. 1
Tunneling curves for YBaCuO with different tip materials. The two lower curves have sharper gap features because aluminum tips were use instead of tungsten (1st curve). Note that their gap values are 44.8 mV and 85 mV, after correction with depairing effect [7]. These values fall within the ranges we discussed in this paper.

We rarely observed curves with more than two gaps up to a bias of +/- 250 mV. In the case of double gap, the second peak was twice as large as the first one (Fig. 2). This does not follow the peak relationship $\Delta = 1\Delta_o, 3\Delta_o, 5\Delta_o$, ... given by the charging mechanism [10]. We interpret this double gap feature as a combination of SIN and SIS junctions in parallel. For those curves with a single sharp peak, we believe only one junction is involved. All of our

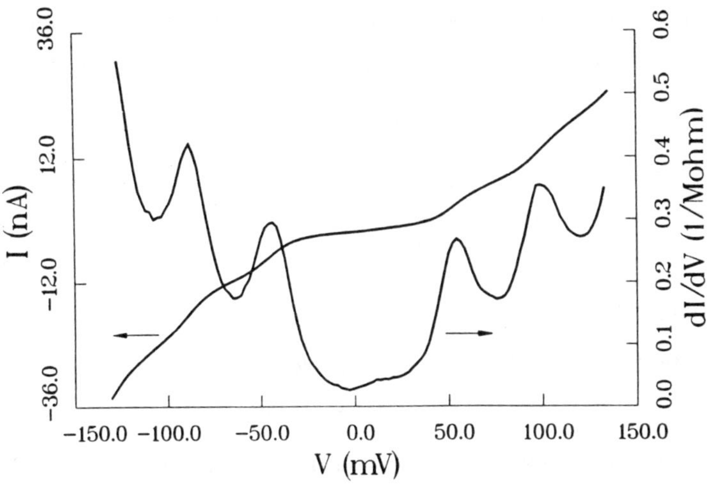

Fig. 2 One of our tunneling curves for YBaCuO showing the double gap feature.

good quality curves have a single gap, very low and flat zero-bias conductance. It is unlikely that small particle tunneling is playing an important role in these curves.

Different gap values with large variation have been observed, but interestingly, the gap values occur more frequently in the ranges of 40 to 45 mV and 80 to 90 mV. If we interpret the lower range to be the gap values and the higher range as a result of SIS tunneling, $2\Delta/kT_C$ will be about three times as large as the BCS theory. Although our values are large in comparison with other authors, it is in agreement with Kapitulnik et al. [11] and Iguchi et al.[3], taking T_C=48K in the latter data.

In conclusion, we obtained different values of the gap of YbaCuO, mostly larger than 40mV, occurring frequently in the ranges 40-45 mV and 80-90 mV. It is possible for other mechanisms such as small particle tunneling to produce a gap-like effect. Before we can provide direct evidence that the gap is related to charging effects or superconductivity, one can only judge according to the curve quality, namely, low conductance within the gap, sharp peaks at the gap, and small curvature at high bias.

ACKNOWLEDGEMENTS

This work is supported by the Texas Advanced Technology Project and by NSF (DMR 8353305) through a Presidential Young Investigator Award (A. L. de Lozanne) with a matching funds from IBM, Bell Communication Research (Bellcore), Texas Instruments, Kodak, and the Microelectronics and Computer Technology Corporation (MCC). LSCO samples were prepared by J. M. Tarascon and L. H. Greene (Bellcore). YBaCuO samples were prepared by A. J. Panson and J. Talvacchio (Westinghouse).

REFERENCES

1. J. S. Tsai, Y. Kubo and J. Tabuchi, Phy. Rev. Lett. 58 1979 (1987)

   J. Moreland, L. F. Goodrich, J. W. Ekin, T. E. Capobianco and A. F. Clark Jap. J. of Appl. Phy. 26 suppl 26-3 pp 999-1000 (1987)

2. J. Moreland, J. W. Ekin, L. F. Goodrich, T. E. Capobianco, A. F. Clark, J. Kwo, M. Hong and S. H. Liou Phys. Rev. B 35 8856 (1987)

3. I. Iguchi, H. Watanabe, Y. Kasai, T. Mochiku, A.

Sugishita and E. Yamaka Japan J. Appl. Phys. 26 L645 (1987)

4. K. E. Gray, M. E. Hawley and E. R. Moog in Novel Superconductivity pp 611-625, edited by S. A. Wolf and V. Z. Kresin, Plenum Press (1987)
5. J. R. Kirtley, R. T. Collins, Z. Schlesinger, W. J. Gallagher, R. L. Sandstrom, T. R. Dinger, D. A. Chance Phys. Rev. B 35 8846 (1987)
6. S. Pan, K. W. Ng, A. L. de Lozanne, J. M. Tarascon and L. H. Greene Phys. Rev. B 35 7220 (1987)
7. K. W. Ng, S. Pan, A. L. de Lozanne, A. J. Panson and J. Talvacchio Jap. J. of Appl. Phy. 26 suppl 26-3 pp993-994
8. M. Naito, D. P. E. Smith, M. D. Kirk, B. Oh, M. R. Hahn, K. Char, D. B. Mitzi, J. Z. Sun, D. J. Webb, M. R. Beasley, O. Fischer, T. H. Geballe, R. H. Hammond, A. Kapitulnik and C. F. Quate Phys. Rev. B 35 7288 (1987)
9. S. T. Ruggiero and J. B. Barner Phys. Rev. B 36 8870 (1987)
10. P. J. M. van Bentum, R. T. M. Smokers and H. van Kempen Phys. Rev. Lett. 60 2543 (1988)
11. A. Kapitulnik, B. Oh, M. Naito, K. Char, A. D. Kent, N. Missert, E. Hellman, S. Arnason, J. W. P. Hsu, M. R. Hahn, P. Rosenthal, R. Barton, M. R. Beasley, T. H. Geballe and R. H. Hammond Proceedings of "High T_C Superconductors" ICTP, Trieste, Italy. Published by World Scientific Publishing, Inc.

CHARACTERIZATION OF PEROVSKITE-BASED SUPERCONDUCTORS BY SURFACE ANALYSIS

Kenneth D. Bomben and William F. Stickle

The Perkin-Elmer Corporation, Physical Electronics Laboratory, 1161-C San Antonio Road, Mountain View, California 94043 USA

ABSTRACT

Perovskite-based Y-Ba-Cu oxides are known to exhibit superconducting transition temperatures near 90 K. Superconducting and non-superconducting samples have been examined using X-Ray Photoelectron Spectroscopy (XPS), Auger Electron Spectroscopy (AES) and Secondary Ion Mass Spectrometry (SIMS). The core- and valence-level spectra will be compared to standard oxides. High resolution XPS spectra reveal subtle changes in the electronic structure between superconducting and non-superconducting ceramics. Carbonates are observed in the carbon 1s regions and a mechanism of the degradation of these materials is thereby suggested. High spatial resolution AES can distinguish the different phases and stoichiometries of the heterogeneous materials that are produced in the preparation of superconductors. The level of impurities shown in the SIMS data have an unknown role on the superconducting transition.

INTRODUCTION

Cava et al.[1] report that the ternary oxide system of BaO, CuO and Y_2O_3 forms a single phase, orthorhombic structure with the formula $YBa_2Cu_3O_{7-\delta}$. This compound exhibits a superconducting transition between 90 and 95 K. We have studied the effects of compositional variation and quenching temperature on the superconductive and microstructural behavior of the ternary system. Our results show that while $YBa_2Cu_3O_{7-\delta}$ forms a single phase orthorhombic structure, a slight change in the yttrium concentration leads to a multiphased material. In the presence of non-superconducting phases, the superconducting transition is decided by percolation effects.

An understanding of the electron transport properties and the electronic structure of oxygen-deficient $YBa_2Cu_3O_{7-\delta}$ perovskites is of considerable interest because of the superconducting properties of these compounds. The superconducting transition temperature, the transition width and the extent of orthorhombic distortion of these materials are highly sensitive to the oxygen concentration. The amount of oxygen available during sintering and cool-down establishes the oxygen vacancies in the perovskite lattice. An additional objective of this work was to address the effects of a gradual change in the the oxygen concentration of $YBa_2Cu_3O_{7-\delta}$ on its electron transport properties, the chemical bonding states of copper and the superconducting transition temperature.

EXPERIMENTAL

Superconducting and non-superconducting samples of Y-Ba-Cu oxides were prepared by grinding $BaCO_3$, Y_2O_3 and CuO together and heating the mixture at 950°C for 24 hours. The mixture was reground and cold pressed into 0.3cm. thick, 2.5cm diameter pellets. The pellets were heated at 950 °C for 24 hours in flowing oxygen, followed by cool-down to room temperature over a two hour period. The oxygen flow was maintained during the cool-down. X-ray diffraction analysis confirmed that the samples were single phase with an orthorhombic perovskite unit cell (a = 3.821Å, b = 3.901Å and c = 11.698Å). Oxygen deficient samples were created by rapid cooling on a thick copper plate during the cool-down cycle . The temperature from which the sample was quenched varied between 500 and 1000 °C.

Electrical resistivity measurements were carried out with a four-point probe with copper wires as current and voltage leads over a temperature range of 63 to 250 K. Temperatures below 77 K were produced by lowering the pressure in the liquid nitrogen Dewar. X-Ray Photoelectron Spectroscopy (XPS, also known as Electron Spectroscopy for Chemical Analysis, ESCA) data were taken with a Perkin-Elmer model 5400 ESCA spectrometer equipped with an Omni Focus™ lens. The data were collected with a spectrometer resolution of less than 0.9 eV FWHM relative to Ag $3d_{5/2}$ using Mg K_α radiation (1253.6 eV). High-resolution secondary electron micrographs of a cross section of the pressed pellets show differences between the morphology of the sample's outer surface and its interior. For this reason the samples were fractured *in vacuo* and data were taken from the center of the fracture area (approximately 1mm in diameter). The deviation in the oxygen concentration from the equilibrium superconducting composition ($YBa_2Cu_3O_{6.9}$), caused by the quenching was established by thermogravimetric analysis (TGA). The TGA was done using a Perkin-Elmer model TGA 7 Thermo-Gravimetric Analyzer. The specimen was heated at 40 °C/minute to 200 °C, held isothermally for one hour, heated at 40 °C/minute to 1000 °C, then held isothermally for two hours. Auger Electron Spectroscopy (AES) and Secondary Ion Mass Spectrometry (SIMS) analyses were performed using a Perkin-Elmer model 660 Scanning Auger Microprobe with a model 3600 SIMS II add-on.

RESULTS AND DISCUSSION

Variations in the Stoichiometry

Figure 1 shows the variation of the electrical resistance for samples with different Y:Ba ratios. Samples A (Y:Ba = 1:4) and B (Y:Ba = 1:2) show a well defined T_c at 93 K. Sample C (Y:Ba = 1:1) shows a sudden drop in resistance at approximately 95 K, but does not reach a superconducting state (to 63 K). Also, unlike the other samples, the temperature coefficient of resistance of Sample C is negative. This semiconductor-like behavior and the absence of a fully superconducting state suggest that a large fraction of the sample consists of an insulating phase.

X-ray diffraction analysis was conducted to identify the crystalline phases present in these samples. In accordance with the observations of Cava et al.[1], Sample B consisted of a single phase material with an orthorhombic unit cell[2]. Sample A has prominent peaks in the diffraction patterns originating from $BaCuO_2$ as well as $YBa_2Cu_3O_{7-\delta}$ while Sample C shows Y_2BaCuO_5 and CuO peaks as well. Comparing the x-ray diffraction data with the electrical resistivity measurements, it is evident that the absence of a fully superconducting state and the negative temperature coefficient of resistance in Sample C is due to the presence of a Y_2BaCuO_5 compound. Although Sample A also contains a non-superconducting phase ($BaCuO_2$), its zero resistance state below 83 K suggests that the fraction of $BaCuO_2$ is not enough to close all the superconducting percolation paths of $YBa_2Cu_3O_{7-\delta}$ in the material. In this case, the non-superconducting phase increases the width of the superconducting transition.

SIMS shows the presence of significant amounts of strontium, iron and chromium. Lithium, sodium, potassium, calcium, and aluminum were also detected along with the expected barium, copper and yttrium. The impact of the contaminants in the superconducting material on its properties is unknown.

The XPS spectra of the samples, referenced to adventitious carbon at 284.6 eV, show changes in the chemistry of the yttrium and barium, and of the chemistry reflected in the valence bands. The chemistry of the copper, as seen in the copper 2p region, does not vary significantly from sample to sample. The analyzed areas result from intergranular fractures and, hence, the observed chemistry does not necessarily reflect the bulk chemistry or stoichiometry of the sample. The atomic concentrations for the three samples with different stoichiometries are tabulated in Table I.

Table I
Atomic Concentrations (%)

Sample	C	O	Y	Ba	Cu
Sample A	17	51	7.7	9.5	15
Sample B	17	46	9.8	8.2	19
Sample C	24	52	2.7	10	10

For Sample B the barium and yttrium are in essentially one chemical environment. The barium, Figure 2, appears as the carbonate despite the heat treatment during sample preparation. A small amount of barium oxide is also present. The other samples exhibit differences in the barium chemistry, in Sample C the barium appears as both an oxide and a carbonate, while in Sample A the barium is carbonate only. The yttrium 3d shown in Figure 3 compares the single oxide of Sample C with the hydroxide and oxides observed in Sample A.

A significant amount of carbon, as carbonate and hydrocarbon, is observed on the cleaved surfaces. In the C 1s spectrum of Figure 4 the carbonate is the high binding energy peak while the hydrocarbon is observed at lower binding energy. This suggests that the porous nature of the material allows the atmosphere to percolate through and provides a means by which the material can degrade, as observed in the large amount of carbonate within the sample.

Additional XPS spectra were taken on Sample B at approximately 80 K. No changes in the chemistry of any of the elements below transition temperature were observed. The XPS data of the copper observed in the surface were essentially the same for each sample. The copper 2p of Sample B, Figure 5, appears to show the copper in multiple chemical environments; the curve-fit suggests three atomic environments. The data from these samples, however, are the same as the XPS spectrum of Cu(II) oxide powder. No indication of the existence of either Cu(I) or Cu(III) is found. Furthermore, prolonged exposure of the sample to the vacuum caused the copper to be reduced, presumably due to the loss of oxygen from the lattice. The reduction is readily observed in the copper 2p spectra shown in Figure 6 where the shake-up satellites associated with Cu(II), the features of curve a at about 944 and 964 eV, are missing.

Variations in the Quenching Temperature

Figure 7 shows the electrical resistivity measurements of a slow-cooled, superconducting sample (curve a) and of a series of rapidly quenched samples. The slow-cooled sample has a temperature coefficient of resistance characteristic of a metal with the onset of superconductivity at 93 K and the zero resistance state at 89 K. Destruction of the superconducting transition is observed with increasing quenching temperatures. Samples quenched from 500, 600 and 700 °C show metallic conduction; however, their resistivity increases with the quenching temperature. Samples quenched from 700 °C do not show any indication of the onset of superconductivity (to 77 K). Upon quenching at 800 °C, the resistivity between 77 and 120 K shows an upturn. Samples quenched from 900 °C or above have resistivities that are characteristic of insulators.

The range of quenching temperatures between 900 - 950 °C, which results in conductivity between 40 $\Omega^{-1}cm^{-1}$ and 15 $\Omega^{-1}cm^{-1}$, marks a metal-insulator transition of this system. The conductivity of the insulating samples is characteristic of thermally activated conduction. However, the activation energy remains temperature dependent for all samples. This behavior indicates a phonon-assisted quantum mechanical tunneling of charge carriers in localized states at the Fermi energy.

A TGA scan of a normally cooled superconductor has a total weight loss of 4%. Assuming that the initial chemical formula of the superconductor was $YBa_2Cu_3O_{6.9}$, and that all the weight loss comes from oxygen in the lattice, the observed 4% weight loss results in a formula of $YBa_2Cu_3O_{5.24}$.

TGA scans were run on the quenched samples with the same scan conditions as for the superconducting sample. Figure 8 shows the total weight loss of the quenched samples as a function of temperature at quenching, assuming the weight loss is from oxygen in the lattice. The oxygen concentration varies from 6.2 to 5.9. By comparing this data with the results of electrical conductivity measurements, it is evident that the onset of the insulating behavior corresponds to a loss of approximately 0.5 oxygen per formula unit. This happens to the samples quenched from temperatures over the range of 850 to 925 °C. X-ray diffraction studies reveal that the extent of orthorhombic distortion (a - b) diminishes on increasing the quenching temperature. These changes in the a and b lattice parameters are accompanied by a progressive expansion of the lattice along the c axis. The samples quenched from 800 °C were essentially tetragonal in character. Since the copper atoms in the perovskite structure have an octahedral coordination, the observed equalization of the a and b axes and the expansion along the c axis have the net effect of enlarging the crystal field splitting of the d_{z^2} and $d_{x^2-y^2}$ components of the e_g band of the copper atoms[3]. The TGA studies indicate that the transition is caused by the loss of oxygen from the material as has been observed by Cava et al[1]. The change in the electrical conduction behavior from metal to insulator around 900 °C quenching temperature appears to be due to the enlarged crystal field splitting and a varying degree of localization in the copper 3d-oxygen 2p hybridized orbitals.

XPS spectra of these samples show variations in the chemistry of the yttrium, and barium, and of the chemistry reflected in the valence bands. The chemistry of the copper, as seen in the copper 2p, does not vary significantly from sample to sample. The surfaces that were examined resulted from the intergranular fracture and hence the observed chemistry does not necessarily reflect either the bulk chemistry or the stoichiometry of the sample. The atomic concentrations of the quenched samples are shown in Table II.

Table II
Atomic Concentrations (%)

Sample	C	O	Y	Ba	Cu
Superconductor	17	51	7.7	9.5	15
Expected		54	7.7	15	23
1000 °C	15	52	5.4	11	17
900 °C	21	51	5.7	9.4	13
800 °C	15	55	5.6	11	14
700 °C	17	51	7.4	8.5	15

The starting material, a superconductor, has barium and yttrium in essentially one chemical environment each. The barium appears as the carbonate despite the heat treatments during sample preparation. A small amount of oxide is also present. The quenched samples exhibit differences in the barium[3] chemistry but not with identifiable trends. For example, in the 700 °C quench sample the barium appears as an equal mixture of oxide and carbonate while the 900 °C sample is predominantly carbonate. The 1000 °C sample shows a mixture of oxides while the 800 °C sample is again predominantly carbonate. The yttrium 3d, shown in Figure 9, shows little change except for the 1000 °C quench sample where essentially only one oxide is evident rather than the oxide-hydroxide mixture observed in the other samples. The copper spectra appear to show copper in a multiplicity of chemical environments, however, it should be noted that these data are not different than the Cu 2p spectrum taken from Cu(II) oxide powder.

A significant amount of carbon is observed on the cleaved surfaces. The carbon, again, appears as both carbonate and as hydrocarbon. The presence of carbonates raises some questions about the interpretation of the TGA data. Sputter etching of the samples show the carbonaceous material to be less than 100Å thick. Hence, the loss of CO_2 will only slightly skew the oxygen-loss numbers.

High-resolution scanning AES images of the surface show significant variation in the composition of surface of the individual grains. In addition to numerous different morphologies, there is no evidence of uniform atomic distribution within individual grains. This would seem to correlate with a recent report indicating that superconductors do not exhibit superconductivity in their uppermost surfaces[4].

SUMMARY

The measurements of bulk properties indicate that changes in the stoichiometry result in changes in T_c, even to the point of making the material a semiconductor. Further, the quenching temperature and, hence, the oxygen vacancy concentration of a superconducting material also causes changes in T_c, even to the point of making the material an insulator. The XPS data suggest that exposure of the superconducting material to the air causes a certain amount of degradation of the material over a period of time by the formation of carbonates, oxides, and hydroxides of barium which accumulate at the grain boundaries. This suggests a degradation mechanism which has implications for the long-term stability of these materials. The TGA experiments show increasing weight loss with decreasing quenching temperature and can be correlated to the transition from an orthorhombic unit cell to a tetragonal unit cell.

ACKNOWLEDGEMENTS

We wish to thank Colin Williams of Perkin-Elmer, Instrument Group (Irvine, California, USA) for the interpretation of the TGA analysis. The samples were prepared by Ramesh Budhani of Roitan Bunshah's group in the Materials Sciences Department of the University of California, Los Angeles, USA.

REFERENCES

[1]R. J. Cava, B. Batlozz, R. B. Van Dover, D. W. Murphy, S. Sunshine, T. Siegrist, J. P. Remeitha, E. A Rictman, S. Gazahuak, and G. P. Espinosa, *Phys. Rev. Lett.*, **58**, 1676 (1987).
[2]R. C. Budhani, S.-M. H. Tzeng and R. F. Bunshah, *Phys. Rev. B*, in press.
[3]M. O. Eatouzh, D. S. Ginley, B. Morosin, E. L. Venturini, *Appl. Phys. Lett.*, **51**, 376, (1987).
[4]R. Pool, *Science*, **240**, 25 (1988).

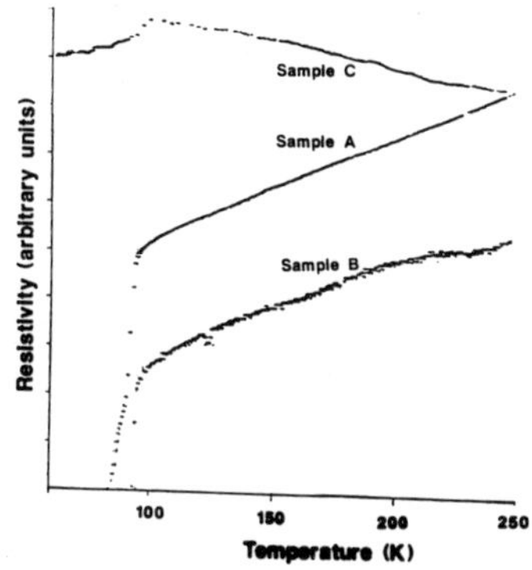

Figure 1. Resistivity of samples with compositional variation as a function of temperature.

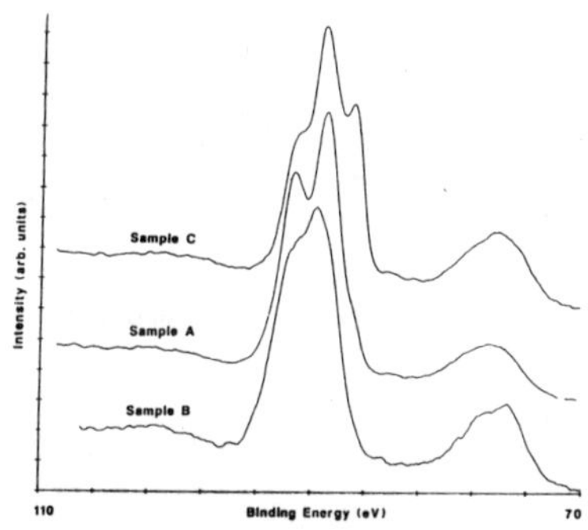

Figure 2. Barium 4d spectra of Samples A, B and C.

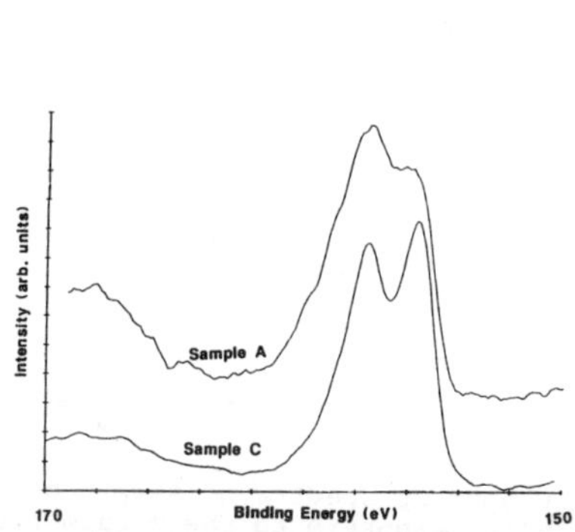

Figure 3. Yttrium 3d spectra of Samples A and C.

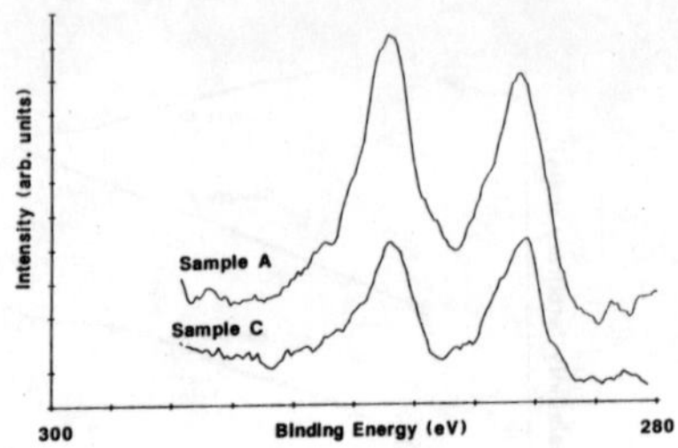

Figure 4. Carbon 1s spectra of Samples A and C.

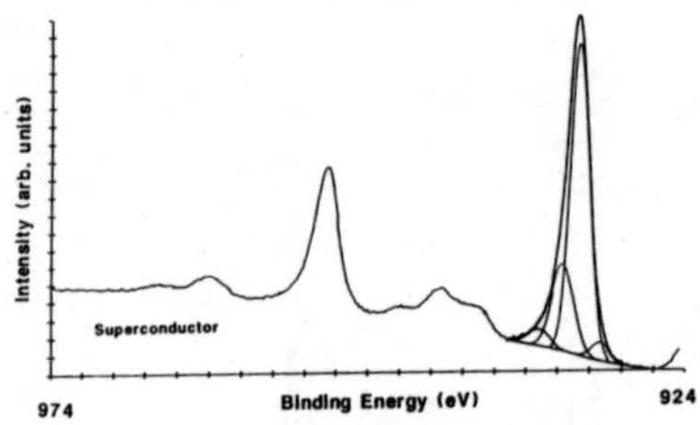

Figure 5. Copper 2p spectrum of Sample B with curve-fit.

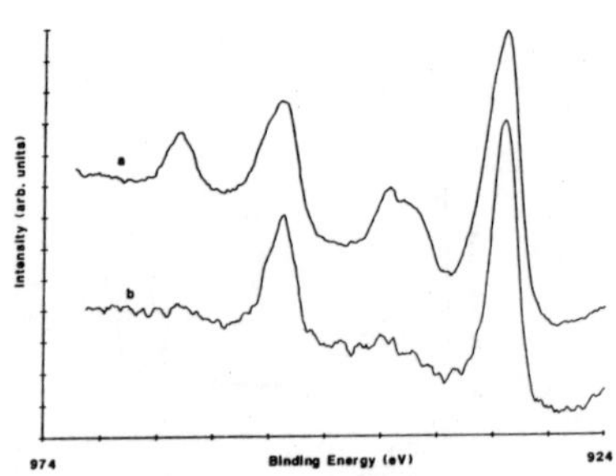

Figure 6. Copper 2p spectra of Sample B "as received" (a) and after exposure to the vacuum for six hours (b).

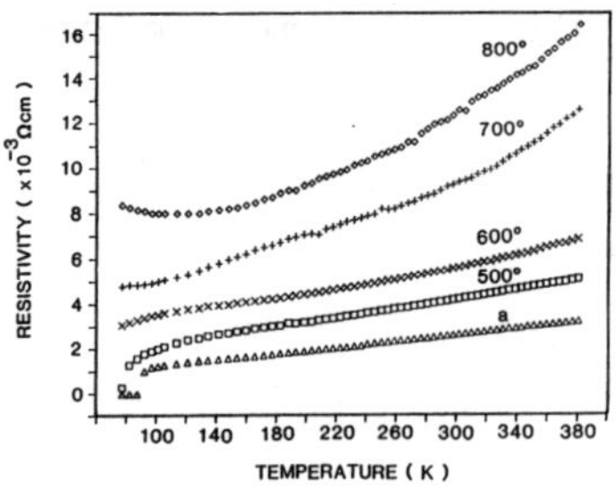

Figure 7. Resistivity of slow cooled (a) and rapidly quenched samples.

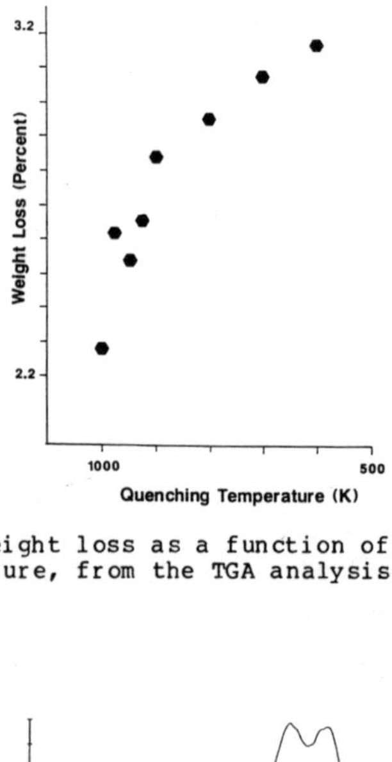

Figure 8. Total weight loss as a function of quenching temperature, from the TGA analysis.

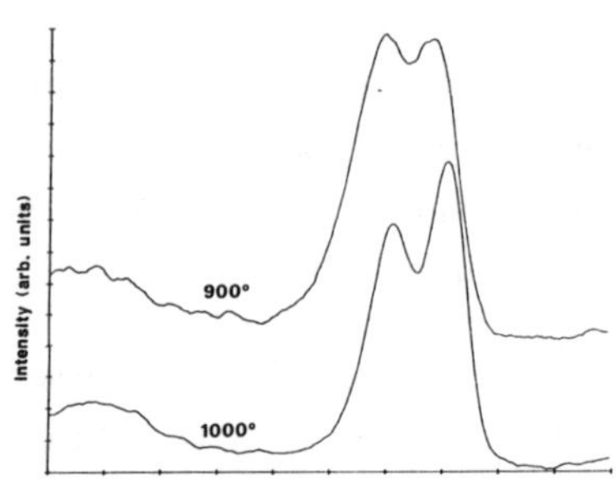

Figure 9. Yttrium 3d spectra of the 1000 ^{o}C quenched sample showing oxide, and the oxide-hydroxide mixture of the 900 ^{o}C quenched sample.

SUPERCONDUCTIVITY IN SPIN ONE HALF DOPED MOTT INSULATOR

by M. CYROT[+], D. MAYOU[++], J.P. JULIEN[++], F. CYROT-LACKMANN[++]

[+]Laboratoire Louis Néel, [++]Laboratoire d'Etudes des Propriétés Electroniques des Solides, C.N.R.S., 166 X, 38042 Grenoble-Cedex, France.

ABSTRACT - We first consider the "parent" compounds of the high T_c superconductors. We describe their magnetism using a strong correlation approach. Arguing that doping cannot screen completely correlations, we describe the high T_c materials as spin one half doped Mott insulators. We find s wave superconductivity and a behavior similar to the B.C.S. one.

1. INTRODUCTION

Since the discovery by Bednorz and Muller of high T_c superconductors, a number of theoretical ideas have been put forward [1]. Among them, Anderson [2] has suggested that correlations between electrons are important and has proposed the resonant valence bond model. We have proposed [3] in a series of papers a somewhat different model although starting from the same hypothesis of strong correlations. We developed the idea that a spin one half doped Mott insulator is a high T_c superconductor.

The high-T_c superconductors results from doping cupric oxydes. The elements introduced give to copper atom a formal valence greater than two. In a first part, we will study the undoped copper oxydes which form the parent compounds and will show that the properties of these perovskites can be explained by using the hypothesis of strong correlations. We will then argue that these correlations must persist in the superconducting phase. Assuming that superconductivity is due to the antiferromagnetic superexchange, we will show that superconducting materials are restricted to those which contains CuO_2 planes and cannot be found in three dimension perovskites. In a second part, following this idea of strong correlations, we then show that one can obtain superconductivity which has all the properties of B.C.S. one. Our model is thus different from the R.V.B. one which is usually discussed in this limit for two main reasons. First our starting point is the usual Heisenberg antiferromagnet for the parent compounds and not a non magnetic R.V.B. state. Second superconductivity is due to the superexchange interaction and not to the simultaneous transfer of an holon and a spinon from one plan to another.

2. MAGNETISM OF THE PARENT COMPOUNDS

Let us first recall how the magnetic properties of these type of perovskites have been discussed using the hypothesis of strong correlations [4]. Cu^{2++} is a Jahn-Teller ions and we have to deal with a lattice a Jahn-Teller ions. The e_g hole of copper in the parent compounds can be considered to be in a doubly degenerate hubbard band if the copper atom is in perfect octahedral environment. Its hamiltonian can be writen :

$$H = H_0 + H_1$$

where H_1 describe the hopping of the hole through the oxygen anion and H_0 the Coulomb repulsion on a site. In the limit of strong repulsion U compared to the hopping integral t, one can derive an effective hamiltonian to second order in t :

$$H_{eff} = - H_1 \frac{1}{H_0 - E_0} H_1$$

For a non degenerate Hubbard hamiltonian, one obtains the Anderson's superexchange hamiltonian :

$$H = \sum_{ij} \frac{2 t_{ij}^{\,2}}{U} \vec{S}_i . \vec{S}_j$$

In the case of orbital degeneracy the corresponding energy resulting from the virtual transition of the d electrons to neighbouring sites depends not only on the magnetic structure but also on the particular orbitals that are occupied at the neighbouring sites. Thus the effective hamiltonian leads not only to magnetic ordering but also to an orbital one determined uniquely. The magnetic state and the crystallographic distorsion results from this orbital superlattice. The effective hamiltonian for a perovskite structure has been first computed by Khomskii and Kugel [4]. We write their results neglecting the Hund's coupling and introducing a pseudo τ to describe the orbital superlattice : $\tau^{\xi} = ½$ corresponding to the d_{z^2} orbital and $\tau^{\xi} = -½$ to the $d_{x_2-y_2}$ one. The arbitrary superposition of orbitals is described in the $\tau^{\eta}\tau^{\xi}$ plane.

$$H_{eff} = \frac{t^2}{U} \sum_{\langle ij\rangle_z} 8 S_i.S_j[\tau_i^{\xi}\tau_j^{\xi} + 1/2\,(\tau_i^{\xi} + \tau_j^{\xi}) + 1/4] + 2\,\tau_i^{\xi}\tau_j^{\xi} - \tau_i^{\xi} - \tau_j^{\xi}$$

$$+ \frac{t^2}{U} \sum_{\substack{\langle ij\rangle_x \\ \langle ij\rangle_y}} 2 S_i.S_j[\tau_i^{\xi}\tau_j^{\xi} - \tau_i^{\xi} - \tau_j^{\xi} + 1 + \sqrt{3}(\tau_i^{\eta}\tau_j^{\xi} + \tau_i^{\xi}\tau_j^{\eta}) \mp \sqrt{3}(\tau_i^{\eta} + \tau_j^{\eta}) + 3\,\tau_i^{\xi}\tau_j^{\eta}]$$

$$+ 1/2\,[\tau_i^{\xi}\,\tau_j^{\xi} + \tau_i^{\xi} + \tau_j^{\xi} \pm \sqrt{3}(\tau_i^{\eta}\tau_j^{\xi} + \tau_j^{\xi}\tau_i^{\eta}) + 3\,\tau_i^{\eta}\tau_j^{\eta} \pm \sqrt{3}(\tau_i^{\eta} + \tau_j^{\eta})]$$

$\langle i,j \rangle_x$, $\langle i,j \rangle_y$, $\langle i,j \rangle_z$ denote summation over the nearest neighbords located along the corresponding axis. The upper sign corresponds to the pair along the x axis and the lower, to the pair along the y axis.

We first consider the two dimensional situation. Assuming different spin structures and optimising the orbitals one, we showed [3] that the lowest energy is obtained for a ferromagnetic orbital lattice i.e. one has $\tau_i^{\xi} = -1/2$ on each site, or a $d_{x^2-y^2}$ orbital. The corresponding spin structure is antiferromagnetic as the effective hamiltonian reduces :

$$H_{eff} = \frac{3t^2}{U} \sum_{ij} \vec{S}_i \cdot \vec{S}_j$$

This is the way we have predicted the antiferromagnetism of La_2CuO_4. We now consider the three dimensional situations. The result has been found by Kugel and Khomskii and figure 1 shows their result. The occupied orbital for the hole is alternatively the $d_{z^2-y^2}$ and $d_{z^2-x^2}$. In that case, the magnetic interaction in the ab plane is

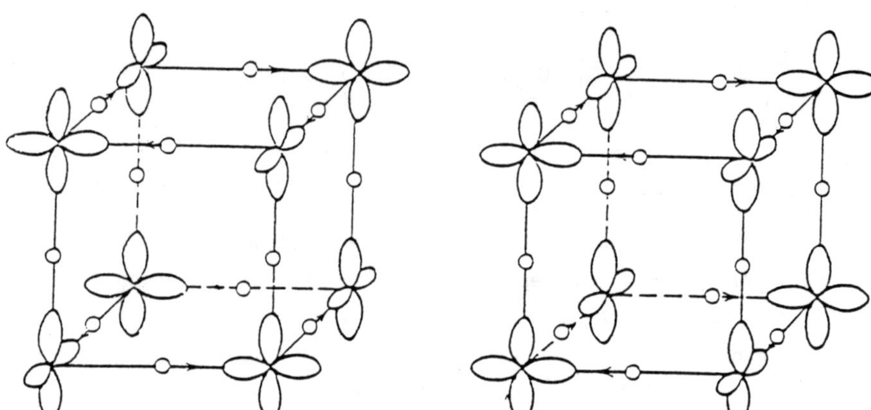

a b

Fig. 1 - Orbital superlattice in 3-D perovskite

ferromagnetic. Along the c axis, it is antiferromagnetic. This result has very important consequences if one believes that superconductivity is due to the antiferromagnetic superexchange. This means that doping the plane in order to have moving carriers can lead to superconductivity but doping the three dimensional structures will not lead to superconductivity.

We think that this is the basic reason why some three dimensional perovskites of copper obtained in the I.B.M. group are not superconducting [5]. This group prepared $La_4BaCu_5O_{13}$ and $La_5SrCu_6O_{15}$ where the formal valence of copper is slightly larger than two. However no superconductivity has been found because the structure is mainly three dimensional and the orbital superlattice leads to ferromagnetic interaction.

3. THE SUPERCONDUCTING STATE

Our basic starting hamiltonian describes carriers in a nearly filled Hubbard lower subband which are coupled by a superexchange interaction through the oxygens. Thus :

$$H = \sum_{i,j} t_{ij}(1-n_{i-\sigma})\, C^{+}_{i\sigma}\, C_{j\sigma}(1-n_{j-\sigma}) + \sum_{ij} J\, \vec{S}_i \vec{S}_j \qquad (1)$$

where i and j are first nearest neighbours sites and t_{ij} is an effective hopping integral between the copper sites in the CuO_2 plane.

In this model, J is not only related to t_{ij} and the Coulomb correlation U. As the superexchange is via the oxygen anion J is also related to the difference between the d and p atomic levels of copper and oxygen. The way the kinetic energy is treated leads to very different results. This kinetic energy is usually treated by a Gutzwiller approximation [9] i.e. one introduces a mass enhancement $1/\delta$ where δ is the number of holes in the subband. Thus we are led to the total hamiltonian :

$$H = - \sum_{ij\sigma} \delta t C^{+}_{i\sigma} C_{j\sigma} + \sum_{ij} J \vec{S}_i . \vec{S}_j \qquad (2)$$

This Gutzwiller approximation requires that we have to deal with a Fermi liquid. But we think that a doped Mott insulator is not a Fermi liquid. Indeed it is well known that the Fermi liquid description leads to serious drawbacks. The most important is that there are four states per site leading to an entropy RLn4 instead of RLn2 as the double occupancy is impossible as long as the temperature is smaller than U. This serious drawback has been recognized as the origin of the breadkdown in the calculation of the susceptibility. To overcome this difficulty and to take into account that the number of available states in the lower Hubbard subband is only one per site, we use a statistics with only one state per k vector with spin up or down. The volume of the Fermi surface is thus doubled and the Luttinger theorem is violated in the Mott state. Our approach is closely related to a phenomenological one developed recently by Spalek and Wojcik [7]. These authors assume that the double occupancy in a Mott insulator is forbidden not in the real space but in the reciprocal space. Thus if the kσ state is occupied the k-σ state is empty.

This result has a very important effect on the possible superconductivity. In a Fermi liquid state, one obtains a d-wave superconductivity which presents absence of gap in some directions of the k space. Properties of this superconductivity are different from the B.C.S. one. In the case where one assumes a non Fermi liquid behaviour, one obtains a s-wave superconductivity which has all the properties of a B.C.S. one.

In previous paper, we have shown that superconductivity can be obtained by peforming a B.C.S. linearisation of the term $S^{+}_{i}S^{-}_{j}$ + $S^{-}_{i}S^{+}_{j}$. The critical temperature and chemical potential are given by the two self consistent equations :

$$\delta-\tfrac{1}{2} = \frac{1}{2}\int N(\varepsilon)\ \mathrm{th}\ [\frac{\beta(\varepsilon-\mu)}{2}]\ d\varepsilon \qquad (3)$$

$$\frac{2}{J} = \int \tilde{N}(\varepsilon)\ \mathrm{th}\ [\frac{\beta(\varepsilon-\mu)}{2}]\ \frac{d\varepsilon}{\varepsilon-\mu} \qquad (4)$$

Equation (3) gives the chemical potential μ as a function of the filling of the band $n = 1-\delta$, for a normal metal. Here $N(\varepsilon)$ is the square lattice density of states for a hopping integral $t\delta$.

Equation (4) is analogous to the B.C.S. equation for T_c. However in the present theory the integration is made over the whole band and not for $|\ \varepsilon-\mu\ |$ less than the Debye energy. Furthermore the density of states is replaced here by an effective density $\tilde{N}(\varepsilon)$ (Figure 2) which is different for s wave symmetry and d wave symmetry.

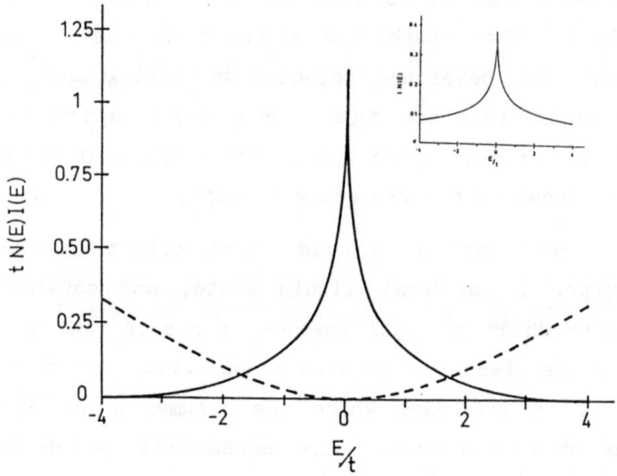

Figure 2 - Effective density of states $\tilde{N}(E)$ as a function of energy for s (----) and d (____) symmetry ; in the inset the square lattice density of states N(E).

The properties of our model, and in particular the critical temperature, can be analysed in a simple way if $N(\varepsilon)$ and $\tilde{N}(\varepsilon)$ are nearly constant for $|\varepsilon-\mu|$ less than a few kT_c. This approximation is valid except for very small δ ($\delta \leq 0,05$ for our two choices of parameters $t/J = 10$, 2.5). This is due to the fact that the bandwidth is $8t\delta \sim 1-0,1$ eV much greater than $kT_c \leq 10^{-2}$ eV (~ 120 K). Thus $N(\varepsilon)$ and $\tilde{N}(\varepsilon)$ can vary appreciably in a region of a few kT_c around μ, only if the chemical potential is very close to the band edge where $N(\varepsilon)$ and $\tilde{N}(\varepsilon)$ are discontinuous. This corresponds to a very small number of holes δ for which the ground state is an antiferromagnetic Néel state. The critical temperature is given by a B.C.S. like formula :

$$kT_c = 1.14\, W(\mu) \exp\left[-\frac{1}{J\,\tilde{N}(\mu)}\right] \qquad (5)$$

The expression for $W(\mu)$ which replaces the Debye frequency is complicated but it is proportional to the hopping integral t for a given δ.

In our model, the band contains only one electron and the range of interest for the doping corresponds to a chemical potential μ between $\mu/t\delta \sim 2$ and $\mu/t\delta = 4$ (band edge). Thus we see from figure (2) that $\tilde{N}(\mu)$ is larger for s type than for d type superconductivity. As a consequence, we predict an s type superconductivity as opposed to many authors (8) which predict a d symmetry. We note that in their theory, the band is nearly half filled (it contains two electrons) and $\tilde{N}(\mu)$ is indeed larger for d type than for s type symmetry.

We have made a complete numerical study of the critical temperature for s and d symmetry which confirms the above prediction. The result is shown in figure (3) for two values of the parameter t/J but from the work of Mila et al (9), it seems that the best choise is $t/J \simeq 2.5$. We note however that the superconductivity is favored when t/J decreases.

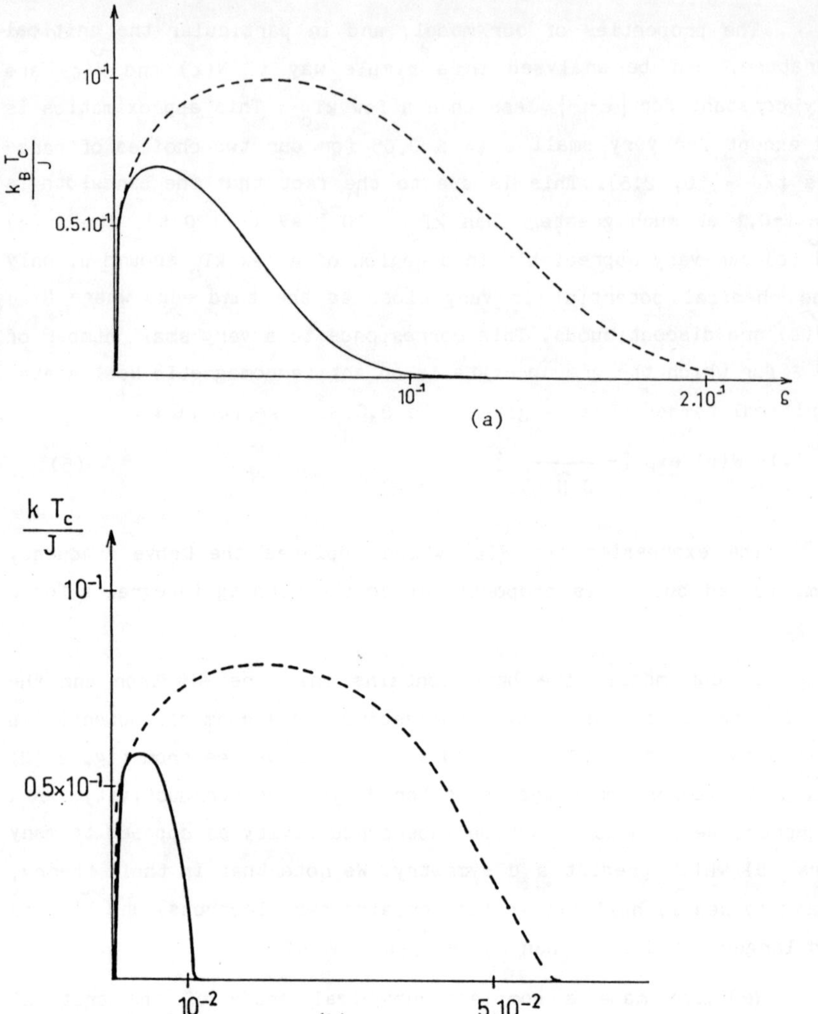

<u>Figure 3</u> - Critical temperature for the square lattice as a function of doping or two values of the parameter t/J :
t/J = 2.5 (----) ; t/J = 10 (——) ;
a) s symmetry ; b) d symmetry

In conclusion our model predicts only an s type superconductivity whose critical temperature has the good order of magnitude. The critical temperature tends to zero for small doping.

The gap $\gamma J\Delta$ has exactly the same temperature dependence in our theory than in the B.C.S. one :

$$\frac{\gamma J\Delta(T)}{kT_c} = f_{B.C.S.}(T/T_c) \qquad (6)$$

In particular at zero temperature $\gamma J\Delta = 1.76\ kT_c$.

REFERENCES

[1] For a review see T.M. Rice, Phys. B. 67, 141 (1987).

[2] P.W. Anderson, Science 235, 1196 (1987)
G. Baskaran, Z. Zou and P.W. Anderson, Solid State Com., 63, 973 (1987)

[3] M. Cyrot Solid State Com., 62, 821(1987), Ibid, 63, 1015 (1987) Nature, 330, 115 (1987),
M. Cyrot, J.P. Julien, D. Mayou, Proceeding of the HTS MS^2 Conference Interlaken 1988

[4] C. Lyon-Caen and M. Cyrot, J. Phys. C 6, L 247 (1973), J. de Physique, 36, 253 (1975) and C.4, 183 (1976).
D.I. Khomskii and K.I. Kugel, Sov. Phys. JETP, 37, 725 (1973), Sol.Stat.Comm., 13, 763 (1973), Sov.Phys. Solid. State, 17, 285 (1975).

[5] J.B. Torrance, Y. Tokura and A. Nazzal, Chemtronics (1988)

[6] M.C. Gutzwiller, Phys. Rev. Lett., 10, 159 (1963) ; Phys. Rev., 137A, 1726 (1965)
W.F. Brinkman and T.M. Rice, Phys. Rev., B2, 4302 (1970)

[7] J. Spalek and W. Vojcik, Phys. Rev., B.37, 1532 (1988)

[8] F.C. Zhang, C. Gros, T.M. Rice and H. Shida, Superconductor Science and technology, 1 (1988).
A. Ruchentein, P. Hirschfield, J. Appl Phys. Rev., B 36, 857 (1987)

[9] F. Mila, F.C. Zhang and T.M. Rice, Proceeding of the HTS MS^2 Conference Interlaken (1988).

Superconductivity and Electronic Structure of High T_c Cu-Oxides

M. Schluter
AT&T Bell Laboratories
Murray Hill, New Jersey 07974

ABSTRACT

The electronic structure of copper oxide superconductors is analyzed. The band picture is contrasted to the Mott-Hubbard picture. The parameters for the Hubbard model are derived from total energy calculations and from a mapping onto the bandstructure. They indicate a "marginal" Mott insulator situation. The different doping mechanisms for the different compounds are considered and possible mechanisms leading to high T_c superconductivity are discussed.

INTRODUCTION

The discovery of high temperature superconductivity[1],[2] in Cu-oxide compounds prompted an enormous activity in trying to understand the underlying physics both from the experimental and theoretical point of view. The situation is, however, still rather unsettled. In this talk I review the basic electronic structure of the Cu-oxide compounds, from which models for superconductivity will have to be derived. It was originally proposed for the (214) La_2CuO_4 compound,[3] later extended to the (123) $YBa_2Cu_3O_7$ compound,[4] and now also confirmed for the new Aurivillius phases $CaBi_2Sr_2Cu_2O_8$[5] and $CaBi_2Tl_2Cu_2O_8$[6] that in this class of compounds the for high T_c superconductivity essential structural elements consist of 2D square Cu-O networks. Differences in structural environment leading to e.g. crystal field splittings and variations in carrier concentrations, etc. are also of relevance in determining the superconducting transition temperature.

The conventional band picture using the local density approximation (LDA) has been applied to all these compounds[7]–[10] and essentially confirms the primary importance of the Cu-O sheets. In particular, the Fermi level is found about halfway up in a broad (W≈5eV) σ^* antibonding band consisting mainly of Cu d(x^2-y^2) and O p(x,y) orbitals. Other occupied (π^*) levels are nearby and some other wavefunction admixtures may be of some relevance. Concentrating on the (214) case for simplicity, the band picture predicts a metallic situation for the stoichiometric compound La_2CuO_4, which, because of extreme Fermi surface nesting properties[7] should be unstable at exactly half filling. In fact, LDA calculations predict[11] a CDW instability involving a breathing mode (oxygen) distortion which, however, is not experimentally observed. The observed antiferromagnetic (AF) insulator,[12] on the other hand, cannot be stabilized within spin density (LSD) band theory.[13] Moreover, the observed magnetic moment associated with the AF phase is too large to be accounted for by a SDW distortion, thus indicating the limitations of the bandpicture in the present case.

Starting from the opposite point of view, namely the localized orbital picture, a multiband Hubbard Hamiltonian has been proposed as appropriate description.[14] The many electron parameters in this Hamiltonian can approximately be derived from spectroscopic data[15] or can be calculated from total energy considerations.[16] This is done by applying the constrained density functional formalism (CDFT) to evaluate the energy associated with impurity-like local charge fluctuations within the LDA scheme. The results are then mapped onto a mean-field solution of the appropriate Hubbard Hamiltonian and the interaction parameters are extracted in the static limit. The results are that Coulomb repulsion energies are important on both copper (U≈9-11eV) and oxygen (U≈5-7eV), but that they are of order of the bandwidth W. The intersite Coulomb energy is found to be significantly smaller (V≈2eV). These findings suggest that the Cu-O sheets in La_2CuO_4 are "marginal" Mott insulators, where both banding and correlation effects are important. The magnetic properties of the non-superconducting parent compounds La_2CuO_4, $YBa_2Cu_3O_6$, etc. can be quantitatively derived from this situation.

The one electron on-site energies of the Hubbard model are extracted by comparing its mean field solution to the LDA band-structure results. Here it is found[7] that the "screened" site energies of copper and oxygen are roughly degenerate. This situation is particular to Cu and O and is in contrast to early transition metal oxides where the oxygen p-level lies well below the metal d-level. The near degeneracy indicates strong covalency in the (ionic) Cu-O bond and it dominates the nature of the low energy charge fluctuation spectrum. It was first suggested by Emery,[14] that the doping holes, which render e.g. $La_{2-x}Sr_xCuO_4$ superconducting, should be of oxygen p-character rather than copper d-character. The Hubbard model calculations confirm this, at least in the W=0 limit. There has recently been interest in a more detailed description of the chemical nature of the dopant holes,[17] i.e. oxygen pxy(σ^*-type), pxy (π^*-type), pz (σ^*-type for out-of-plane oxygens) etc. Although more refined calculations are needed to settle these questions, there is no direct evidence for the importance of these orbitals on the basis of Hubbard model considerations.

While in La_2CuO_4 the conducting holes are introduced by chemical doping on the donor "A"-lattice (Sr^{2+} or Ba^{2+} in place of La^{3+}) in the (123) compound $YBa_2Cu_3O_{7-\delta}$ hole carriers are presumably created by variations in oxygen stoichiometry (within the chains). Here, however, the situation is somewhat obscured by the fact that the 1D Cu-O chain electronic states also lie near the Fermi level. The recently discovered high T_c Aurivillius phases based on Bi or Tl metals allow for yet another possible mechanism of hole "self-doping" in the Cu-O sheets. Band calculations[9],[10] show that in these cases the interleaving Bi-O (or Tl-O) layers dip themselves slightly below the Fermi level creating small electron pockets at the expense of carrier transfer from the Cu-O sheets. This happens in the pure, stoichiometric compounds. Non-stoichiometry can, of course, also provide dopant holes. The Aurivillius phases can be viewed as intercalated structures of Cu-O sheets and slightly semi-metallic Bi(Tl)-O sheets with their Fermi levels lined up electrostatically. The observed increase in T_c with increasing number of Cu-O sheets per unit cell can be due to a variety of reasons: a simple increase in the density of states near E_F, a higher density of carriers per Cu-O sheet or an increased metallicity in the Bi(Tl)-O layers giving better 3D coupling. Although more work is needed to clarify these questions, it is believed that in each case the superconductivity occurs in the strongly correlated Cu-O sheets which are doped away from the insulating half-filled band situation.

The relevance of these electronic structure details for the understanding of the basic superconductivity mechanisms remains to be seen. Anderson's RVB model[18] is based on a simple one-band Hubbard description, which he argues can be viewed as a renormalized multiband description. Emery's,[14] Hirsch's,[19] Varma's,[20] Weber's,[21] Chen's,[22] Stechel's[23] and many other models[17] make explicit use of the extra degrees of freedom in the multi-band Hamiltonian. More importantly, it is still unclear whether spin - or charge fluctuations (or possibly strongly coupled hybrids) provide the predominant coupling mechanism that leads to high superconducting transition temperatures. After about two years since the original discovery of these exciting materials our theoretical understanding is still rather fragmentary and the big challenge remains.

References

1. Bednorz, J. G. and Muller, K. A., Z. Phys. B 64, 189 (1986).
2. Chu, C. W., Hor, P. H., Meng, R. L., Gao, L., Huang, Z. J. and Wang, Y. Q., Phys. Rev. Lett. 58, 405 (1987).
3. Uchida, S., Takagi, H., Kitazawa, K. and Tanaka, S., Jpn. J. Appl. Phys. 2, 26, LI (1987).
4. Cava, R. J., Batlogg, B., vanDover, R. B., Murphy, D. W., Sunshine, S., Siegrist, T., Remeika, J. P., Rietman, E. A., Zahurak, S. and Espinosa, G. P., Phys. Rev. Lett. 58, 1676 (1987).
5. Maeda, H., Tanaka Y., Fukotomi, M. and Asano, T., Jpn. J. appl. Phys. Lett. (to be published).
6. Sheng, Z. Z., Hermann, A. M., Ali, A. E., Almasan, C., Estrada, J., Datta, D. and Matson, R. J., Phys. Rev. Lett. 60, 937 (1988).
7. Mattheiss, L. F., Phys. Rev. Lett. 58, 1028 (1987).

8. Mattheiss, L. F. and Hamann, D. R., Sol. Stat. Commun. 63, 395 (1987).

9. Hybertsen, M. S. and Mattheiss, L. F., Phys. Rev. Lett. 60, 1661 (1988).

10. Hamann, D. R. and Mattheiss, L. F., to be published.

11. Weber, W., Phys. Rev. Lett. 58, 1371 (1987).

12. Vaknin, D., Sinha, S. K., Moncton, D. E., Johnston, D. C., Newsam, J., Safinya, C. R., and King, H. E., Phys. Rev. Lett. 58, 2802 (1987).

13. Leung, T. C., Wang, X. W., and Harmon, B. N., Phys. Rev. B 37, 384 (1988).

14. Emery, V. J., Phys. Rev. Lett. 58, 2794 (1987).

15. Shen, Z., Allen, J. W., Yeh, J. J., Kang, J. S., Ellis, W., Spicer, W., Lindau, I., Maple, M. B., Dalichaouch, Y. D., Torikachvili, M. S., and Sun, J. Z., Phys. Rev. B 36, 8414 (1987).

16. Schluter, M., Hybertsen, M. S. and Christensen, N. E., Proc. Intl. Conf. High T_c Superconductivity, Interlaken, 1988.

17. Proceedings of the Intl. Conf. High T_c Superconductivity, Interlaken, 1988.

18. Anderson, P. W., Science 235, 1196 (1987).

19. Hirsch, J. E., Phys. Rev. Lett. 59, 228 (1987), Hirsch, J. E., Tang, S., Loh, E. and Scalapino, D. J., Phys. Rev. Lett. 60, 1668 (1988).

20. Varma, C. M., Schmitt-Rink, S. and Abrahams, E., Sol. Stat. Commun. 62, 681 (1987).

21. Weber, W., Z. Phys. B, to be published.

22. Chen, G. and Goddard, III, W. A., Science, 239, 899 (1988).

23. Stechel, E. B. and Jennison, D. R., Phys. Rev. B, to be published.

Theory of High Temperature Superconductors: Field-Theoretic and Monte-Carlo Studies

W. Hanke*, L. Lilly, A. Muramatsu*
Physikalisches Institut der Universität Würzburg
8700 Würzburg, FRG
*also at: Max Planck Institut für Festkörperforschung
7000 Stuttgart 80, FRG

Abstract

On the one hand, a simple CuO_2 plane plus CuO chain model with short range (extended Hubbard) repulsive interactions is presented that naturally explains many of the differences between $YBa_2Cu_3O_{7-x}$ and $La_{2-y}Sr_yCuO_4$-type superconductors. We obtain (a) a substantial enhancement of T_c mediated by density fluctuations in the chains and (b) a superposition of s- and d-wave symmetry in the gap function, which is consistent with recent experiments of an orthorombic distortion around T_c. On the other hand, we have used the the Quantum Monte-Carlo technique for the 2-D Hubbard model to study the stability of the various groundstates (CDW, SDW, superconducting) against "frozen-in" phonon lattice displacements, and to clarify the role played by the lattice displacements for the pair correlation functions. For the frozen-in lattice displacements on-site (breathing) lattice vibrations are considered for half- as well as non-half-filled band cases.

1 Part I

1.1 Introduction

The discovery of the high temperature superconductors (HTSC) [1] – [3] led to an intensive and still exciting theoretical search for new pairing mechanisms. Models based on a strong on-site Coulomb repulsion on the Cu-atoms like the resonating valence bond (RVB) model [4] and others [5] – [8] are of special interest; they are supported by the presence of an antiferromagnetic state with rather high Nèel temperatures [9] – [11] and large antiferromagnetic fluctuations [12]. Most investigations focused just on the pairing mechanism on the CuO_2 planes, while only very little theoretical efforts were directed towards more specific aspects of the HTSCs. We give an explanation for the increase in critical temperature (T_c) in the Y-Ba-Cu-O compounds compared to the La-Ba-Cu-O ones. This question is obviously of special interest in the search of even higher T_cs.
Here we present our investigation of a model for the 1-2-3 HTSCs ($YBa_2Cu_3O_7$) with assumes extended Hubbard type interactions for both planes and chains. A layer of chains is coupled to an underlying plane. We found that for an appropriate choice of parameters the charge density (CD) fluctuations in the chain lead (a) to a substantial enhancement of T_c when double occupation of the Cu-sites in the planes is suppressed [4] – [8], and (b) to a gapfunction displaying a mixture of s- and d-wave symmetry. The increase in T_c is a general feature that is found both in models like in Emery's model [6], [7], [8],

where the dynamics on the O-atom is taken into account explicitely (holes on O-atoms), or not, as was suggested by Anderson's RVB [4], or [5]. The result (b) might explain recent suggestions to explain experimental observations of an orthorombic distortion at the superconducting transition in $YBa_2Cu_3O_7$ [13].

1.2 Model Hamiltonian

The CuO_2 plane as well as the CuO chains are described by extended Hubbard models (EHM):

$$H_p = -t_p \sum_{\substack{<i,j> \\ \sigma}} c^{\dagger}_{i\sigma} c_{j\sigma} + U_p \sum_i n_{i\uparrow} n_{i\downarrow} + \frac{V_p}{2} \sum_{\substack{<i,j> \\ \sigma,\sigma'}} n_{i\sigma} n_{j\sigma'} \tag{1.1}$$

(For the chain Hamiltonian replace $p \to c$). p (c) denotes the plane (chain) parameters and operators. To make things easier, we will work in the "superexchange picture" [4], [5] where the oxygen atoms are eliminated. We will comment below that the mechanism proposed also enhances T_c when the dynamics on the O-atoms is considered explicitely [6], [7], [8]. We assume only Coulomb coupling between chains and planes, i.e. that hopping between chains and planes is negligible [14], [15]. The interaction is given by:

$$H_{int} = V_{cp} \sum_{\substack{l,i \\ \sigma,\sigma'}} n^{p}_{i,\sigma} n^{c,l}_{i,\sigma} \tag{1.2}$$

where l numbers the chains, i the sites in the x direction (along which the chains extend).

1.3 Chain Dynamics

First we consider the low-lying excitations in the chains, because they will have the strongest influence on the ground state properties of the planes. They are extracted by a series of well-known transformations in the physics of one dimensional (1-D) systems [16]. We will just briefly enumerate them: First we pick out the states close to the Fermi surface via the Kogut-Susskind scheme. Afterwards, the continuum version is bosonized and a trivial canonical transformation yields decoupled charge- and spin-density fluctuations, each one being described by a sine-Gordon Hamiltonian [16]:

$$\begin{aligned} H_\rho &= A_\rho \int dx \left\{ \frac{1}{2}[\Pi_\rho^2 + (\nabla_x \phi_\rho)^2] \right\} + \\ &+ \int dx \left\{ \frac{(U_c - 2V_c)t_c}{\pi^2} \cos[\sqrt{8\pi}\beta_\rho \phi_\rho - \pi\gamma_\rho] \right\} \end{aligned} \tag{1.3}$$

where ϕ_ρ is a canonical Bose field, Π_ρ is its conjugate and we used the following constants:

$$\begin{aligned} A_\rho &= \left(1 - \frac{V_c}{2\pi t_c}\right)^{\frac{1}{2}} \left(1 + \frac{U_c + 5V_c}{2\pi t_c}\right)^{\frac{1}{2}} \\ \beta_\rho &= \left(\frac{2\pi t_c - V_c}{2\pi t_c + U_c + 5V_c}\right)^{\frac{1}{4}} \\ \gamma_\rho &= N_{1\uparrow} + 2N_{1\downarrow} + N_{2\uparrow} \end{aligned} \tag{1.4}$$

Applying the same transformations to the 1-D operators in equation (1.2), we obtain for the l-th chain:

$$H_{int}^{l} = V_{cp}\sqrt{\frac{2}{\pi}}\beta_{\rho}\int dx n_{p}(x)\nabla_{x}\phi_{\rho}(x) \tag{1.5}$$

Now we perform a Gaussian expansion in the fluctuations of ϕ_ρ around the classical (mean field) solution of the sine Gordon system. Going to a path integral formulation, the quadratic form is integrated out. Thus we arrive at the following renormalization of the parameters in the plane:

$$U_{p}^{R} = U_{p} - 2\gamma(1 - \frac{\eta}{2}) \tag{1.6}$$

$$V_{p,x}^{R} = V_{p} + \gamma\eta exp(-2\eta) \tag{1.7}$$

with

$$\gamma = \beta_{\rho}^{2}\frac{V_{cp}^{2}}{2t_{c}A_{\rho}} \tag{1.8}$$

$$\eta = \left(\frac{\beta_{\rho}^{2}(U_{c} - 2V_{c})}{2\pi t_{c}A_{\rho}}\right)^{\frac{1}{2}} \tag{1.9}$$

The low-lying CD-fluctuations in the chain decrease the on-site Coulomb repulsion and increase the nearest-neighbor interaction in x-direction in the plane through excitonic effects. Furthermore this renormalization introduces an anisotropy that is much larger compared to the one stemming from the difference in lattice constants a and b. This anisotropy results in a mixture of s- and d-wave symmetry as we will show in the next section.

1.4 Anisotropic Superconductivity

For the 2-D system we assume the ground state to be dominated by the limit where double occupancy of the Cu-sites is not allowed [4] – [7]. We perform a mean-field evaluation of the effective Hamiltonian in this limit [5].

Confined to the subspace of singlet pairing, the gap function displays a superposition of extended s- and d-wave symmetry:

$$\Delta_{k} = f_{s}(k)\Delta_{s} + f_{d}(k)\Delta_{d} \tag{1.10}$$

where

$$f_{s}(k) = \cos k_{x}a + \cos k_{y}a \tag{1.11}$$

$$f_{d}(k) = \cos k_{x}a - \cos k_{y}a \tag{1.12}$$

In figure 1.1.a) we plotted the critical temperature T_c as a fuction of doping δ.

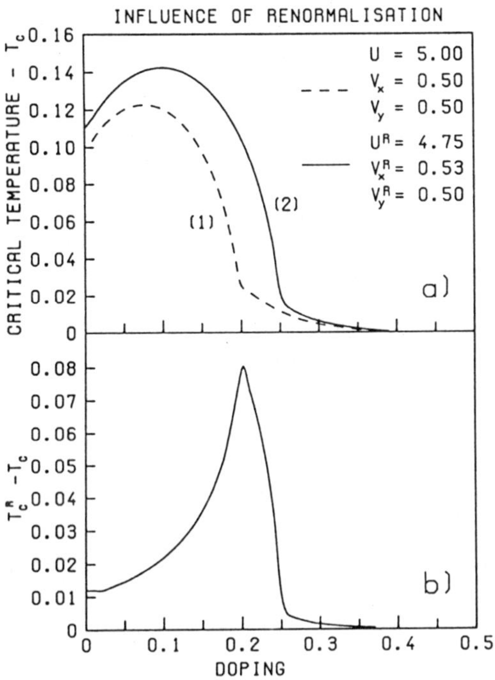

Figure 1.1

a) Curve (1) is the T_c for a plane only (no chains attached), as a function of doping. Curve (2) corresponds to the renormalized parameters by coupling the chains to the plane.

b)Difference in T_c as a function of doping.

In curve (1) we plotted the results for a plane without coupling to the chains. The s- and d-wave parts are decoupled in this isotropic case. The curve (2) shows the critical temperature when including the renormalization of the bare parameters of the plane by the chains as derived in equations (1.6) and (1.7). Now a mixture of s- and d-wave symmetry appears. For low concentration of holes the s-wave character wins while the d-wave character is stronger up for higher concentrations. The anisotropy in the order parameter gives a quite natural description for an additional orthorombic distortion at T_c which was found experimentally [13]. Figure 1.1.b) plots the difference between the curve (1) and (2), with a maximum for hole count $\delta \sim 0.2$. That means, in the frame of our model a maximum improvement of T_c can be achieved by doping. Furthermore the value of $\delta \sim 0.2$ is in reasonably good agreement with the doping concentration obtained in experiments [17]. A more detailed description of the presented calculations will be given elsewhere.

2 Part II

2.1 Introduction

The discovery of high temperature superconductivity [1], [2] in cuprates has generated a renewed interest in the two-dimensional (2-D) Hubbard model. Of particular importance for some mechanisms [4] – [5] is the stability of the 2-D Hubbard model against lattice displacements at and near the half-filled case. For the HTSC [18], [19], attention has been focused on the breathing mode of the oxygen ions vibrating against the Cu ions. In view of the relatively strong electron-phonon coupling and the nesting features of the 2-D band [18], [19], this mode may play a decisive role in the structural stability. In this paper we present Monte-Carlo calculations, which study the structural stability of the 2-D Hubbard model against these "breathing" lattice displacements.

2.2 Model and Method

We start by considering the Hamiltonian:

$$H = -t \sum_{\substack{<i,j> \\ \sigma}} c^{\dagger}_{i\sigma} c_{j,\sigma} - 4V \sum_{i,\sigma} u_i c^{\dagger}_{i,\sigma} c_{i,\sigma} + U \sum_i n_{i\uparrow} n_{i\downarrow} + \frac{K}{2} \sum_i (x_i^2 + y_i^2) \tag{2.1}$$

with

$$u_i = \frac{1}{4}(x_{i_x,i_y} - x_{i_x-1,i_y} + y_{i_x,i_y} - y_{i_x,i_y-1}) \tag{2.2}$$

which has already been proposed by Prelovśek, Rice and Zhang [18] (PRZ). The displacements x_i, y_i correspond to both O-atoms in the unit cell ($i = (i_x, i_y)$). Like in PRZ, the phonon modes are taken into account in the adiabatic approximation. However, the electronic part is treated essentially exactly, within the frame of Quantum Monte-Carlo (MC) simulation. The algorithm used is the same as developed by Hirsch [20], [21] and others [22]. Following PRZ, we introduce the dimensionless electron-phonon coupling $\lambda = \frac{V^2}{Kt}$. Here we present results with λ ranging from 0.15, a weak coupling situation, upto 0.6 (intermediate coupling). Furthermore, a dimensionless quantity $\tilde{V} = 4V\frac{u}{t}$ is defined as a measure of the lattice distortion for a given coupling V (typically $u \sim 0.04 - 0.16$Å). In our simulations, we concentrate on $U = 4t$, since it was previously found, that around this value an enhancement of pairing correlation functions takes place.

2.3 Results and Interpretation

Figure 2.1 displays the total energy for a 6×6 lattice in the half- filled band case. For all values of λ an increase in energy is observed when the "displacement" $\tilde{V}$ is increased, showing that the system is stable against a CDW formation. According to mean field [23], this should be the case for $U > U_0 = 8\lambda$. However, our MC simulation shows a departure from this criterion as the strength of U is increased.

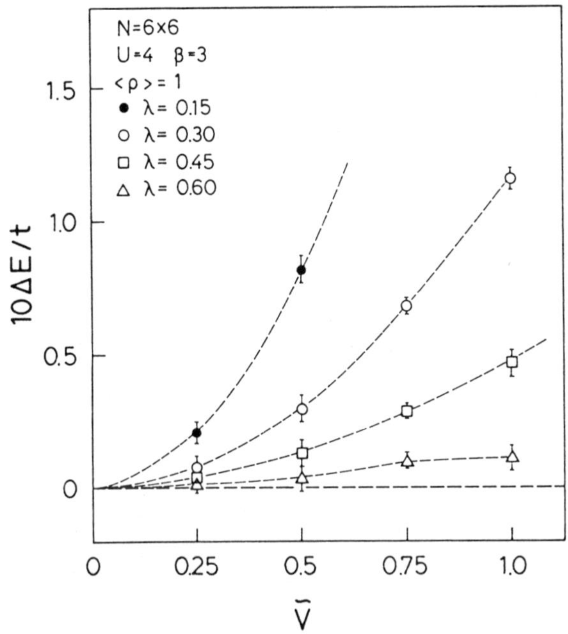

Figure 2.1

Change ΔE in total energy of H in eq. 2.1 for a 6×6 lattice in $\frac{1}{2}$ filled band case, for $U = 4t$ and $\beta^{-1} = k_B T = \frac{t}{3}$ The system is stable for all values of the electron-phonon coupling strength λ, ranging from weak coupling to strong coupling.

For $U = 2t$ (and $\tilde{V} = 0.5$), we obtain a critical value of $\lambda_c = 0.27$ and the mean field value $U_{cMF} = 2.16t$ in good agreement with the actual value of U. For $U = 4t$ we find $\lambda_c = 0.71$, the corresponding mean field value is $U_{cMF} = 5.68t$, showing a departure of more than 40% from the actual value.

The situation changes completely upon doping. Figure 2.2 shows the change in total energy for the same parameters as before, but now in the presence of holes. ($< \rho > \sim 0.84$). A general softening of the lattice is observed and, in particular for $\lambda \geq 0.45$, the distortion is energetically favored. A stable minimum develops for displacements $\tilde{V} \simeq 0.5$. For displacements larger than $\tilde{V} \simeq 0.75$, the computed curves deviate from a parabolic behavior, an indication of the importance of anharmonic effects.

The fact that the lattice instability sets in away from half-filling demonstrates clearly that it is not a consequence of the nesting properties of the 2-D lattice. Instead, the instability is due to an interplay between doping and Coulomb correlation; the added holes tend to move to the top of the alternating Cu potentials, thereby lowering the total energy. Thus the instability may be viewed as being due to the formation of local hole-polarons, as the doping is increased.

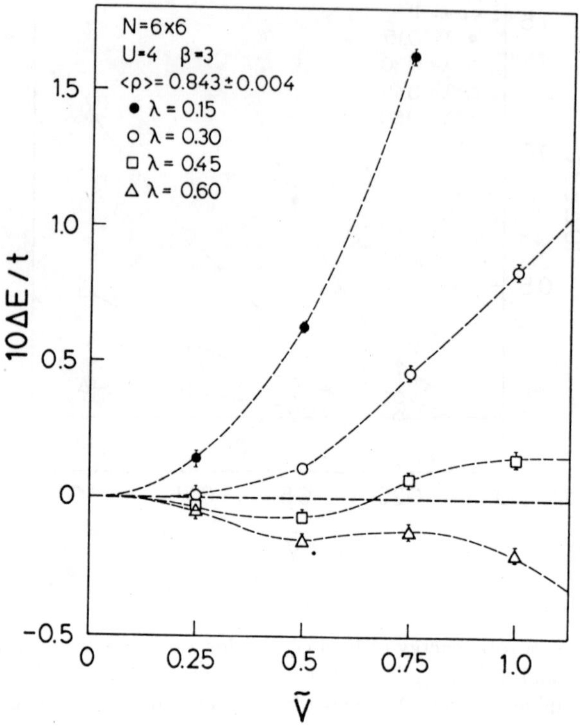

Figure 2.2

Change ΔE in total energy for the same parameters as in Fig. 2.1, but now in the presence of holes ($< \rho >= 0.843$). A CDW lattice instability occurs for values of $\lambda \geq 0.45$.

Summarizing, we have presented quantum Monte-Carlo studies of the 2-D Hubbard model with frozen-in "breathing" lattice displacements. We found that for intermediate values of U, nesting is not effective in inducing a lattice distortion. Moreover, the mean-field results overestimate the tendency to a CDW instability when compared to the MC simulation. This deviation from mean- field increases for increasing U.
Adding holes significantly changes the physical picture, in that now a CDW superlattice becomes stable at a critical value of the electron-phonon coupling. This can be understood in terms of the formation of local polarons, with holes moving towards the raised Cu potentials. Pairing and SDW correlation functions display only a weak dependence on the lattice displacements. We would like to add that simulations upto 12×12 lattice were performed without qualitative changes in the results. A more complete account of our

results will be published elsewhere [24].

References

[1] G. Bednorz, K.A. Müller, Z.Phys.B 64, 189 (1986)

[2] C.W. Chu et.al., Phys.Rev.Lett. 58, 405 (1987)

[3] M.K. Wu et.al., Phys.Rev.Lett. 58, 908 (1987)

[4] P.W. Anderson, Science 235, 1196 (1987); ibid. in "Frontiers and Borderlines in Many Particle Physics", International School of Physics "Enrico Fermi", ed. J.R. Schrieffer and R.A. Broglia (North Holland) to be published and references therein.

[5] A. Ruckenstein, P. Hirschfeld, J. Appel, Phys.Rev.B 36, 857 (1987)

[6] V.J. Emery, Phys.Rev.Lett. 58, 2794 (1987)

[7] J.E. Hirsch, Phys.Rev.Lett. 59, 228 (1987)

[8] A. Muramatsu, R. Zeyher and D. Schmeltzer (unpublished)

[9] D. Vaknin et.al., Phys.Rev.Lett. 58, 2802 (1987)

[10] Y.J. Uemura et.al., Phys.Rev.Lett. 59, 1045 (1987)

[11] J.M. Tranquada et.al., Phys.Rev.Lett. 60, 156 (1988)

[12] G. Shirane et.al., Phys.Rev.Lett. 59, 1613 (1987)

[13] P.M. Horn et.al., Phys.Rev.Lett. 59, 2772 (1987)

[14] S. Massidda et.al., Phys.Lett.A 122, 198 (1987)

[15] F. Herman, R.V. Kasowsky und W.Y. Hsu,Phys.Rev.B 36, 6904 (1987)

[16] V.J. Emery, in "High Conducting One-Dimensional Solids", ed. by J.T. Devreese, R.P. Evrard and V.E. van Doren (Plenum, New York, 1979)

[17] Z.Z. Wang et.al., Phys.Rev.B 36, 7222 (1987)

[18] P. Prelovśek, T.M. Rice, F.C. Zhang, J.Phys.C 20, L229 (1987)

[19] W. Weber, Phys.Rev.Lett. 58, 1371 (1987)

[20] J.E. Hirsch, Phys.Rev.B 31, 4403 (1985)

[21] J.E. Hirsch, Phys.Rev.Lett. 54, 1317 (1985)

[22] R. Blankenbecler, D.J. Scalapino, R.L. Sugar, Phys.Rev.D 24, 2278 (1981)

[23] A. Prêtre and T.M. Rice, J.Phys.C 19, 1363 (1986)

[24] A. Muramatsu, W. Hanke, Phys.Rev.B. Rapid Comm. (to be published)

MAGNETIC PROPERTIES OF YBaCuO CRYSTALS: SUPERCONDUCTIVE GLASS OR FLUX PINNING?

A. P. Malozemoff, Y. Yeshurun[+], L. Krusin-Elbaum,
T. K. Worthington, D. C. Cronemeyer, T. Dinger, F. Holtzberg and
T. R. McGuire
IBM Thomas J. Watson Research Center
Yorktown Heights, NY 10598

and

P. H. Kes,
Kamerlingh Onnes Laboratory,
State University Leiden, P. O. Box 9506, 2300RA Leiden, The Netherlands

ABSTRACT:

We review various magnetic properties of YBaCuO crystals: hysteresis loops, field-cooled and zero-field-cooled moments, remanent moments, ac susceptibility and magnetic relaxation. Some of these results point to a conventional Type II flux pinning and critical state model while others, notably the appearance of an irreversibility line, have been taken to indicate superconductive glass behavior. We argue that the most consistent interpretation favors the flux pinning model once it is recognized that the irreversibility line can be understood in terms of giant flux creep caused by the high temperatures and low pinning potentials in these superconductors.

INTRODUCTION

Magnetic properties of the new high temperature superconductors are of interest for many reasons. The phenomena are complex and interesting in their own right. Magnetic measurements also permit determination of many of the key superconducting parameters, like coherence lengths and penetration depths, which are vital for understanding the superconducting mechanism. Furthermore, magnetic properties control the superconductor's performance in applications, particularly because of the relation between magnetic hysteresis and the critical current density.

Although a tremendous amount of data has already appeared on the new materials, many aspects of these data have been puzzling and have led to a number of competing interpretations. In this paper we discuss this problem in the context of our data[1-9] on YBaCuO crystals[1,10] and show how the flux pinning picture, suitably extended, appears to provide a good framework for understanding the data.

THE IRREVERSIBILITY LINE

Perhaps the key problem was identified by Müller, Takashige and Bednorz in their early work[11] on the magnetic properties of the new high temperature superconductors. They performed dc SQUID measurements of the magnetization, first cooling in zero applied field, then turning on the field and measuring the flux exclusion with increasing temperature. This is called the "zero-field-cooled" (zfc) measurement. At low fields for high density ceramics or crystals, as illustrated in Fig. 1 for an YBaCuO crystal, this measurement usually shows close to complete "flux exclusion" or diamagnetic shielding ($M/H = -1/4\pi$) at low temperatures. But as the temperature is increased close to the transition temperature T_c, flux begins to penetrate and -M decreases in magnitude.

Next, cooling in the same field from above the transition temperature, one measures the "field-cooled" (fc) moment as the temperature is decreased. The "Meissner fraction" $4\pi M/H$ obtained from this measurement at low temperatures is usually significantly less than that obtained in the zfc experiment and corresponds to "flux expulsion" (see Fig. 2). Of particular interest to Müller et al. was the region of overlap between these two measurements near T_c, indicating reversible magnetic behavior (see Fig. 1). They defined a

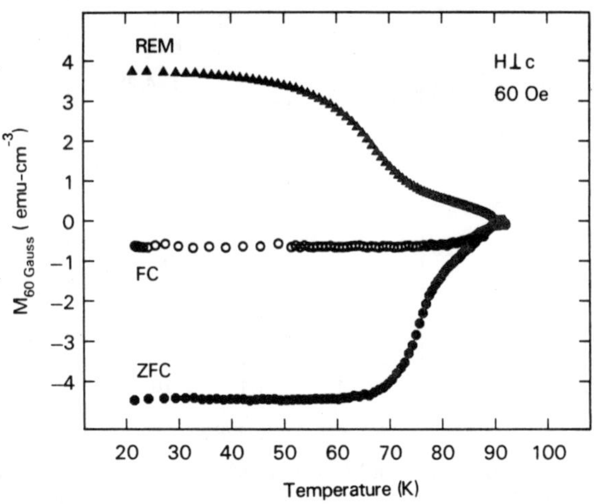

Fig. 1. Zero-field-cooled (zfc), field-cooled (fc) and remanent (rem) moments of an YBaCuO crystal[10] with an applied field of 60 Oe oriented perpendicular to the c-axis. The low temperature value of M_{rem} is accurately equal to the difference $M_{fc} - M_{zfc}$. After Malozemoff et al.[9]

temperature, which we here call T_{irr}, below which the irreversibility sets in. Plotted as a function of field, these temperatures follow a simple power law

$$1 - t = aH^q , \tag{1}$$

where t is the reduced temperature T_{irr}/T_c relative to the true zero-field superconducting transition temperature T_c, and where the exponent q ranges from about 2/3 to 3/4 in different measurements. Müller et al. found q=2/3 in their original measurements on ceramic LaBaCuO. By analogy with spin glasses where similar measurements give an irreversibility line with a 2/3 power, they called this a "quasi-de-Almeida-Thouless" line; we call it more generally an "irreversibility" line. They also reported non-exponential, typically logarithmic, time dependence of the zero-field-cooled and remanent magnetizations,[11-12] which are similar to relaxation in spin glasses.

Müller et al. suggested that these phenomena could be interpreted in terms of the superconductive glass model introduced principally by Ebner and Stroud.[13] In this model the superconductor is viewed as granular, with Josephson elements connecting the different grains. Randomness in the grain

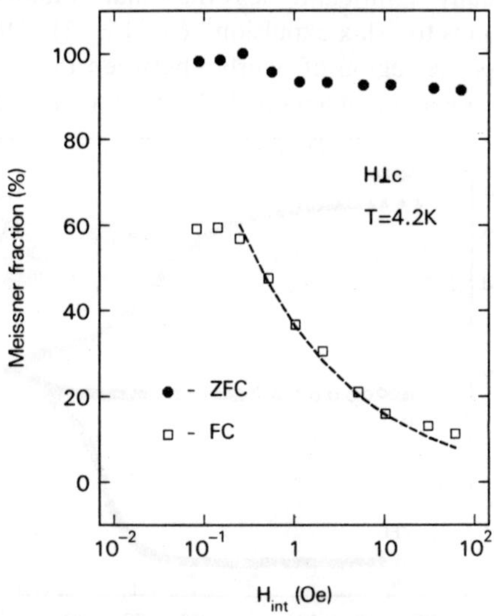

Fig. 2. Zero-field-cooled and field-cooled Meissner fractions as a function of field for an YBaCuO crystal.[10] The dramatic increase of the field-cooled fraction at low fields shows that a measurement at a single field cannot be used as a reliable measure of the quality of a superconducting sample. The dotted line is a theoretical prediction as described in the text.[6] After Krusin-Elbaum et al.[6].

positions implies a random "gauge field" or phase factor in the Josephson relation, in the presence of an applied field. This leads in turn to the presence of currents of random polarity around different loops of neighboring grains, leading to randomly oriented moments. The effect disappears in the absence of field, and a characteristic field scale for the complete disordering of the moments is

$$H \simeq \Phi_0/a^2 \ , \tag{2}$$

where Φ_0 is the flux quantum 2×10^{-7}Gcm^2, and where a is a characteristic spacing between the grains. The field for the onset of nonlinearities in the moment is considerably lower, a small fraction of that in Eq. 2, as shown by simulations.[13,14] Thus the reduced Meissner fraction in field-cooled experiments could be interpreted as arising from such random polarity reversal of local moments in the superconductive glass model.

Müller et al.[11,14] pointed out that the field scale for observation of such effects implies, through Eq. 2, length scales much smaller than the characteristic 10 μm grain size of typical ceramics in which the original measurements had been done. For example, a field of 1000 Oe corresponds to only 0.14 μm. This they took as an indication that the glassy structure could be intrinsic to the grains themselves, and therefore should be characteristic of crystals also. Indeed, Deutscher and Müller[15] suggested possible Josephson elements in the crystals, namely the twin boundaries, which are generally found in YBaCuO crystals with spacings in the range of 10's to 1000's of nanometers.

While there is very strong evidence[16] for the existence of Josephson elements at grain boundaries in YBaCuO, the presence of such elements in crystals is less clear. Given the complexity of the phenomena in polycrystalline material,[17–24] in which granular behavior overlays contributions from within the grains, it is clearly of importance to first identify the behavior of the bulk (though possibly twinned) material. Understanding of bulk material would then form a basis for analyzing data on polycrystalline ceramics. This is the aspect which we have addressed so far in our work and which we review below.

In fact our data on YBaCuO crystals[1,10] with extremely sharp susceptibility transitions show many of the phenomena observed in the ceramics. The overlap region visible in Fig. 1 illustrates the existence in crystals of an irreversibility line, which, for one case with $H \parallel c$, gives the dependence[7] $1 - t \simeq 5 \times 10^{-4} H^{2/3}$. The Meissner fraction in field-cooling, shown in Fig. 2, is indeed suppressed[5,6] with increasing field strength, and, as suggested by the glassy model, it increases towards full flux expulsion at very low fields.

(Incidentally, this observation is also of importance because in the literature the fraction of superconducting material in a sample is sometimes evaluated from the size of the Meissner effect; our results show that this evaluation is meaningless since the Meissner effect depends so strongly on the field strength, and saturating values are attained at extremely low fields available only in specially designed apparatus like our non-commercial SQUID magnetometer.[5,6,9]) The observation[7] of logarithmic relaxation in the magnetization, with a temperature dependence of dM/dln(t) illustrated in Fig. 3, recalls the roughly logarithmic time depences reported in spin glasses.

Nevertheless a number of other observations point toward an alternative "flux pinning" picture of the magnetic properties. Hysteresis loops[1] show a characteristic shape familiar in conventional superconductors and consistent with the Bean model of the critical state.[25,26] This picture considers a Type II superconductor with a conventional Abrikosov vortex or flux line lattice.[27] Inhomogeneities in the material cause pinning of these vortices in potential valleys of height U_0. Such pinning prevents motion of vortices in the presence of current and thus controls the critical current density J_c . Because of the Maxwell relation $dh/dx = 4\pi J/c$ (in cgs units) between the gradient of the local induction field h and the current density J, the penetration of flux is determined by the critical current density J_c. Thus simple relations can be derived[25,26] between the critical current and the magnetic irreversibility ΔM obtained in a hysteresis loop (here ΔM can be defined as the remanent moment after removal of a large field exceeding H_{c1}). For a cylinder of radius r, one ha s $J_c = 30\Delta M/r$, or for a slab of thickness d, one has $J_c = 40\Delta M/d$,

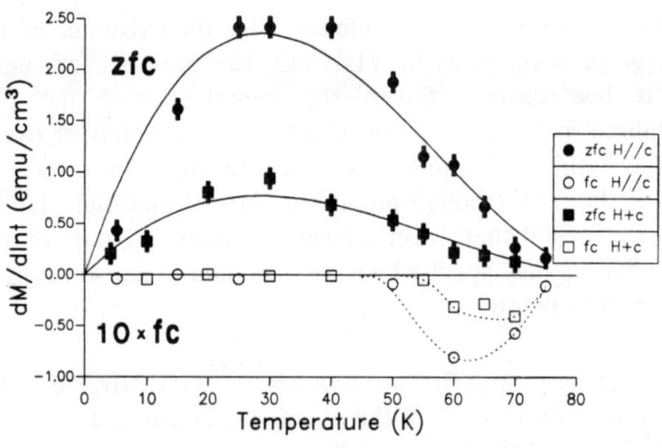

Fig. 3. Magnetic relaxation, zero-field-cooled and field-cooled, as a function of temperature for an YBaCuO crystal.[1] The small relaxation in the field-cooled moment indicates the presence of irreversible effects near the irreversibility line. The size of the relaxation is very large compared to conventional superconductors. After Yeshurun and Malozemoff.[7]

where J_c is in Amps/cm^2, M is in emu/cc and r or d are in cm. Using this treatment, the hysteresis loops can be understood quite well,[22] and the critical current density can be extracted.[1,31]

Another measurement which may distinguish between the models is of the dc remanent moment obtained by turning off the magnetic field after field-cooling. The resulting positive moment is measured with increasing temperature, giving results like those shown in Fig. 1 and labeled "rem". The magnitude of this remanent moment is found to equal quite closely the difference between the field-cooled moment and the zero-field-cooled moment at low fields and temperatures.[5,9,18] This observation is easily explained in the flux-pinning picture.[9] The field-cooled moment consists of a surface diamagnetic screening signal decreased by the effect of positive trapped flux. When the field is turned off, the surface screening current disappears but the trapped flux remains, giving the remanent moment. By contrast this effect is problematic for the superconductive glass model because if there are canceling random polarity local moments after field-cooling, the remanent moment need bear no precise relationship to that field-cooled moment.[9]

Perhaps the clearest contradiction to the glassy model in crystals comes from the direct observation of the Abrikosov vortex lattice[27] by a magnetic decoration technique.[28] Since the observed vortex density changes linearly with field strength, consistent with Eq. 2, it is clear that the vortex structure corresponds not to a fixed structure of Josephson elements in the crystal but rather to a conventional Type II vortex lattice.

A possible explanation for the apparent disappearance of the vortex lattice near, but below, T_c in these studies[28] comes from the concept of vortex "melting", possibly due to a Kosterlitz-Thouless transition. This idea also offers an interpretation of the irreversibility line, which presumably corresponds to the melting phenomenon since melting implies that vortices are mobile and can find their true reversible equilibrium configuration. We do not consider this interesting proposal further because it rests on a statistical mechanical mechanism for the irreversibility line which would depend only on fundamental parameters of the material. The weight of our evidence, however, points to different irreversibility lines in different crystals, depending upon the nature of their defects. Further work is required to demonstrate this conclusively. In what follows we develop a mechanism for the irreversibility line depending on defects and flux pinning in the crystals.

GIANT FLUX CREEP

While many articles have pointed out one or another of the features described above, a comprehensive view of the magnetic properties has been lacking. We attempt such a comprehensive view here by pointing out that the features originally taken as evidence for glassy behavior, namely the irreversibility line and the logarithmic relaxation, can be explained at least in a preliminary way by a simple extension of the flux pinning picture. Since the other observations have not been explained by the glassy model, the weight of evidence leans in our view towards the flux pinning model, at least in the crystals. We emphasize that glassy effects are more likely in ceramics and polycrystalline material, but further work needs to be done to demonstrate them conclusively.

The key extension required for the flux pinning picture in crystals is to recognize that in the new superconductors, because of the high temperature and low pinning barriers, flux creep can occur on a scale far beyond what has been observed in conventional low temperature superconductors. This is consistent with the data in Fig. 3, where the moment relaxation is extremely large by comparison with usual materials. This "giant" flux creep underlies our proposal for the origin of the irreversibility line.

To develop a semi-quantitative theory of this effect,[7] we follow the flux creep theory of Anderson and Kim[29] as developed by Campbell and Evetts[26]. In this theory, thermal activation of vortices over an energy barrier U_0 is modulated by a Lorentz force on the vortices proportional to the critical current J_c and leads to a reduction of the measured critical current below its value J_{c0} in the absence of thermal activation[26]

$$J_c = J_{co}[1 - (kT/U_0)\ln(f_0/f)] , \tag{3}$$

where f_0 is a characteristic attempt frequency of order 10^{10} Hz.

Since Maxwell's equations can be used to relate J_c to the magnetization, one can directly derive the predicted magnetization relaxation[30] of a cylinder of radius r, valid for the limit of $H > H_{c1}$ and high temperature

$$dM/d\ln t = (rJ_{c0}/3c)(kT/U_0) , \tag{4}$$

where U_0 is the activation energy in the absence of a gradient. For the case of low field and low temperature where the flux does not penetrate entirely through the sample, an alternative formula can be derived (assuming J_c independent of H):

$$dM/d\ln t = (cH^2/16\pi^2 rJ_{c0})(kT/U_0) . \tag{5}$$

These results predict logarithmic relaxation of the type observed experimentally[9] and thus offer an alternative to the superconductive glass model.

These equations describe the qualitative data trends very well. The rise and fall with temperature can be interpreted in terms of a competition between the explicit linear T term and the implicit dropoff in J_c to zero at the critical temperature. (U_0 is also temperature dependent but usually more weakly, see below.) Eq. 5 explains the low temperature field dependence to be reported elsewhere.

With earlier estimates[1] for J_c, one can extract U_0 of order 0.1 eV from comparison of the relaxation data with these equations. These values are an order of magnitude smaller than in typical low temperature superconductors.[30] This can be understood semiquantitatively from the simple scaling suggested by Anderson and Kim[29] $U_0 \propto H_c^2 \xi^3 / 8\pi$, where H_c is the thermodynamic critical field and ξ is the superconducting coherence length. For conventional type II materials this usually comes out to several eV, in agreement with earlier experiments.[30] For YBaCuO we take $\xi^3 = \xi_{ab}^2 \xi_c$ with published parameters[2] and obtain $U_0 \simeq 0.15$ eV at low temperatures, roughly as observed experimentally.

Next we show how this giant flux creep can also explain the irreversibility line.[7] Conventionally the thermal activation term in Eq. 3 is a negligible correction[26] to J_c, and even though the activation energy U_0 drops to zero at T_c, the temperature difference over which one might be able to observe an effect on J_c near T_c is exceedingly small. This is no longer true in the high temperature superconductors, where the temperature T is an order of magnitude larger than in conventional superconductors and where the activation energy U_0 is an order of magnitude lower, as we have shown. This suggests an explanation of the irreversibility line because now the temperature difference over which J_c is measurably affected by flux creep is much larger. In particular $J_c = 0$ in Eq. 3 corresponds roughly to the temperature T_{irr} where the critical current becomes immeasurably small and where magnetic irreversibility, which depends on a finite J_c, disappears.

The field and temperature dependence of this line is determined by the scaling of U_0 with field and temperature. Many models of pinning in superconductors have been proposed.[26,31] However at this point the pinning mechanisms in the new high temperature superconductors are just beginning to be studied, and so we confine ourselves here, as in the earlier work,[7,8] to a very simple scaling argument which gives only one of many possible scaling forms. The basic idea is that the activation energy U_0 scales as the condensation energy per volume $H_c^2/8\pi$ times a characteristic excitation volume. For sufficiently large fields, that volume is limited laterally by the area which a single flux quantum occupies in the flux line lattice. This area is approximately a_0^2, where $a_0 = 1.075\sqrt{\Phi_0/B}$ is the flux line spacing in field B.

In the third direction, along the applied field, the minimum possible extent of the activation volume is ξ, giving $U_0 \propto (H_c^2/8\pi)a_0^2\xi$. According to the Ginsburg-Landau theory, H_c scales as 1-t and ξ scales as $(1 - t)^{-1/2}$; so U_0 scales as $(1 - t)^{3/2}/B$. Substituting in Eq. 3, approximating T by T_c, and solving for 1-t with $t = T_{irr}/T_c$, one finds an expression for the irreversibility line valid with $f << f_0$

$$1 - t = p[BkT_c \ln(f_0/f)]^{2/3} , \qquad (6)$$

where p is a constant to be determined experimentally. This derivation of the irreversibility line of Eq. 1 is key because it gives an alternative to the superconductive glass explanation. Estimates of the constant p give order-of-magnitude agreement with experiment.[7]

Understanding the irreversibility line in the context of flux creep clarifies several other interesting effects. The field-dependent flux exclusion (Meissner fraction) of Fig. 2 can now be understood from the fact that during field-cooling, the flux and hence the magnetization are frozen in when the irreversibility line is reached. As illustrated in Fig. 4, the upward curvature

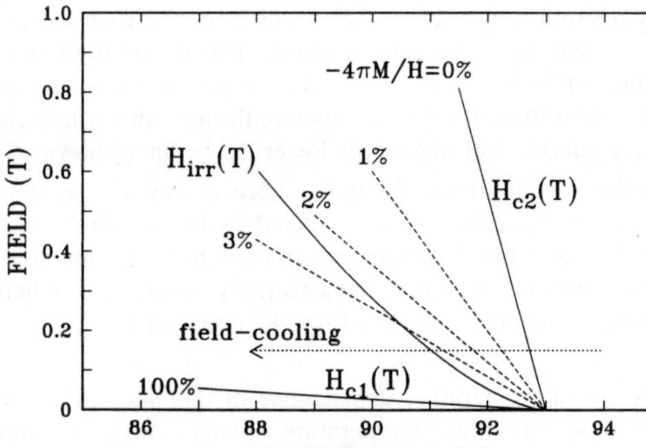

Fig. 4. Schematic field-temperature diagram for a superconductor, showing the upper and lower critical fields, the irreversibility line, and the equilibrium moment fractions predicted by the Ginzburg-Landau and Abrikosov theory for Type II superconductivity. The dotted line indicates the process of field-cooling, and the intercept with the irreversibility line determines the Meissner fraction at low temperatures. After Krusin-Elbaum et al.[6]

of the irreversibility line in the H-T plane causes it to intersect the equilibrium lines of $4\pi M/H$ at increasingly high ratios for lower fields. By using the conventional Abrikosov theory for $4\pi M/H$, and combining it with the measured irreversibility line, we have been able to make a quantitative prediction[6] of the Meissner fraction, as shown by the dashed line in Fig. 2 (The vertical scale is set by the rather imperfectly known Ginzburg Landau κ ratio, which has been used as a fitting parameter here; details will be given in later work.) Thus we see that the unusual field dependence of the Meissner fraction appears to be a consequence of the irreversibility line and hence of the pinning in the material.

Another surprising prediction comes from Eq. 6: the irreversibility line should be frequency-dependent. We can in fact derive this dependence by differentiating Eq. 6 to obtain

$$dT_{irr}/d\ln(f) = (2pT_c/3)(BkT_c)^{2/3}[\ln(f_0/f)]^{-1/3} , \qquad (7)$$

Eqs. 6,7 show that the apparent transition temperature **increases** with increasing frequency. This is physically plausible because at longer measurement times there is more time for flux lines to relax to an equilibrium configuration, and thus reversibility can be achieved at lower temperatures. The effect is close to logarithmic in frequency, although the -1/3 power in Eq. 7 causes a small upward curvature.

These predictions clarify[8] problems which have emerged in the study of the upper critical field H_{c2} by pinning, ac susceptibility and resistivity measurements.[2,3,31,32] Ac susceptibility data[3] shown in Fig. 5 follow the power law of Eq. 1 with exponents close to 2/3 and 3/4 for field along the two crystal directions. What is more, these results exhibit the roughly logarithmic frequency dependence[8] shown in Fig. 6, which can be fit to Eq. 7 with the choice $f_0 = 10^9$ Hz. These results suggest[8] that the ac measurements, originally interpreted as the upper critical field, are in fact a measurement of the irreversibility line. Disturbing discrepancies[31] between these measurements and pinning or resistance measurements of H_{c2} can now also be understood in terms of the frequency effect.[8] The puzzling broadening[32] of the resistance-versus-temperature curve in field can now be understood in terms of flux-flow resistance.

What do these results imply for the true H_{c2}'s and the fundamental parameters like the coherence lengths deduced from them? One cannot simply extrapolate the irreversibility line to higher frequencies to determine H_{c2}, although surely the true H_{c2}'s lie above the irreversibility lines, and so coherence lengths are smaller than those previously reported.[1,2] Reversible magnetization measurements are needed and have been reported by Fang et al.[33] but show a puzzling downward curvature (the opposite tendency to that in Fig. 5). This has been attributed[33] to effects of twin boundaries, but more work needs to be done on this problem before convincing coherence lengths can be extracted.

In summary "giant flux creep" explains properties previously interpreted in the superconductive glass context and clarifies several other surprising effects like the field-dependent Meissner fraction and the frequency dependent ac susceptibility transition. These preliminary successes support a flux pinning model for the magnetic properties of YBaCuO crystals. Clearly more work needs to be done to explore different aspects of this model in detail as well as to extend the work to ceramics where the conclusions are at this point less clear.

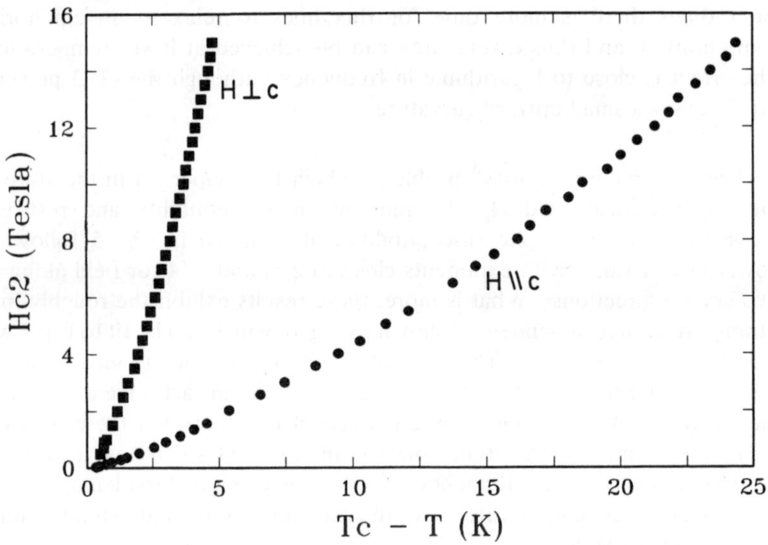

Fig. 5. Field dependence of the transition temperature determined from the peak in the out-of-phase ac susceptibility on an YBaCuO crystal.[10] Originally interpreted as a measure of the upper critical field $H_{c2}(T)$, this data should now be interpreted as a measure of the irreversibility line in the MHz frequency range. After Worthington et al.[3]

The authors thank K. A. Müller, W. J. Gallagher, I. Morgenstern, T. Schneider, R. Greene, J. Evetts, M. B. Salamon, D. Pines, W. W. Webb, G. J. Dolan, J. A. Mydosh, J. Clem, D. K. Finnemore, and P. M. Horn for helpful discussions. P. H. K. thanks A. A. Menovsky, J. van den Berg and R. J. Wijngaarden for their cooperation.

+Permanent address: Dept. of Physics, Bar-Ilan University, Ramat-Gan, Israel.

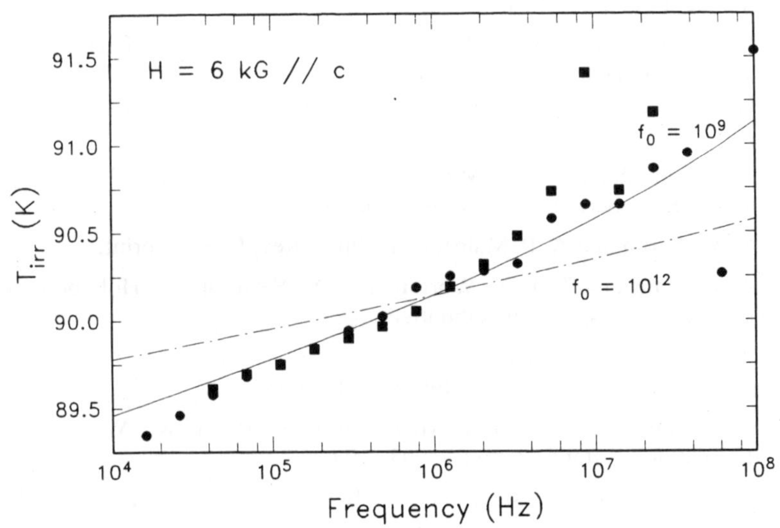

Fig. 6. Logarithmic frequency dependence of the apparent transition temperature in the ac susceptibility of the crystal of Fig. 5, at a field of 0.6 T oriented parallel to the c-axis. These data give strong evidence against the original interpretation in terms of the upper critical field and point towards the interpretation in terms of flux creep and the irreversibility line. After Malozemoff et al.[8]

References

1. T. R. Dinger, T. K. Worthington, W. J. Gallagher and R. L. Sandstrom, Phys. Rev. Lett. **58**, 2687 (1987).

2. T. K. Worthington, W. J. Gallagher and T. R. Dinger, Phys. Rev. Lett. **59**, 1160 (1987); W. J. Gallagher, T. K. Worthington, T. R. Dinger, F. Holtzberg, D. L. Kaiser and R. L. Sandstrom, Physica **148B**, 228 (1987); T. K. Worthington, W. J. Gallagher, T. R. Dinger and R. L. Sandstrom, in **Novel Superconductivity**, ed. S. A. Wolf and W. Z. Kresin (Plenum Press, New York, 1987), p. 781.

3. T. K. Worthington, W. J. Gallagher, D. L. Kaiser, F. H. Holtzberg and T. R. Dinger, Physica, to be published (Proceedings of the Interlaken Conference, Switzerland, Feb. 29-Mar. 4, 1988).

4. T. R. McGuire, T. R. Dinger, P. J. P. Freitas, W. J. Gallagher, T. S. Plaskett, R. L. Sandstrom and T. M. Shaw, Phys. Rev. B **36**, 4032 (1987); T. R. McGuire, F. Holtzberg, D. L. Kaiser, T. M. Shaw and S. Shinde, to be published.

5. L. Krusin-Elbaum, A. P. Malozemoff and Y. Yeshurun, in **High Temperature Superconductors**, ed. M. B. Brodsky, R. C. Dynes, K. Kitazawa and H. L. Tuller (Materials Research Society, Pittsburgh PA, 1988), Vol. 99, p. 221.

6. L. Krusin-Elbaum, A. P. Malozemoff, Y. Yeshurun, D. C. Cronomeyer and F. Holtzberg, Physica, to be published.

7. Y. Yeshurun and A. P. Malozemoff, Phys. Rev. Lett., in print.

8. A. P. Malozemoff, T. K. Worthington, Y. Yeshurun, F. Holtzberg and P. H. Kes, Phys. Rev. B, submitted.

9. A. P. Malozemoff, L. Krusin-Elbaum, D. C. Cronemeyer, Y. Yeshurun, F. Holtzberg and C. Tsuei, Phys. Rev. B, submitted.

10. D. L. Kaiser, F. Holtzberg, M. F. Chisholm and T. K. Worthington, Journal of Crystal Growth **85**, 593 (1987).

11. K. A. Müller, M. Takashige and J. G. Bednorz, Phys. Rev. Lett. **58**, 1143 (1987).

12. A. C. Mota, A. Pollini, P. Visani, K. A. Müller and J. G. Bednorz, Phys. Rev. B **36**, 401 (1987).

13. C. Ebner and D. Stroud, Phys. Rev. B **31**, 165 (1985).

14. I. Morgenstern, K. A. Müller and J. G. Bednorz, Z. Phys. B **69**, 33 (1987).

15. G. Deutscher and K. A. Müller, Phys. Rev. Lett. **59**, 1745 (1987).

16. P.Chaudhari, J. Mannhart, D. Dimos, C.C. Tsuei, J. Chi, M. Oprysko and M. Scheuermann, Phys. Rev. Lett. **60**, 1653 (1988).

17. H. Maletta, A. P. Malozemoff, D. C. Cronemeyer, C. C. Tsuei, R. L. Greene, J. G. Bednorz and K. A. Müller, Solid State Commun. **62**, 323 (1987).

18. D. C. Cronemeyer, A. P. Malozemoff and T. R. McGuire, in **High Temperature Superconductors**, ed. M. B. Brodsky, R. C. Dynes, K. Kitazawa and H. L. Tuller (Materials Research Society Pittsburgh PA, 1988), Vol 99, p. 837.

19. C. Giovannella, G. Collin, P. Rouault and I. A. Campbell, Europhys. Lett. **4**, 109 (1987).

20. Y. Wolfus, Y. Yeshurun and I. Felner, Phys. Rev. B **37**, 3667 (1988).

21. M. Tuominen, A. M. Goldman and M. L. Mecartney, Phys. Rev. B **37**, 548 (1988). These authors point out that the remanent moment does not disappear at T_{irr}, as illustrated in Fig. 1. They argue that this is inconsistent with the concept of an irreversibility line. A possible resolution to this contradiction is to recognize that remanent moment relaxation is sensitive to surface conditions, where the pinning could be different from that in the bulk.

22. U. Atzmony, R. D. Schull, C. K. Chiang, L. J. Schwarzendruber and L. H. Bennett, J. Appl. Phys. **63**, 4179 (1988).

23. S. Senoussi, M. Oussena, M. Ribault and G. Collin, Phys. Rev. B **36**, 4003 (1987).

24. E. L. Venturini, D. S. Ginley, J. F. Kwak, R. J. Baughman, J. E. Shirber and B. Morosin, Proceedings of the MRS Symposium, Anaheim CA, April 20-23, 1987, p. 97.

25. C. P. Bean, Phys. Rev. Lett. **8**, 250 (1962).

26. A. M. Campbell and J. E. Evetts, Adv. Phys. **21**, 199 (1972).

27. A. A. Abrikosov, Soviet Phys.-JETP **5**, 1174 (1957).

28. P. L. Gammel, D. J. Bishop, G. J. Dolan, J. R. Kwo, C. A. Murray, L. F. Scheemeyer and J. V. Waszczak, Phys. Rev. Lett. **59**, 2592 (1987).

29. P. W. Anderson, Phys. Rev. Lett. **9**, 309 (1962); Y. B. Kim, Rev. Mod. Phys. **36**, 39 (1964).

30. M. R. Beasley, R. Labusch and W. W. Webb, Phys. Rev. **181**, 682 (1969).

31. P. H. Kes, Physica, to be published; J. van den Berg, C. J. van der Beek, P. H. Kes, J. A. Mydosh, A. A. Menovsky and M. J. V. Menken, Physica, to be published ; R. J. Wijngaarden, K. Heeck, R. Griessen, A. A. Menovsky and M. J. V. Menken, Physica, to be published. (Proceedings

of the Interlaken Conference, Switzerland, Feb. 29-Mar. 4, 1988).

32. Y. Iye, T. Tamegai, H. Takeya and H. Takei, Jpn, J. Appl. Phys. Part 2, **26**, L1057 (1987); also Physica, to be published; B. Oh, K. Char, A. D. Kent, M. Naito, M. R. Beasley, T. H. Geballe, R. H. Hammond and A. Kapitulnik, to be published.

33. M. M. Fang, V. G. Kogan, D. K. Finnemore, J. R. Clem, L. S. Chumbley and D. E. Farrell, Phys. Rev. B, to be published; L. A. Schwartzkopf, M. M. Fang, L. S. Chumbley and D. K. Finnemore, Physica, to be published.

THE WEAK LINKS AND THE MAGNETIC AND TRANSPORT PROPERTIES OF THE HIGH T_C SUPERCONDUCTORS

S. Senoussi, M. Oussena, J.B. Bieri, J. Arabski and R. Reich,*
Laboratoire de Physique des Solides, Bâtiment 510
Université de Paris Sud, 91405 ORSAY (France)
* Fuller Theological Seminary Pasadena, California 91182, USA *

PACS : 74.70, 74.60J, 74.30C

ABSTRACT

Magnetic and transport measurements on $YBa_2Cu_3O_7$, $La_{1-x}Sr_xCuO_4$ and BiSrCaCuO are reported. The low-field region ($H < H_{c1}$) of the magnetic cycle of granular materials is characterized by two threshold fields H^w_{c1} and H^w_{c2} related to the transport critical current by a Silsbee-like criterion. However, no such fields are seen in single crystals. These low-H data are ascribed to the weak-links and are interpreted quantitatively using the Ferrell-Prange and the Ambegaokar-Baratoff equations at low-fields ($H < H^w_{c2}$) a granular London-like picture at intermediate fields ($H^w_{c2} < H < H^w_{c1}$) and finally the familiar critical state model at higher fields ($H^w_{c1} < H < H_{c2}$).

The magnetic and transport properties of granular superconductors are very complex and exhibit several regimes of magnetic behaviors as a function of field (H) [1-5], temperature (T) [6,7] or current (I) [2,8,9].
An interesting feature of these materials is that, at high field the magnetic cycle (H > 1kG at 4.2 K) is well described by the critical state model provided that the sample is assumed to be a collection of decoupled filaments of the same radius as the grains [1-5].
Other striking features which will be discussed here are : (1) the anomalous low-value of the transport critical current density J_t ($\leq 10^3$ A/cm^2) and its high sensitivity to any applied field [2,8,9], (2) the presence of an unusual low-field hysteresis loop (H < 100 G at 4.2 K) characterized by two threshold limits H^w_{c1} and H^w_{c2} marking the onset respectively the vanishing of a regime of low irreversibility effects [2,3], (3) A large drop (10 to 30 %) [2,3] of the magnetic susceptibility (-m/H) when the field increases from the region $H < H_{c1}$ to the region $H^w_{c2} < H < H_{c1}$ where H_{c1} is the familiar first critical field for the grains. (4) It turns out that for sufficiently sintered and compacted specimens the transport critical current density of granular materials is related to the low-H characteristic fields of the magnetic cycle by a Silsbee like equation of the form :

$$H^{w}_{c_1} < \frac{2\pi J_t R}{10} < H^{w}_{c_2} \qquad (1)$$

where the units are in A/cm^2 and in Gauss (practical units).
Here R is the radius of the specimen assumed to have a cylindrical geometry with no demagnetizing field. The above formula suggests strongly [2,3] that J_t is to some extent imposed by the magnetic field (H_{ind}) induced by the current itself. As a consequence the transport current J_t should depend strongly on the field whether self-induced or external. Then, as the self field at the surface of the specimen tends to zero with R ($H_{ind} \alpha J_t R$) we expect that for sufficiently thin samples the transport critical current would be limited by some intrinsic features of the barriers (i.e. the weak-links) between the grains composing the material. In this paper we report new data together with an improved theoretical analysis based on the Ferrell-Prange-Josephson equations [10] and on the Ambegaokar-Baratoff [11] formulae. We find that these equations (hereafter refered to as ABFPJ equations) extend the well known critical state model which seems to work reasonably well at high fields but is insufficient (within its familiar formulation) to describe the low-H magnetic behavior. The characterization and the preparation of the $YBa_2Cu_3O_7$ material followed the same procedures as in refs 2,3. As in these references the magnetic measurements were performed using a vibrating sample magnetometer.

RESULTS

LOW-FIELD HYSTERESIS CYCLE

A detailed investigation of the M vs H relationship at 4,2 K for $YBa_2Cu_3O_7$ cooled in zero field is presented in Fig. 1 (sintered specimen) and Fig. 2 (powder) at various field scales showing three different magnetic regimes :
1. It is clear from the inspection of Fig. 1 and 2 that, initially the curve (virgin branch) is fairly linear and reversible up to a first threshold value H^{w}_{c1} which depends on the sintering conditions. Moreover, in this field region the screening is almost complete ($-4\pi m/H \approx 95$ %) for the highly sintered specimen (Fig. 1).
2. As H is further increased (i.e. $H > H^{w}_{c1}$) a regime of weak irreversibilities sets in abruptly and holds up to a second, but ill-defined threshold field H^{w}_{c2} which also depends on the sintering conditions. It is believed /2,3/ that H^{w}_{c1} is the field at which the flux starts penetrating the sample through the weak links connecting the grains whereas H^{w}_{c2} would indicate that most of the weak-links (those parallel to the field) have been broken.
At this point we note that the size of the low-H cycle (as measured by H^{w}_{c1} and H^{w}_{c2}) is notably smaller for the powder than for the sintered specimen. This together with the fact that no low-H cycle is seen in single crystals /12/ suggests that the number of weak-links is much smaller in the powder and that the single grains composing the powder are not single-crystals but include many smaller crystallites. We also find that embedding the powder in silicon oil leads to no detectable changes in the m-H curves. The differential susceptibility of $YBa_2Cu_3O_7$ in various crystalline and granualr states is illustrated in Fig. 3 which presents the three above magnetic regimes in a different way.

3. For still higher fields ($H > H^W_{c2}$) the m-H curve (Figs 1, 2, 3)becomes again reversible and linear (but with a reduced slope) up to a third critical field $H^g_{c1} \approx H_{c1}$ beyond which strong irreversibilities take place gradually : it is probable that H^g_{c1} marks the penetration of the vortex lines within the grains. Consequently, H^g_{c1} would be equal to the usual first critical field H_{c1} if the surface barriers opposing the entry of flux quanta within the grains are negligible.

TRANSPORT-CRITICAL CURRENT AND V-I CHARACTERISTICS

We have measured the voltage (V) as a function of current (I) of about ten $YBa_2Cu_3O_7$ samples prepared in different experimental conditions. The shape of the V(I) cycles varies strongly from sample to sample.
Figure 4 represents an example of the V versus I characteristics measured at 77 K after the sample has been cooled in $H_{ext} = 0$ to that temperature. As for the magnetic cycle we often observe several electrical regimes.
1. Starting from H = 0 just after zero-field cooling (Virgin branch labelled "1") and increasing the field we observe a marked kink at some critical current I_{c1} above which V increases very suddenly. This is followed by some irreversibility effects which persist up to an upper current I_u the induced field of which (H_{ind}) (at the surface of the specimen) seems to be of the order of the field H^W_{c2} of the magnetic cycle (formulae 1).
2. If the current is subsequently reduced from I_u (branch 2) we observe a second kink at a different current I_{c3} below which the voltage drops rapidly toward zero.
3. Finally, sweeping again the field from H = 0 to H_u (third branch) we find a third kink with a higher critical field I_{c2}. If the field is cycled again up and down one always passes by branches 2 and 3.

HIGH-FIELD HYSTERESIS CYCLE

The magnetic cycles of $YBa_2Cu_3O_7$ and $La_{1.85}Sr_{0.1}CuO_4$ in relatively high fields are displayed in Figs 5 and 6 for several temperatures.

It is worth noticing that both LaSrCuO and BiSrCaCuO ceramics [1,13] exhibit qualitatively the same m(H) relationships as $YBa_2Cu_3O_7$ system. In particular, for these three oxydes and for high enough H and T the m-H curves approach once again a reversible and roughly linear behavior above some threshold field $H_r(T)$ probably marking a transition toward a resistive (or flux-flow) state [14].

DISCUSSION

The data reported here are too complex to be interpreted rigourously within the framework of any existing models for granular superconductors [15-18]. However, we believe that the magnetic cycle can be interpreted consistently over the whole field range using three different pictures each of which being respectively appropriate to one of the magnetic regions considered in Figs 1, 2, 3 and defined successively by $H \leq H^W_{c1}$, $H_{c1} < H < H_{c2}$ and $H_{c2} < H < H_{c1}$.

ABFPJ-EQUATIONS, $H \leq H^W_{c2}$

We beleive that most of the low-H magnetic and transport properties of the new high-Tc oxydes can be interpreted qualitatively in terms of Ambegaokar-Baratoff and Ferrell-Prange equations [10,11]. Accordingly, a Josephson-junction of thickness d is characterized by an effective penetration depth λ_j, an effective field H_j and an upper current density J_j given respectively by (Gaussian units) :

$$\lambda j = \left(\frac{h\,c^2}{8\pi e J_j(2\lambda + d)} \right)^{1/2}$$

$$H_j = 8\pi J_j \lambda_j / c \tag{3}$$

Here J_j is an upper limit of the critical current density of the junction whereas λ is the usual London penetration depth for the grains on both sides of the junction.

In this discussion we assume that J_j is defined by the Ambegaokar-Baratoff equation [11] :

$$J_j = \frac{\pi}{2e} \frac{\Delta(T)}{R_j(T)} \,\mathrm{Tanh}\, \frac{\Delta(T)}{k_B T} \tag{4}$$

where $\Delta(T)$ is the energy gap and R_j the normal state resistance (per cm^2) of the junction.

To get an order of magnitude for the junction-resistance R_j we propose that the residual resistivity R_0 of the whole sample as extrapolated from the high T regions is principally due to the grain boundaries. Then assuming that the sample is formed by (n+1) grains of radius r_j each and by n junctions in series we get :

$R_j(T) = R_0/n$ with $n = 1/(2\, r_j)$

As an example we take $R_0 = 10^{-3}\ \Omega\, cm^2$, $n = 10^3$ and thus $R_j \approx 10^{-6}\ \Omega\, cm^2$. Using results from the literature for λ, we also get $\lambda \approx 1/2\ (\lambda_\perp + \lambda_{||}) \approx 1/2\ (2 10^{-5} + 8 10^{-5}) = 5 10^{-5}$ cm, $d \approx 0$ (generally $d \approx \xi << \lambda$) and $\Delta(0) \approx 2k_B T_c$ (≈ BCS value) ≈ 200 K.

Putting the above numerical values in equations 2, 3, 4 gives $J_j \approx 5 10^4$ A/cm^2, $\lambda_j \approx 1\ \mu m$ and $H_j \approx 5$ G. These values are not very sensitive to the exact choices for λ and R_j. The most significant result to be emphasized is that the critical field H_j calculated just above is quite close to the threshold field H^W_{c1} (< 2O G)deduced from the magnetic measurements. It also has the same physical meaning as the one ascribed [2,3] to H^W_{c1} :

Indeed, according to Ferrell and Prange [10], and Josephson, λ_j is a "London"-like penetration depth for the junction whereas H_j can be regarded as a first critical field marking the beginning of flux penetration into the core of the junction. This finite penetration depth may also explain why the transport current (≈ 10^3 A/cm^2 at 4.2 K) of our present granular ceramic sample is much lower than the one calculated ($J_j \approx 5 10^4$ A/cm^2) from eq. 4. Again, following Ferrell and Prange and Josephson, the transport and magnetic behaviors of a

Josephson barrier can be explained qualitatively by analogy with type II "superconductors". There are three interesting situations :

1. The junction is ideal but small compared with λ_j ($\lambda_j >> r_j$). The global current density and the transport current are of order Jj ($J_t \approx J_j$) whereas the field dependence of this global current is given by a Fraunhofer diffraction pattern and thus decreases markedly with H (in few tens Gauss). At this point, we stress that for real-granular samples there is a large distribution in the junction parameters (shape, dimension...) with no phase coherence from junction to junction. As a result, the diffraction pattern will be reduced to its central lob.
2. The junction is large compared to λ_j ($r_j >> \lambda_j$) and still ideal in the sense that it has no defects to pin the vortex. Then the Josephson current is a "surface-like current" confined to a fringe of width λ_j at the junction's edge. In other terms the interior does not contribute to the super-current reducing the apparent density by a factor of order 2 λ_j/r_j. Besides, if the field (either external or self induced by the current) exceeds H_j a second order phase transition occurs and quantized flux lines start penetrating into the barrier leading to a resistive state as the vortex feel no pinning centers.
Nevertheless, as the two ideal situations considered above are not expected to lead to any hysteresis effects they can not explain the irreversibilities of the m(H) curve for $H \leq H^W_{C2}$, since irreversibilities require inhomogeneities.
3. The barriers are "inhomogeneous" with strong pinning forces acting on the Josephson-vortex-lines. It is possible in this case for the current to be carried by the interior of the junction just as in superconductors of the second kind. By analogy with the latter, it is expected that the critical state model could (in this limit) describe both the magnetic and transport behaviors of the junction. The critical current density would approach J_j if the pinning forces are very strong. This model can also explain the irreversilitities and the m(H) curve for $H < H^W_{C2}$.
From the above discussion it appears that the transport current would depend strongly on the junction characteristics and would be governed by the two relationships below :

$J_j \geq J_t \geq 2\,Jj\,(\lambda_j/r_j)$ (with $r_j > \lambda_j$) (5)

and $J_t \approx 15\,(M_+(0) - M_-(0))/r_j$ (in Bean's practical units) (6)

Here $M_+(0)$ and $M_-(0)$ refer to the magnetizations at H = 0 for the low-H cycle of Fig. 1. We find that the radius r_j of the junction is of the ordere of the sample radius R. Note also (Fig. 1) that $M^+(0) - M^-(0) \approx (2H^W_{c1})/4\pi$. Putting this value in formula 6 leads to Silsbee rule (within a factor of two) :

$H^W_{c1} = 2\pi J_t R/15$ (in practical units).

<u>INTERMEDIATE H REGION : $H^W_{C2} \leq H \leq H_{C1}$</u>

We have already noted that one of the most puzzling features of the low-H m(H) curve is the marked drop of the ratio :

$$\alpha(H,T) = \frac{(m/H)_{H<H_{c_1}}}{(m/H)_{H=0}}$$

of the susceptibility at H such that $H^W_{c2} \leq H \leq H_{c1}$ to the initial susceptibility ($H < H^W_{c1}$). The factor α(H,T) measures the screening capability of the sample in increasing field as depicted in Fig. 3. In the case considered in Fig. 3 α is almost constant up to H^W_{c1}, drops steeply in a narrow field region and then

becomes again almost constant in the interval $[H^W_{c2}, H_{c1}]$.
According to the analysis developed in the precedent paragraphs and in refs 2, 3 the linearity and the reversibility of the m(H) curve for $H^W_{c2} < H < H_{c1}$ would mean that most of the weak links connecting the strongly superconducting grains (those parallel to the field H) have been broken by the field and that the grains have then become decoupled. Assuming that for $H > H^W_{c2}$ the sample behaves like a collection of independent spherical grains of radius r we can calculate the screening factor $\alpha(H,T)$ and show that it takes the form :

$$\alpha(H,T) = 1 - \frac{3(2\lambda + d)}{2r} \coth \frac{2\lambda + d}{2r} + \frac{3(2\lambda + d)^2}{4r^2} \tag{8}$$

We recall that λ is the London penetration depth of the grain whereas d is the average thickness of the grain barriers. The above expression is in fact derived from an equivalent expression giving the susceptibility of "very-fine-grained-preparation of colloidal mercury", as calculated by F. London [19] some 50 years ago and has a straightforward physical meaning : we have already mentioned that for $H \le H^W_{c1}$ the screening is almost complete (compact sample) so that the whole volume of the sample appears as superconducting. On the other hand for $H > H^W_{c2}$ the effective superconducting volume is reduced and it is equal to the initial volume minus that of the weak-links which have become normal. But this is not the whole story : it was shown by F London that as long as the magnetization is concerned and because of the penetration of the field in the grain the effective volume and therefore the effective radius of an isolated (or fully decoupled) grain is reduced by λ and this adds a "dead layer of volume" $4\pi r^2\lambda$ to that of the weak link reducing globally the effective volume of each grain of the sample by a reduced factor $\approx 4\pi r^2 (\lambda + d/2)/(4/3\, \pi r^3) = 3(\lambda + d/2)/r$.

It is to be stressed that formulae 8 applies only when the grains are fully decoupled in the directions (x,y,z) of the space (three degrees of freedom). We believe that the grains are certainly fully decoupled in the x,y plane perpendicular to the direction of H, but it is not clear for us whether this is also true for the z direction. This is because it is much more easy to destroy a junction when it is parallel to H than when it is perpendicular to it. In anyway, if the grains are decoupled in the x-y plane only (i.e. in the plane perpendicular to H) then the sample behaves like a collection of perfectly diamagnetic filaments or cylinders of radius $(r - \lambda - d/2)$. Extending again the calculation of London to the cylindrical case we find (for $r >> \lambda + d/2$)

$$\alpha(H,T) \approx \left(1 - \frac{\lambda + d/2}{r}\right)^2 \tag{9}$$

It is interesting to see that the efective reduction $2\lambda + d$ of the grain diameter is just equal to the effective thickness of the Josephson-junction entering F-P equations 2, 3.

THE CRITICAL STATE MODEL $H_{C1} << H << H_{C2}$

According to the Bean model [15] there exists a unique limiting macroscopic current density J_c that a hard superconductor can carry and that any local electromotrice force however small will induce such a J_c to flow fully. The direction of J_c at any point being imposed by the last e.m.f variation at that point. The physical origin of J_c was not clear in the original work of Bean but has been clarified later with the emergence of the critical state concept [16-18] which states that J_c is obtained when the pinning force density P(B) retaining the vortex lattice equilibrates exactly and everywhere the Lorentz force density ($\approx J_cB$) acting on the same lattice.

As already mentioned, we find that at high enough field ($H > H_{c1}$) and at low temperatures the magnetic cycle is reasonably well described by the above critical state model. Actually, more detailed inspection of the experimental curves reveals that the latter differ in many details from the theoretical prediction. Some of the departure, from the theory can probably be accounted for by : (1) the highly anisotropic properties of the grains [20], (2) the surface current of these grains the role of which should increase notably with the ratio (λ/r_J) defined before and (3) as a result of the fact that J_c could depend strongly on H [21]. However, as long as J_c is concerned, the later complicating effect (i.e. effect of H) can, to a first approximation, be eliminated by assuming that the relatioship between J_c (in A/cm^2), the radius r (in cm) of the grain and the magnetization (in emu/cm^3) are related by the relationship [18] :

$$J_c = 15\,(M_+(H) - M_-(H))/r \qquad \text{(in practical units)} \qquad (10)$$

where $M_+(H)$ and $M_-(H)$ refer to the algebric values of the magnetization at the same H on the ascending, respectively the descending branches of the cycle and for $H \gg H_{c1}$.

Using equation 10 and returning back to Fig. 5 and 6 it is now possible to understand the meaning of the field H_r above which the high-H cycle becomes reversible and roughly linear : it is the field beyond which the critical current density J_c of the grains becomes zero within the sensitivity and the time scale (1 mn) of the experiment. It is equally interesting to remark that H_r would also indicate that the pinning forces have become vanishingly small.

In conclusion, we have shown that the whole shape of the magnetic cycle of the new high Tc superconductors can be accounted for (1) in the framework of the F.P. equations at very low H, (2) using a simple London-like approach at intermediate fields and (3) the critical state model at the highest fields. The interpretation of the data can probably be improved significantly by taking into account the anisotropic properties [20] of the material and the contribution of the surface currents to the diamagnetic response. Finally, it is worth noticing that the irreversible behaviour has also been interpreted in terms of a spin-glass-like picture [22,23]. It is possible that the glassy picture and the present approach provide two complementary descriptions of the same physical effects.

We are grateful to I Campbell and to P Monod for many helpful discussions.

REFERENCES

1) Oussena, M., Senoussi, S. and Collin, G., Europhys. Lett. 4, 625 (1987)
2) Senoussi, S., Oussena, M., Ribault, M. and Collin, G., Phys. Rev. B 36, 4003 (1987)
3) Senoussi, S., Oussena, M. and Hadjoudj, S., J.M.M. (to appear in Journal of Appl. Phys.
4) Pureur, P. and Schaf, J., J.M.M.M. 69, L 215 (1987)
5) Mc Henry, M.E., Mc Kittrick, J., Sassayama, S., Kwapong, V., O'Handley, R.C. and Kalonji, G., Phys. Rev. B 37, 623 (1988)
6) Civale, L., Safar, H., de la Cruz, F., Esparza, D.A. and D'ovidio, C.A. Solid State Communication, 65, 129 (1988)
7) Cronemeyer, D.C., Malozemoff, A.P. and Mc Guire, T.R., (preprint)
8) Jin, S., Shenwood, R.C., Tiefel, T.H., Van Dover, R.B., Johnson, D.N., Grader, Jr. and G.S., Appl. Phys. Lett. 51, 855 (1987)
9) Newcomb, S.B., Growacki, B.A., Campbell, A.M., Evetts, J.E. and Stobbs, W.M., British Ceramic proceedings 40 (1988)
10) Ferrell, R.A. and Prange, R.E., Phys. Rev. Lett. 10, 479 (1963)
11) Ambegaokar, V. and Baratoff, A., Phys. Rev. Lett. 10, 486 (1963)
12) Senoussi, S., Oussena, M. and Collin, G., Phys. Rev. B (May 1988)
13) The only significant exception concerns the fields H^{W}_{c1} and H^{W}_{c2} which in the present conditions of preparation of Pi-based perovskites are not yet well defined. More detailed data will be reported soon.
14) We feel that near $H = H_r(T)$ both m(T,H) and J(T,H) would be strongly dependent on the time scale of the experiment and this makes it difficult to define $H_r(T)$ rigourously. Time effects measurements are underway.
15) Bean, C.P., Phys. Rev. Lett. 8, 250 (1962)
16) Anderson, P.W., Phys. Rev. Lett. 9, 309 (1962)
17) Friedel, J., de Gennes, P.G. and Matricon, J., Appl. Phys. letters 2, 119 (1963)
18) Fickz, W.A. and Webb, W.W., Phys. Rev. 178, 657 (1969)
19) London, F. : Superfluids, Macroscopic theory of superconductivity "Dover Publications, Inc." N.Y. page 35-36
20) Markiewcz, R.S., Chen, K. and Jaggi, N., (Phys. Rev. B, June 1988)
21) Ravi Kumar, G. and Chaddah, P., preprint
22) Morgenstern, I., Müller, K.A. and Bednorz, J.G., Z. Phys. B, Condensed Matter 69, 33 (1987)
23) Giovannella, C., Chappert, C. and Beauvillain, P., Europhys. Lett., 5, 535 (1988)

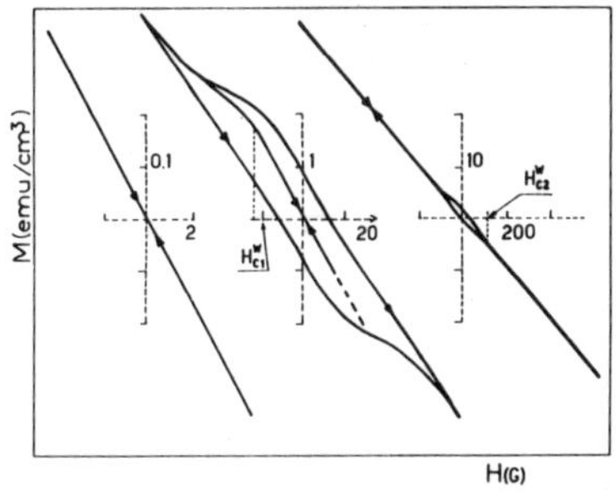

Figure 1 : The low-field magnetic cycle of $YBa_2Cu_3O_7$ at 4.2 K and at various expanded-field scales. The associated sample is very compact (≈ 95 %) and highly sintered.

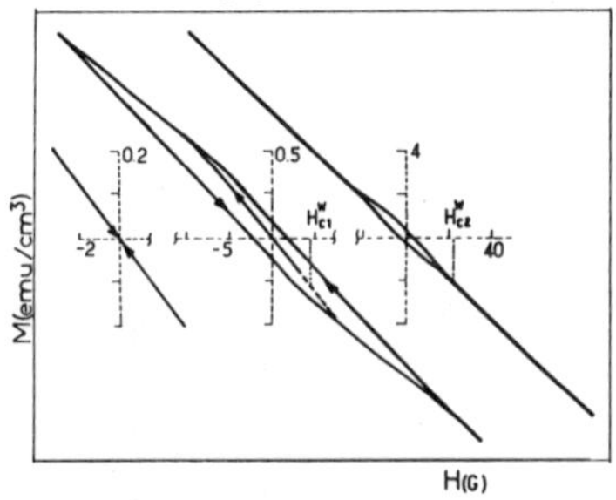

Figure 2 : The same as Figure 1 but now the sample is a powder 10 to 40 μm diameter. Microscopic observations show that the "individual" grains are not single crystals but include many smaller grains. Embedding the powder in oil leads to no detectable change in the m(H) curve.

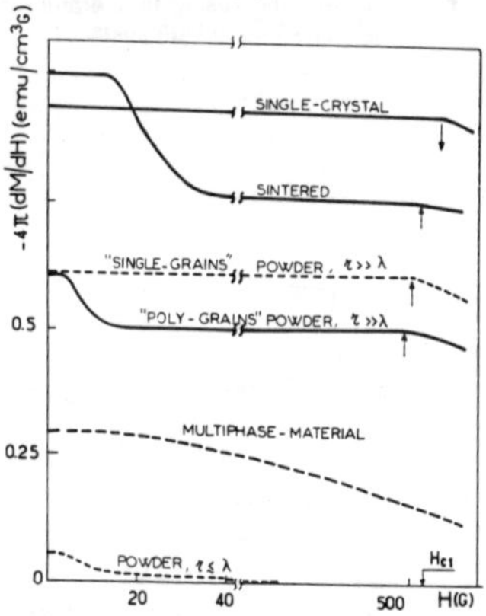

Figure 3 : Field dependence of the differential susceptibilities of $YBa_2Cu_3O_7$ at 4.2 K in the indicated crystalline states. The individual grains of the powder represented by the full line are not single grains but agregates including many smaller crystallites. The dashed curves refer to the expected behaviour if each grain was a single crystal with radius r either greater (up) or smaller (bottom) than the London penetration depth λ. The uncertainty on H_{c1} is very large due to demagnetizing and surface pinning effects.

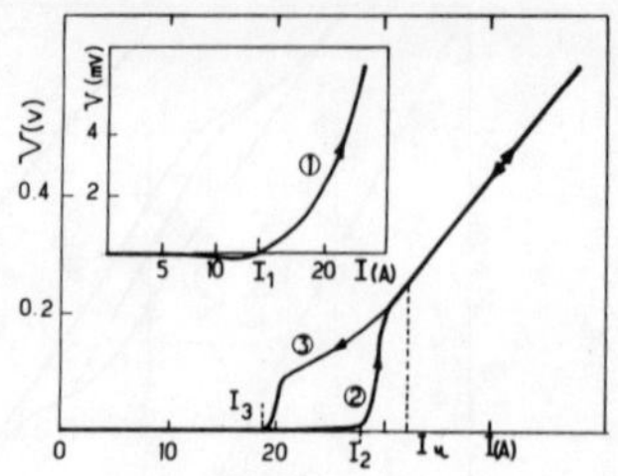

Figure 4 : The voltage versus current characteristic at 77 K for the same $YBa_2Cu_3O_7$ specimen as in Fig. 1. Branch 1 was traced just after zero-field-cooling. Note the different scales between the insert and the figure.

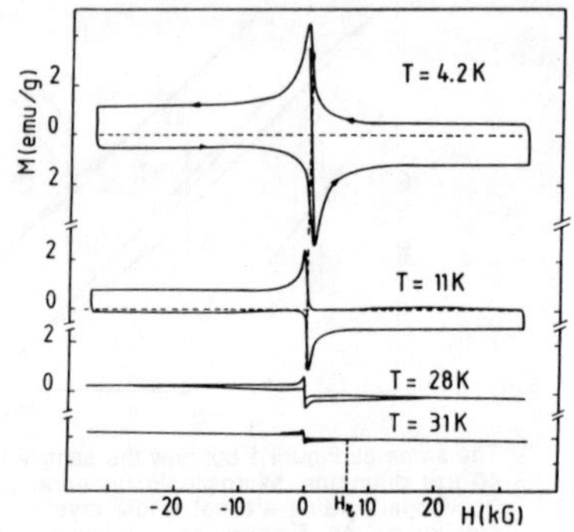

Figure 5 : Evolution with increasing temperature of the magnetic cycle of $La_{1.85}Sr_{0.15}CuO_4$ in high fields.

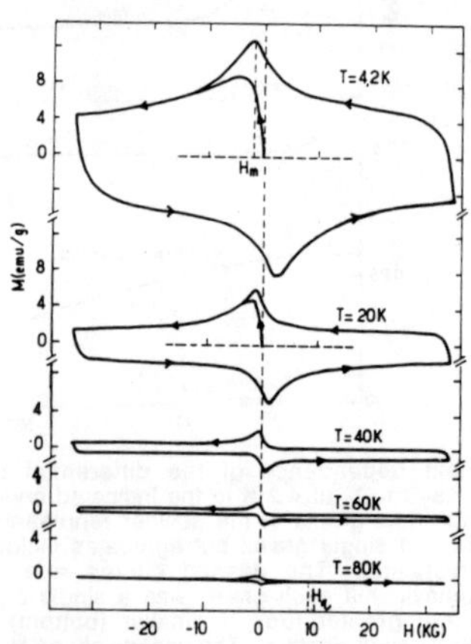

Figure 6 : The same as in Fig. 5 but for $YBa_2Cu_3O_7$.

ELECTRIC AND MAGNETIC MEASUREMENTS ON THE HIGH-T_c OXIDE SUPERCONDUCTORS $YBa_2Cu_3O_7$, $EuBa_2Cu_3O_7$ and $GdBa_2Cu_3O_7$

P. Pureur, J. Schaf and J. V. Kunzler
Instituto de Física - Universidade Federal do Rio Grande do Sul
Av. Bento Gonçalves, 9500
91500 Porto Alegre, RS, Brasil

ABSTRACT

We perform electrical resistivity and DC magnetic susceptibility experiments on the sintered oxide superconductors $RBa_2Cu_3O_7$ (R = Y, Eu, Gd). The temperature derivative of the resistivity at the normal-superconducting transition shows a two peaks structure, suggesting a double transition characteristic of granular systems. The magnetic measurements reveal broken ergodicity phenomena analog to those found in spin glasses.

1. INTRODUCTION

The inhomogeneous character of the superconducting state of the new high-T_c oxide superconductors has been recognized since the early stages in the experimental work[1]. The most accepted picture is that of a disordered granular superconducting network where the overall coherence is established through weak coupling between grains. A clear identification of the grains, which might occur inside a single crystal[2], as well as the nature of the links, which could be set either by Josephson tunneling or proximity effect, has to be determined. However, the consequences of this inhomogeneous superconducting state have been clearly revealed in physical properties as electrical resistance[3], magnetization[1,2,4] and microwave absorption[5]. In particular, Müller et al[1] were the first to show that the magnetic properties observed in the LaBaCuO superconductor are very similar to those observed in spin glasses. This is obviously related to its disordered network structure where frustration is expected to occur when a magnetic field is applied[6].

In this work we present electric resistivity and magnetic measurements in sintered samples of $YBa_2Cu_3O_7$, $EuBa_2Cu_3O_7$ and $GdBa_2Cu_3O_7$ which tend systematically to corroborate the assumption of an inhomogeneous and frustrated superconducting state for characterizing the new high-T_c oxide superconductors.

2. EXPERIMENTAL

Polycrystalline samples were prepared by the solid-state reaction technique[7]. Calcination was done at 950 °C in O_2 atmosphere. The resulting pellets were powdered, pressed into a disk and sintered at 950 °C in flowing O_2 for 16 hours, then slowly cooled to room temperature. Cooling in the vicinity of the tetragonal-orthorhombic transformation was done quite carefully. X-ray analysis has shown the expected phase. The electrical measurements were performed with a lock-in AC-technique with current densities in the range 50 to 100 mA/cm^2. Pressure contacts of copper-berillium were used with low level noise. Accuracy in measuring both resistance and temperature is high enough to determine the temperature derivative of the resistivity near T_c using the procedure described in ref. (8). The magnetic measurements were performed with a Foner magnetometer with sensitivity of about 10^{-4} emu in fields up to 7 kOe. For these experiments the temperature was determined with a carbon-glass sensor with accuracy of 0.1 K in the temperature range near T_c.

3. RESULTS

3.1 Electric Measurements

Figure 1.a shows the measured electric resistivity of one of our $YBa_2Cu_3O_7$ samples. Figure 1.b shows the temperature derivative of the resistivity near the normal-superconductor transition. A two peaks structure is clearly resolved. A pronounced maximum is centered at 91.75 K and a weaker maximum is centered at 90.69 K. The zero resistance state is reached at 90.22 K. This result may be interpreted as a two stages transition typical of a granular system. The higher temperature peak corresponds to the superconducting transition occurring inside the grains whereas the lower temperature peak indicates that the junctions become superconducting. Recently, Dubson et al[3] have shown that the resistance of YBaCuO is non-ohmic in the temperature range corresponding roughly to the lower peak in fig. 1.b. This strongly suggests that this regime is indeed dominated by weak links which are expected to be quite sensitive to temperature, current intensity and magnetic field. Figures 2.a and 3.a show the measured resistivities for $EuBa_2Cu_3O_7$ and $GdBa_2Cu_3O_7$, respectively. Parts b in these figures show the resistivity derivatives near T_c. For $EuBa_2Cu_3O_7$, again a two peaks structure is apparent. The higher temperature peak occurs at 90.2 K and the lower temperature one at 84.5 K. For this sample, the heights of the two peaks are similar, which reinforces the interpretation of a two stage transition. For GdBaCuO, $d\rho/dT$ shows only one peak centered at 93.34 K. However, closely examining, one observes that this peak is asymmetric with respect to its maximum. One possible interpretation of this feature is the superposition of two close peaks as sketched in fig. 3.b.

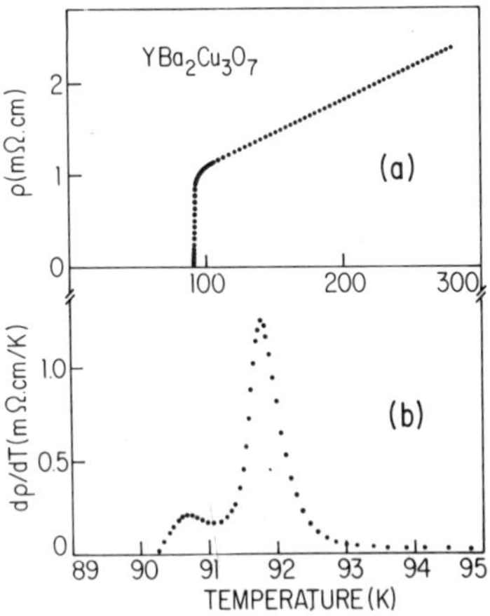

Figure 1. (a) Temperature dependence of the electrical resistivity of $YBa_2Cu_3O_7$ and (b) temperature derivative of the resistivity near T_c.

All of our resistivity results are thus suggesting the granular picture for describing superconductivity in the high-T_c oxide systems. In consequence, one may wonder about the occurrence of percolation phenomena at the transition to the zero resistance state. Studying the approach of the resistivity to the zero value, we have found that a

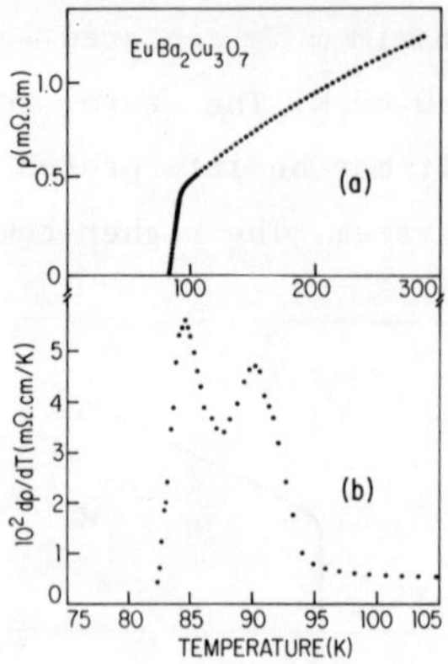

Figure 2

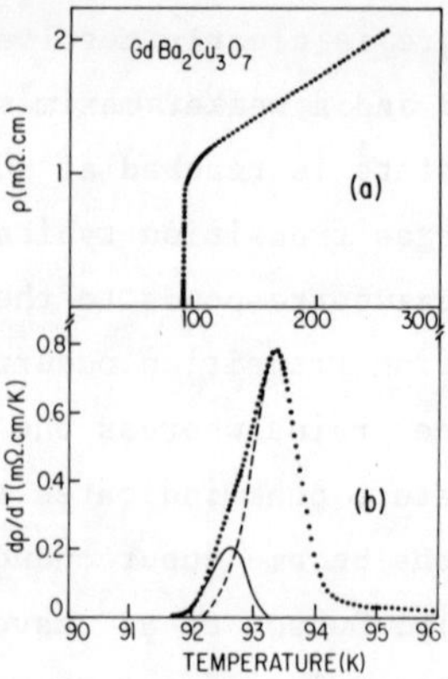

Figure 3

(a) Temperature dependence of the electrical resistivity of $EuBa_2Cu_3O_7$ and $GdBe_2Cu_3O_7$; (b) temperature derivative of the resistivity. In fig. 3.b, the two peaks drawn with dashed and continuous lines are obtained from the experimental points (see text).

power law of the type

$$\rho \cong \varepsilon^{s'} \tag{1}$$

where $\varepsilon = (T-T_{co})/T_{co}$, with T_{co} being the zero resistance state temperature, is obeyed for all of our samples in the range $\varepsilon < 0.01$, with the same exponent $s' \cong 1.45$. This is indeed suggestive of an universal critical behaviour. However, our s' value seems too high if compared to values obtained from a percolation type calculation in resistor networks, which gives

$$\rho \cong (p_c - p)^s \tag{2}$$

where p_c is the threshold superconductor concentration and $s \cong 1.1$ for 2D systems and $s \cong 0.5$ for 3D systems[9]. On the other hand, recent calculations for resistive transition in 3D Josephson networks[10] give $s' \cong 0.4$.

3.2 Magnetic Measurements

Müller et al[1] first proposed that the LaBaCuO superconductor presents a non-ergodic behaviour, reminiscent of spin glasses, which is induced by disorder and frustration below a characteristic temperature denoting the splitting of the zero field cooled (ZFC) and field cooled (FC) diamagnetic susceptibilities. The Hamiltonian proposed to describe the properties of the "glassy" granular superconductors is written as[6]:

$$H = -\sum_{i,j} J_{ij} \cos (\theta_i - \theta_j - A_{ij}) \tag{3}$$

where the sum is generally restricted to nearest-neighbors, J_{ij} is the coupling energy between grains i and j, θ_i and θ_j are the phases of the superconducting order parameter in grains i and j. The phase factor $A_{ij} = K_{ij} H$ depends on the geometrical distance between grains. Frustration is introduced in the model by this random phase factor, when a magnetic field H is applied to clusters containing closed loops of coupled grains[6]. In this situation there is no configuration of a phase coherent state which simultaneously minimizes the energies of all the couplings. In consequence, the system acquires a large number of inequivalent configurations which are nearly degenerate in energy and close to the ground state value. This leads to non-ergodic behaviour[11] very much as in spin glasses below the freezing temperature.

Our results in the three studied samples tend to support the glass-like superconducting picture for describing the magnetic responses of the ceramic high-T_c systems. In particular, we have found strong differences in our DC magnetic susceptibility measurements according if the samples are zero field cooled or field cooled to temperatures much lower than T_c, as exemplified in figure 4 for $EuBa_2Cu_3O_7$. The FC susceptibilities are totally reversible but the ZFC results are time dependent and not reproducible for heating or cooling the sample. We observe that both FC and ZFC susceptibilities are strongly suppressed by increasing the field. In the ZFC curves, this effect is particularly pronounced when the temperature approaches T_c. This characteristic ZFC-FC behaviour has been interpreted as resulting from a "glassy" superconducting state[1,8], although explanations in terms of pinned flux may also be proposed[12]. We have performed other magnetic experiments as hysteresis cycles and measurements of the trapped flux as a function of time and temperature which are also indicative of non-

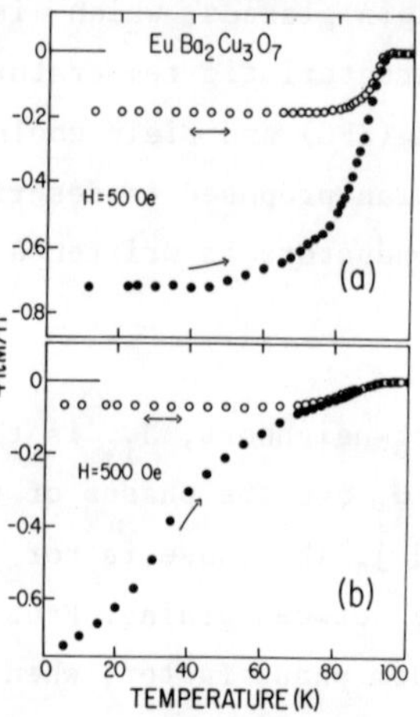

Figure 4. ZFC (full circles) and FC (open circles) magnetic susceptibilities of $EuBa_2Cu_3O_7$ as a function of temperature, measured in fields of (a) 50 Oe and (b) 500 Oe. The ZFC measurements are performed while warming up the sample. The results are corrected for the magnetizing field.

ergodic behaviour in our samples. These results will be reported elsewhere.

In order to study in detail the ergodicity breaking phenomena in our samples we measured carefully the DC-susceptibility in temperatures near T_c (see fig. 4), for determining the temperature which marks the bifurcation of the ZFC and FC branches as a function of H. Results for the three studied samples follow a quite general trend and are shown in figs. 5, 6 and 7. In spin-glasses, the temperatures where the ZFC and FC susceptibilities split are assumed to separate ergodic and non-ergodic regimes in the H-T phase diagram. Using a mean-field theory for Ising spins, de Almeida and Thouless[13] derived a line given by

$$H^{2/3} \sim [1 - T_g(H) / T_g(0)] \qquad (4)$$

where $T_g(H)$ corresponds to the onset of the irreversibilities in the field H and $T_g(0)$ is the freezing temperature at zero field. The so-called de Almeida-Thouless line, eq. (4), is generally found in many experiments in spin-glasses and other glass-like systems[14]. Particularly, Müller et al[1] showed that a de Almeida-Thouless-like line separates reversible from non-reversible behaviour in the DC susceptibility of a LaBaCuO superconductor. As indicated in figs. 5-7, we have found that such a line may be fitted through our experimental points in the low field region for all of the three compounds.

On the other hand, it should be noted that departures from the de Almeida-Thouless line similar to those appearing systematically in figs. 5 to 7, have been observed in some spin-glass systems[15]. This has been ascribed to a crossover from a high-field Heisenberg regime to a low field Ising regime induced by anisotropy forces. Indeed, for Heisenberg

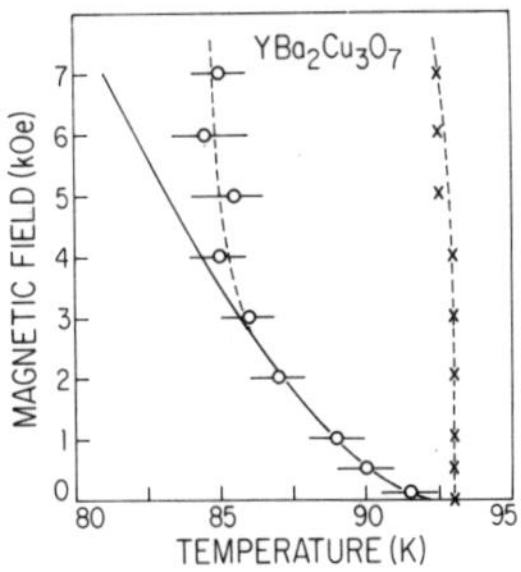

Figure 5

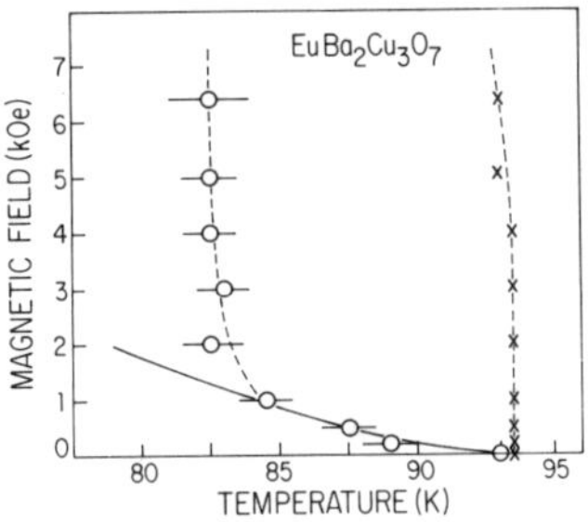

Figure 6

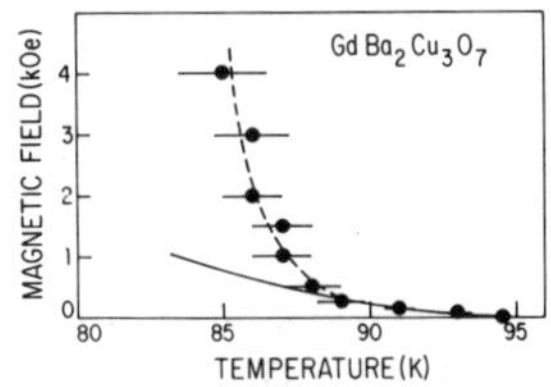

Figure 7

Experimentally determined points which mark the onset of the ZFC-FC splitting in the susceptibilities of the three studied superconducting compounds. These points define lines which separate ergodic from non-ergodic regions in the H-T plane. The solid lines correspond to fits with a de Almeida-Thouless-like line, Eg. (4). Dashed lines are guides to the eye. In figs. 5 and 6, the points denoted by x determine T_c on-set, defined as the temperature where the superconducting diamagnetic susceptibility first becomes apparent.

and XY spins, mean field theory of spin-glasses predicts a phase transition occurring at the much less field dependent Gabay-Toulouse line[16]:

$$H^2 \sim [1 - T_g(H) / T_g(0)] \tag{5}$$

We remark that the Hamiltonian (3) is the superconducting analog of the frustrated XY spin model. Therefore, if we venture to extend the analogies between oxide superconductors and spin-glasses so much, we would expect a Gabay-Toulouse-like rather than a de Almeida Thouless-like behaviour for $T_g(H)$. Thus, it may be possible that the low field de Almeida-Thouless-like line, systematically observed in our samples as well as in the LaBaCuO sample of Müller et al[1], results from a XY-Ising crossover analog to the one found in spin-glass transitions[15]. The origin of the anisotropy forces which induces such a crossover in the high-T_c superconductors has to be clarified. However, it may be related to the very anisotropic character of the bulk superconducting state of these oxide systems[17].

In conclusion, our electric and magnetic experiments in the high-T_c $RBa_2Cu_3O_7$ (R = Y, Eu, Gd) superconductors reveal systematic features which are coherent with a description of these systems in terms of a granular superconducting network, where disorder and frustration play a major role.

REFERENCES

1. Müller, K.A., Takashige, M. and Bednorz, J.G., Phys. Rev. Lett. 58, 1143 (1987).
2. Senoussi, S., Orsay preprints vol. 6 (1987).
3. Dubson, M.A., Herbert, S.T., Calabrese, J.J., Harris, D.C., Patton, B.R. and Garland, J.C., Phys. Rev. Lett. 60, 1061 (1988).
4. Esparza, D.A., D'Ovidio, C.A., Guimpel, J., Osquiguil, E., Civale, L. and de la Cruz, F., Solid State Commun. 63, 137 (1987).
5. Durny, R., Hautala, J., Ducharme, S., Lee, B., Symko, O.G., Taylor, P.C., Zheng, D.J. and Xu, J.A., Phys. Rev. B36, 2361 (1987).
6. Ebner, C. and Stroud, D., Phys. Rev. B30, 134 (1984).
7. Cava, R.J., Batlogg, B., van Dover, R.B., Murphy, D.W., Sunshine, S., Siegrist, T., Remeika, J.P., Rietman, E.A., Zahurak, S. and Espinosa, G.P., Phys. Rev. Lett. 58, 1676 (1987).
8. Pureur, P. and Schaf, J., J. Mag. Mag. Mat. 69, L-215 (1987).
9. Kirkpatrick, S., Phys. Rev. B15, 1533 (1977).
10. Cai, M. and Straley, J.P., Phys. Rev. B36, 3556 (1987).
11. Palmer, R.G., Adv. Phys. 31, 669 (1982).
12. Wolfus, Y., Yeshurun, Y. and Felner, I., Phys. Rev. B88, 3667 (1988).
13. de Almeida, J.R.L. and Thouless, D.J., J. Phys. A11, 983 (1978).
14. See review by Binder, K. and Young, A.P., Rev. Mod. Phys. 58, 801 (1986).
15. de Courtanay, N., Fert, A. and Campbell, I.A., Phys. Rev. B30, 6791 (1984).
16. Gabay, M. and Toulouse, G., Phys. Rev. Lett. 47, 201 (1981).
17. Dinger, T.R., Worthington, T.K., Gallagher, W.J. and Sandstrom, R. L., Phys. Rev. Lett. 58, 2687 (1987); Worthington, T.K., Gallagher, W.J. and Dinger, T.R., Phys. Rev. Lett. 59, 1160 (1987).

BOSON EXCHANGE MECHANISMS, BOUNDS AND ASYMPTOTIC LIMITS

J.P. CARBOTTE
Physics Department, McMaster University
Hamilton, Ontario, Canada L8S 4M1

ABSTRACT

Select physical properties of superconductors are calculated as a function of $T_c/\omega_{\ell n}$. Here T_c is the critical temperature and $\omega_{\ell n}$ an appropriately defined average boson exchange energy. In the BCS limit ($T_c/\omega_{\ell n} \ll 1$), universal properties are predicted, but not when $T_c/\omega_{\ell n}$ becomes larger as for the conventional superconductors. A critical review and comparison with theory of the relevant data for the high T_c oxides leads us to the conclusion that, besides a phonon contribution, there is an additional mechanism which involves a large $\omega_{\ell n}$ value. While the phonon contribution is dominant in La Sr Cu O it is not in Y Ba Cu O. Using a BCS type approximation to describe the second mechanism the expected physical properties are calculated and discussed. Finally, the idea of functional derivatives is introduced and used to obtain bounds on superconducting properties. Asymptotic limits are also established for the case $T_c/\omega_{\ell n} \to \infty$.

I - INTRODUCTION

The Eliashberg equations, with the electron-boson exchange spectral density identified as due to the electron-phonon interaction, can explain the observed physical properties of all known conventional superconductors.[1)] In most cases the properties deviate markedly from universal BCS behaviour and reflect the retarded nature of the interaction which can be characterized, in a first approximation, by a single parameter $T_c/\omega_{\ell n}$.[2,4] Here T_c is the critical temperature and $\omega_{\ell n}$ is an appropriately defined average boson energy describing the virtual exchange between two electrons of the pairing excitations. It is only when $T_c/\omega_{\ell n}$<<1 that BCS results and the physical properties of all super- conductors are universal and so no information about mechanism can be obtained from thermodynamic properties.

In this paper we are concerned with the variation in the thermodynamic and in other properties, as $T_c/\omega_{\ell n}$ is increased without restriction.[5-6] We will also not want to specify the origin and exact nature of the kernels appearing in the Eliashberg equations. They could be due to the exchange of any boson such as phonons, plasmons, excitons, paramagnons, etc. As a first approximation, we assume that, even if Migdal's theorem is not valid in some of the above mentioned cases, the Eliashberg equations can still be retained as phenomenological equations with effective kernels. While these effective kernels may be hard to relate to the fundamental microscopic interaction, this is not a major concern here since we intend to model them.

The properties considered in section II are the ratio of the gap edge Δ to T_c, the jump in the specific heat at T_c ($\Delta C(T_c)$) normalized to the normal state value γT_c with γ the Sommerfeld constant and the slope at T_c of $\Delta C(T)/\gamma T_c$. In principle these quantities, as well as others, carry information on mechanism and we will give a brief discussion of comparison with experiment in the specific case of La Sr Cu O and Y Ba Cu O. It will be found that definitive conclusions are not possible at present because of extreme variability in the available data. Even with this reservation, however, it is nevertheless quite clear that another mechanism (in addition to phonons) with large value of $\omega_{\ell n}$ is needed to explain the data.[7,8] For La Sr Cu O, it is possible that phonons make a very substantial contribution, but for Y Ba Cu O they do not. While we suggest one particular possibility for the additional contribution[7-9] to the pairing potential, many others have been suggested.[10] As long as they correspond to a value of $\omega_{\ell n}$ which is large, compared to T_c, they can all be adequately modeled by a constant BCS pairing potential which, in our formalism, corresponds to taking a negative value for the Coulomb repulsion pseudopotential μ^*.

After presenting some results (section III) for a combined phonon and BCS pairing mechanism (of unspecified origin), we turn to a discussion of functional derivatives[11-12] and bounds (section IV) on superconducting properties.[13-15] Finally, some asymptotic limits (section V) are given which apply when the electron-boson exchange spectral density is assumed to be arbitrarily large. Brief concluding remarks are to be found in a last section VI.

II - SUPERCONDUCTING PROPERTIES AS A FUNCTION OF $T_c/\omega_{\ell n}$

The Kernels of the Eliashberg equations are an electron-boson exchange spectral density $\alpha^2F(\omega)$, which describes the effective interaction between two electrons at the Fermi energy due to the exchange of some boson (phonon, plasmon, exciton, paramagnon, etc.) and the Coulomb repulsion pseudopotential μ^*. There are three moments of $\alpha^2F(\omega)$ which are central to the theory. They are the electron mass enhancement parameter $\lambda \equiv 2 \int_0^\infty [\alpha^2F(\omega)/\omega]d\omega$ which gives the renormalized to bare electron mass ratio $m^*/m = 1+\lambda$, the area under the spectral density $A = \int_0^\infty \alpha^2F(\omega)d\omega$, which characterized its overall strength[16-18] and the characteristic boson energy $\omega_{\ell n}$ given by

$$\omega_{\ell n} = \exp\left\{\frac{2}{\lambda}\int_0^\infty \frac{\alpha^2F(\omega)\ell n(\omega)}{\omega}\, d\omega\right\}. \tag{1}$$

B.C.S. theory, which predicts universal laws for the physical properties of superconductors, results in the $T_c/\omega_{\ell n} \ll 1$ limit in which case $2\Delta/K_BT_c = 3.54$ and $(\Delta C(T_c)/\gamma T_c) = 1.43$. For conventional superconductors $T_c/\omega_{\ell n}$ is finite but still small (say $\lesssim 0.25$)[4] and it is then possible to make expansions of the Eliashberg gap equations and of the attendant free energy formula, in the small parameter $T_c/\omega_{\ell n}$, to get relations such as[3]

$$\frac{2\Delta}{K_BT_c} = 3.54\left[1+12.5\left[\frac{T_c}{\omega_{\ell n}}\right]^2 \ell n\left[\frac{\omega_{\ell n}}{2T_c}\right]\right]. \tag{2}$$

and[4]

$$\frac{\Delta C(T_c)}{\gamma T_c} = 1.43\left[1+53.0\left[\frac{T_c}{\omega_{\ell n}}\right]^2 \ell n\left[\frac{\omega_{\ell n}}{3T_c}\right]\right]. \tag{3}$$

These relationships are satisfied remarkably well by almost all conventional superconductors and allow one to extract from a measurement of either of these quantities a value of $\omega_{\ell n}$ and so get some information on mechanism. It is also clear from equation (2) and (3) that Eliashberg theory implies a definite relationship between $\Delta C(T_c)/\gamma T_c$ and $2\Delta/K_BT_c$. This is shown in Fig. 1 where we have plotted the trend curve (solid line) implied by Eliashberg theory as well as placed on the figure results (solid dots) for various real materials. A few words of explanation are in order. The solid points are actual results of calculations based on the tunneling derived values of the spectral density $\alpha^2F(\omega)$.[19] The experimental results would, in all cases, be close and so confirm the predictions of Eliashberg theory. The deviations from the trend curve in a few cases reflect features of $\alpha^2F(\omega)$ not completely captured by the single parameter $T_c/\omega_{\ell n}$ but these fluctuations are not important for the discussion here. For comparison, the dashed line applies to a delta function spectrum rather than to an observed one.

We now present results for arbitrary values of $T_c/\omega_{\ell n}$ and so do not restrict ourselves to the small values seen in conventional cases. To accomplish this, we need to specify some shape for the spectral density $\alpha^2F(\omega)$. Two choices will be made. In one case, we use the results calculated by W. Weber[20] for La Sr Cu O which can produce the measured T_c of 36K and, as a second example, we use the tunneling results for Pb.[19] Each of these base spectral densities, which we denote by $\alpha_0^2F(\omega)$, are then scaled by constants B and b to get our model[21,4]

$$\alpha_M^2F(\omega) = B\,\alpha_0^2F(b\omega). \tag{4}$$

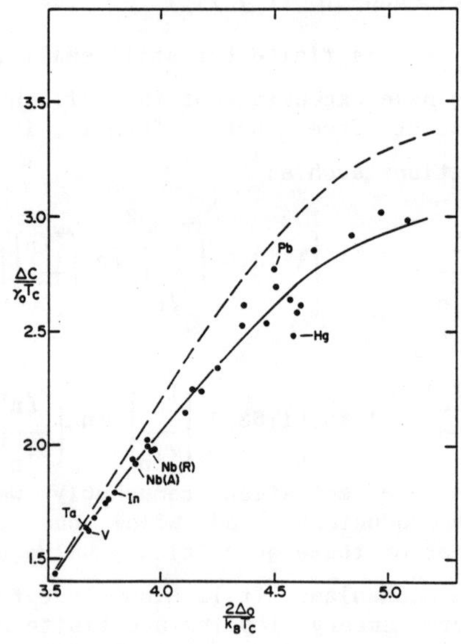

Fig. 1 The dimensionless ratio $\Delta C(T_c)/\gamma T_c$ against $2\Delta/K_BT_c$ for many real conventional superconductors (solid dots).

<u>Table 1</u> La Sr Cu O Gap to critical temperature ratio.

$\frac{2\Delta_0}{K_B T_c}$ 2.4, 1.6-2.7, 2.5, 2.9-4.5, 5.2-9.1, <4.5, 5.0-8.7, 4.07-4.78, 4.5-5.8, 3.0-6.0, 3.5-4.0, 0.7-2.7, 8.0-18.0, 2.6, 1.3±0.2, 5.0, 4.7, 7±2, 2.0(c-axis)

It is possible to show that as b changes, all important quantities simply scale if the Coulomb pseudopotential μ^* does not play a large role. For example $\omega_{\ell n} = \frac{1}{b}\omega_{\ell n0}$, $T_c = \frac{1}{b}T_{c0}$, $\Delta = \frac{1}{b}\Delta_0$, etc., and universal ratios remain invariant.[21] By varying B, curves for any quantity of interest can be traced out for arbitrary $T_c/\omega_{\ell n}$ values.[5,6] Results for the dimensionless gap to critical temperature ratio are presented in Fig. 2. While these curves depend somewhat on the shape of the base spectral density, the general trend is clear. The ratio rises steadily with increasing $T_c/\omega_{\ell n}$ with some indication of saturation in the limit $T_c/\omega_{\ell n} \to \infty$. We will return to this point later. The implied values of λ are shown in Fig. 3 for both base spectra and in one case (La Sr Cu O) for two values of μ^*. We note that λ rises to very large values (probably not consistent with

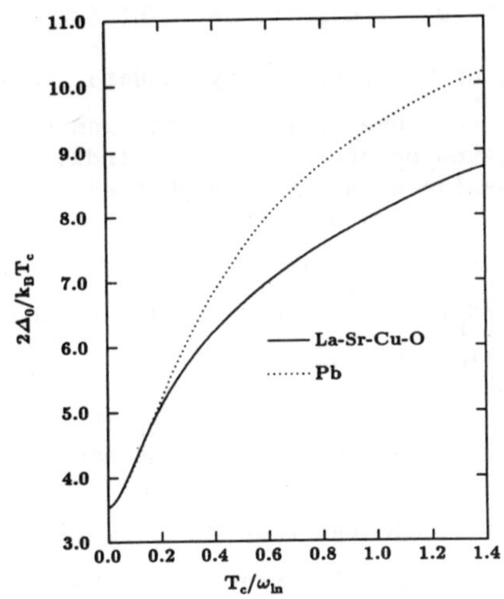

Fig. 2 The dimensionless ratio $2\Delta/K_B T_c$ vs $T_c/\omega_{\ell n}$.

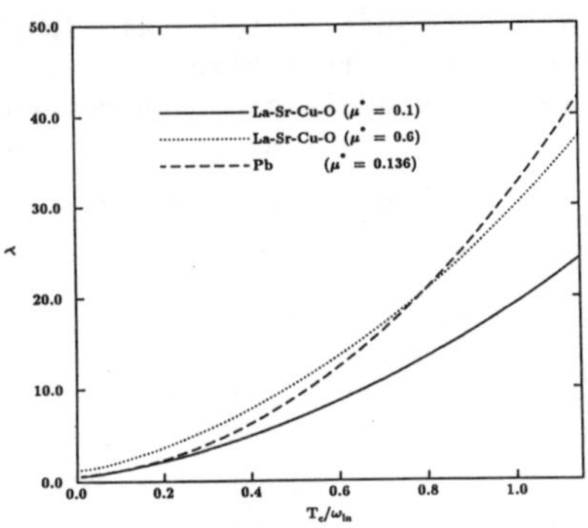

Fig. 3 The mass enhancement parameter λ vs $T_c/\omega_{\ell n}$.

Table 2 Y Ba Cu O Gap to critical temperature ratio.

$\frac{2\Delta_0}{K_B T_c}$ 3.2±0.3, 3.5±0.3, 2.0, 3.5, 3.7-5.6, 1.6-3.4, 2.3-3.5, 2.5-4.2, 1.3±0.2, 3.3, 3.2, 11.0, 4.5-6.0, 3.9-4.8, 4.8, 3.8-4.5, 10.0, 13.0, 3.9, 3.2±0.4, 7.0-13.0, 3.5, 5.0±0.2, 3.4±1.5, 4.8±0.5, ~8.0, 7.5, 3.9, 3.8

lattice stability if the $\alpha^2F(\omega)$ is due to phonons) as $T_c/\omega_{\ell n}$ grows beyond one.

It is of some interest to apply these results to the high T_c oxides with the aim of obtaining some information on mechanism through comparison with experiment. In Table 1[22] we show a compilation of data on $2\Delta/K_BT_c$ for La Sr Cu O. For reference to the literature on which this table is based, the reader is referred to the thesis of F. Marsiglio (McMaster 1988).[22] It is quite clear from this table that $2\Delta/K_BT_c$ varies unacceptably from paper to paper. Note particularly, that infrared values are on the whole smaller than the tunneling values although there is some overlap in some cases. Weber's spectrum for $\alpha^2F(\omega)$ coming from the electron-phonon interaction gives a value of $\lambda = 2.6$, $\omega_{\ell n} = 14.0$ meV and $T_c/\omega_{\ell n} = 0.23$ for a T_c of 36K and predicts a $2\Delta/K_BT_c = 5.3$.[23] Because of the variability in data that we have just noted, it is hard to make a firm conclusion but, tentatively, this high value is not favoured and the data indicates that some additional non-phonon mechanism with a much higher value of $\omega_{\ell n}$ is playing some role and that the electron phonon interaction is overestimated. The same conclusion applies, but in stronger form, for Y Ba Cu O with $T_c/\omega_{\ell n} = 0.6$ and $2\Delta/K_BT_c \simeq 7$[4] (we have assumed that $\omega_{\ell n}$ remain the same to a first approximation as in La Sr Cu O). Comparison with the data of Table 2[22] indicates clearly that the value of $2\Delta/K_BT_c = 7.0$ is too large although there is no consensus as to its exact value. The data is clearly more consistent with a small value of $T_c/\omega_{\ell n}$ than a large one. This indicates that there is little room for a large phonon contribution and that high energy excitations are involved. Very much the same conclusion can be reached if the specific heat jump at T_c is considered. Detail analyses leads us to believe that the electron-phonon contribution in La Sr Cu O is probably large, with some small contribution from another high $\omega_{\ell n}$ mechanism, while for Y Ba Cu O the phonons contribute little. This conclusion is consistent with a significant isotope effect observed for La Sr Cu O [24] and none (or almost none) for Y Ba Cu O.[25]

Before leaving this section, we show in Fig. 4 our results for the normalized specific heat jump $(\Delta C(T_c)/\gamma T_c$ and in Fig. 5 for its slope $\frac{d}{dt}(\Delta C(T)/\gamma T_c)$ at $t = T/T_c = 1$.[26] All conventional superconductors described by equation (3) would fall on the rising part of the $\Delta C(T_c)/\gamma T_c$ for $T_c/\omega_{\ell n} \lesssim 0.25$. The maximum exhibited above this value was unexpected as was the trend down towards values <u>smaller</u> than BCS in the region of large $T_c/\omega_{\ell n}$. Similar remarks hold for Fig. 5. It is seen that the slope first becomes steeper as $T_c/\omega_{\ell n}$ increases from value zero, then it has an extremum and reverses itself. For very large values of $T_c/\omega_{\ell n}$, the slope can be smaller than in B.C.S. Both this small slope and small value of the jump are signatures of the extreme strong coupling regime. Of course, it is important to keep in mind that this regime may never be attained in real cases.

III - COMBINED PHONON AND BCS PAIRING MECHANISMS

Many suggestions have been made as to possible mechanisms[10] for the superconductivity in the oxides and there is some evidence from optical data[9] for excitonic superconductivity. If an Eliashberg formulation can be contemplated (as a first approximation) and the bosons exchanged by the electrons are of high energy (electronic instead of phonons) then it can be modeled by a constant BCS pairing contribution in the kernels. This means using a negative μ^* which now represents the sum of attractive mechanism plus Coulomb repulsions.

In Fig. 6, we show preliminary results for a joint, phonon plus B.C.S. contribution.[26] What is plotted is $2\Delta/K_BT_c$ as a function of $T_c/\omega_{\ell n}$ for various negative values of μ^*. In all calculations the phonon contribution is modeled by a delta function of weight B centered at the Einstein frequency $\omega_E = 8.27$ meV and the cutoff on $\mu^*(\omega_c)$ is 400 meV. The solid curve corresponds to $\mu^* = 0$. The dotted curve corresponding to a $\mu^* = -0.1597$ gives a $T_{c0} = 10K$ when the phonons are neglected. The remaining curves are for larger values of negative μ^* corresponding to cases with a larger limiting T_c value for the non-phonon part. It is clear that as the negative μ^* value is made more important the phonon contribution is less effective in increasing the gap to critical temperature ratio for a given value of $T_c/\omega_{\ell n}$. This is due, of course, to the ever decreasing value of B that is required.

Results for the specific heat jump $\Delta C(T_c)/\gamma T_c$ are presented in Fig. 7. It is clear that the large maximum observed at low $T_c/\omega_{\ell n}$ values in the case of positive μ^* (see Fig. 4) is rapidly reduced as μ^* is made negative. In particular for $\mu^* = -0.2317$, which corresponds to a T_{c0} = 70K (from the second mechanism alone) the peak is completely missing and all values fall below BCS, as is the case for the very strong coupling limit of the pure phonon case (Fig. 4). These results are particularly interesting and indicate that the signature for the existence of a phonon contribution to $\Delta C(T_c)/\gamma T_c$ in addition to a significant BCS contribution is an anomalously small value for this quantity (below BCS).

IV - BOUNDS ON SUPERCONDUCTING PROPERTIES

We now turn to the idea of functional derivatives and bounds on superconducting properties which exist in Eliashberg theory independent of the shape, size, or origin of the spectral density. We will start with the critical temperature T_c and ask how an infinitesimal change (ϵ) in $\alpha^2F(\omega)$ at a specified frequency ω_0 changes (ΔT_c) the value of T_c. By definition the functional derivative is

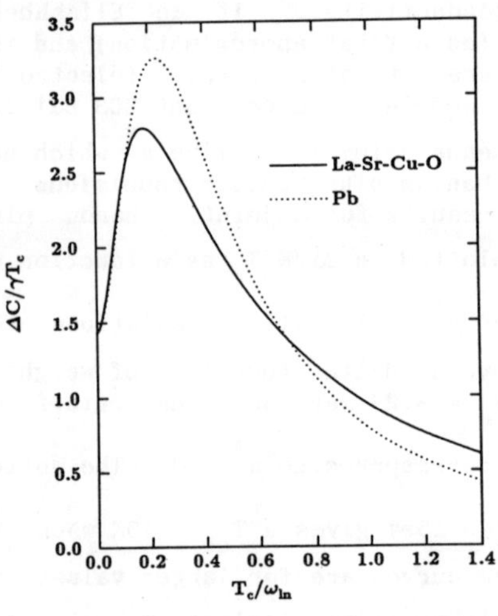

Fig. 4 The normalized specific heat jump $\Delta C(T_c)/\gamma T_c$ vs $T_c/\omega_{\ell n}$.

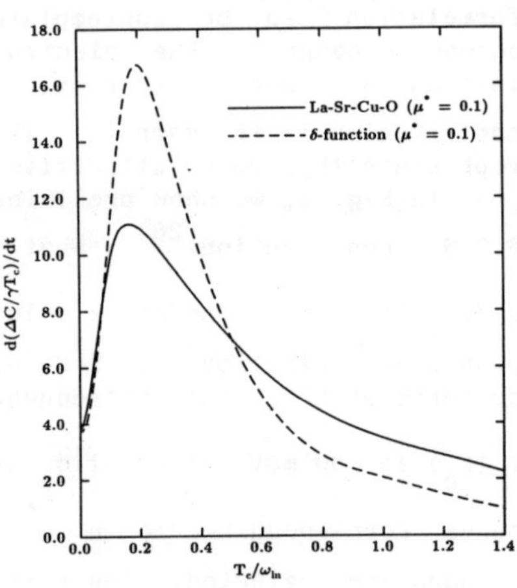

Fig. 5 The slope of the normalized specific heat jump at T_c $(d(\Delta C(T)/\gamma T_c)/dt)/_{t=1})$ vs $T_c/\omega_{\ell n}$.

$$\frac{\delta T_c}{\delta \alpha^2 F(\omega_0)} = \lim_{\epsilon \to 0} \frac{\Delta T_c}{\epsilon}. \tag{5}$$

The work of Bergmann and Rainer[11)] and others has established that while the absolute value of this functional derivative depends significantly on the base spectrum used, its shape is fairly universal. For instance, it goes to zero like ω at $\omega \to 0$ and like $1/\omega$ at $\omega \to \infty$ and exhibits a single broad maximum in between.[19)] The position of the maximum, which defines the optimum phonon frequency, is always around $8\ K_BT_c$. This suggested to Leavens[17)] that it may be possible to maximize T_c for a given strength of spectral density (taken to be the area (A) under $\alpha^2F(\omega)$) by placing all available weight at a specific optimum frequency ω_E^* which should also be the position of the maximum in its own functional derivative. This suggests that a delta function $\alpha^2F(\omega) = A\delta(\omega-\omega_E^*)$ will maximize T_c and that all other shapes will lower it. Leavens was able to prove that for a delta function T_c/A is a function only of ω_E/A for a fix value of μ^* and that it becomes maximum (see figure 8) at some frequency $\omega_E^* = Ad(\mu^*)$. The value at maximum is $T_c = Ac(\mu^*)$ with $d(\mu^*)$ and $c(\mu^*)$ universal numbers completely determined from the Eliashberg equations

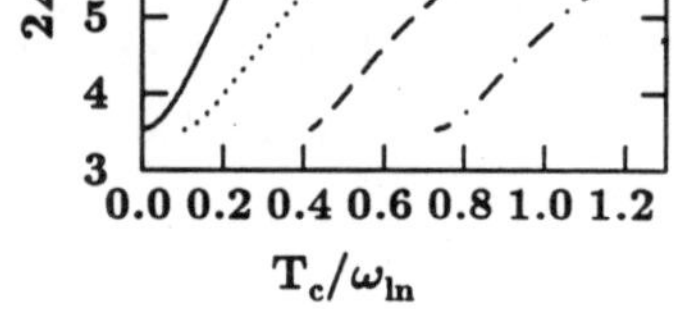

Fig. 6 The dimensionless ratio $2\Delta/K_BT_c$ vs $T_c/\omega_{\ell n}$ for a joint mechanism.

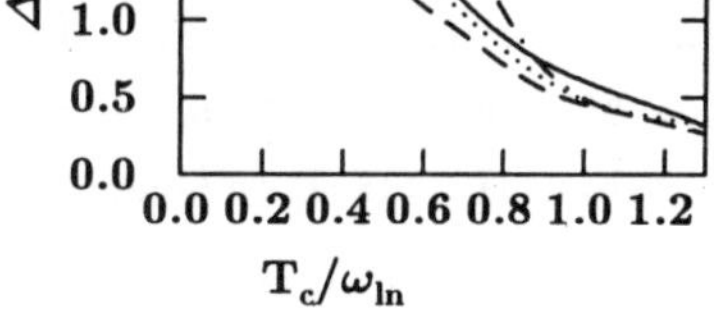

Fig. 7 The normalized specific heat jump $\Delta C(T_c)/\gamma T_c$ vs $T_c/\omega_{\ell n}$ for a joint mechanism.

themselves. Leavens[17)] further suggested that for any other shape

$$T_c \leq Ac(\mu^*). \tag{6}$$

This relationship is tested in Fig. 9 where we have plotted $c(\mu^*)$ as a function of μ^*[17,18)] and also placed in the same figures results (solid dots) for real materials. In all cases, T_c/A falls below the maximum value as expected. Note in passing that the relation (6) with the equality sign holding for a delta function, says that T_c is not limited in size by the mathematical structure of the Eliashberg equations themselves. As A grows, T_c also grows indefinitely. For an electron-phonon mechanism, however, T_c will certainly be limited by conditions for lattice stability and A cannot grow forever. Similar criteria must also apply in the case of other boson exchange mechanisms.

In Fig. 8, we showed the variation of T_c/A as a function of ω_E/A for several values of μ^* based on a delta function (upper frame). In the lower frame, we show the functional derivative of (T_c/A) for the base delta function which maximizes T_c/A for $\mu^* = 0.1$. It is seen that this functional derivative is negative definite everywhere with a zero value exactly at the frequency appearing in the base delta function $\omega_E^* = Ad(\mu^*=0.1)$. This shows that this delta function represents a local maximum. Adding weight to $\alpha^2F(\omega)$ exactly at ω_E^* leaves T_c/A unchanged while adding it anywhere else reduces this ratio.

We have obtained similar results for many other superconducting properties. For example, in the next section we will show that the gap to critical temperature ratio is limited to be less than 13 for $\mu^*=0$. Figure 4 indicates that the normalized specific heat jump peaks at intermediate values of $T_c/\omega_{\ell n}$ and it was shown[13)] that it takes on an absolute maximum of 3.74 for $\mu^*=0.15$ with some small variation with μ^*. This bound must be satisfied by any Eliashberg superconductor in the isotropic limit independent of spectral function $\alpha^2F(\omega)$.

V - ASYMPTOTIC LIMITS

Here we retain the delta function spectrum of the previous section

$$\alpha^2F(\omega) = A\delta(\omega-\omega_E) \qquad (\lambda = 2A/\omega_E) \tag{7}$$

and consider the limit as $\omega_E \to 0$ for some fixed A which corresponds to the mass enhancement parameter $\lambda \to \infty$. Carbotte et al.[27] were able to show that in this case $T_c \to 0.25\sqrt{A\omega_E} = \frac{1}{4}\sqrt{\frac{\lambda}{2}}\,\omega_E$ a result already known from the work of Allen and Dynes[2] and that $\Delta \to 1.6\sqrt{A\omega_E} = 1.6\sqrt{\frac{\lambda}{2}}\,\omega_E$ which is close to an analytic approximate estimate by Kresin.[28] Our limit was determined numerically. The above results combine to give

$$\lim_{\substack{\omega_E \to 0 \\ (\lambda \to \infty)}} \frac{2\Delta}{K_B T_c} \equiv 13 \tag{8}$$

Using functional derivative techniques it is possible to prove that this asymptotic limit represents a local maximum[14] and that results

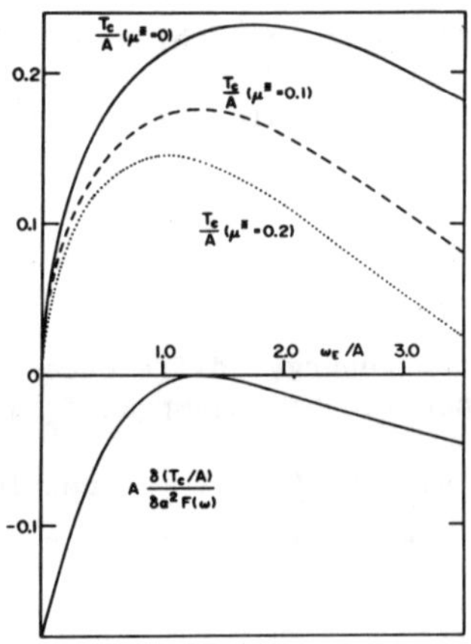

Fig. 8 Variation of T_c/A with ω_E/A for three values of μ^* based on a delta function spectrum (upper frame). The functional derivative with $\alpha^2F(\omega)$ of T_c/A for the $\mu^*=0.1$ case with base delta function placed at the optimum frequency $\omega_E^* = Ad(\mu^*)$.

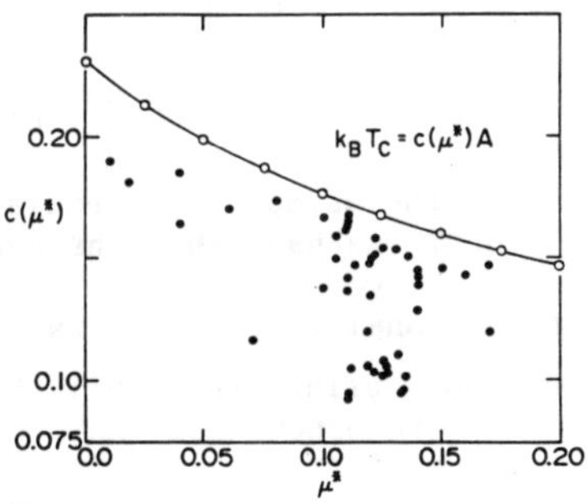

Fig. 9 The maximum T_c/A as a function of μ^*.

for any other shape of spectral density should fall below this limit for an isotropic Eliashberg superconductor. The saturation of $2\Delta/K_BT_c$ when the characteristic phonon frequency $\omega_{\ell n}\to 0$ can already be seen in Fig. 2 where results are presented for two spectral shapes. In the large $T_c/\omega_{\ell n}$ limit the curves bend over and become flat although the asymptotic value would appear to depend on spectral shape.

Other asymptotic limits can also be established. For example, the free energy difference ΔF as a function of reduced temperature $t = T/T_c$ can be shown to vary as

$$\lim_{\substack{\omega_E\to 0\\(\lambda\to\infty)}} \Delta F(T) = A\omega_E \Delta f(t) \tag{9}$$

where $\Delta f(t)$ is a universal function of t applicable to the case of any delta function spectral density as $\omega_E\to 0$. Equation (9) can be rewritten as $\Delta F\sim A^2/\lambda$. It is important to make clear that here, we are fixing A and varying ω_E. If instead A is increased indefinitely for fix ω_E, λ would again go to infinity but would then think of $\Delta F \sim \lambda\omega_E^2$. It follows immediately that the thermodynamic critical field $H_c(t)$ has the asymptotic limit

$$\lim_{\substack{\omega_E\to 0\\(\lambda\to\infty)}} H_c(t) = \frac{A}{\sqrt{\lambda}} h_c(t) \tag{10}$$

From the asymptotic formula for the free energy difference and that for T_c we have that the normalized specific heat jump at T_c will be proportional to $(1+\lambda)^{-1}$ as will its slope at T_c. Numerical evaluation using the full Eliashberg theory formulated on the imaginary frequency axis (Matsubara formalism) gives

$$\frac{\Delta C(T_c)}{\gamma T_c} = \frac{19.88}{1+\lambda} \quad \text{and} \quad \frac{d}{dt}\left[\frac{\Delta C(t)}{\gamma T_c}\right]\Bigg|_{t=1} = \frac{39.2}{1+\lambda} \tag{11}$$

These results are consistent with the $1/\lambda$ drop observed in Fig. 3 and 4 as $T_c/\omega_{\ell n}$ gets large. Note, however, that these latter results are still not quite in the asymptotic limit and that they apply to real spectral shapes namely that of La Sr Cu O and Pb respectively while our asymptotic limits are established for a delta function shape.

VI - CONCLUSIONS AND DISCUSSION

Starting from the Eliashberg equations with arbitrary strength, shape, or origin of the electron-boson exchange spectral density we have calculated several superconducting properties with a view to understanding how mechanism gets reflected in such physical quantities. It was found that the properties of interest depend mainly on the single strong coupling parameter $T_c/\omega_{\ell n}$ and less importantly on other details of $\alpha^2F(\omega)$ not captured in the effective boson energy $\omega_{\ell n}$. Comparison of our results for large $T_c/\omega_{\ell n}$ with experiment in the high T_c oxides La Sr Cu O and Y Ba Cu O proved disappointing because a very large scatter exists in the data at the present time. Nevertheless, it could be concluded that a non-phonon mechanism corresponding to large boson exchange energy (electronic energy) is operative in addition to a phonon part. For La Sr Cu O the phonon contribution is possibly dominant but for Y Ba Cu O it is not.

We have given preliminary results of calculations of superconducting properties for a combined phonon and other unspecified pairing mechanism which is assumed to involve high energies and is therefore treated in a BCS approximation. In our formalism, this is accomplished by introducing a negative contribution in the usual Coulomb repulsion pseudopotential μ^*. Results are presented for a fix BCS contribution and increasing phonon contribution. The physical properties can differ very significantly from the pure BCS case even for modest values of electron-phonon mass renormalization (λ).

In a final section (V) we considered asymptotic limits and gave some results for the large λ limiting values of the critical temperature, the energy gap and the ratio $2\Delta/K_BT_c$. We also show that the jump in the normalized specific heat and its slope at T_c go like $(1+\lambda)^{-1}$ and give numerical values for the proportionality constants. The unpublished part of this work was done with R. Akis, F. Marsiglio, and P. Williams.

1) Daams, J.M. and Carbotte, J.P., Jour. Low Temp. Phys. 43, 263 (1981).
2) Allen, P.B. and Dynes, R.C., Phys. Rev. B12, 905 (1975).
3) Mitrovic, B., Zarate, H.G. and Carbotte, J.P., Phys. Rev. B29, 184 (1984).
4) Marsiglio, F. and Carbotte, J.P., Phys. Rev. B33, 6141 (1986).
5) Marsiglio, F., Akis, R. and Carbotte, J.P., Phys. Rev. B36, 5245 (1987).
6) Marsiglio F. and Carbotte, J.P., Phys. Rev. B36, 3633 (1987).
7) Marsiglio, F. and Carbotte, J.P., Phys. Rev. B36, 3937 (1987).
8) Marsiglio, F., Akis, R. and Carbotte, J.P., Solid State Comm. 64, 905 (1987).
9) Kamaras K. et al., Phys. Rev. Lett. 59, 919 (1987).
10) Numerous reference in Novel Superconductivity (edited by S.A. Wolf and V.Z. Kresin, Plenum Press New York 1987).
11) Bergmann, G. and Rainer, D., Z. Phys. 263, 59 (1973).
12) Rainer D. and Bergmann, G., J. Low Temp. Phys. 14, 50 (1974).
13) Blezius J. and Carbotte, J.P., Phys. REv. B36, 3622 (1987).
14) Carbotte, J.P. in Novel Superconductivity (edited by S.A. Wolf and V.Z. Kresin, Plenum Press New York 1987) p. 73.

15) Akis, R., Marsiglio, F., Schachinger, E. and Carbotte, J.P., Phys. Rev. (in press).
16) Leavens, C.R. and Carbotte, J.P., J. Low Temp. Phys. 14, 195 (1974).
17) Leavens, C.R., Solid State Comm. 17, 1499 (1975).
18) Leavens C.R. and Carbotte, J.P., Ferro Electrics, 16, 295 (1977).
19) McMillan, W.L. and Rowell, J.M. in Superconductivity (edited by R.D. Parks, Marcel Dekker New York 1969) p.
20) Weber, W., Phys. Rev. Lett. 58, 1371 (1987).
21) Coombes, J.M. and Carbotte, J.P., Jour. Low Temp. Phys. 63, 431 (1986).
22) For references see Marsiglio, F. Ph.D. thesis McMaster University (1988) unpublished.
23) Schossmann, M., Marsiglio, F. and Carbotte, J.P., Phys. Rev. B36, 3627 (1987).
24) Batlogg, B. et al., Phys. Rev. Lett. 58, 2333 (1987).
25) Batlogg, B. et al., Phys. Rev. Lett. 59, 912 (1987).
26) Akis, R. and Carbotte, J.P. (unpublished).
27) Carbotte, J.P., Marsiglio, F. and Mitrovic, B., Phys. Rev. B33, 6135 (1986).
28) Kresin, V.Z., Solid State Comm 6, 725 (1987).

EXOTIC NORMAL AND SUPERCONDUCTING PROPERTIES OF HIGH T_c OXIDES AND SMALL E_f

V. Z. Kresin

Lawrence Berkeley Laboratory
University of California
Berkeley, CA 94720

G. Deutscher

School of Physics and Astronomy
Tel-Aviv University
Ramat-Aviv, Israel

S. A. Wolf

Naval Research Laboratory
Washington, DC 20375-5000

A key feature of the high T_c oxides is a very small value of the Fermi energy. This feature leads to peculiar thermodynamic and transport properties in the normal and superconductive state.

Introduction. A small value of the Fermi energy in the new high T_c oxides [1, 2] and its consequences form the essence of this paper. Of course, it means that these new materials can be described by general Fermi liquid approach with the concept of a Fermi surface. Recent positron annihilation data [3] indeed show the existence of the Fermi surface in Y-Ba-Cu-O. Furthermore, the results obtained in [3] are in agreement with our previous theoretical analysis [4].

Fermi Energy. In this section we are going to focus on the value of E_f. It will be shown that the new oxides, indeed, have a small value of E_f relative to conventional superconductors. In conjunction with a high T_c this leads to a number of peculiar features. Our approach is based on the analyses of heat capacity, critical behavior, thermal conductivity, etc.

New high T materials are highly anisotropic systems. For example, $\sigma_\perp / \sigma_\parallel \simeq 10$ for La-Sr-Cu-O [5]. The dispersion relation $E(\vec{p})$ is anisotropic and as a result, the Fermi surface $E(\vec{p})=E_f$ has a cross section that is far being spherical.

The Fermi surface of La-Sr-Cu-O has an almost cylindrical shape [4]. Small deviations from this shape are connected with interlayer transitions.

The Fermi energy can be calculated from the expression [4]:

$$E_f = \frac{\pi \hbar^2 d_c n}{m^*(0)} \qquad (1)$$

Using experimental data on heat capacity [6] which allow us to evaluate m*[4] and Hall effect data for n, we obtain:

$$E_f \simeq 0.1 \text{ eV}. \qquad (2)$$

This value is very small (usually for metals $E_f \sim 5$–10 eV). The small value of E_f is due to a small carrier concentration and a large value of the effective mass ($m^* \simeq 5\ m_e$, see [4]).

In conventional superconductors the parameter (T_c/E_f), or the related parameter $E(0)/E_f$ is very small ($<10^{-4}$), here E(0) is the energy gap at T=0 K. In the cuprates the situation is different. In La-Sr-Cu-O the parameter $E(0)/E_f \sim 0.1$; it means that a large fraction of carriers is paired [4]; the coherence length appears to be small (~20Å).

These features lead us to conclude that the unusual critical behavior of the new high T_c superconductor is similar to that of liquid He. Such a conclusion has been stated by one of the present authors [7] (see also [8]). The role of the fluctuations is very important; of course, such behavior manifests itself stronger in Y-Ba-Cu-O.

The analysis of the heat capacity date in the region near T_c carried out in [9-11] has allowed us to observe the effect of fluctuations. Indeed, the temperature behavior of the heat capacity appears to be similar to the observations for liquid helium.

Based on the expression $\tau < [(13/\eta)(T_c/E_f)]^{1/4}$ (see [7]), $\tau = T - T_c / T_c$, $\eta = 2E(0)/3.5\ T_c$, with the use of the experimental data. One can estimate the value of E_f for Y-Ba-Cu-O and obtain $E_f \simeq 0.1 - 0.3$ eV, that is, E_f has the same order of magnitude as for La-Sr-Cu-O.

We think that a small value of E_f, along with strong anisotropy, the key features of new systems responsible for many peculiar, normal, and superconducting properties. We will discuss below some of the consequences of the small value of E_f.

Heat Capacity. In this section we will discuss a peculiar behavior of the heat capacity due to the small value of E_f. As is well known, the electronic heat capacity C_{el} is usually small relative to the total value C, except the region $T \to 0$; this small value is connected with the large value of E_f, because C_{el} represents a series expansion in the parameter $(T/E_f) \ll 1$. Usually $C_{el} \ll C_{ph}$ and only in the region $T \to 0$ the sharper decrease of C_{ph} allows us to separate the electronic contribution.

A small value of E_f makes the situation in cuprates entirely different. Contrary to the usual metals, the inequality $T \ll E_f$ does not hold in the region $T \gg \omega_D$. In this region C_{ph} depends slowly on T, because of anharmonicity, see e.g. [9, 10]; the temperature dependence C(T) is caused, mainly, by the electronic part $C_{el}(T)$, which is not small in cuprates. Of course, a small carrier concentration is a diminishing factor, but, nevertheless, the closeness T and E_f makes the contribution

C_{el} noticeable. Thus, for these unusual materials, the electronic contribution can be separated in the high, rather than low, temperature region.

Let us evaluate C_{el} for a layered system. The total energy for a cylindrical Fermi surface can be written in the form $E=\nu \int d\varepsilon\varepsilon \times [\exp(\varepsilon-\mu)/T+1]^{-1}$ where $\nu=m^*(\pi\hbar^2 d_c)^{-1}$ is the density of states. In addition to the direct temperature dependence, it is necessary to take into account the temperature dependences of the chemical potential $\mu(T)$ and the effective mass $m^*(T)$. The dependence $m^*(T)$ is due to the electron-phonon coupling [12-15]. As for the temperature dependence of $\mu(T)$, this can be determined from the equation describing a number of carriers. As a result, we obtain $C_V = T\gamma F(T)$; here γ is the Sommerfeld constant for a cylindrical Fermi surface, and

$$F(T) = G(\tilde{E}_f/T)-1/2\ G'(\tilde{E}_f/T)(\tilde{E}_f/T)[1+(T/m^*(T))dm^*(T)/dT];\quad \tilde{E}_f=(m^*(0)/m^*)E_f \qquad (3)$$

The function $G(t)$ is equal to $G(t) = \int dx\ x(e^x e^{-tr(t)}+1)^{-1}$.

If $T \ll E_f$, we obtain a linear law $C = \gamma T$. The temperature region, $\omega_D < T < E_f$, is of particular interest. As was noted above, the lattice term in this region C_{ph} slowly depends on temperature (in fact, this dependence can be determined from thermal expansion data, see e.g. [9, 10]). The Fermi energy (see above) leads to a situation when the ratio (T/E_f) is not small in the region $T>\omega_D$ (for La-Sr-Cu-O, $\omega_D \sim 350K$, $E_f \simeq 10^3 K$). As a result, one can separate a noticeable electronic temperature-dependent contribution $C_{el}(T)$.

In the region $E_f>T>\omega_D$, C_{el} (see Fig. 1) is not described by a linear law. This is important in order to obtain C_V from the measurement of C_p, because the difference $C_p - C_V$ does depend linearly on T.

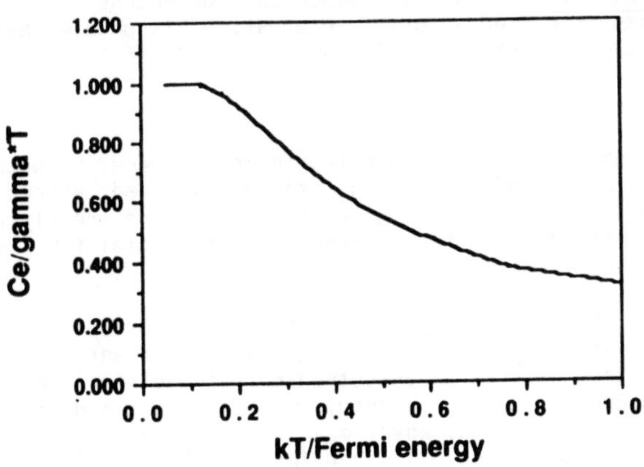

Fig. 1

The measurements of the electronic heat capacity allow us to determine an important parameter, namely, the strength of the electron-phonon coupling λ. The usual method for evaluating λ is based on tunneling spectroscopy. At present, there are no experimental tunneling spectroscopy data on the cuprates. It is also known that the short coherence length, interface problems, etc., make such spectroscopy very complicated. This is why it is very important to explore possibilities of determining λ.

The heat capacity data allows us to determine the value of the effective mass m* (see [15]). It is essential that the effective mass depends on temperature. In the region $T \to 0$, $m^*=m_b(1+\lambda)$, where m_b is the band value. Hence, in the low temperature region the quantity measured experimentally, is the value renormalized by the electron-phonon coupling. The intensive thermal motion in the region $T>\omega_p$ leads to an absence of the renormalization and hence, we measure m_b. A direct comparison $m^*|_{T \to 0}$ and $m^*|_{T \geq \omega_D} = m_b$ allows us to determine the value of λ.

We can determine γ from low temperature data [16]. Then the band value of m_b can be expressed as $m_b = (3\hbar^2/\pi)\, K_B^{-2}\, d_c\, \gamma(1+\lambda)^{-1}$. In the high temperature region $m^* = m_b$. C_e (T) depends parametrically only on (the last term in Eq. (3) is negligibly small in the region $T>\omega_p$), and 1 can be determined from heat capacity data.

Note that heat capacity data were used in [10] in order to determine λ, and his efforts were caused by a lack of the tunneling spectroscopy data by that time. Our motivation aimed on determination λ of the cuprates is similar.

The determination of λ is important for the understanding of the origin of high T_c. It would be very interesting to carry out the corresponding experiments.

Normal Conductivity. Unusual temperature dependence of one normal conductivity has attracted a lot of interest. This problem has been discussed in [17]. We think that a linear dependence of the resistance on temperature is due to a strong anisotropy of the system and to a small carrier concentration.

Carrier's subsystem of the cuprates forms a layered structure. This leads to an anisotropy of the phonon spectrum. Indeed, according to the adiabatic theory, the phonon's degrees of freedom are greatly affected by the electronic motion. Electronic terms form potential for the ionic motion. (See e.g. the review [18]).

As a result, the electronic term is highly anisotropic and an analysis based on the crystal's anisotropy (see [19]) is relevant to the cuprates. The phonon spectrum of such an anisotropic system differs drastically from the usual 3D picture. Name, one should distinguish two different Debye temperatures. Θ_{max} and Θ_{min}, where $\Theta_{min} = \eta\, \Theta_{max}$, $\eta \simeq d_a/d_c$ where d_c is the interlayer distance, and d_a is the interatomic distance in the plane, $\Theta_{max} \simeq 350K$.

Consider the temperature region $\eta\, \Theta_{max} \ll T \ll \Theta_{max}$. The major contribution to the lattice dynamics comes from the branch: $\omega = s\gamma_q^2$ [19];

based on the relation $\sigma = n\, e^2 \tau / m^*$ one can obtain the dependence $\tau \sim T^{-1}$, and, hence $\rho \sim T$. Indeed, a number of phonons with dispersion relation $\omega \sim q^2$ is proportional to T. In addition, the Fermi momentum and the phonon momentum have the same order of magnitude. As a result, the additional temperature dependence of does not appear (as is well known, the small value $q \ll P_f$ leads to an additional dependence $\tau \sim T^{-2}$). Hence, $\tau \sim T^{-1}$ and we obtain the linear dependence of the resistance.

The situation is different in the region $\eta^2 \,\Theta_{max} \ll T \ll \eta \Theta_{max}$. For La-Sr-Cu-O system it corresponds to the interval $25K \le T \le 10^2 K$. The interlayer interaction is noticeable and the dispersion relation should be written in the form (see [19]): $\omega \sim q$.

In this case $N_{ph} \sim T^2$ and $\tau \sim T^{-2}$. However, it is necessary to take into account the dependence of $m^*(T)$. The effective mass decreases almost linearly in this region (see [12]). As a result, we obtain, that in the region $\eta^2\, \Theta_m \ll T \ll \eta\, \Theta_m$ ρ also display linear dependence; $\rho \sim T$. In the region $T \sim \eta\, \Theta_m$ there is a change of one slope in the function of $\rho \sim T$.

Note that in the region $T > \Theta_D$ one should take into account the electron-electron scattering. For the ordinary metals with a large value of the Fermi energy this mechanism of relaxation does not play a noticeable role, because it is proportional $\sim (T/E_f)^2 \ll 1$, if $T \ll E_f$. For the cuprates the temperature in the region $T > \Theta_D$ is not small relative to E_f. As a result, the temperature dependence of ρ_{ee} is different from being quadratic, and, in addition, the contribution of this channel become important. A more detailed analysis of ρ_{ee} will be given elsewhere.

Thermal Conductivity. The behavior of the thermal conductivity in cuprates is affected strongly by a small value of the Fermi energy. As is known, the total thermal flow consist of two parts: $\kappa = \kappa_{el} + \kappa_{latt}$. In ordinary metals $\kappa_{el} \gg \kappa_{latt}$, that is, the thermal conductivity is dominated by the electrons (see e.g. [20]). Even for dirty metals with a small mean free path, the lattice contribution does not exceed several percent.

Thermal conductivity of the new oxides has been studied experimentally in [21] and the picture appears to be entirely different. Namely, main contribution to the thermal flow comes from phonons, the electronic part is only ~ 20% of the total thermal conductivity.

Such abnormal behavior is due to a small value of the Fermi energy. Indeed, the electronic term κ_e depends on E_f: $\kappa_e \sim E_f$ (see e.g. [22]). The value of the Fermi energy in cuprates is ~ two order of magnitude less than in ordinary metals and it leads to a decrease of κ_e observed experimentally.

An experimental investigation of the dependence $\kappa(T)$ is the region T/T_c has resulted in a very important observation. Namely, $\kappa(T)$ increases with decreasing temperature below T_c; and at the same value T_0 one can observe κ_{max}, and then $\kappa \to 0$ if $T \to 0$.

Such behavior is not unusual. As is known (see e.g. [22]) there are six relaxation mechanisms determining the behavior of the thermal conductivity. An increase of κ below T_c corresponds to the lattice contribution due to scattering of phonons by electrons. It is described by the same electron-phonon interaction which is the major mechanism of BCS theory. In the region $T \sim T_0$ the phonon mean free path $l_{pe} \sim l_{p,imp}$ and the subsequent decrease in T leads to a decrease of κ.

Such a manifestation of the electron-phonon coupling, which appear to be a dominant mechanism of the thermal conductivity, is very important from a point of view of the understanding of the mechanisms of high T_c. The described behavior of the thermal conductivity means that the electron-phonon coupling plays an important role in the cuprates.

Note that the strength of the electron-phonon coupling also is connected with the value of the Fermi energy. As is known, the electron-phonon coupling is a deviation from the adiabatic picture (see e.g. the review [18]) and this non-adiabaticity is described by the ratio ω_D/E_f. A small value of E_f means a large non-adiabaticity and, hence, a strong electron-phonon coupling.

Conclusion. A small value of the Fermi energy (~0.1eV) is a key feature of the new high T_c oxides. A large value of T_c/E_f leads to a noticeable contribution of fluctuation. Such a small value of E_f allows us to separate the electronic contribution to the heat capacity in the high temperature region $E_f > T > \Theta_D$ to determine the value of the electron-phonon coupling constant λ.

The linear temperature dependence of the normal resistance is mainly due to a large anisotropy of the system. A small value of E_f leads to the situation when the lattice contribution to the thermal conductivity plays a dominant role.

A strong electron-phonon coupling is manifested in the increase of the thermal conductivity in the region $T < T_c$ and the appearance of such coupling is also connected with a small value of E_f.

REFERENCES

1. J. Bednorz and K. Muller, Z. Phys. B 64, 189 (1986).

2. M. Wu et al., Phys. Rev. Lett. 58, 908 (1987).

3. L. Smedskjaer et al. (preprint).

4. V. Kresin and S. Wolf, a) Solid State Comm. 63, 1141 (1987); b) in Novel Superconductivity, ed. by S. Wolf and V. Kresin, p. 287, Plenum, NY (1987); and c) J. of Superconductivity, 1, 143 (1986).

5. Y. Midaka et al., Jpn. J. Appl. Phys. 26, L33 (1987), J. Worthington et al., Ref. 4b, p. 781.

6. N. Phillips et al., Ref. 4b, p. 739; B. Batlogg et al., Phys. REv. B 35, Ref. 4b, 5340 (1987).

7. G. Deutcher, Ref. 4b, 5340 (1987) p. 293.

8. G. Deutcher, Physica C [Proc. of Int. Conf., Interlaken, (1988)], in press.

9. N. Phillips, et al., Ref. 8 (in press).

10. G. Knapp and B. Brockhouse, Phys. Rev. B 6, 1761 (1972).

11. A. Muller and B. Brockhouse, Can. J. Phys. 49, 704 (1971).

12. V. Kresin and G. Zaitsev, Sov. Phys. - JETP, 47, 983 (1978).

13. K. Kihlstrom et al., Ref. 4b, p. 95; Phys. Rev. B (in press).

14. G. Grimvall, The Electron-Phonon Interaction in Metals, North-Holland (1981).

15. P. Allen and M. Cohen, Phys. Rev. B 1, 1329 (1970).

16. N. Phillips et al., Ref. 4b, p. 739; B. Batlogg et al., Phys. Rev. B 35, 5340 (1987).

17. M. Gurvitch and A. Fiory, Ref. 4b, p. 663.

18. V. Kresin and W. Lester, The Adiabatic Method in the Theory of Many-Body systems, in Mathematical Analysis of Physical Systems, Van Nostrand, NY (1985).

19. L. Landau and E. Lifstutz, Statistical Physics, p. 203 Pergamon, NY (1980).

20. R. Berman, Thermal Conduction in Solids, Clarendon (1976).

21. A. Jezowski et al., Phys. Lett. A 122, 431 (1987); C. Uher et al., Phys. Rev. B 36, 5680 (1987).

22. B. Geilikman and V. Kresin, Kinetic and Non Steady--State Effects in Superconductivity, Wiley, NY (1974).

23. B. Geilikman, Sov. Phys. - Solid State 18, 54 (1971).

24. V. Kresin and M. Morawitz, J. of Superconductivity 1, 89 (1988).

PHONON-PLASMON MECHANISM OF HIGH T_c SUPERCONDUCTIVITY

V. Z. Kresin

Lawrence Berkeley Laboratory
University of California
Berkeley, California 94720

H. Morawitz

IBM Research Division
Almaden Research Center
San Jose, California 95120-6099

The observed high superconducting transition temperatures are caused, in our view, by the coexistence of strong electron-phonon coupling and the plasmon mechanism. A generalized Eliashberg equation describing the effects of phonons and plasmons on the pairing is presented. The structure of the plasmon bands in conducting, highly anisotropic layered materials is studied.

Introduction. The new high T_c materials [1-3] display BCS-type superconductivity. This implies: (1) pairing due to exchange of excitations of some intermediate field and (2) the presence of an energy gap. According to the usual BCS theory, which describes with great success the properties of hundreds of conventional superconductors, the superconducting state is due to an effective electron-electron interaction via exchange of virtual phonons. As regards the new high T_c materials, we think that in addition to phonons, a very important role is played by plasmons. As is known, plasmons arise from the collective motion of the carrier subsystem, whereas phonons describe the quantized excitations of the ionic lattice.

The role of plasmons in high T_c superconductivity has been pointed out and described in our papers [4,5]. Later, various aspects of the electron plasmon interaction and plasmon mechanisms were discussed in [6].

The question on the origin of high T_c by virtual exchange of some intermediate field may be further broken down into two distinct parts of each of which has to be met to be taken seriously:

(1) The physical demonstration of the existence of the particular excitation invoked for the mechanism.

(2) The coupling of the carriers to be paired to this excitation.

For phonons and plasmons the answer to (1) is positive. Quasi-2D plasmons have been observed experimentally in [7] (infrared spectroscopy, electron losses). For excitons there seems to be no evidence for a strong optical excitation up to an upper limit of 5-6 eV. For magnetic excitations and magnetic (anti-ferromagnetic) ordering there are experimental observations with Neel temperatures of several hundred degrees K in both undoped and slightly doped La_2CuO_4 and Y $Ba_2Cu_3O_{6+\delta}$ [8] in the insulating phase. The Neel temperatures observed decreased with increased doping or increasing δ suggesting that the anti-ferromagnetic and conducting states of the system are unable to coexist. On the other hand, a detailed polarized neutron scattering result [9] on the superconducting $YBa_2Cu_3O_7$ showed no evidence of magnetic excitations in the expected energy region probed by inelastic neutron scattering.

First of all, we would like to stress that the high T_c in the new materials is due not to the plasmon mechanism by itself, but to its coexistence with strong electron-phonon coupling. On the other hand, there is experimental evidence that the electron-phonon interaction is very important, but it is not sufficient to lead to the observed high transition temperatures.

Speaking of exchange by phonons and plasmons, one should note that these are not hypothetical, but real excitations. At the same time, the high anisotropy of the new oxides, caused by the presence of layer and chain structures, makes the plasmon mechanism very favorable. The behavior of plasmons in quasi-low-dimensional materials, their dispersion relations, etc., differs in a striking way from those in the usual 3D systems. The interaction of an electron with the rest of the Fermi sea is not only responsible for the usual screening, but, in addition, it contains a dynamic part. This dynamic part corresponds to the collective motion of electrons with respect to the lattice or to the relative motion of two groups of carriers in a system with two energy bands. Plasmons are the quasi-particles describing such collective motion, and, in this case, they are similar to phonons which are the quasi-particles describing the collective excitations of the lattice.

Interaction. Consider a highly anisotropic system with a layered structure. The z-axis is chosen to be perpendicular to the layers. Our approach assumes the applicability of the Fermi liquid description of the system. Such an approach has been used by S. Wolf and one of the other authors [10] in order to evaluate the major parameters of the oxides. A recent positron annihilation data [11], indeed, have resulted in the obtaining the Fermi surface of Y-Ba-Cu-O system, in agreeing with [10].

The self-energy part $\Delta(\vec{\kappa}, p_z, \omega_n)$ ($\vec{\kappa}$ is a 2D momentum, $\omega_n=(2n+1)\pi T$) is described by the equation

$$\Delta(\vec{\kappa}, p_z, \omega_n)Z = T \sum_{\omega_n'} \int d\vec{\kappa}' \, dp_z' \times$$

$$\times \Gamma(\vec{\kappa} - \vec{\kappa}'; p_z; p_z'; \omega_n - \omega_n') F^+(\vec{\kappa}', p_z', \omega_n'). \qquad (1)$$

Here F^+ is an anomalous Green function, Z is the renormalization functions, and Γ is the total vertex describing the effective electron-electron interaction.

The total vertex can be written in the form $\Gamma = \Gamma_{ph} + \Gamma_{pl}$. Γ_{ph} describes the electron-phonon interaction: $\Gamma_{ph} = g^2 D$, where D is the phonon Green's function and g the electron-phonon coupling constant. Note that the region of the crossing of the optical phonon and plasmon branches should be considered separately, then it is necessary to take into account the hydridization of the branches.

Plasmons in Anisotropic Metals. The pure 2D dispersion relation was obtained in [10] and analyzed in detail by us in [4,5).

The existence of a low-lying plasmon branch, which does not have an energy gap at q=0, is a major difference between 2D and 3D plasmons. It is important that this branch exists even for a single energy band. Appearance of the additional "demon" branch is caused by overlap at partially-filled bands.

2D dispersion relation can be obtained in the random phase approximation (RPA). As is known, RPA can be modified by introducing the special function G(q) [13]: $G(q)=A[1-e^{-B(q/p_f)^2}]$, where A and B depend on r_s. This modification leads to a change in the dispersion relation. As a result, we obtain the same plasma branch $\omega \sim q^{1/2}|_{q\to 0}$ [12], but with a modified coefficient of the dispersion relation.

Interlayer interaction leads to a noticeable modification of the pure 2D dispersion relation, namely to the formation of plasmon bands. Nevertheless, anistropy results in a picture which differs in a striking way from the isotropic 3D case. In addition, the maxima of the plasmon density of states are shifted towards the boundaries of the plasmon branches, which enhances the electron-plasmon coupling considerably. In our view, it is the layered nature of the superconducting cuprates which confines the collective motion of the electrons predominantly to the planes and leads to the unusual features of its plasmon spectra.

Two factors due to interlayer interactions affect the plasmon dispersion relation. Firstly, the interlayer Coulomb interaction plays an important role. Secondly, one should consider interlayers transitions of the electrons. (The importance of this factor increases with decreasing interlayer distance.) The new high T_c materials are characterized by high anistropy of dispersion relation comes from the first factor, that is, from the Coulomb interaction. The Fermi surface of a layered structure has a cylindrical shape [13] with the electron dispersion relation $\varepsilon(\vec{k})$, where $\vec{k}$ is the 2D quasi-momentum. A small deviation from the cylindrical shape is a manifestation of interlayer transitions.

Layered crystals are characterized by a plasmon band $\Omega = f(\vec{k},k_z)$, where k, k_z are wave vectors in the planes and perpendicular to them. The z-axis is chosen to be perpendicular to the layers.

The values of are restricted to lie between the upper and lower branches. These branches correspond to k_z=0 and $k_z\pi/c$ for the uppermost and the lowest branches, respectively. In addition, we note that these boundary modes correspond to in-phase motion of electron on different planes (k_z=0) and out-of-phase motion ($k_z=\pi/c$) on adjacent planes. Note that the layer spacing c is equal to half the lattice constant c_0 for $La_{2-x}Sr_xCuO_4$.

In Fig. 1, we show the dispersion relations for the layered electron gas in a cylindrically symmetric system -- the excitations are both electron-hole pairs and the various plasmon branches for the in-plane wave vector, k, and the wave vector perpendicular to the conducting sheets, p_z.

The upper branch corresponds to three-dimensional behavior, and the dependence of $\Omega(\vec{k},0)$ is similar to the behavior of the usual three-dimensional sample. On the other hand, the behavior of the lowest branch $(k,\pi/c)$ is entirely different. It is important to note that in the limit $k_z >> 1/c$, the interlayer interaction does not play an important role, and we are dealing with a two-dimensional dispersion relation. There is a crossover from 3D to 2D behavior in the region $k_z \sim 1/c$. This crossover corresponds to a maximum in the density of states. This can be seen by considering the dependence of the plasmon frequency on one less variable and hence, the derivative with respect to this variable goes to zero. This leads to an effective increase in the density of states in this region.

Hence, interlayer interaction leads to the formation of a highly anisotropic plasmon band $\Omega(k,k_z)$. A very important feature of this band is the nonuniform distribution of the density of states.

One can show [5], that the plasmon density of states has two peaked regions. The sharp increase of the density of states near the lower boundary is particularly important for the superconducting state.

Plasmon-Optical Phonon Hybridization. In this section, we want to draw attention to the potential hybridization of electronic charge density oscillations with ionic charge motion such as polar modes [5]. This case corresponds to the crossing of the plasmon and optical phonon branches. Then the total vertex cannot be written as a sum, and this means that it is impossible to separate the excitations.

As a result of this hybridization, the two branches acquire characteristics of each other as in standard polarition theory. The modes do not cross, but are split off from each other by an amount dependent on the strength of the coupling. This coupling in the lowest approximation is proportional to the product $\sim\rho_{pl}\rho_{ph}$ where $\rho_i=\lambda_i[1+\lambda_{ph}+\lambda_{ph}]^{-1}$, i=pl,ph. A large value of λ_{ph} may result in a noticeable splitting. The lower branch strongly disperses in the intermediate frequency region ($100\ cm^{-1}<\omega<1000\ cm^{-1}$). As a consequence, the plasmon density of states behaves similarly to the usually phonon density of states.

The function $g_{pl}(\omega)$ is, therefore, characterized by a strong peak in the intermediate frequency range. This peak has a structure similar to the usual phonon peak and can be determined by tunneling spectroscopy. Because of the mixing of electronic and ionic degrees of freedom, it seems plausible that optical measurements in this frequency region will be enhanced over purely ionic oscillator strength. There is recent experimental evidence for very strong absorption in the phonon region in both the La-Ba-Cu-O and Y-Ba-Cu-O systems [14].

Superconducting State: Coexistence of the Phonon and Plasmon Mechanisms. As was noted in the Introduction, our approach is based on the concept of coexistence of the phonon and plasmon mechanisms. At present, this concept is receiving various experimental support.

Experimental data on thermal conductivity [15], heat capacity [16], sound attenuation [17], etc., show that the electron-phonon coupling plays an important role.

The increase of the thermal conductivity $\kappa(T)$ at temperatures $T < T_c$ observed in [15] means (see e.g., [18]) that the phonons make a major contribution to the total thermal flow, and the electron-phonon interaction (EPI) is a main relaxation mechanism. EPI was analyzed in [16] with the use of the heat capacity data. The method [16] is based on the dependence of the Sommerfield's constant on temperature (see [19]). As a result, the value $\lambda \simeq 2$ was obtained.

Strong electron-phonon coupling is very important for high T_c but, nevertheless, it is not sufficient and there is a need for an additional mechanism. Indeed, let us consider the question: how large a value of λ_{ph} in La-Sr-Cu-O is required in order to provide $T_c \simeq 40K$. In order to answer this, it is convenient to use the expression for T_c obtained in [20],

$$T_c = \frac{0.25\,\tilde{\Omega}}{[e^{2/\lambda_{eff}} - 1]^{1/2}} \quad . \tag{2}$$

Here $\tilde{\Omega} = \langle\Omega^2\rangle^{1/2}$, $\langle\Omega^2\rangle = \int d\Omega\alpha^2(\Omega)F(\Omega)\Omega$, $F(\Omega)$ is the phonon density of states, $\alpha^2(\Omega)$ describes the electron-phonon interactions; $\lambda_{eff}=(\lambda - \mu^*)(1 + 2\mu^* + \mu^*\lambda + \mu^*\lambda t(\lambda))^{-1}$; the function $t(\lambda)$ is defined in [20]. Equation (3) was obtained directly from the Eliashberg equation and is valid for any value of λ_{ph} (for more detailed analysis and limiting cases, see [20]).

Let us apply Eq. (2) to the La-Sr-Cu-O system in order to estimate the value of λ_{ph}. According to data obtained by neutron inelastic spectroscopy [21], the phonon density of states $F(\Omega)$ in the $La_{1.8}Sr_{0.2}CuO_4$ system exhibits two sharpe peaks at $\Omega_1 \simeq 100K$ and $\Omega_2 \simeq 200K$. The function $\alpha^2(\Omega)$ can be determined by the tunneling spectroscopy technique (these measurements have not yet been carried out) but usually $\alpha^2(\Omega)$ is a relatively smooth function. As a result, it is reasonable to put $\tilde{\Omega} \simeq 150K$. Assuming that $\mu^* \simeq 0.1$ (larger values of μ^* lead to an increase of λ_{ph}, see below), we obtain from Eq. (2): $\lambda_{ph} \simeq 5$. This means that the electron-phonon interaction can account for the high $T_c \simeq 40K$ in La-Sr-Cu-O compound, but this requires very strong coupling.

The value of λ_{ph} for different materials can be measured with the use of tunneling spectroscopy, but the data are unavailable at present. However, the necessary information can be obtained from another set of data, namely from measurements of the energy gap. Indeed, the quantity $\beta = 2\Delta(0)/T_c$ depends on the value of λ_{ph}. For example, $\beta_{BCS}=3.52$ (weak coupling) $\beta_b \simeq 4.3(\lambda \simeq 1.5)$, $\beta_{Hg} \simeq 4.5$ $(\lambda \simeq 1.6)$, $\beta_{pb-Bi} \simeq 5$ $(\lambda\ \ 2)$.

The magnitude of the energy gap has been determined by infrared spectroscopy and various types of tunneling (see e.g. [3]). Although there is some variation in the data that have been reported, the majority corresponds to $\beta \leq 5$ (our analysis corresponds to the single-gap picture, or to the situation when $l < \xi_0$). This means that λ_{ph} for the La-Sr-Cu-O system does not exceed the value of $\lambda_{ph} = 2$ (cf. [16], see above). However, this is not sufficient to provide the experimentally observed $T_c \simeq 40K$. This temperature requires $\lambda_{ph} = 5$ (see above).

Hence, we come to the conclusion that the electron-phonon interaction plays an important role in La-Sr-Cu-O, as manifested by various experimental data (see above), but, nevertheless, there is need for an additional mechanism of attraction between the carriers. We think that the electron-plasmon interaction is an additional mechanism which, jointly with the strong electron-phonon coupling, is responsible for high T_c superconductivity.

Plasmon branches corresponding to the poles of $\Gamma_{pl}(\omega,q)$ appear in the region $\omega > qv_F$. The region $\omega < qv_F$ does not contain any singularities of $\Gamma_{pl}(\omega,q)$. In this region, one can use static approximation and $\lambda_{pl}(\omega,q)$ represents the usual screening. As a result, we obtain:

$$\Delta(p_z,\omega_n)Z + T\sum_{\omega_{n'}}\int d\Omega\left[\lambda_{ph}(\Omega,p_z,p_z')\frac{\Omega^2}{\Omega^2+(\omega_n-\omega_n')^2} - V_c\theta(|\omega_n|-|\omega_o|)\right.$$
$$\left. + \lambda_{pl}(\Omega,p_z,p_z')\frac{\Omega^2}{\Omega^2+(\omega_n-\omega_n')^2}\right]\frac{\Delta(\omega_n',p_z')}{|\omega_n'|}\Bigg|_{T=T_c}. \tag{3}$$

The concept of coexistence means that the electron-phonon interaction plays an important role. The Coulomb repulsion is overcome mainly by the electron-phonon interaction. As for the plasmon contribution one can see directly from Eq. (3) that the electron-plasmon interaction provides an additional mechanism of electron-electron attraction and in the presence of electron-phonon interaction it leads to an additional increase in T_c.

Equation (3) can be written in the form

$$\Delta(p_z,\omega_n)Z=T\sum_{\omega_n'}\int d\Omega\left[\lambda_{ph}(\Omega,p_z,p_z')\frac{\Omega^2}{\Omega^2+(\omega_n-\omega_n')^2} - V_c\theta(\omega_n-\omega_o)\right]\frac{\Delta(\omega_n;p_z')}{|\omega_{n'}|}\Bigg|_{T=T_c} \tag{4}$$

where $\lambda = \lambda_{ph} + \lambda_{pl}$. The Eq. (4) is a generalization of usual Eliashberg's equation. The plasmons will appear as a modification of the usual function λ_{ph}.

One can see directly from Eq. (4) that the anisotropy of the Fermi surface [13] and the interaction lead to the anisotropy of the energy gap.

The behavior of the anisotropic energy gap and T_c in the presence of the anisotropic Fermi surface and several coupling mechanisms is a complex problem and will be analyzed in detail elsewhere. For a rough estimate one may use the expression obtained in [4], (assumed that $\lambda_{ph} \neq 0$):

$$T_c = T_c^{ph}\left(\frac{\tilde{\Omega}_{pl}}{T_c^{ph}}\right)^a$$

$$a = \frac{\lambda_{pl}}{\lambda_{pl} + \lambda_{ph}} \qquad (5)$$

where T_c^{ph} is the critical temperature in the absence of the plasmon mechanism. One can see that the large value of the plasmon energy $\tilde{\Omega}_{pl}$ makes the plasmon contribution noticeable, even for small λ_{pl}. For example, if $\lambda_{ph} \simeq 2$ (this value corresponds to $2\,\Delta(0)/kT_c \simeq 5$ (see above) and we use our model of a polar phonon-plasmon hybridized density of states (5) $\lambda_{pl} \simeq 0.3$, $\Omega_{PO} \simeq 60$ meV, we obtain $T_c^{ph} \simeq 22K$ and $T_c \simeq 38K$. In the absence of phonon-plasmon hydridization and using an experimental value of 1 eV for the bulk plasmon frequency which is a lower limit for the 3D to 2D crossover, we estimate a $\lambda_{pl} \simeq 0.2$ for $T_c \simeq 38K$. It is clear that an increase of Ω_{pl} results in a decrease of λ_{pb}.

Plasmon-induced pairing can, therefore, make a noticeable change in T_c relative to T_c^{ph} even for small values of λ_{pl} and it arises from the large value of the plasmon frequency.

Although some factors such as exchange and vertex correction [22] (in this work an isotropic 3D jellium model is considered, different from the case studied here) may result in some decrease of electron-plasmon coupling, we would like to stress that our concept of coexistence of strong electron-phonon coupling and the plasmon mechanism (see below) can explain the observed high transition temperature with weak electron-plasmon coupling.

In addition, the presence of different energy bands with different effective masses results in the appearance of additional plasmon bands breaches, the "demon" states.

Conclusion. In this paper we describe our approach to the theory of high T_c superconductivity. The dynamic part of the interaction between electrons in layered systems differs drastically from that in the usual 3D system. The collective excitations form a plasmon band, and the lowest branch corresponds to the maximum of the density of states.

Note that the plasmon edge measured by infrared technique corresponds to the upper plasmon branch (see Fig. 1), whereas the pairing is affected mainly by the lower branch. We studied also the hybridization of polar phonon and plasmon branches. The phonon-plasmon mechanism is described by the generalized Eliashberg equation. The large scale of the plasmon energy leads to a noticeable effect of plasmons upon T_c even for weak electron-plasmon interaction.

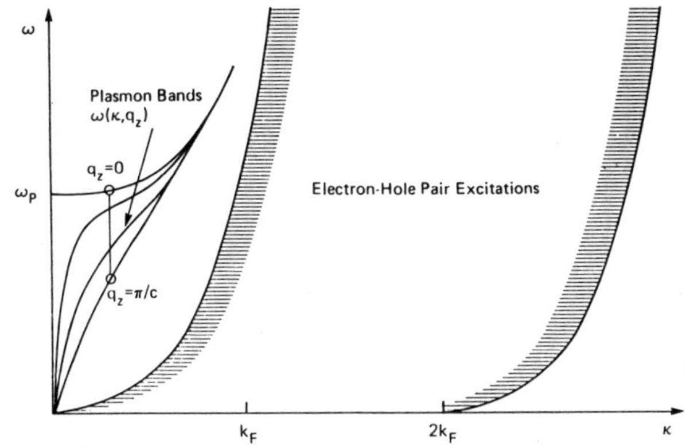

Fig. 1. Plot of the various plasmon branches and electron hole pair excitations as a function of k (absolute value of inplane wave vector) and q_z for a layered electron gas.

The work of one of the authors (V.Z.K.) was supported, in part, by the Office of U. S. Naval Research under Contract No. N 00014-86-F00015 and carried out at the Lawrence Berkeley Laboratory under contract No. DE-AC03-76SF00098.

REFERENCES

1. J. Bednorz and K. Muller, Z. Phys. B 66, 189 (1986).

2. M. Wu, et al., Phys. Rev. Lett. 58, 908 (1987).

3. For a summary see for example "Novel Superconductivity," S. Wolt and V. Kresin, eds., Proceedings of the Berkeley Conference (Plenum, New York, 1987); or "High Temperature Superconductivity," reprints from Phys. Rev. Lett. and Phys. Rev. B (American Physical Society, American Institute of Physics, New York, January-June 1987).

4. V. Z. Kresin, Phys. Rev. V, p. 1 (American Physical Society Meeting, March 1987, New York), Proc. Materials Research Soc. Meeting, Anaheim, April 1987, p. 19), Phys. Rev. B 35, 8716 (1987); Ref. 3, p. 309.

5. V. Z. Kresin and H. Morawitz, Ref. 3, p. 445; J. of Superconductivity, 1, 89 (1988); Phys. Rev. B 37 (in press); Physica C. L. Prac. of Int. Conf., Interlaken, 1988), in press.

6. J. Ruvalds, Phys. Rev. B 35, 8869 (1987); Ref. 3, p. 455, J. Jhm and D. Lee, Ref. 3, p. 451; J. Ashkenazi, et al., Solid State Comm. 63, 1144 (1987) A. Griffin (preprint).

7. J. Bassat et al., Phys. Rev. B 35, 7126 (1987); J. Fink et al., Physica C (in press).

8. D. Vaknin et al., Phys. Rev. Lett. 58, 2801 (1987); G. Shirane, et al., Phys. Rev. Lett. 59, 1613 (1987); 60, 156 (1988).

9. T. Brueckel, H. Capellmann, W. Just, O. Schaefer, S. Kemmler-Sack, R. Kiemel and W. Schaefer, Europhys. Lett. 4, 1189 (1987).

10. V. Z. Kresin and S. A. Wolf, Ref. 3, p. 287, J. of Superconductivity, 1, 143 (1988).

11. L. Smeds Jaer, et al., (preprint).

12. F. Stern, Phys. Rev. Lett. 18, 546 (1987).

13. D. Vashista and K. Singwi, Phys. Rev. B 6, 875 (1972).

14. D. Bonn, et al., Phys. Rev. Lett. 58, 2249 (1987).

15. A. Jezowski, et al., Phys. Rev. Lett. A 122, 431 (1987); D. Morelli, et al., Phys. Rev. B 36, 3917 (1987), C. Uher et al., Phys. Rev. B 36, 5676, 5680 (1987).

16. G. Panova et al., ZETF Lett. 46, 79 (1987).

17. M. Levy, et al. (preprint).

18. B. Geilikman and V. Z. Kresin, Non-Steady and Kinetic Effects in Superconductors, Wiley, New York (1974).

19. V. Z. Kresin and G. Zaitsev (Sov. Phys.-YETP, 47, 983 (1978)).

20. V. Z. Kresin, Phys. Lett. 122A, 434 (1987).

21. A. Masaki, et al., Jpn. J. Appl. Phys., 20, L405 (1987); A. Ramirez, et al., Phys. Rev. B 35, 8883 (1987).

22. O. Entin-Wohlman and H. Gutfreund, J. Phys. C17, 1071 (1984); H. Gutfreund, Ref. 3, p. 465; M. Rietschel and L. Sham, Phys. Rev. B 28 5100 (1983); M. Grabowski and L. Sham, Phys. Rev. B 29, 6132 (1984).

THE ENERGY GAP OF $YBa_2Cu_3O_{7-\delta}$: TUNNELING, FAR INFRARED REFLECTION AND ANDREEV REFLECTION.

H. van Kempen, P.J.M. van Bentum, H.F.C. Hoevers, L.E.C. van de Leemput, L.W.M. Schreurs, R.T.M. Smokers, P.A.A. Teunissen, Research Institute for Materials, University of Nijmegen, Toernooiveld, NL-6525 ED Nijmegen, The Netherlands.

A basic parameter for the description of the superconducting state is the energy gap 2Δ. Three different methods to determine the gap value are described. Quasi particle tunneling using a low temperature scanning tunneling microscope and far infrared spectroscopy give consistent values for the gap in $YBa_2Cu_3O_{7-\delta}$. It is found that $2\Delta/k_BT_c \approx 3.6$ ($2\Delta = 28 \pm 4$ meV), compatible with a weak coupling BCS state. The existence of Andreev reflection at a silver $YBa_2Cu_3O_{7-\delta}$ interface shows the presence of a zero momentum (k,-k) ground state. In addition the Andreev experiment gives a lower limit on the gap of 25 meV.

1. INTRODUCTION.

The discovery of the high T_c superconductors has resulted in an unprecedented worldwide research effort. Nevertheless there is still no consensus on the theoretical origin of the phenomenon. To make more progress, one needs quite basic information like: How does the energy spectrum of the excitations look like, is there an energy gap and if so, what is the ratio of the gap value to T_c, can the density of states be described by a BCS model, what is the nature of the ground state, and is the ground state a zero momentum paired ground state. To answer these questions many experiments have been performed already, but the nature of the high T_c superconductors is still very unclear.

In this paper three methods to gain information on the energy gap and the ground state will be discussed: tunneling, Andreev reflection and infrared spectroscopy. The emphasize of the paper will be on the first two mentioned methods.

2. TUNNELING.

Tunneling is one of the most commonly used methods to study the density of states around the Fermi surface and the electron phonon interaction. The results published so far on the high T_c superconductors show a considerable scattering in the obtained results. Several effects can contribute to this situation, such as:

1) Anisotropy of the gap.
2) Intergrain superconductor-insulator-superconductor tunneling.
3) Stoichiometric deviations at grain boundaries.
4) Spreading resistance in 2 point geometry.
5) Pressure effects.
6) Coulomb gap.
7) Single electron charging.

In this paper we will concentrate on the points 2, 5, 6, and 7.

2.1 SIN And SIS Junctions.

For nearly all of our experiments we used a normal metal needle to form a Superconductor-Insulator-Normal metal (SIN) junction. In many cases however, we do find I-V curves which clearly show the characteristics of a Superconductor-Insulator-Superconductor (SIS) junction. For comparison, the theoretical (BCS model) I-V curves for SIN and SIS junctions are shown in Fig. 1. Experimental curves measured on $YBa_2Cu_3O_{7-\delta}$, showing both types of behaviour are shown in Fig. 2. The cause for the unexpected occurrence of the SIS curves might be that the point has to be pushed quite strongly into the samples to get a stable junction. Under these conditions it is feasible that a metallic contact is made between the needle and a grain of the sample such that a SIS junction is formed between this grain and the rest of the sample.

Because the I-V curves are often somewhat smoothed, it is not always possible to make a distinction between the two cases which leads to an uncertainty of a factor of two in Δ. So for the determination of $2\Delta/k_BT_c$ we only used I-V curves which are sharp enough to distinguish the two cases. Even with this precaution we found for $La_{1.85}Sr_{0.15}CuO_{4-\delta}$ a large scattering in the results. The main cause will be discussed in the next section.

2.2 Pressure Effects.

Many tunneling measurements on the high T_c superconductors have been performed by the point contact method. Because the sample material is hard and also the point is often made of a hard material, the contact area can be quite small even when a force is present. This means that the pressure can become very high. In Fig. 3 an example is given of a set of measurements on $La_{1.85}Sr_{0.15}CuO_{4-\delta}$, where the force was adjusted in a controllable way by using a piezo element to move a flexible point relative to the sample. From Fig. 3 it can be seen that the Δ can be increased by nearly a factor of three before a saturation is reached. The saturation is probably due to yielding of the needle material at an estimated pressure of the order of 100 kbar. We found that the La-compound was much more pressure dependent than the Y-compound, in accordance with the results of Chu et al.[1,2]. Extrapolating to low pressures, we found $2\Delta/k_BT_c \approx 3.6$ ($2\Delta = 14 \pm 2$ meV) for $La_{1.85}Sr_{0.15}CuO_{4-\delta}$.

2.3 Single Electron Tunneling.

The next effect to be discussed is not directly related with superconductivity, but is an effect generally present when the capacitance of a tunnel junction is very small. For very small tunnel junctions the charging energy $e^2/2C$ (e is the electron charge and C is the capacitance) is large compared to k_BT. In that case it is energetically unfavourable to tunnel as long as the capacitor is charged to less than e/2 ("Coulomb blockade"). This leads to tunneling of the electrons one by one at regular time intervals:

Single Electron Tunneling (SET) oscillations induced by the Coulomb blockade. This effect strongly influences the I-V curves[3,4]. Around the origin there is a quadratic voltage dependence while for large voltages the curve approaches an asymptote displaced from the origin with $\Delta V = e/2C$ (Fig. 4). The full curve can be calculated with a semiclassical Monte Carlo simulation[5,6].

The SET does not have anything to do with superconductivity. This is illustrated in Fig. 5, where similar results are shown for a small junction formed between a tungsten needle and a ceramic $YBa_2Cu_3O_{7-\delta}$ sample as well as a stainless steel sample.

As is obvious from the figures it is possible to confuse the SET curves with SIN curves. A distinction however is that the SET characteristics have a continuous rising derivative. Also, the apparent gap for the SET curves can be increased at will by lowering of the capacitance (increasing the tip-sample distance). Note however that changing the distance tip-sample can change the local pressure and so can change the I-V relation of a SIN junction via the pressure dependence of the gap value. Another question is what the combined effect will be, when both the Coulomb blockade effect and a superconductivity gap are present at the same time. Preliminary results of the Monte Carlo simulations indicate that in that case the apparent gap will be larger than the real one.

2.4 Incremental Charging.

Another effect, related to the SET effect discussed in 2.3, shows up when a small particle is situated between tip and sample[4,6]. Then there are two small capacitances in series and the tunnel current can only flow if sufficient voltage is present to charge the particle with one or more electrons. In that case, the total charge on the common electrode is no longer a classic variable, but becomes quantized in units of the electron charge. If the net charge of the grain is zero, then the potential is a geometric mean of the potential difference between tip and sample. In addition, the potential of the grain increases stepwise if the number of additional charge carriers increases. The maximum charge on the particle is limited by the applied voltage on the tip, and the tunnel current increases stepwise if an additional electron can be accommodated on the grain. This characteristic shape of the I-V curve is often referred to as the Coulomb staircase. More in particular, no current will flow until a sharply defined threshold voltage is reached. This threshold voltage is determined mainly by the larger capacitance of the two series junctions, and may easily be misinterpreted as an energy gap in the superconductor. A more extended treatment can be found in van Bentum et al.[7]. Figs. 6a and 6b show the similarity between a measurement on Aluminium and on $YBa_2Cu_3O_{7-\delta}$.

Relatively small particles with diameters of a few nanometers may give rise to step structures at sufficient large voltage spacing to give confusion with multiple superconducting energy gaps.

A clear discrimination between the charging effect and a superconducting multiple gap structure can be obtained again by observing the distance (and so the capacitance) dependence of the I-V relation: In the case of incremental charging the apparent gaps increases for larger distances. Summarizing our tunneling results and taking the effects discussed in the sections 2.1, 2.2, 2,3 and this section into account, we do find a $2\Delta/k_BT_c \approx 3.6$ ($2\Delta = 28 \pm 4$ meV) for $YBa_2Cu_3O_{7-\delta}$.

3. INFRARED SPECTROSCOPY.

A second more or less classical technique that is used to determine the energy gap in superconductors is based on the absence of electromagnetic absorption if the photon energy is too small to break up Cooper pairs. The far infrared reflection measurements on polycrystalline $YBa_2Cu_3O_{7-\delta}$ indicate an increase of the normalized reflectivity[8-13] R_s/R_n above unity for frequencies below approximately 250 cm^{-1}. The far infrared results for the $2\Delta/k_BT_c$-value scatter roughly between 2.5 and 3.5. In view of the uncertainty in these values, this is not inconsistent with the weak-coupling BCS prediction. More recent experiments on single crystals, however, give quite conflicting results. See for example references 14 and 15. It is clear that from the far infrared measurements no firm conclusions can be drawn yet.

4. ANDREEV REFLECTION.

A third method to determine the energy gap of the high T_c superconductors is the superconductors is the Andreev reflection. When an electron, with energy smaller than the gap energy Δ, reaches a superconductor-normal metal interface, it cannot enter the superconductor as a quasiparticle, since there are no states below $E=\Delta$. Instead, in the "classical" superconductors, it can condense, together with a second electron of opposite spin and momentum from the normal metal, to form a Cooper pair in the superconductor. The hole, or missing electron, that results in the normal metal will move back in exactly the direction where the incident electron came from (Andreev reflection[16)]). So the observation of the Andreev retro-reflection is an indication of the presence of impulse compensated pairs in the superconducting ground state. To observe the energy dependence of the Andreev reflection, we use a point-contact geometry. The point contact is used to inject electrons into a normal metal layer. When Andreev reflection is present holes will travel back through the point contact and will give rise to the so called excess current. In other words, for electron energies below Δ, one expects an increase in the current and so a decrease of the point-contact resistance. When a very small point contact with a contact radius of approximately 10 nm (Sharvin type[17)]) is used, the electrons can be injected with energies up to eV (where V is the voltage over the point contact) and the energy dependence of the Andreev reflection can be determined and so the gap value can be found (Fig. 7). The Andreev reflection has already been measured using both single[18,19)] as well as double[20,21)] point contact techniques on "classical" superconductors like Pb.

A typical result is displayed in the upper part of Fig. 8. The point-contact resistance at zero voltage is higher than that at low bias voltages; at voltages higher than approximately 10 mV the resistance starts to increase and finally becomes more or less constant above 15 mV.

For comparison, we have also plotted a typical experimental result obtained for a Ag-Pb bilayer (lower curve b).

There are two mechanisms which can contribute to the observed energy dependence, the energy dependence of the Andreev reflection itself and the energy dependence of the electron-phonon scattering in the normal metal. Concerning the last effect, the mean free path $\ell(\varepsilon)$ of the injected electrons in the normal metal consists of an energy independent part, ℓ_{imp}, and the energy dependent phonon-emission length $\ell_{eph}(\varepsilon)$. Measurements of the energy-dependent electron-phonon coupling constant α^2F of Ag[22] indicate that $\ell_{eph}(\varepsilon)$ will decrease rapidly just in the region of the expected gap energy of the $YBa_2Cu_3O_{7-\delta}$. When $\ell(\varepsilon)$ decreases drastically the excess current becomes smaller and this effect can dominate the energy dependence of the Andreev reflection. Therefore, the inferred gap value $\Delta \approx 12$ meV mainly represents a lower limit for the superconductive gap in $YBa_2Cu_3O_{7-\delta}$. Additional experiments with the Ag replaced by Cu, which has the electron-phonon coupling at higher energies than Ag, may give information on the effect of $\ell(\varepsilon)$.

5. CONCLUSIONS.

We conclude that we have observed the excess current due to Andreev reflection, with a point contact on a thin film Ag-$YBa_2Cu_3O_{7-\delta}$ bilayer. The observation of Andreev reflection is in itself a clear manifestation of the fact that the superconducting ground state in $YBa_2Cu_3O_{7-\delta}$ consists of a paired zero-momentum (k,-k) state. From these preliminary measurements, a lower limit of $\Delta = 12.5 \pm 2$ meV of the $YBa_2Cu_3O_{7-\delta}$ is found. Tunneling measurements on $YBa_2Cu_3O_{7-\delta}$ indicate a slightly larger gap $\Delta = 14 \pm 2$ meV. The results are not incompatible with the weak-coupling BCS value $2\Delta/k_BT_c = 3.5$. The combined results from the Andreev reflection and tunneling are compatible with a BCS $(k\uparrow,-k\downarrow)$ pairing.

6. ACKNOWLEDGEMENTS.

This work was part of the research programme of the Stichting voor Fundamenteel onderzoek der Materie (FOM) and was made possible by the financial support from the Nederlandse organisatie voor Zuiver Wetenschappelijk Onderzoek (ZWO) and from Philips Research Laboratories. We would like to thank A.G.J. Schellingerhout of the Technical University Delft for the evaporation of the thin films and C.J. Beers, J.W. Gerritsen and J.G.H. Hermsen for technical assistance.

REFERENCES

1 Chu, C.W., Hor, P.H., Meng, R.L., Gao, L., Huang, Z.G. and Wang, Y.Q., Phys. Rev. Lett. 58, 405, (1987).

2 Hor, P.H., Gao, L., Meng, R.L., Huang, Z.J., Wang, Y.Q., Foster, K., Vassilious, J., Chu, C.W., Wu, M.K., Ashburn, J.R. and Torng, C.J., Phys. Rev. Lett. 58, 911 (1987).

3 Averin, D.V. and Likharev, K.K., J. Low Temp. Phys. 62, 345 (1986).

4 Ben-Jacob, E. and Geffen, Y., Phys. Lett. 108A, 289 (1985).

5 van Bentum, P.J.M., van Kempen, H., van de Leemput, L.E.C. and Teunissen, P.A.A., Phys. Rev. Lett. 60, 369 (1988).

6 Mullen, K., Ben-Jacob, E., Jacklevic, R.C. and Schuss, Z., Phys. Rev. B37, 98 (1988).

7 van Bentum, P.J.M., Smokers, R.T.M. and van Kempen, H. Phys. Rev. Lett. 60, 2543 (1988).

8 Bonn, D.A., Greedan, J.E., Stager, C.V., Timusk, T., Doss, M.G., Herr, S.L., Kamaras, K. and Tanner, D.B., Phys. Rev. Lett. 58, 2247 (1987).

9 Thomas, G.A., Ng, H.K., Millis, A.J., Bhatt, R.N., Cava, R.J., Rietman, E.A., Johnson, D.W., Espinosa, G.P. and Vandenberg, J.M., Phys. Rev. B36, 846 (1987).

10 Wrobel, J.M., Wang, S., Gygax, S., Clayman, B.P. and Peterson, L.K., Phys. Rev. B36, 2368 (1987).

11 Wittlin, A., Genzel, L., Cardona, M., Bauer, M., Köning, W., Garcia, E., Baharona, M. and Cabanas, M.V., preprint.

12 Cardona, M., Genzel, L., Liu, R., Wittlin, A., Mattausch, Hj., Garcia-Alvarado, F. and Garcia-Gonzalez, E., Solid State Comm. 64, 727 (1987).

13 Perkowitz, S., Carr, G.L., Lou, B. Yom, S.S., Sudharsanan, R. and Ginley, D.S., Solid State Comm. 64, 721 (1987).

14 Schlesinger, Z., Collins, R.T., Kaiser, D.L., Holtzberg, F., Phys. Rev. Lett. 59, 1958 (1987).

15 Timusk, T., Herr, S.L., Kamaras, K., Porter, C.D., Tanner, D.B., Bonn, D.A., Garret, J.D., Stager, C.V., Greedan, J.E. and Reedyk, M., Physica C153-155, 1744 (1988).

16 Andreev, A.F., Zh. Eksp. Fiz. 46, 1823 (1964) [Sov. Phys.-JETP 19, 1228 (1964)].

17 Sharvin, Yu.V., Zh. Eksp. Teor. Fiz. 48, 984 (1965) [Sov. Phys.-JETP 21, 655 (1965)]. A Sharvin junction is a junction between two metals in the Knudsen limit, i.e. $\ell \gg b$, where ℓ is the electron mean free path and b the contact radius.

18 Benistant, P.A.M., van Gelder, A.P., van Kempen, H. and Wyder, P., Phys. Rev. B32, 3351 (1985).

19 van Son, P.C., van Kempen, H. and Wyder, P., Phys. Rev. Lett. 59, 2261 (1987).

20 Bozkho, S.I., Tsoi, V.S. and Yavkolev, S.E., Pis´ma Zh. Eksp. Teor. Fiz. 36, 123 (1982) [JETP Lett. 36, 152 (1983)].

21 Benistant, P.A.M., van Kempen, H. and Wyder, P., Phys. Rev. Lett. 51, 817 (1983).

22 Jansen, A.G.M., Mueller, F. and Wyder, P., Phys. Rev. B16, 1325 (1977).

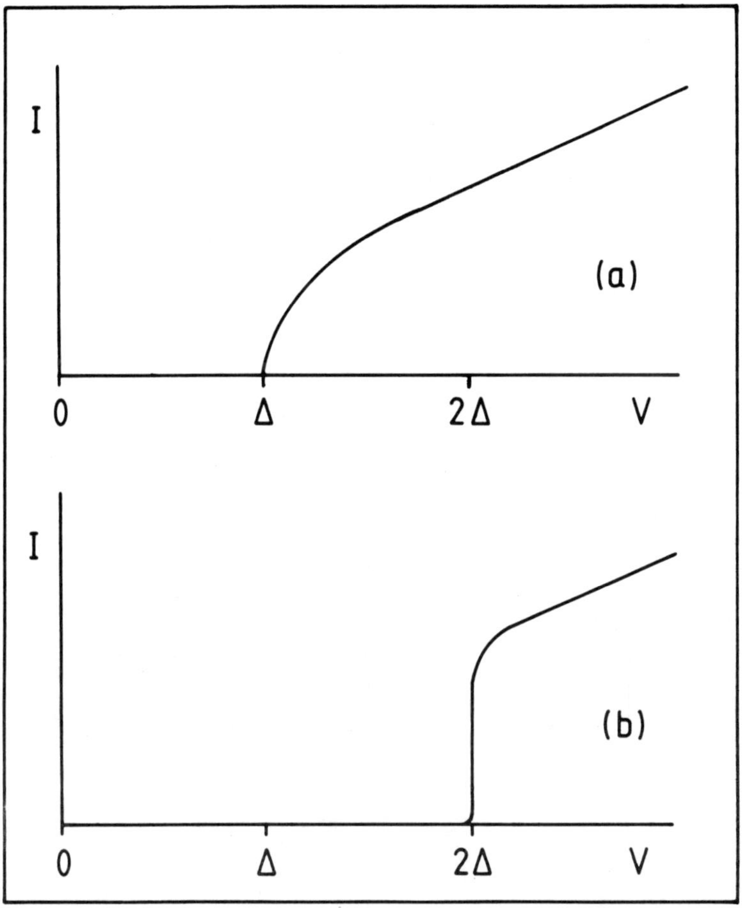

Fig. 1. Calculated BCS tunneling characteristics for a SIN junction (a) and a SIS junction (b).

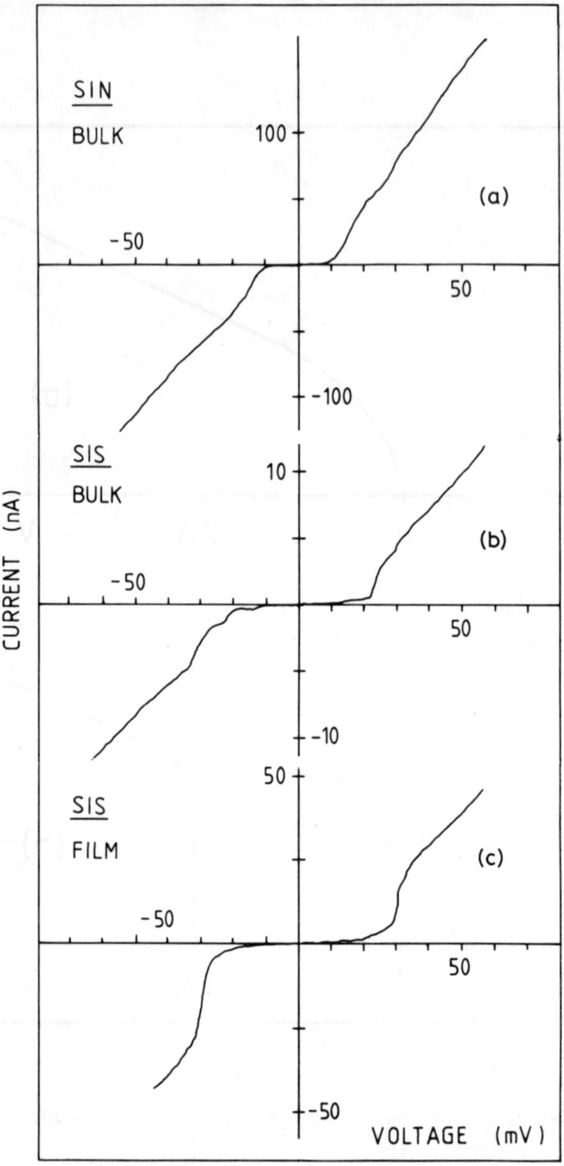

Fig. 2. I-V curve for a SIN junction between a tungsten tip and a sintered $YBa_2Cu_3O_{7-\delta}$ sample (a), I-V characteristic for a SIS junction between two grains embedded in the sample (b), and I-V characteristic of a SIS juction observed on a thin $YBa_2Cu_3O_{7-\delta}$ film (c).

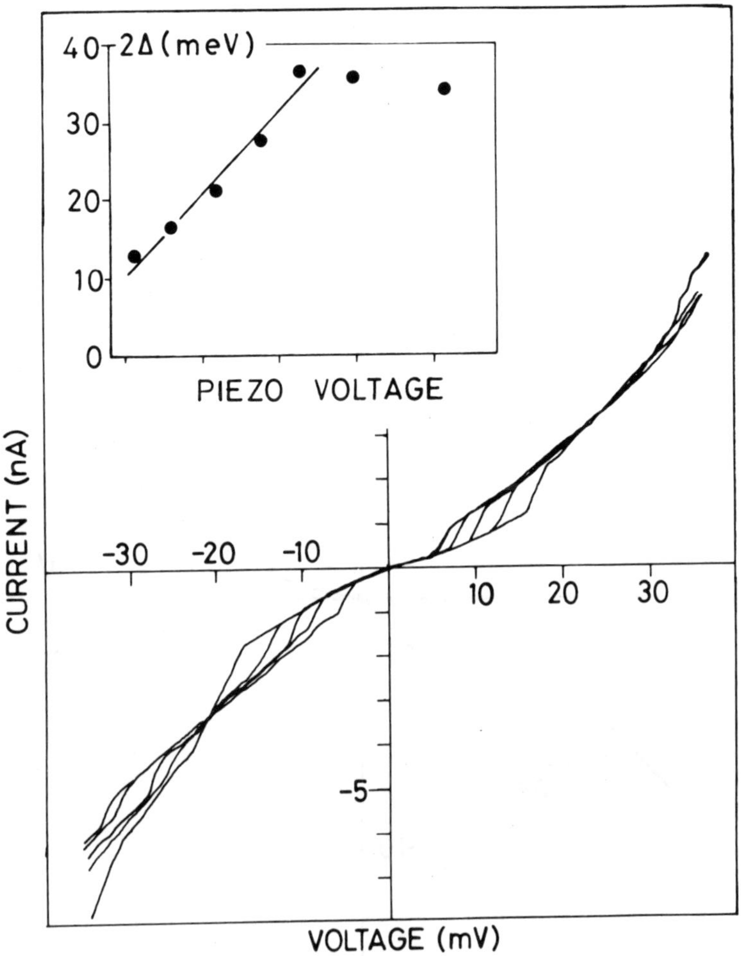

Fig. 3. I-V curves for a SIN junction on $La_{1.85}Sr_{0.15}CuO_{4-\delta}$ for different values of the pressure, exerted by the microprobe. The inset shows the energy gap 2Δ as a function of the voltage over the piezo, which is proportional to the force exerted by the tip.

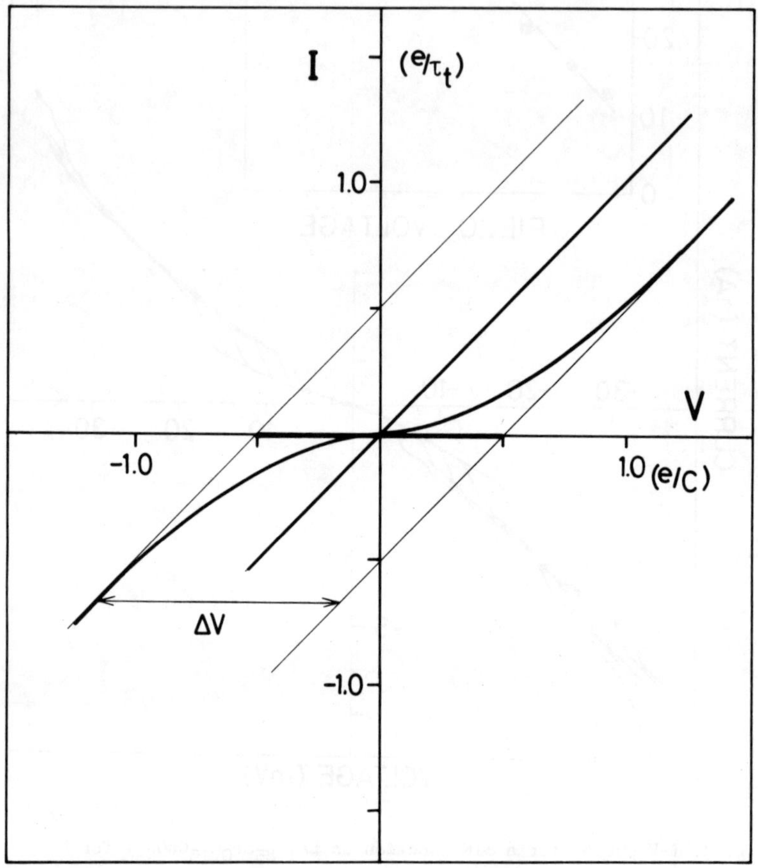

Fig. 4. Expected I-V curve for SET. The current is expressed in units of e/RC

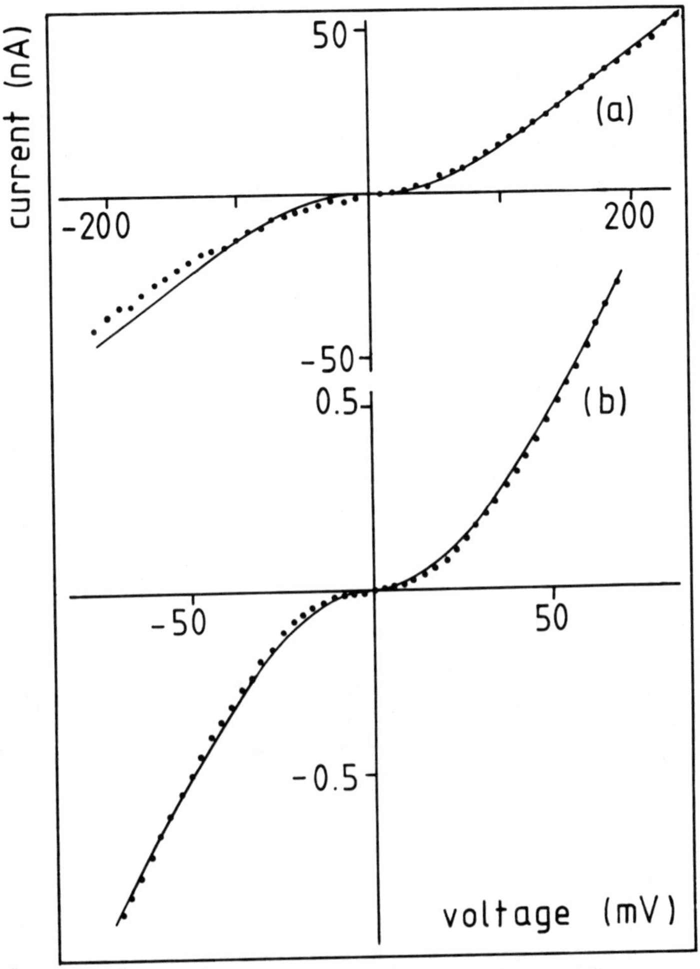

Fig. 5. I-V characteristic of a tunnel junction between a tungsten tip and a stainless steel surface (solid dots), the continuous curve is a calculation from Monte Carlo simulations with $C=3\times10^{-18}$F and R=3.3 MΩ (a). I-V characteristic of a juction between a tungsten tip and an $YBa_2Cu_3O_{7-\delta}$ surface, compared with a Monte Carlo simulation with $C=3.2\times10^{-18}$F and R=52 MΩ (b).

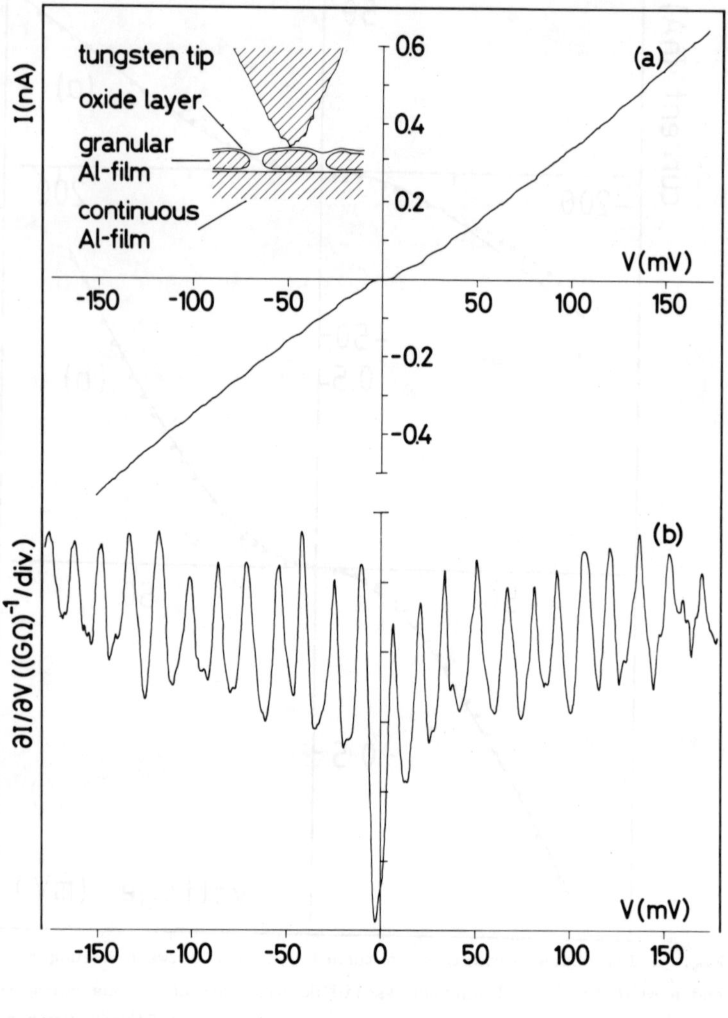

Fig. 6a. I-V (a) and dI/dV-V characteristics of a point contact on a sample consisting a granular Al film (35 Å) deposited on an oxidized continuous Al film (500 Å). The inset shows an artists impression of the configuration.

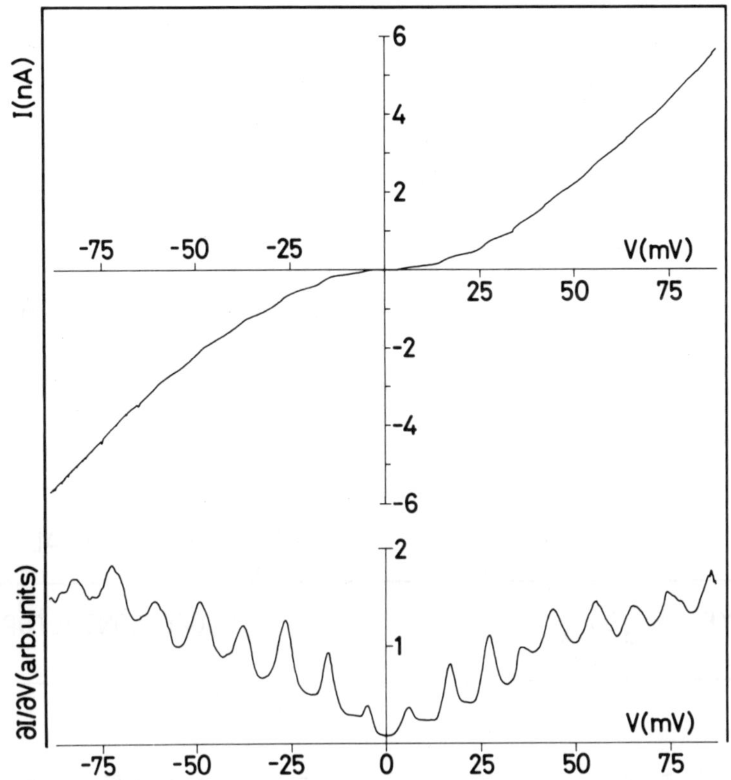

Fig. 6b. I-V and corresponding dI/dV-V characteristics of a point contact on a single crystal $YBa_2Cu_3O_{7-\delta}$ sample, polluted with isolated conducting particles.

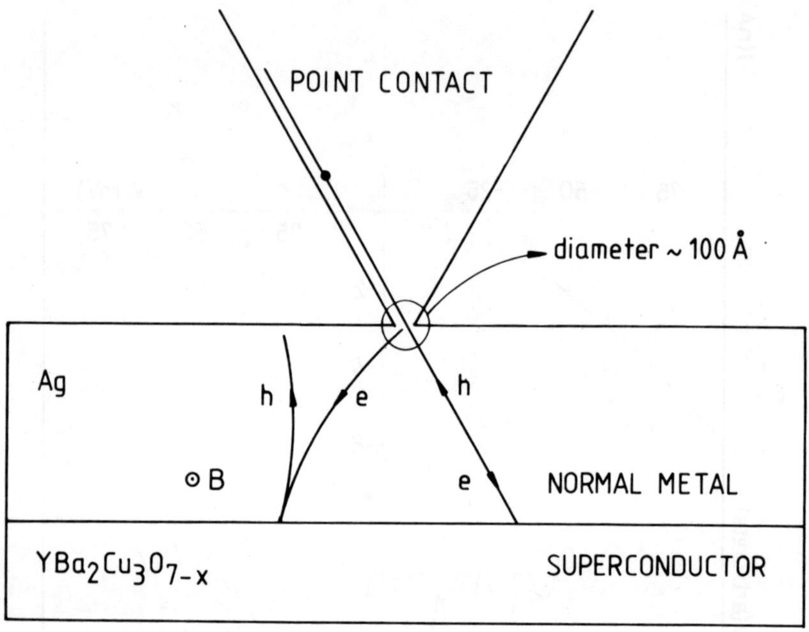

Fig. 7. Geometry of the single point contact Andreev reflection experiment.

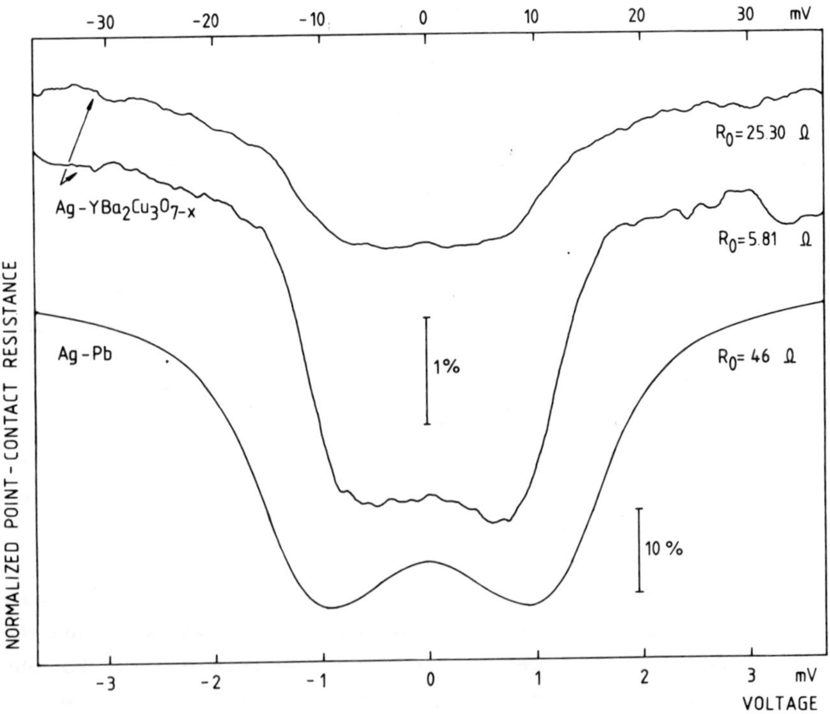

Fig. 8. Differential resistance of a point contact versus voltage. The upper curves show two measurements with different point-contact resistances. The measurements are made on a 0.25-1 μm thick Ag-$YBa_2Cu_3O_{7-\delta}$ bilayer. The lower curves shows, for comparison, a measurement for a Ag-Pb bilayer. Note the different voltage scale for the $YBa_2Cu_3O_{7-\delta}$ and the Pb measurement.

ELECTRON TUNNELING IN CERAMIC SUPERCONDUCTORS.

R. Escudero and F. Morales.
Instituto de Investigaciones en Materiales.
Universidad Nacional Autónoma de México. Apdo. Postal
70-360. México ,D.F. 04510 México.

ABSTRACT

Tunneling experiments were made on different types of tunnel junctions. In particular, for Bi-Sr-Ca-Cu-O break junctions we observed interesting features that may be related to the gap or subharmonics of the energy gap. The temperature dependence of these features shows a BCS-like behaviour. We discuss the dI/dV - V characteristics in terms of different possible contributions to superconducting tunneling.

Electron tunneling is one of the most important experimental techniques to investigate the microscopic characteristics of the superconducting state, and it is one of the most direct ways to obtain the energy gap of superconductors. Its wide use in the field produced results in the last few years that were essential for the understanding of important theoretical aspects, and today it is clear that tunneling is a very valuable tool to obtain detailed information about the interactions that give rise to the superconducting state [1].

Recently, the tunneling experiments in the new ceramics superconductors shown unusual features, probably associated to the intrinsec granularity of the material. As a consequence, all sorts of dV/dI - V characteristics are produced [2-6], meaning that the tunneling data present a wide range of different contributions to the current in the device. For instances, some of these contributions can be due to superconducting-superconducting tunneling, superconducting-normal tunneling, normal-normal tunneling, normal- normal contact, normal-superconducting contact, superconducting - superconducting contact,and of course charging effects that can be important in granular superconducting tunneling. The importance of each contribution is difficult to evaluate and, of course, the tunnel characteristic is distorted and its interpretation is not a straightforward matter. Thus, taking into account all those possible differents mechanisms, one can assume that the best way to observe better data is to fabricate small area or point contact tunnel junctions, and to select only the characteristics more according with what its expected in thin film superconducting tunnel junctions [1]. In this paper we present some tunneling results in the Bi-based ceramic superconductor of 80K. We discuss some relevant aspects related to the dI/dV - V data, and the evolution of some features with temperature.

A number of different tunnel junctions were fabricated with the new Bi based superconductors such as break junctions [7], point contacts, normal thin films - bulk ceramic, etc. The samples used for tunneling measurements were characterized by measurements of resistivity to determine the transition temperature (Tc), X-ray diffraction pattern to determinate phase purity, and were made starting from powders of their respective oxides and carbonates. These were pelletized, fired and sintered using the normal procedures already reported in the literature by several authors [8,9].

The tunnel junctions were made with samples of Bi-compound [10], which showed a Tc=70K with onset at 78K and with almost no anomaly at 107K.(Fig. 1b) In Fig. 1a the X-ray diffraction pattern shows that the predominant phase is the 80K (2212) phase, as reported by Hazen [11]. A break junction was formed by breaking a pellet into two pieces. which were mounted on a saphire holder with high vaccum epoxy. The two pieces were joined together with a small press, while the epoxy dries off overnight. Care was taken in order to avoid epoxy on the two joined pieces. this method leads to values of the differential resistance (dV/dI) always in a reasonable range for tunneling measures with low noise. With this method a very stable junction is obtained, as can be verified by its dV/dI curve as a function of temperature (Fig 1b), this curve was measured with a bias current of 1 μAmp.

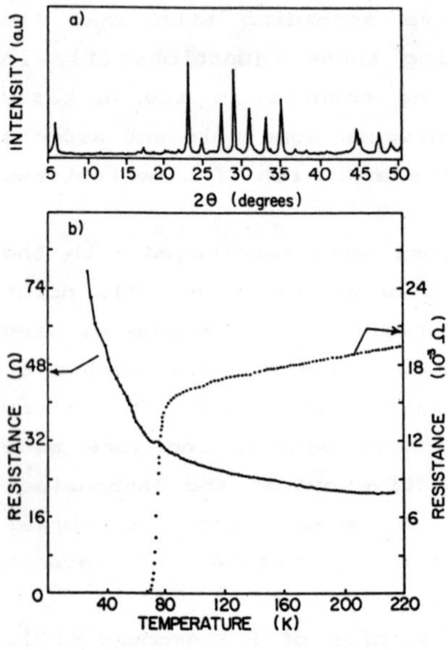

Fig. 1a) X-ray powder diffraction pattern of a sample of BiSrCaCuO.
b) Resistance-Temperature of the pellet, (dotted line) and of the break junction made with it (full line), the bias current was 1 μAmp.

The tunneling measurements were carried out by taking dV/dI, d^2V/dI^2 - V characteristics from 10K to room temperature, using the usual modulation technique [12]. The amplitud modulation used in each measure was adjusted as to optimize the signal to noise ratio, without smearing out the features. Fig. 2 shows the (dI/dV) vs V characteristic taken at various temperatures. At temperatures above Tc the junction shows the typical characteristic of tunneling between normal metals. However, below 80K some interesting features start to be clear:

1) A decrease of the resistance at zero bias, which disappears at Tc, that could reflect the presence of Josephson current or trapped flux in the superconducting grains. This feature has been noted by others, but still not fully explained .

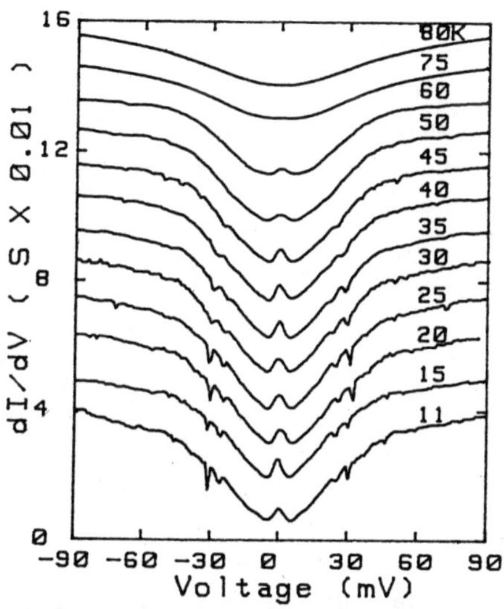

Fig. 2 dI/dV vs. V characteristic of the tunnel junction taken at various temperatures, indicated by the numbers on the right of the each curve.

2) Clear peaks at 23.9 and 29.1 mvolts, which have been observed sistematically by several authors and which are not fully understood at present. Some information about them can be extracted by investigating their temperature dependance, that is

discussed below.

3) Presence of a very dim peak at 11.5 mvolts (clearly noted in the d^2V/dI^2 - V characteristic Fig. 3) , and two small peaks at 46 and 72 mvolts. It is worth noticing that the energies of all peaks could be thought as harmonics or subharmonics of a fundamental. For instance, 11.5 mvolts is almost half of 23.9 mvolts, and 46 and 72 mvolt are multiples of 23.9 mvolt as well.

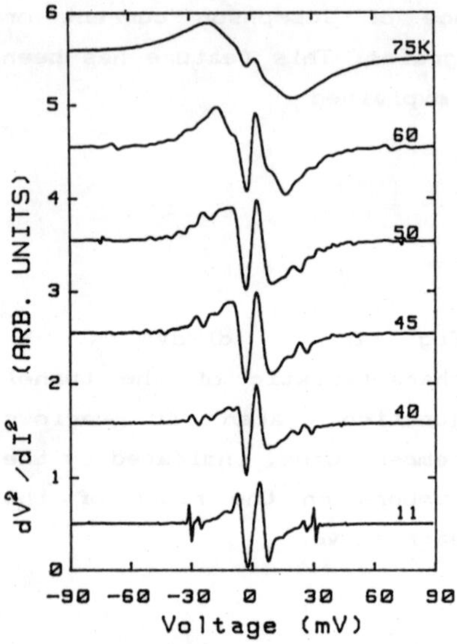

Fig. 3 dV^2/dI^2 vs. V characteristic of the tunnel junction taken at various temperatures indicated by the numbers on the right of each curve.

We also observe that all peaks move with temperature in a continuous way and they tend to fade out as the temperature increases. The temperature dependence of all these peaks was measured using the second derivative curves,(Fig.3) which provide a more accurate way to see the volt peak position. Fig. 4 shows the 23.9 mvolt peak vs temperature, the experimental data are shown as crosses, the value of 23.9/KTc is 3.56 provided one takes Tc= 78K. The BCS prediction is shown in the figure as a full line in order to compare it to the experiment. If one normalizes the temperature using Tc= 60K the results are the dots with error bars in fig. 4. Notice the remarkable agreement between the experimental data and the BCS prediction. Normalizing to a effective Tc=60K can be physically explained if one consideres that the coherence length of these ceramics is very short and therefore the order parameter should be depressed very rapidly near the surface of the superconductor [13]. We found that all the peaks follow the same behaviour respect to temperature. The ratio Vpeak/KTc (Tc= 78K) for the 29.1, 46 and 72 mvolts peaks is 4.34, 6.86 and 10.7 respectively.

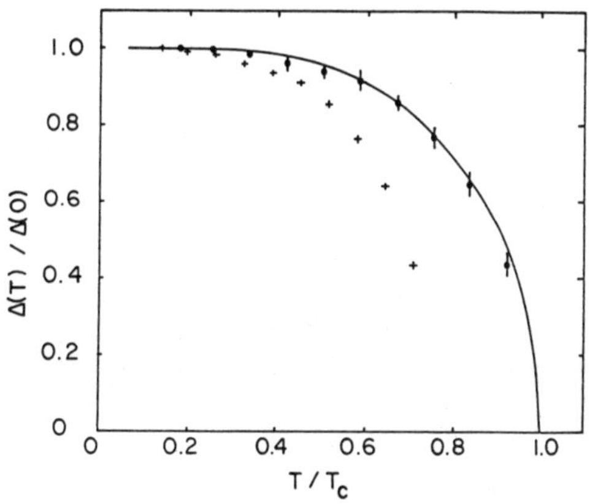

Fig. 4 BCS universal curve the temperature dependence of the gap (full line). The experimental data, normalized with T_c=78K are shown by crosses. When one normalizes the same data with T_c=60K the dots are obtained.

There is no doubt that all these harmonics like peaks have to do with a phenomenon similar to a BCS-like energy gap. Usually one observes subharmonics of the gap when there are Andreev reflexions in microbridges or point contacts weak links. Therefore one is tempted to assure that similar configurations are inevitably present in this kind of tunnel devices, which are imperfect. There is a need of improving tunnel experiments in order to disminish all unwanted contributions to the current other than straighforward superconducting tunneling.

Acknowledgment: We would like to thank professor. R. Barrio for helpful discussions, F. Estrada for assistance with tunneling measurements, E. Chavira for the sample preparation. This work was supported by the Proyecto Universitario de Investigaciones sobre Superconductores de Alta Temperatura de Transición, UNAM and by CONACYT.

REFERENCES

1) W.L. McMillan and J.M. Rowell. in Superconductivity Edited by R.D. Parks. (Marcel Dekker, N.Y., 1969). Chapter 11.

2) J.R. Kirtley, C.C. Tsuei, Sung I. Park, C.C. Chi, J. Rozen and M.W. Shafer. Phys. Rev. B35 ,7216(1987).

3) K.E. Gray, M.E. Hawley, and E.R. Moog in Novel Superconductivity Edited by Stuart A. Wolt and Vladimir Z. Kresin. 1987. Page 611

4) N. Hohn, R. Koitun, H. Schmid, S. Blumenröder, H. General, G. Güntherodt, and D. Wohlleben. Z. Phy. B Condensed Matter 69 , 173(1987)

5) H. Enomoto, K, Gotoh, K. Iida, Y. Takano, N. Mori and H. Ozaki Mat. Res. Soc. Symp. Proc. Vol 99, Page 853, 1988

6) S. Span, K.W. Ng and A.L. de Lozanne Phys. Rev. B35 , 7220(1987).

7) J. Moreland and J.W. Ekin J. Appl. Phys. 58 , 3888 (1985)

8) J.M. Tarascon, L.H. Greene, B.G. Bagley, W.R. McKinnon,

P. Barboux and G.W. Hull in Novel Superconductivity Edited by Stuart A. Wolf and Vladimir Z. Kresin. 1987.

9) J.S. Swinnea and H. Steinfink J. Mater. Res. 2(4), 424(1987).

10) E. Chavira, R. Escudero, D. Rios-Jara and L.M. Leon This Proceeding.

11) R.M. Hazen et. al. Phys. Rev. Lett., 60 , 1174(1988).

12) J.G. Adler, and J.S. Rogers, Phys. Rev. Lett. 10 , 217(1963); J.G. Adler, T.T. Chen, and J.G. Strauss, Rev. Sci. Instrum. 42 , 326(1971).

13) J.A. Blackburn, B.B. Schwartz, and A. Baratoff, J. Low. Temp. Phys. 20 , 523(1975).

MAGNETISM IN THE HIGH-T_c RELATED COMPOUNDS

S. K. Sinha

Exxon Research and Engineering Company
Annandale, New Jersey 08801 USA

ABSTRACT

We review what has been learned about antiferromagnetic ordering in the high-T_c family of compounds, using neutron scattering, magnetic susceptibility and muon spin resonance techniques. The "parent" compounds La_2CuO_4 and $YBa_2Cu_3O_6$ appear to be antiferromagnetic insulators with a three-dimensionally ordered Neel state. This long-range order disappears rapidly with doping and is replaced by short-range 2D antiferromagnetic correlations. The magnetic phase diagrams for the $La_{2-x}Sr_xCuO_{4-y}$ and $YBa_2Cu_3O_{7-y}$ family of compounds are presented. Some comments are made about possible implications of these results regarding the mechanism for superconductivity.

INTRODUCTION

What is fascinating about the new high-T_c superconducting family of compounds is that after over a year of unprecedented worldwide research on these materials and the uncovering of copious clues as to their microscopic behavior, we still have no consensus about the fundamental mechanism for superconductivity. This is probably because they happen to be metallic oxides, a class of materials which are still poorly understood from the theoretical point of view.

The occurrence of antiferromagnetism in these compounds in the undoped or oxygen-deficient phases points to some of the difficulties in theoretically describing these materials. The parent compounds La_2CuO_4 and $YBa_2Cu_3O_6$ are antiferromagnetic (AF) insulators, while local-density-functional based band theoretical calculations predict a non-magnetic metallic state.[1)] Such theories also predict strong electron-phonon induced phonon softening which is not observed.[2)] On the other hand, Resonating-Valence-Bond (RVB) type theories predict that the parent compounds should be insulators where the Neel state is preempted in favor of short-range AF spin correlations.[3)] La_2CuO_4 and $YBa_2Cu_3O_6$ can be thought of as corresponding to the situation where the Cu-O planes are half-filled, i.e., where we have Cu^{++}, O^{--} ions, corresponding to one hole on every Cu site. If the Hubbard correlation energy U corresponding to a double occupation of holes on the Cu site, or the formation of Cu^{+++}, is large, current ideas of superexchange (which provides an effective Cu-Cu nearest neighbor exchange of ~ - $t^2/4U$, t being the effective Cu-Cu hopping matrix element), would predict a Neel insulating state, as observed. Thus, the parent compounds appear to belong to the family of Mott-Hubbard antiferromagnetic insulators. Matters become considerably more complex when holes are introduced by doping and as we move towards the superconducting phases in these materials, and this will be discussed later. Even if it turns out that magnetism does not provide the basic superconducting mechanism in these materials, it is clear that magnetic-type correlations must be properly taken into account in a description of the electronic state of the cupric oxide superconductors.

The La_2CuO_4-Based Compounds

Early combined magnetic susceptibility $\chi(T)$[4)] and neutron diffraction[5)] experiments established the occurrence of long-range AF ordering at temperatures $T_N \lesssim 300K$ in the parent compound La_2CuO_{4-y}. While some samples did not exhibit AF ordering down to fairly low temperatures, this was later established as being due to the extreme sensitivity of T_N to the oxygen deficiency y.[6,7)] Weak superlattice peaks seen in powder samples at (100) and equivalent reciprocal lattice points were analyzed to yield the AF spin structure for La_2CuO_{4-y}, illustrated in Figure 1, where only the Cu atoms are shown for clarity.[5] The Neel state occurs in the orthorhombic (CmCa) phase, and the ordered moments lie along the c-axis in the Cu-O planes, with a saturated moment of ~ 0.4 μ_B/Cu at low temperatures in a sample with $T_N \doteq 300K$ and $y \doteq 0.03$. The magnetic origin of the superlattice peaks was confirmed by polarized neutron diffraction measurements[8] and by muon spin resonance (μSR) measurements.[9]

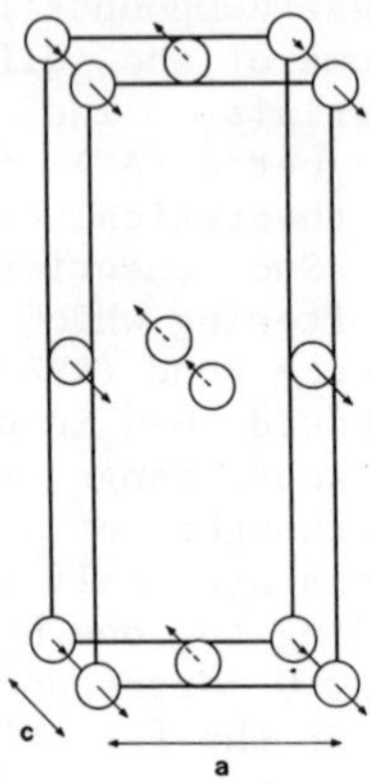

FIGURE 1
AF Spin Structure for La_2CuO_4 (only the copper atoms are shown for simplicity).

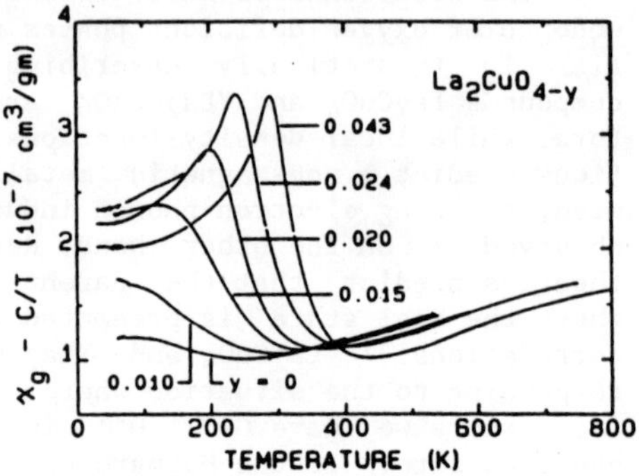

FIGURE 2
Corrected magnetic susceptibility χ_g - C/T for several samples of $La_{2-z}CuO_{4-y}$ vs. temperature.

Figure 2 shows that the peak in the susceptibility $\chi(T)$ (corresponding roughly to T_N) rapidly moves to lower temperature with decreasing oxygen deficiency y.[7] The measurements were made by heating La_2CuO_{4-y} samples <u>in situ</u> in a Faraday magnetometer, thus continuously changing the oxygen content. T_N also rapidly decreases on doping with Sr, Ba, etc.[7] If stoichiometric La_2CuO_4 is regarded as the half-filled band limit of the Mott-Hubbard insulator for which T_N is expected to be maximum, Figure 2 suggests that the samples looked at must have had a small La deficiency of $\doteq 0.02$, so that $y \doteq 0.03$ actually corresponds to the (compensated) half-filled band limit. The susceptibility measurements of Johnston et al.[7] also showed a strong dependence of the tetragonal-orthorhombic transition temperature on oxygen deficiency. It is tempting to assume that the phase diagram depends only on n_H, the number of holes as we dope away from this limit (although this might not be true in detail). Figure 3 shows such a phase diagram obtained from the above susceptibility studies of Johnston et al. combined with other measurements. (The dashed line represents earlier measurements of the Bell Laboratories group[10]) and appears displaced from the present data, perhaps because of differences in La or O stoichiometry.) The "spin-glass" phase is observed by μSR data[9]) which indicate static but disordered spins below ~ 10K. These μSR data also reveal that the Cu moment is nearly independent of T_N, in contrast to the ordered AF moment found from neutron diffraction which increases approximately linearly with T_N.[6]) This suggests that the volume fraction over which long-range AF exists increases ~ linearly with T_N.

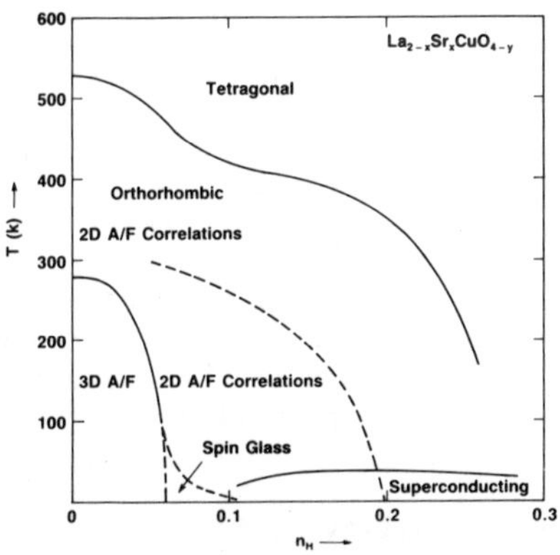

FIGURE 3
Phase diagram of $La_{2-x}Sr_xCuO_{4-y}$ system as discussed in text.

The high-temperature behavior in Figure 2 is suggestive of the behavior of a two-dimensionally ordered Heisenberg antiferromagnet with a 2D ordering temperature of ~ 1000K. With subsequent measurements on large single crystals at Brookhaven National Laboratory,[11]) it became clear that the 3D Neel state is simply the "tip of the iceberg". Above T_N, strong 2D correlations were observed to at least 550K. 2D fluctuations were also observed below T_N, consistent with the fact that not all the Cu moments participate in the static 3D ordered structure. The in-plane instantaneous correlation length becomes at least 200Å at T_N. The interesting fact about these 2D correlations is that they are not static but have fluctuations associated with them on a rather large energy scale. The energy dependence of the magnetic scattering as studied with triple-axis spectrometry was consistent with a very steep gapless spinwave dispersion curve emanating from the magnetic superlattice point and corresponding to an in-plane nearest neighbor exchange of 1300K.[11)] Magnetic excitations at energies of ~ 300 cm^{-1} have also been observed by light scattering in single crystals of La_2CuO_4[12)] and been ascribed to two-magnon processes arising from these in-plane spin fluctuations. Recent experiments indicate that the 2D AF correlations persist with doping, even into the superconducting phase, the correlation length decreasing approximately as the average spacing between holes.[13)]

From very general principles, one knows that a 2D Heisenberg antiferromagnet cannot have long-range order at finite temperatures. Thus, the (3D) Neel state establishes itself by virtue of the (weak) interlayer coupling. It is an interesting consequence of the crystal structure of La_2CuO_4 that in the tetragonal phase, the AF coupling (via superexchange) between any Cu spin and its neighboring spins in adjacent Cu-O planes exactly cancels. The orthorhombic distortion removes this cancellation and so turns out to be extremely important in bringing about 3D order. The net coupling between planes is extremely weak and the relatively large 3D Neel temperature is due to the fact that when $T \sim T_N$, the in-plane correlation lengths are quite large.[14)] The disorder (and consequent decrease in the global average of the static 3D-ordered moment) with T_N may be associated with the fact that there is frustration in the interlayer coupling if local regions of the crystal are transformed into the orthorhombic phase along different axes, i.e., due to twinning effects.

The peak in the susceptibility $\chi(T)$ in the 3D ordered samples turns out, however, not to be due to the 2D-3D transition (at which theory does not predict a maximum in $\chi(T)$) but rather due to a subtle in-plane ferromagnetic coupling between Cu spins due to the Moriya-Dzyaloshinski antisymmetric exchange.[15)]

The $YBa_2Cu_3O_{7-y}$ System

$YBa_2Cu_3O_6$ can be thought of as the half-filled band case, if we think of the filled Cu-O planes as being made up of Cu^{++} and O^{--} ions, and the Cu atoms in the empty planes to be Cu^+. Thus, one might think of this material as being an AF insulator. After considerable effort, powder neutron diffraction data finally did show the antiferromagnetic ordering in the oxygen deficient compounds.[(16)] This was soon confirmed by several groups.[(17)] The AF structure is shown in Figure 4. The spin direction cannot be determined from the powder data (other than that it lies in the basal plane) owing to the tetragonal symmetry. The Cu-O layers are

AF ordered as in La_2CuO_4, but the stacking of the AF layers is different. For $y \simeq 1$ and high temperatures, the Cu atoms in the "empty" Cu planes show no ordered moment consistent with the Cu^+ state for these ions. Some single crystal experiments with higher oxygen content indicate other low temperature phases with a small ordered moment on these Cu sites and a different stacking sequence of layers.(17) There may also be some sensitivity to details of the sample. Neutron diffraction measurements as a function of oxygen content yield the phase diagram shown in Figure 5.

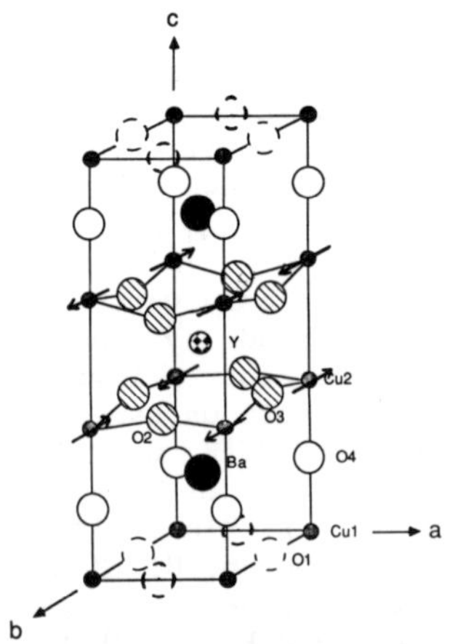

FIGURE 4
AF spin structure for $YBa_2Cu_3O_6$, as determined in Ref. 16. The O1 oxygen sites are empty. The spin direction in the basal plane is arbitrary.

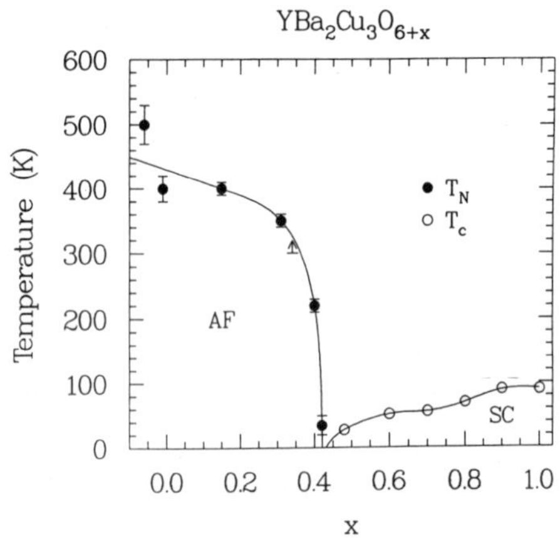

FIGURE 5
Phase diagram of $YBa_2Cu_3O_{7-y}$ determined in Ref. 16.

The Neel state appears to be associated with the tetragonal and insulating part of the phase diagram, while the superconductivity has been determined to exist in the orthorhombic phase. For $YBa_2Cu_3O_6$, the low temperature ordered AF moment on the Cu ion is $0.5\ \mu_B$. While there is no cancellation of the interlayer exchange coupling as in La_2CuO_4, the layering (and lack of bridging oxygens every third layer) makes the interlayer coupling weak, so one would expect strong in-plane AF correlations above T_N and perhaps in the superconducting phase. These may already have been observed.(18) Doping with oxygen will presumably add holes to the Cu-O layers resulting in a rapid decrease of T_N near $y \simeq 0.6$, similar to what occurs in La_2CuO_4. Exactly how holes disrupt AF order in a 2D AF insulator is currently the subject of much theoretical work.

There are a variety of models for the superconductivity based on an attractive interaction in the presence of AF order.[19] These include models where the holes perturb the AF order around them (either by suppressing AF spin fluctuations or by introducing local ferromagnetic interactions between their adjacent Cu spins) and thus lower their mutual energy as they approach each other. As discussed above, there is no long-range AF order in the superconducting phases of these materials, but since the superconducting coherence lengths are fairly small (~ 4-20Å), it may be possible that short-range magnetic order can bring about the same coupling. This has not yet been firmly established, however. It is also not yet clear whether the peculiar quantum ground state formed by holes in the highly correlated insulator when no long-range magnetic order exists is best described by the RVB state, or rather by Heisenberg-based spin models, or whether these end up being very similar entities. Clearly, many fascinating questions are still to be answered in this area.

The neutron diffraction experiments described here were carried out at the High Flux Brookhaven Reactor, operated by the U.S. Department of Energy at Brookhaven National Laboratory. I wish to acknowledge with pleasure my collaborators in various aspects of this work: M. Alvarez, M. H. Cohen, A. S. Cooper, D. E. Cox, J. E. Fischer, T. Freltoft, A. Goldman, G. Grest, B. N. Harmon, D. Harshman, A. J. Jacobson, D. C. Johnston, H. E. King, Jr., W. Kunnmann, J. T. Lewandowski, S. Mitsuda, D. E. Moncton, H. Moudden, J. M. Newsam, J. P. Remeika, C. R. Safinya, G. Shirane, C. Stassis, M. Suenaga, J. M. Tranquada, D. Vaknin, W. Xue, and P. Zolliker. I have also benefitted from many stimulating conversations with many others.

REFERENCES

1) Leung, T. C., Wang, X. W., and Harmon, B. N., Phys. Rev. B37, 384 (1988); Yu, J., Freeman, A. J., and Xu, J. H., Phys. Rev. Lett. 122, 198 (1987); Mattheis, L. F., Phys. Rev. Lett. 58, 1028 (1987).

2) Weber, W., Phys. Rev. Lett. 58, 1371 (1987).

3) Anderson, P. W., Science 235, 1196 (1987); Kivelson, S. A., Rokhsar, D. S., and Sethna, J. P., Phys. Rev. B35, 8865 (1987).

4) Johnston, D. C., Stokes, J. P., Goshorn, D. P., and Lewandowski, J. T., Phys. Rev. B36, 4007 (1987).

5) Vaknin, D. et al., Phys. Rev. Lett. 58, 2802 (1987).

6) Freltoft, T. et al., Phys. Rev. B36, 826 (1987); Yamada, K. et al., Sol. St. Comm. 64, 753 (1987).

7) Johnston, D. C., Sinha, S. K., Jacobson, A. J., and Newsam, J. M., Proc. Int'l. Conf. on High-T_c Superconductors and Materials and Mechanisms of Superconductivity, Interlaken, Switzerland (Muller, J. and Olsen, J. L., Eds.), Physica B (to be published) 1988.

8) Mitsuda, S. et al., Phys. Rev. B36, 822 (1987).

9) Uemura, Y. J. et al., Phys. Rev. Lett. 59, 1045 (1987); Harshman, D. et al. (unpublished).

10) Fleming, R. M. et al., Phys. Rev. B35, 7191 (1987).

11) Shirane, G. et al., Phys. Rev. Lett. 59, 1613 (1987); Endoh, Y. et al., Phys. Rev. B (to be published) (1988).

12) Lyons, K. P. et al., Phys. Rev. B37, 2353 (1988).

13) Birgeneau, R. J., et al. (to be published).

14) Xue, W., Grest, G., Sinha, S. K., and Cohen, M. H., Phys. Rev. B (to be published) 1988.

15) Thio, T., et al., Phys. Rev. B (to be published) (1988).

16) Tranquada, J. M., et al., Phys. Rev. Lett. 60, 156 (1988); Tranquada, J. M., et al., Phys. Rev. B (1988), to be published.

17) Brewer, J. H., et al., Phys. Rev. Lett. 60, 1073 (1988); Rossat-Mignod, J., et al., Physica C152, 19 (1988); Kadowaki, H., et al. (to be published).

18) G. Shirane, private communication.

19) Schrieffer, J. R., Wen, X.-G., and Zhang, S.-C., Phys. Rev. Lett. 60, 944 (1988); Emery, V. J., Phys. Rev. Lett. 58, 2794 (1987); Aharony, A., et al., Phys. Rev. Lett. 60, 1330 (1988).

ELECTRONIC BEHAVIOUR IN Cu-O PLANES: CONSEQUENCES TO SUPERCONDUCTIVITY.

R.A. BARRIO, J.A. COGORDAN, J. TAGÜEÑA, and I.ORTEGA-BLAKE

Instituto de Investigaciones en Materiales
Universidad Nacional Autónoma de México
Apartado Postal 70-360, 04510, México, D.F.

ABSTRACT

Ab-initio self-consistent calculations have been performed in molecules simulating the Cu-O arrays in high-Tc superconductors. The changes in the electronic structure were investigated when the molecule is distorted according to the displacements of the atoms in a selected mode of vibration. Particular attention was paid to certain states in which charge is transferred to the oxygen p-orbitals during the motion. The relevance of this type of phenomenon to superconductivity is discussed.

I. INTRODUCTION

The recent appearance of new high-Tc ceramic superconductors, based on Bi [1] and Tl [2], has given some guide to the theoretical thoughts about the possible cause of supeconductivity in these materials. On one hand, discussions about twinning, incommesurability, rare earth magnetic behaviour, orthorhombic to tetragonal transition, linear arrays of copper and oxygen, etc. have been invalidated, since the new compounds do not present these characteristics. On the other hand, it is seen that all compounds present a structure in which transition metal-oxygen planes are stacked on the c-axis. It is observed that when the number of Cu-O planes increases in the unit cell, the critical temperature is enhanced. Theories trying to predict this correlation, due to changes of the density of states at the Fermi level, have been put forward [3]. Therefore, it seems that the important part of the structure, as far as superconductivity is concerned, is the Cu-O planar array.

There have been several calculations of various types, in Cu-O square lattices [4,5], which show that the ground state of the system is insulating, and that an attractive interaction between added holes is possible when large interatomic interactions are considered.

It is clear by now that a pure BCS description of the pairing is not satisfactory, and that some other excitation (not only phonons) is responsible for the high critical temperature. Carbotte [6] has repeatedly stated that phonons are certainly playing a rôle, but they are not the only excitations responsible for superconductivity. These statements are based on calculations using the Eliashberg formalism.

Therefore, a detailed investigation of the electron-phonon interaction and of the possible appearance of electronic excitations in these materials is needed. Here, a study aiming to this goal is presented.

II. PHYSICAL CONSIDERATIONS

In order to model the electronic behaviour of the system, let us take into account some important experimental facts:

1) The low dimensionality of the system is crucial, despite the tridimensional character of supeconductivity it is seen that these materials are extremely anisotropic. The anisotropy increases as Tc is more elevated, for instance, the ratio between the resistance along the c-axis and the resistance along the ab axes is about 20 to 40 in the La compounds, and the same quantity is about 10^4

in the Bi compounds [7]. This means that the electronic scattering in the Cu-O planes is much different from the one perpendicular to the planes.

2) The undoped material is an antiferromagnetic insulator. This is somewhat surprising, because band calculations, using various approximations [8,9], predict the appearance of charge density wave instabilities and not antiferromagnetism. On the other hand, Hubbard like Monte Carlo calculations [10] predict the antiferromagnetism correctly. This means that correlations are important in these materials, and that they are Mott insulators. However, they are peculiar, since the Coulomb interaction U is not much larger than the intersite integral W [9].

3) Charge is carried in pairs and <u>there is</u> a gap. There has been much confusion about this fact because infrared and tunneling experiments give values for the gap (2) that are spread over a wide range of energies. However, from our point of view, this must be due to problems of measurement and the very existence of the gap should not be a matter of discussion. In the 1-2-3 compounds values for $2\Delta/K_BT_c$ range from 1 to 12, or even greater [11]. In the Bi compounds recent measurements give exactly the BCS value and the temperature dependence of the gap follows, almost exactly, the BCS universal curve (Fig.1).

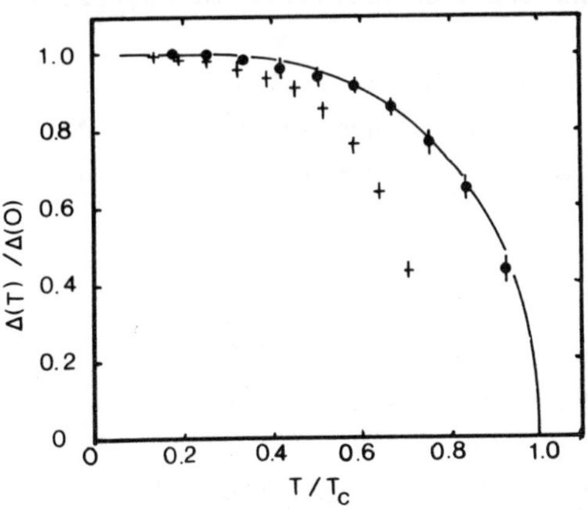

Figure 1. Temperature dependence of the gap in the 80 K Bi compound, (from ref. [12].

In the figure the crosses are the row data and the dots with error bars are the experimental points normalized to Tc=60K instead of the measured Tc=78K. This can be justified by considering the very short coherence length in this material, and that near the tunnel junction the order parameter is depleted significantly over a range of a few Amstrongs near the surface. More details concerning this experiment can be found in the paper by Escudero et al. [12].

4) There are other experimental quantities that change abruptly around Tc. Since we are interested in phonons, let us concentrate on Raman scattering experiments. There are four modes at the gamma point which are totally identified as belonging to the 1-2-3 structure [13]. Of these, there are two modes (at 335 and 500 cm^{-1}) which soften anomolously at Tc, particularly the first one. This is an indication that those modes are involved in the mechanism for superconductivity, although using the Eliashberg equations, they do not give such a high Tc, due to their low frequency. Therefore, one should look for other ways by which these modes could play a rôle. We have performed a phonon calculation in the 1-2-3 structure usin a Born-type Hamiltonian [14]. After fitting the main features of the spectrum with neutron scattering experiments, we have identified an Ag mode at 500 cm^{-1}, which is depicted schematically in Fig. 2. It can be seen that this mode is peculiar, since only the oxygen atoms in the Cu-O planes move substantially.

All these four facts must be taken into account in order to study the electronic behaviour of the metallic planes in these materials.

III. MOLECULAR CALCULATION

The cluster of atoms considered in order to model the Cu-O planar array is the one shown in Fig. 3. This configuration of seven oxygen atoms surrounding two copper atoms is the minimum one that contains a copper and an oxygen atom not in

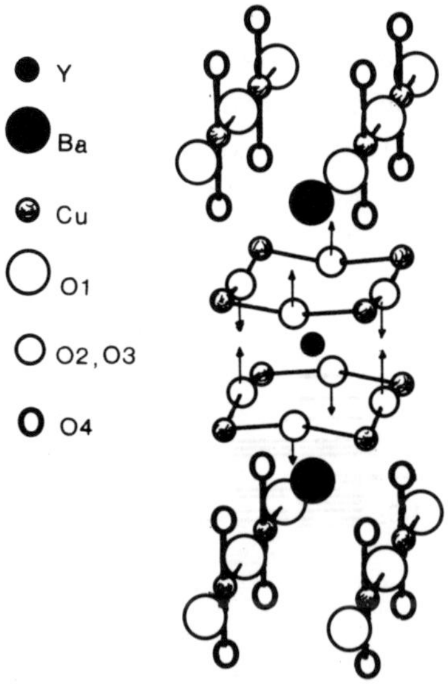

Figure 2. Displacements of the atoms in the 1-2-3 structure during the motion of the selected mode of vibration. Only the O displacements are indicated by arrows, since all others are less than 1%.

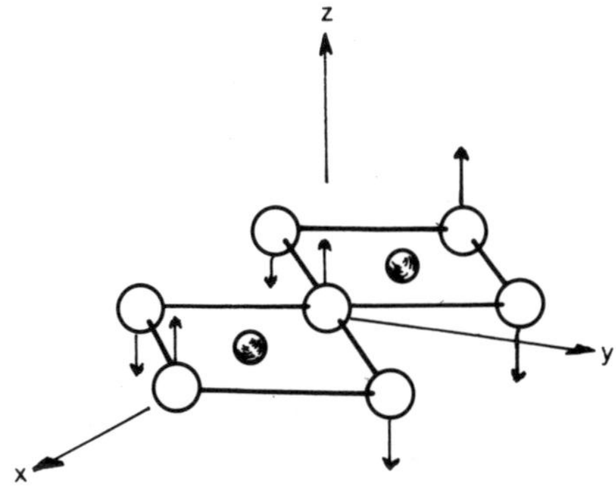

Figure 3. Molecule in which the calculations were made. The displacements in order to obtain the distorted configuration are indicated by arrows.

the surface of the molecule. The molecular calculations were made using the PSHONDO program [15], which essentially uses a Hartree-Fock approach that includes pseudopotentials for the inner electrons. These pseudopotentials were extracted from reference [16], and the basis sets used were 2-Z, optimized in the molecule. A total number of 92 basis functions were considered, that is the copper s,p and d electrons and the oxygen s and p. Although this type of calculation is only quantitatively reliable when examining total energies, the qualitative trends of charge transfer, relative level position and eigenvectors is correct. Calculations were performed in three systems, the undistorted neutral molecule of Fig. 3, the distorted molecule, in which the relative displacements along the z-axis were of 0.5Å, and the undistorted configuration with two holes, that is a doubly ionized molecule. The results of the calculations are shown in Fig. 4 for the three cases mentioned above.

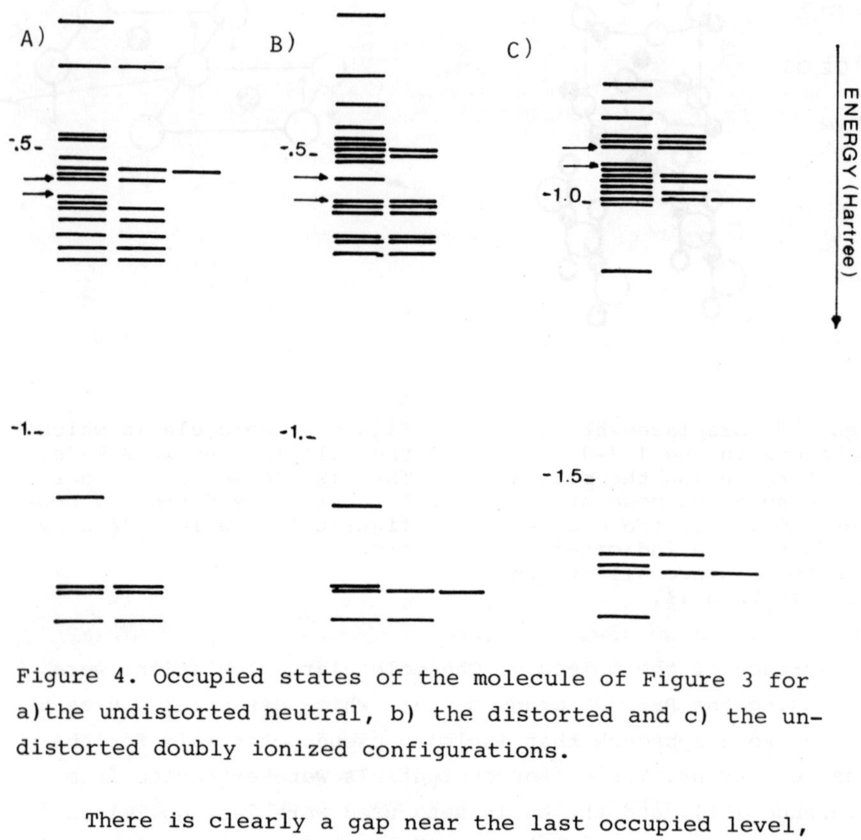

Figure 4. Occupied states of the molecule of Figure 3 for a)the undistorted neutral, b) the distorted and c) the undistorted doubly ionized configurations.

There is clearly a gap near the last occupied level, which can be thought as the molecular version of the solid state bands calculated for transition metal oxides with a simple self-consistent one-electron approach [17]. In this type of band calculation it is found a set of filled oxygen bands, and unfilled bands that are a covalent mixture of transition metal d-orbitals and a small amount of oxygen p-orbitals. The upper gap edge is practically all d and the lower one is almost p. In the molecular calculation of the undistorted neutral configuration these edges are indicated by arrows in Fig 4a). The upper one has an eigenvector Cu($3d_{\pi x}^{.77}$ O($2p_z^{.31}$), where the superscripts indicate the value of the coefficient of the atomic orbital in the linear combination to form the eigenvector. The lower "gap edge" is a totally even admixture of all orbitals.

When the molecule is distorted (Fig.4b) these states change to Cu($3d_\sigma^{.20} 3d_{\pi x}^{.56} 3d_{x^2}^{.31}$)O($p_z^{.12}$) and O($p_x^{.80}$) respectively. Only the components larger that .10 are written. One can notice two facts: 1) The nature of the states changes to the expected one in the solid, that is the upper level becomes almost Cu-d and the lower almost O-p; and 2) There is a charge transfer to the oxygen p_z orbital of about 1/3 of electron during the motion. This was detected with a Mulliken population analysis of the orbitals.

It is interesting to see that adding holes to the molecule produces almost exactly the same effect on the electron configuration. In Fig. 4c the eigenvectors for the doubly ionized configuration are Cu($3d_{\pi y}^{.82}$) for the upper level and O($p_x^{.80}$) for the lower one.

IV. DISCUSSION

The results of the present calculation are relevant because they point towards a possible origin of a polarization wave excitation of the type described by Varma [17]. Preliminary results from a Configuration Interaction calculation on the same molecule in order to study the excited states, indicate that the first excited state is obtained by taking out one electron from the lower edge of the gap. If this turns out to be true, one is left with the following picture: The phonon produces motions of the ions in the plane in such a way that the electrons in the oxygen p_z orbitals move in and out, since this motion is coherent along the crystal, this produces a polarization wave which could scatter the conduction electrons. This interaction in turn might become attractive, resulting in eventual pairing at some temperature. It is important to notice that the electronic excitation has to do with fairly localized states and that it is enhanced by the presence of holes. This fact removes the problems of separating arbitrarily the electrons in two non-interacting gases in excitonic superconductors [18] in order to avoid annoying vertex corrections. Obviously, this phenomenology is only speculation at the present time, nevertheless, the calculation presented here is valid in its own right and might help clarifying the ideas about new mechanisms for superconductivity.

REFERENCES

1) H.Maeda et al. Jpn. J. Appl. Phys. (1988) in press.
2) P.M.Grant et al. (preprint)
3) P.M.Grant (this volume)
4) H.Castillo et al. (this volume)
5) J.E.Hirsch Phys. Rev. Lett. 59, No 2, 228(1987)
6) J.Carbotte (this volume)
7) B. Batlogg (this volume)
8) M.Schluter, Proc. IX Winter Meeting on Low Temp. Phys. Mexico, (ed. by J.Heiras, J.Tagüeña and R.A.Barrio) World Sci. Publ. 1988
9) M.Schluter (this volume)
10) J.E. Hirsch Phys. Rev B31, 4403(1985)
11) A. de Lozanne (this volume)
12) R. Escudero et al. (this volume)
13) M. Cardona et al. (to be published in Sol. State Commun.)
14) A. Calles et al. (to be published)
15) M.Dupuis, J. Rys and H.F.King Quantum Chemistry Programme Exchange 11, 338(1977). Pseudopotential adaptation by J.P. Daudey, Université Paul Sabatier (unpublished)
16) J.C.Barthelat, Ph.Durand and A.Serafini, Molecular Physics 33, 159(1977)
17) C.M.Varma (this volume)
18) W.A.Little in "Novel Superconductivity" (ed. by S.A. Wolf and V.Z. Kresin) Plenum N.Y. (1988) p.341

SUPERCONDUCTIVITY AND CHARGE TRANSFER EXCITATIONS IN HIGH T_C MATERIALS

C.A. BALSEIRO*
L.E.P.E.S.-C.N.R.S.
BP 166 - 38042 GRENOBLE CEDEX - France

ABSTRACT

We study the electronic structure of Cu-O high T_c superconductors using a tight binding Hamiltonian with intra and interatomic Coulomb repulsion. We present different numerical and analytical results to show that in these materials charge transfer excitation produces pairing of holes.

It is now generally accepted that conventional electron-phonon interactions can not explain the high values of T_c observed in the new superconductors. Concerning the Cu-O materials it seems clear that the electrons free to move are those contained in the Cu-O planes of the structure. The peculiar magnetic and transport properties of the ceramic superconductors can be interpreted as the evidence of strong correlations. Moreover, most of the proposed non electron-phonon mechanisms include high electronic correlations as a starting point [(1-4)].

Here we show that high correlations in these materials can give rise to pairing between holes due to the electronic polarization of the medium. Our analysis includes what is a common feature of both La-Sr-Cu-O and Y-Ba-Cu-O : Cu-O planes in which O atoms lie between two Cu atoms. We describe the electronic structure of these planes making the two following assumptions : i) Only the Cu $3d_{x^2-y^2}$ and the O 2_{px}, 2_{py} orbitals are relevant and ii) strong intra-atomic and interatomic correlations.

The model Hamiltonian is [(5,6)] :

$$H = \sum_i E_d d^+{}_{i\sigma} d_{i\sigma} + U_d d^+{}_{i\uparrow} d_{i\uparrow} d^+{}_{i\downarrow} d_{i\downarrow}$$

$$+ \sum_j E_p p^+{}_{i\sigma} p_{i\sigma} + U_p p^+{}_{i\uparrow} p_{i\uparrow} p^+{}_{i\downarrow} p_{i\downarrow}$$

$$+ \sum_{ij} t\,(d^+{}_{i\sigma} p_{j\sigma} + p^+{}_{j\sigma} d_{i\sigma}) + G\, d^+{}_{i\sigma} d_{i\sigma}\, p^+{}_{j\sigma'} p_{j\sigma'}$$

here $d^+{}_{i\sigma}$ ($p_{j\sigma}$) creates a hole in the Cu3d (O2p) orbital of energy E_d (E_p). The parameters U_d and U_p are the intra-atomic Coulomb interactions and G represents the interatomic repulsion. In Eq(1) $<i\,j>$ indicates summations over n.n. sites.

The undoped La-Cu-O or stoichiometric Y-Ba-Cu-O correspond to one hole per Cu orbital.

---

*On leave from Centro Atomico Bariloche and Instituto Balseiro - 8400 S.C. de Bariloche, Argentina.

Total energy calculations and spectroscopic experiments indicate that the Cu and O orbitals are nearly degenerate which implies large charge fluctuations in both Cu and O atoms. This seems to be a characteristic shared by all ceramic superconductors, namely Ba-Pb-Bi-O, La-Sr-Cu-O, Y-Ba-Cu-O and the new Ba-K-Bi-O compound [7].

The ground state and the excitation spectrum of Hamiltonian (1) are extremely complicated due to the many body interactions. However, there are two limiting case that can be easily analyzed :

a) Weak coupling limit : if all interactions are small they can be neglected or treated in the Hartree-Fock approximation. The resulting band structure consists of three bands, one bonding, one antibonding and one non bonding O band. The Fermi energy for the stoichiometric compounds lies at the center of the lower hole-band and the systems should be metallic or undergo a CDW instability.

b) Strong coupling limit : if both U_d and U_p are large ($U_p, U_d >> t$) the bands are splitted into two sets of Hubbard bands. In this case, the Fermi level corresponding to one hole per unit-cell lies in a gap. The size of the gap is given by the charge transfer energy ($E_g \sim E_p - E_d$). (Fig. 1). Doping or changing the O concentration adds holes which go mainly into O-like bands.

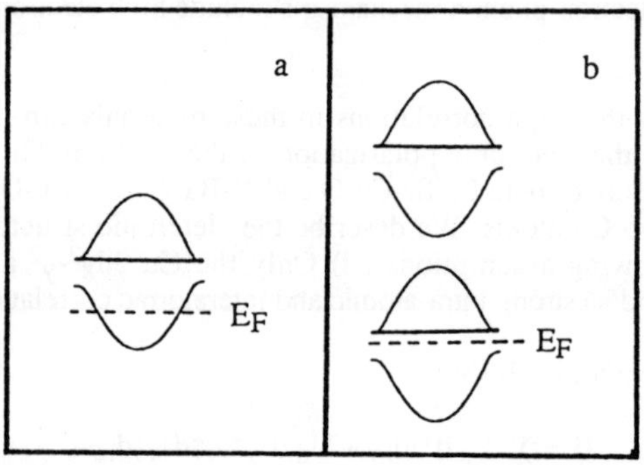

Figure 1 - Schematic band structure for the Cu-O planes. a) weak coupling and b) strong coupling limits

We believe that the weak coupling limit is a good starting point to understand the superconductivity in the ceramics with no Cu (in the Hamilton[1] the S-orbital of Bi replaces the d-orbital of Cu) - while for the Cu compounds the second limit is more appropriate.

Varma et al [8,9] proposed that the interatomic Coulomb repulsion G together with the quasi-degeneracy of the metal and oxygen orbitals leads to superconductivity mediated by charge transfer resonances. They assumed weak intra-atomic correlations. On the other hand it has been shown that charge transfer excitations can produce pairing if the intra-atomic repulsions are large, at least in the metal-ion [3,5,6,10].

In what follows we present a summary of our results which include numerical studies of small clusters and analytical approximations of the strong coupling limit.

First we present results corresponding to the bidimensional lattice of Fig. 2 In order to study the effective interactions between particles we calculate the ground state energies E(n) for different number of particles n in a cluster with periodic boundary conditions. We define the quantity E_b by the following expression [3,5]

$$E_b = E(n + z) + E(n) - 2E(n + 1) \qquad (2)$$

If n is equal to the number of Cu atoms, $E_b > O$ means that when two holes are added to the stoichiometric state it is energetically favorable to accommodate them in different clusters. On the other hand, when E_b is negative the lower energy state is that in which the two extra holes were added to the same cluster. We therefore interpret $E_b > O$ as due to an attractive interaction of the two extra particles $|E_b|$ being the binding energy.

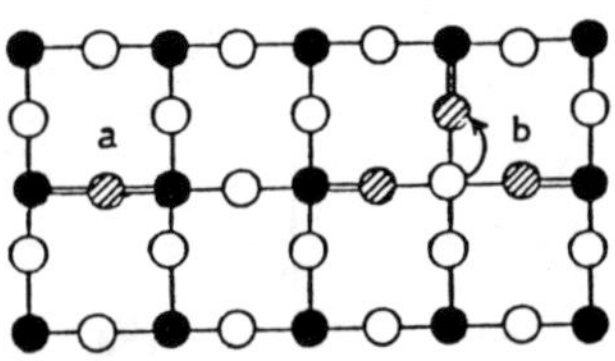

Figure 2 - Cu-O plane structure and charge rearrangement when two holes are added to the stoichiometric state (t = 0).

In fig. 3 we show E_b as a function of G for fixed (E_p-E_d) in two different cases: $U_p, U_d \to \infty$ and $U_p = \frac{U_d}{2} = 4$ t. The former case has been evaluated using an approximation which consist of describing the lower Hubbard bands as a system of spinless fermions.

The results show that the effect of the interatomic repulsion G is to produce an effective attraction between holes. Hirsh et al have found very similar results.

These results can be interpreted by taking $t \to 0$ in Hamiltonian (1). In this limit, adding one hole at an O site to the reference state - which consists of one hole per Cu atom - increases the energy by a quantity $E_p + 2G$. If a second hole is added sufficiently far from the first the energy increases again by the same quantity. If however the second hole is added in one of the O sites next to the first, charge can be rearranged so as to give an energy expense of $3E_p - E_d + 3G$ (see Fig. 2). Thus the energy difference between a bound hole-hole pair and two holes far apart is $E_b = 2(E_\pi - E_d)$-G. If $G < 2(E_\pi - E_d)$ no charge redistribution occurs and E_b is zero.

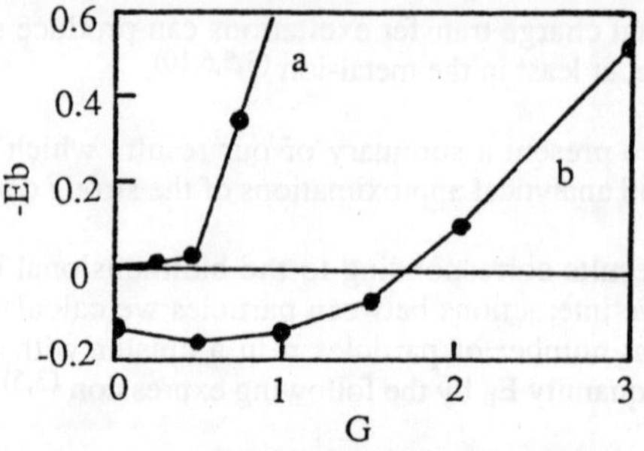

Figure 3 - Eb as a function of G for E_p-E_d = 0.5
a) infinitelly correlated limit
b) $U_p = U_d/2 = 4$. All parameters are measured in units of t.

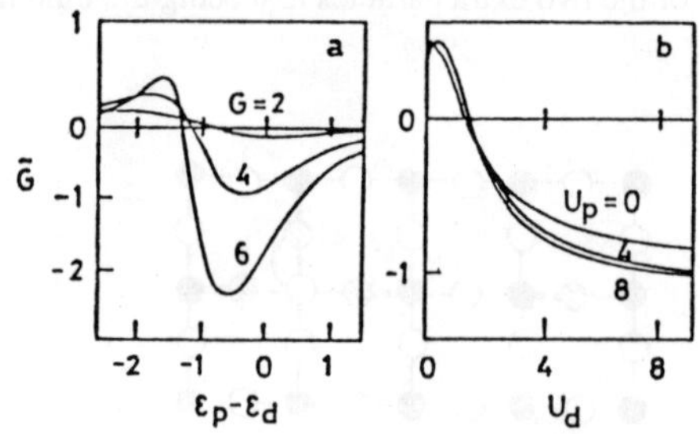

Figure 4 - Effective intercell interaction
a) as a function of E_p-E_d for different values of G and
b) as a function of U_d for various values of U_p.

An effective Hamiltonian which makes evident the existence of an attractive interaction can be obtained by projecting the total Hamiltonian into the lower states of a unit cell [(5)]. Resorting to the ideas of the real space renormalization group we calculate the many body states of a unit cell and defined new cell fermion operators associated with these states. In terms of the new cell-fermion operators the Hamiltonian reads :

$$
\begin{aligned}
H = & \sum_i \tilde{E}\, C^+_{i\sigma} C_{i\sigma} + \tilde{U}\, C^+_{i\uparrow} C_{i\uparrow}\, C^+_{i\downarrow} C_{i\downarrow} \\
& + \sum_{ij} \bar{t}\, C^+_{i\sigma} C_{j\sigma} + \tilde{G}\, C^+_{i\sigma} C_{i\sigma}\, C^+_{j\sigma'} C_{j\sigma'} \\
& + H'
\end{aligned}
$$

Here $C^+_{i\sigma}$ creates a particle in the i-th cell and H' includes other many particles interactions. If however the Hamiltonian is projected into the states of zero and one particle per cell H' can be neglected. The effective interaction $\tilde{G}$ between particles at neighbouring cells is shown in Fig. 4. The effective interaction $\tilde{G}$ is attractive for $E_p - E_d$ and large U_d and is not very sensitive to U_p These results are in qualitative agreement with the numerical results obtained by Hirsh et al.[3,10].

The superconducting critical temperature can be calculated easily if H' is neglected in Eq(3) by treating the intercell interaction in the mean field approximation and taking $\tilde{U}$ large. The results show that for nearly one particle per cell the superconducting order parameter has s-type symmetry. The critical temperature as a function of the umber of particles shows a maximum for nearly one particle per cell which corresponds to the stoichiometric situation[11].

Finally, the attractive forces due to the charge transfer excitations can also be studied in the strong coupling limit by perturbations in the interatomic repulsion G. As mentioned above in the strong coupling limit the total Hamiltonian can be approximated by a spinless Hamiltonian in which the only remaining may body interaction is G. The effective interaction between particles can then be estimated in the RPA using both direct and exchange diagrams [12]. The results show again attractive interactions and make evident that exciton-like excitation play an important role in the effective hole-hole attraction.

In summary, we have presented different type of calculations and both numerical and analytical results indicate that the charge transfer excitations produce pairing between holes in the Cu-O planes of high T_c superconductors.

Our results - in agreement with the results of Ref 3 and 10 - indicate that the attractive interaction decreases as U_d decrease and can even become repulsive for small U_d. How big U_d must be to get attraction is still an open question, the numerical results are size dependent and the cell approximation gives only qualitative results. As mentioned above, Varma et al made a pertubative calculation for $U_d = U_p = 0$ and showed that even in this limit charge transfer excitations give an effective attraction between carriers.

It is important to study in more details the dependence of the attraction with U_d to show in a more rigorous way that the mechanism works in both the weak and the strong coupling limits. This is an important point since ten new K compounds - which do not show any signature of strong coupling - are probable to be considered as High T_c material. Non of the alternative non-phonon theories can explain high T_c superconductivity in both limits and in two and three dimensional structures [1,2,13,14]. This makes the charge transfer mechanism the best candidate at present.

REFERENCES :

(1) P.W. Anderson, G. Baskaran, Z. Zou and T. Hsu, Phys. Rev. Lett. 58, 2790 (1987).

(2) V.J. Emery, Phys. Rev. Lett. 59, 228 (1987).

(3) J.E. Hirsch, S. Tang, E. Loh and D.J. Scalapino, Phys. Rev. Lett. 60, 1668 (1988).

(4) B.K. Chakraverty, D. Feinberg, H; Zheng and M. Avignon, Solid State Comm. 64, 1147 (1987).

(5) C.A. Balseiro, C.R. Gagliano, A. Rojo and B. Alascio, to be published.

(6) C.A. Balseiro, A. Rojo, G.R. Gagliano and B. Alascio, Proceedings of the Interlaken Conference, to be published.

(7) R.J. Cava, B.Batlogg, J.J. Krajewski, R. Farrow, L.W. Rupp, A.E. White, K. Short, W.F. Peck and T. Kometani, Nature 332, 814 (1988).

(8) C.M. Varma, S. Schmitt-Rink and Elihu Abrahams, Solid State Comm. 62, 681 (1987).

(9) C.M. Varma, Proceedings of this Conference and Proceedings of the "International discussion meeting on high T_C Superconductors" Ed. by Weber, to be published.

(10) J. Hirsch, Proceedings of the Interlaken Conference, to be published.

(11) M. Foglio and C.A. Balseiro, Proceedings of this Conference.

(12) H. Castillo, C.A. Balseiro, H. Ceva and B. Alascio, same as (11).

(13) M. Cyrot, Nature 330, 115 (1988).

(14) V.Z. Kresin and H. Morawitz, Phys. Rev. B, 332, 814 (1988).

GRANULARITY AND GLASS-LIKE BEHAVIOUR OF SINTERED CERAMICS.

J.Rosenblatt, A. Raboutou, C. Lebeau and P. Peyral, Laboratoire de Physique des Solides, U. A. 040786, INSA, B.P.14A, F-35043 RENNES-CEDEX, FRANCE.

ABSTRACT: The formal analogy between the paracoherence-coherence transition of granular superconductors and the paramagnetic-ferromagnetic transition of XY ferromagnets provides a quantitative basis to study the diamagnetic response of sintered and powdered superconducting ceramics. This explains in particular the puzzling "incomplete Meissner effect" found in the literature. Measurements of the penetration depth in sintered ceramics and in isolated grains suggest the existence of intragranular structure, or equivalently, multiple levels of granularity. The critical exponents found in the transition to coherence of granular superconductors, including Y-Ba-Cu-O, are close to those found in several spin glasses. This glassy feature exists in the absence of a magnetic field, and therefore of frustration, in the superconductor.

1. INTRODUCTION.

The discovery of high-temperature superconducting ceramics has focused attention on granularity and its relevance for the phenomenology of the new materials. On the other hand, several authors have described a possible spin-glass (SG) state of granular superconductors[1-3], induced by frustration[4], due to frozen current loops when a sufficiently strong magnetic field is applied. The pertinence of this concept for high-T_c superconductors has also been pointed out[5,6]. However, it is hard to see how certain assumptions made in the models apply to "natural" granular superconductors like the new ceramics: proximity to the percolation threshold and long-range interactions (the Josephson coupling between grains is of course short-range) or validity of the Ginzburg-Landau equations. It remains that the physical reality of frustration effects has been proven, particularly in two-dimensional (2D) superconducting networks[7].

Our group has been working for quite a long time on 3D random granular superconductors. The samples were model systems made of 10^6 to 10^8 slightly oxidised Nb or Ta grains of 10 or 50 μm nominal diameter in electrical contact with each other and imbedded in epoxy resin. Among our results, the following are pertinent in this paper:

(i) There is a second order phase transition, from "paracoherence" to coherence among the superconducting phases of the grains[8] at a critical temperature $T_c < T_{cg}$, the superconducting critical temperature of the grains (purely intragrain properties are hereafter indicated by the subscript g).

(ii) The phase transition is isomorphic with the XY model of a ferromagnet and paracoherence is formally analogous to the critical region above the Curie temperature T_c[9].

(iii) Like in other phase transitions, the behaviour of *macroscopic* quantities near T_c is governed by power laws. The order parameter is given by the thermal average $\Psi = \langle \exp(i\phi) \rangle$ where ϕ_i is the superconducting phase in grain i. Then $\Psi \sim \Delta T^{\beta}$, where $\Delta T = |T - T_c|$ and the exponent β can be obtained by measuring the penetration depth $\lambda(T) \propto 1/\Psi(T)$. Similarly the critical decoupling current density $j_c(T) \sim \Delta T^{2\beta+\nu}$ where ν is the exponent describing the singularity of the correlation length, $\xi \sim \Delta T^{-\nu}$. By measuring $\lambda(T)$ and $j_c(T)$ we obtained β and ν[10]. Similarly, the excess conductivity $\sigma_S = \sigma - \sigma_N$, due to supercurrent fluctuations at $T > T_c$, is described by the same exponent γ as the susceptibility of an XY ferromagnet[11]. Finally, the equal-time order parameter-order parameter correlation function at $T = T_c$ decays with distance like $r^{2-\eta-d}$, where d is the dimensionality of space and η is a rather ellusive very small exponent. Ours is the first measurement of η in a 3D XY system[11].

We discuss in section 2 the relevance of granularity for certain superconducting and normal state properties. We show in section 3 that this concept, together with the ferromagnetic analogy, provides an interpretation for the low-field response of sintered Y-Ba-Cu-O. Measurements on isolated grains suggest that several levels of granularity are simultaneously present in these materials. Finally section 4 is devoted to a model of phase transitions in extremely disordered matter, particularly applicable to granular superconductors.

2. GRANULARITY.

The macroscopic (thermal, electrical, mechanical,...) properties of granular matter may differ spectacularly from those of the individual granules. The same is true for homogeneous condensed matter in connection with atoms and molecules. It is this particular relationship that we emphasize here; in other words, we define granularity as the property whereby certain *macroscopic* phenomena depend on a *mesoscopic* average grain size a which plays the role of a *microscopic* unit of length. A simple and sometimes overlooked phenomenological consequence of granularity is that the replacement of angströms by micrometers may change by orders of magnitude the values of certain macroscopic parameters, as discussed below. Disorder in grain sizes will expectedly replace translational invariance by self similarity[12], i. e., if $N(L)$ is the number of grains in a region of size $L \gg a$,

$$N(L/b) = b^{-d}N(L), \quad L > a, \quad b < 1 \tag{1}$$

2.1. Phase Coherence and Granularity.

The XY model transition consists in the ordering in the complex plane, of the vectors representing the superconducting order parameter in individual grains. The coupling between grains (or vectors) is supplied by the Josephson energy, which at $T \ll T_{cg}$ is given, for tunnelling junctions, by[13]

$$J_a = (R_c / 2R_J)\, \Delta(0) \simeq (R_c / R_J)\, k_B T_{cg} , \tag{2}$$

where $R_c = \Phi_0 / 2e \simeq 6450\ \Omega$ is of the order of the charateristic resistance for which weak localization effects are expected in the normal state[14], R_J is the junction normal resistance and $\Delta(T)$ is the superconducting energy gap. Certain straightforward conditions are necessary for granularity to be relevant:

a) Isolated grains are superconducting below T_{cg}. For small classical superconducting particles this implies that the superconducting energy gap must be greater than the distance between electron energy levels[14], i. e.

$$\Delta(0) > W_a = 1/[a^3 \, N(E_F)] \qquad (3)$$

with $N(E_F)$ the density of states (DOS) at the Fermi energy.

b) Paracoherence can be observed in the temperature interval $T_c < T < T_{cg}$. We require that this interval be large enough to be observable, that is $k_B T_c \simeq J_a < k_B T_{cg}$. Defining[14] $a_0^{-3} = N(E_F)\,\Delta$ and taking into account Eqs. (2) and (3) we find the conditions

$$R_c / R_J < 1 \ll 2\,(a/a_0)^{-3} \qquad (4)$$

for the observation of a granular transition to phase coherence. Typically $a_0 \simeq 3$ to 5 nm.

2.2. Granular Localization.

At first sight one would expect the normal resistivity to provide a simple estimate of the junction properties through $\rho_N \cong R_J a$. This is not necessarily the case, in particular for clean small grains. In fact, the resistance R_J depends on the DOS and on the barrier transmission coefficient t, but *not* on the grain size; on the other hand R_a, the effective normal resistance between two grains, does depend on a, as we show now for a free-electron model in the limit $a \ll \ell_\infty$; here ℓ_∞ is the intragranular electronic mean free path. The lifetime of an electron against diffusion out of the grain is the product of the transmission probability t times the attempt frequency, that is $1/\tau_a \simeq t\, v_F/a$. The diffusion coefficient is $D_a \simeq a^2/\,3\tau_a$, and from Einstein's relation $\sigma_a \simeq e^2 D_a N(E_F)$ we obtain the conductivity at the mesoscopic size scale:

$$\sigma_a = 1/R_a a = a\, k_F^{\,2}\, t/3\pi\, R_c \qquad (5)$$

and for the conductance:

$$g_a \,/\, g_c = R_c /\, R_a = a^2\, k_F^2 t\,/3\pi. \qquad (6)$$

When $g_a \simeq g_c$ we expect weak localization to set in. Superconductivity and granularity will disappear simultaneously as $a \to a_0$; for $R_J < R_a \simeq R_c$ for small a and neither condition in Eq. (4) is satisfied. We remark that amazingly high values of $t \simeq 0.1$ (Eq. (6)) are needed to keep the system on the metallic side of the metal-insulator transition when $a \simeq a_0$. On the other hand when $a > a_0$, ℓ_∞, $R_J \simeq R_a$ and Eq. (4) implies rather important localization effects of granular origin together with a granular phase transition to coherence. This is illustrated in the approximate phase diagram of Fig. 1. The figure, like the preceding discussion, does not take into account other effects, like intra- and intergrain charging energies[15], which also become important when $a \simeq a_0$.

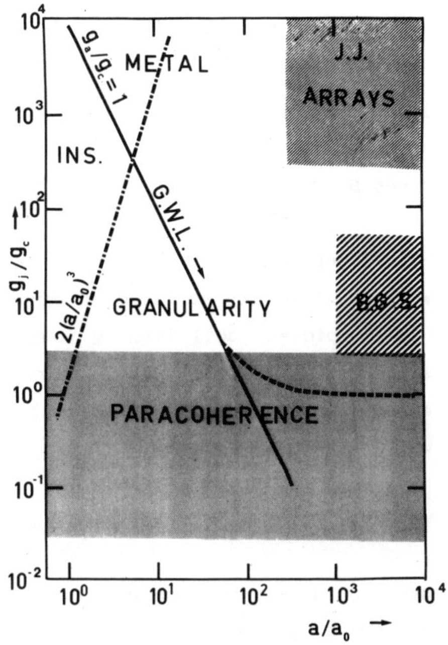

Fig. 1. Schematic phase diagram of a granular superconductor at T = 0. The dot-and-dash line marks the limit between relevant and irrelevant granularity for the superconductor, the full line indicates incipient granular weak localization (GWL) in the normal metal. Hatched regions show qualitatively the expected behaviour of Josephson junction (J.J.) arrays and bulk granular superconductors (B. G. S.)

3. FIELD PENETRATION IN SINTERED CERAMICS.

A few years ago we described an *absolute* measurement of the penetration depth λ of "classical" granular superconductors[10]. By this we mean *not* the superconducting intragrain penetration depth λ_g but the size of the region where intergrain supercurrents flow coherently to screen out an applied field. The "microscopic" unit of length for λ is the grain size $a \simeq 1$ to $10\ \mu m$, so we should expect rather large values of λ. The XY model analogy gives[15]:

$$\lambda \simeq \frac{\Phi_0}{2\pi}\left[\frac{6\ a^{d-2}}{\mu\ z\ J_a |\Psi|^2}\right]^{1/2} \qquad (7)$$

in d dimensions, with z the average coordination number. Since the XY order parameter $\Psi \sim \Delta\, T^{\beta}$, a measurement of λ as a function of temperature gives β.

3.1. Experimental Method.

We apply a purely ac field of varying amplitude H_m and record the output V of a pick-up coil tightly wound around the sample. As long as H_m is small compared with the intragrain penetration field H_{c1g}, the grain magnetization M_g is linear with the applied field. We can then write (MKS units are used throughout):

$$\mathbf{B} = \mu_0\mathbf{H} + \mathbf{M}_g + \mathbf{M}_c = \mu\mathbf{H} + \mathbf{M}_c = \mu\ \mathbf{H}\ f(R/\lambda), \qquad \mathbf{H} \to 0. \qquad (8)$$

Here $\mu < \mu_0$ is a permeability describing the diamagnetic response of *uncoupled* grains; the couplings and the corresponding intergrain screening currents result in the coherent magnetization M_c or equivalently, the global penetration depth λ. The function $f(R/\lambda)$ is the solution of London's equation in the geometry of the sample, characterized by the parameter R (radius of a cylinder, a sphere, etc.). So by measuring the low frequency response $B(H)/H$ we can get $f(R/\lambda)$ and thereby λ *provided* we know the value of μ, i. e. the non coherent response.

3.2. The Penetration Depth.

We expect these materials to display extreme type II behaviour; in particular a Meissner effect and therefore a linear response to the field, described by Eqs. (6) and (7), for fields $H < H_{c1} \ll H_{c1g}$. When $H > H_{c1}$, irreversibilities appear and the response to an ac field becomes non linear: the system describes a hysteresis cycle in each oscillation and its response depends on the field H_m. We then expect a critical state model[17] to apply; vortices come in and out of, say, a cylindrical sample with a maximum gradient $-\partial B_z/\partial r = \mu j_p$, where j_p is the critical *pinning* current. A double integration of this equation shows that the maximum difference between the applied flux $\pi R^2 \mu H$ and the internal flux is $\pm 2\pi R^3 \mu j_p/3$. So if the applied field amplitude $H_m \gg j_p R$, the measured field is $B \simeq \mu H$, giving a second field interval where the response is linear. Actually the effects of pinning should start to saturate when the distance between two vortices becomes of the order of the grain size, $B_m = H_m \simeq \Phi_0/a^2$. This is shown in Fig. 2, where $H_m \simeq 2$ to 10 G gives $a \simeq 2$ to 5 μm. The ratio between the two values of V/H_m in the low- and the high-field plateaux gives $f(R/\lambda)$. The results are very similar to those found for 10 μm Nb grains[10]: λ(77 K) $\simeq$ 0.8 mm for sintered ceramics[18], λ(1.5 K) $\simeq$ 0.6 mm for Nb grains, $\beta \simeq 0.7 \pm 0.1$ in both cases. These abnormally large penetration depths explain the "partial Meissner effect": λ is not negligible compared with the sample diameter, usually a few mm.

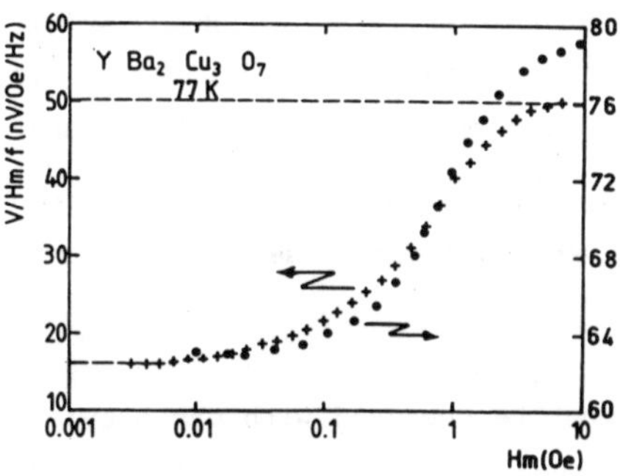

Fig. 2. Response of a coil wound around a cylindrical sample of sintered grains (crosses) and decoupled grains (circles), under an ac field of amplitude H_m.

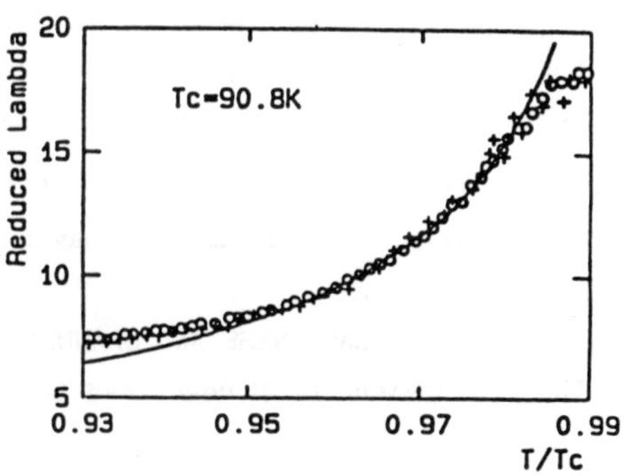

Fig. 3. Reduced penetration depth of two assemblies of uncoupled grains of Y-Ba-Cu-O with different filling factors. The line is a fit to the law $\lambda(T) \propto \Delta T^{-0.7}$. A very similar fit is obtained for sintered ceramics.

Next we repeated our experiment on *uncoupled* grains, which we obtained by grinding a sintered sample[19]. The results were surprisingly similar to those of bulk ceramics (Fig. 2). In both cases a power law fitted the experimental results (Fig. 3) with $\beta \simeq 0.7$ and not the Ginzburg-Landau value 0.5. This is a clear signature of granularity. The striking fact is that now $\lambda(77\ K) \simeq 15\ \mu m$, fifty times smaller than in the bulk but still a macroscopic value.

3.3. Evidence for Multiple Levels of Granularity in Ceramics.

The preceding results indicate that "mechanical" grains of about 60 μm display as much granularity as bulk samples. How far can we carry on with this process? Apparently quite far in view of the very short coherence length $\xi_s(0) \simeq 2$ nm of these materials. Then practically any type of defect - deviations from stoichiometry, grain boundaries, crystallite mismatch, twin boundaries[20,21],...- may be efficient in modifying the superconducting order parameter $\Psi_s(r)$. In the same way, granularity modulates self-similarly (cf. Eq. (1)) $\Psi_s(r)$. So defects may blur the borderline between intra- and intergranular properties. More precisely, in systems like the Nb grains we worked with[10], both couplings and grain sizes could vary a lot, but there was no correlation between them, $\overline{\Delta J_a \Delta a} = 0$. In this sense we may speak of a single level of granularity. For ceramics the correlations between couplings and distance would certainly modify the value of λ depending on the size scale, as we found. On the other hand correlations of the type $\overline{\Delta J_a(r)\Delta J_{a'}(r')} \neq 0$ may change drastically[22] the critical behaviour of second order phase transitions.

4. SCALING, GLASSINESS AND GRANULARITY.

The measurement of β described before must be compared with our previous results on Nb and Ta granular systems; here we obtained four exponents: β, γ, ν and η. The consistency of our results is illustrated by the fact that the exponents satisfy the well-known scaling relations:

$$d\,\nu = 2\,\beta + \gamma, \qquad \tilde{\gamma} = \gamma/\nu = 2 - \eta. \tag{9}$$

However, the measured values $\beta = 0.68 \pm 0.05$, $\nu = 1.3 \pm 0.1$ and $\gamma = 2.6 \pm 0.02$, are about twice what is predicted by calculations[23] on ordered XY systems, while $\eta = 0.027 \pm 0.015$ is close to the theoretical value. Obviously disorder is largely responsible of this behaviour[10,19]. Furthermore the values of the first three exponents, and particularly of ν, are close to those found in spin glasses[24,25]. A few examples: for a 3D Ising spin-glass Montecarlo simulation $\nu = 1.3 \pm 0.1$[26]. For Cu-Mn experimentally $\nu = 1.3 \pm 0.2$[27], $\nu = 1.39$ for a-FeNiPAl[28] and $\nu = 1.26 \pm$

0.2 for $CdCr_2(In)S_4$[29]. This coincidence is rather unexpected, because the theoretically predicted SG state[1,2] requires frustration and therefore the presence of a magnetic field. In our experiments the remanent field is ≤ 10 mOe. So either frustration can be efficiently induced by very low fields or a different phenomenon - randomness for example - can result in a quasi-spin-glass behaviour of granular superconductors.

4.1. Disorder and Critical Exponents.

We recall here very schematically an argument we developed in previous papers[24,30]. A bulk granular superconductor will conceivably display the Meissner effect and infinite conductivity when at least one phase-coherent cluster spans the whole sample. A grain at point r belonging to this cluster will be coupled to its neighbours by an energy $J_r > k_B T$. The transition becomes possible at a temperature T_c such that the grains with $J_r > J_c \simeq k_B T_c$ have a concentration $p = p_c$, the percolation threshold. If the distribution of couplings is such that $J_c > \bar{J}$, the average coupling (and this is a rather weak constraint in 3D), we expect, by the same token, that at $T > T_c$ the critical behaviour will be governed by the strongest couplings, $J_r > J^*(T) > k_B T > J_c$, with concentration $p(J^*) < p_c$; $J^*(T)$ may be sample-dependent, but $\Delta p = |p - p_c|$ is the important parameter here. Actually the relevant couplings at $T > T_c$ form finite (disconnected) percolation clusters of maximum size $\xi_P \sim \Delta p^{-\nu_P} \sim \xi(T) \sim \Delta T^{-\nu}$; here the subscript "P" stands for percolation; the correlation length exponent $\nu_P \simeq 0.82$[31]. Percolation clusters are characterized by the pair connectedness $C(r) \sim r^{-2\tilde{\beta}_P}$, ($\tilde{\beta}_P = \beta_P/\nu_P$), which is the conditional probability that, given that a point, taken as the origin, belongs to a cluster, another point at distance r from the first, also belongs to the same cluster. The number of points belonging to clusters in a region of volume $\xi_P{}^d$ is $\xi_P{}^D$, where $D = d - \tilde{\beta}_P$ is known as the percolation fractal dimensionality. The exponent β_P describes the probability $P \sim \Delta p^{\beta_P}$ that a point belongs to the infinite cluster when $p > p_c$.

Consider now the double sum $\sum_{r,r'} \Delta J_r \Delta J_{r'}$, taken over a volume ξ^d. This should go like the number of pairs of points in the percolation clusters:

$$\sum_{r,r'} \Delta J_r \Delta J_{r'} \sim \xi_P^{\,D} \sum_r C(r) \sim \xi_P^{\,2d-3\tilde{\beta}_P}. \tag{10}$$

On the other hand Weinrib and Halperin[22] have shown that if an *ordered* system has a correlation length exponent ν_0 and disorder is introduced with correlations in the local critical temperatures $T_c(r)$, satisfying

$$\sum_{r,r'} \Delta T_c(r) \Delta T_c(r') \sim \xi_P^{\,2d - b} \tag{11}$$

with $b < 2/\nu_0$, then the exponent of the *disordered* system will be $\nu = 2/b$. The proportionality between critical temperature and coupling justifies relating Eqs. (10) and (11) to obtain $b = 3\tilde{\beta}_P$ and

$$\nu = 2\nu_P/3\beta_P \simeq 1.33 \tag{12}$$

where we have used published values of percolation exponents[31]. We have not yet obtained the value of ν for ceramics. However, the fact that the β exponent appears to be the same as in "classical" bulk granular superconductors, together with the relations $2\beta = (d - 2 + \eta)\nu$ and $\eta \ll 1$, so $\nu \simeq 2\beta$, is a rather strong hint that $\nu \simeq 1.3 - 1.4$ also for ceramics.

4.2. Discussion.

The power laws characteristic of critical behaviour are known to be valid in a small interval near T_c, when $\xi \to \infty$. For ceramics, the existence of several levels of granularity may introduce further complications: ξ grows smaller as the temperature decreases, therefore probing correlations of shorter and shorter range. This is equivalent to an increase in b in Eq. (10); if we were able to predict the corresponding thermal behaviour, we would have a nice tool to explore the statistics of the microstructure.

Our argument invokes only randomness to obtain *one* exponent common to superconducting glasses and spin glasses. By virtue of Eqs. (8) a second one is needed to determine all the others and therefore the class of universality of these systems. For granular superconductors the situation compared with the ordered XY model is rather clear: randomness affects β, γ and ν but $\delta = (d + 2 - \eta)/(d - 2 + \eta)$ and the ratios $\tilde{\beta}$ and $\tilde{\nu}$ are universal.

REFERENCES.

1. Shih, W. Y., Ebner, C. and Stroud, D., Phys. Rev. B30, 134 (1984).
2. John, S. Lubensky, T., Phys Rev. Lett. 55, 1014 (1985).
3. Vinokur, V., Ioffe, L. B., Larkin, A. I. and Feigel'man, M. V., Zh. E. T. F., 93, 342, (1987) ; Mod. Phys. Lett. B1, 27 (1987).
4. Toulouse, G., Commun. Phys. 2, 155 (1977).
5. Müller, K. A., Takashige, M. and Bednorz, J. G., Phys. Rev. Lett. 58, 1143 (1987).
6. Morgestern, L., Müller, K. A. and Bednorz, J. G., Z. Phys. B69, 33 (1987).
7. Pannetier, B., Chaussy, J. and Rammal, R., Proceedings of LT 18, Japn. J. Appl. Phys. 26-3, 1994 (1987).
8. Pellan, P., Dousselin, G., Cortès, H. and Rosenblatt, J., Sol.St. Comm. 11, 427 (1972); Rosenblatt, J., Rev. Phys. Appl. 9, 217 (1974).
9. Rosenblatt, J., Raboutou, A. and Peyral, P., Proc. L.T. 14, edited by Krusius, M. and Vuorio M., North Holland, Amsterdam, 1975, Vol. 2, 361.
10. Raboutou, A., Rosenblatt, J. and Peyral, P., Phys. Rev. Lett. 45, 1035 (1980).
11. Peyral, P., Raboutou, A., Lebeau, C. and Rosenblatt, J., J. Phys. C20, L155 (1987).
12. Mandelbrot, B., "The Fractal Geometry of Nature", Freeman, San Francisco 1982.
13. Ambegoakar V., Baratoff, A., Phys. Rev. Lett. 10, 486 (1963).
14. Imry, Y. in "Percolation, Localization and Superconductivity", edited by Goldman, A. M. and Wolf, A. S., NATO ASI Series, Plenum Press, N. Y., 1984, p. 189.
15. Abeles, B., Sheng, P., Coutts, M. I. and Arie, Y. Adv. Phys. 23, 407 (1975); Simanek, E. Sol. St. Comm. 31, 419 (1979).
16. Rosenblatt, J., in "Percolation, Localization and "Superconductivity", Ref. 14, p. 431.
17. Bean, C. P., Phys. Rev. Lett. 8, 250 (1962); Rev. Mod. Phys. 36, 31 (1964).

18. Raboutou, A., Peyral, P., Rosenblatt, J., Lebeau, C., Peña, O., Perrin, A., Perrin, C. and Sergent, M., Europhysics Letters, 4, 321 (1987).

19. Peyral, P., Rosenblatt, J., Raboutou, A., Lebeau, C., Perrin, C. Peña, O. and Sergent, M., Proc. of the Interlaken Conference, to be published in Physica C, 1988.

20. Deutscher, G. and Müller, K. A., Phys. Rev. Lett., 59, 1745 (1987).

21. Giovannella, C., Chappert, C. and Beauvillain P. Europhysics Letters, 5, 535, (1988).

22. Weinrib, A. and Halperin, B. I., Phys. Rev. B27 (1983).

23. Le Guillou, J. C. and Zinn-Justin, J., Phys. Rev. Lett. 39, 95 (1977).

24. Rosenblatt, J., Raboutou, A., Peyral, P. and Lebeau, C., Proc. of LT18, Japn. J. Appl. Phys., 26, 1421 (1987); Physica, 152B, (1988).

25. Campbell, I. A., Phys. Rev. B37, 9800 (1988).

26. Ogielski, A. T., Phys. Rev. B32, 7384 (1985).

27. de Courtenay, N., Bouchiat, H., Hurdequint, H. and Fert, A., J. Phys. (Paris) 47, 1507 (1986).

28. Lundgren, L., Nordblad, P. and Svendlindh, P., Phys. Rev. B34, 8164 (1986).

29. Vincent, E. and Hammann, J., J. Phys. C. 20, 2659 (1987).

30. Rosenblatt, J., Lebeau, C., Peyral P. and Raboutou, A., "Josephson Effects: Achievements and Trends", edited by A. Barone, World Scientific, 1986, p.320.

31. Stauffer, D. Phys. Rep. 54, 1 (1979).

NOVEL TIME-EFFECTS IN THE MEISSNER REGION OF SUPERCONDUCTORS

A.C. Mota, P. Visani, A. Pollini, and G. Juri
Solid State Physics Laboratory, ETH, 8093 Zürich, Switzerland

The recently discovered high -T_c oxides Ba-La-Cu-O, Sr-La-Cu-O and Y-Ba-Cu-O show anomalous behaviour in their low field magnetization at $T << T_c$ [1,2]. Both for specimens in sintered or powdered form and for single crystals, we observe the following: if a field $H_i < H_{c1}$ is applied after zero-field cooling, the isothermal dc magnetization evolves in time rather rapidly following the law:

$$M(t) - M(t_0) \propto H_i^n \, f(T) \ln(t/t_0)$$

with $n = 3.5 \pm 0.5$ over observation times extending from 1 sec to 10^5 sec. The function f(T) is directly proportional to the temperature for $1\ K < T < 10\ K$ indicating that the relaxation process is thermally activated. For $T < 1\ K$, f(T) becomes almost independent of temperature.

The same type of law is followed by the remanent magnetization after the field H_i is removed. At a given temperature T and $H = 0$, the logarithmic rate $\partial M/\partial \ln t$ is determined by the value of the field, H_i, applied previously.

Similar time effects with even stronger values of $\Delta M/M$ per decade of time than for the high-T_c oxides, have been found in the heavy fermion superconductor $CeCu_2Si_2$ [3] and also in the organic superconductor ß-$(BEDT\text{-}TTF)_2I_3$ by us recently [4].

In order to understand such anomalous relaxation of the magnetization we have made investigations in several conventional type II superconductors. Although they also show logarithmic decays of the remanent magnetization at H = 0, the logarithmic rates are about two to three orders of magnitude smaller than those in the superconducting oxides and occur only after cycling to fields above H_{c1}.

REFERENCES

1. Müller, K.A., Takashige, M, and Bednorz, J.G., Phys. Rev. Lett. 58, 1143 (1987).
2. Mota, A.C., Pollini, A., Visani, P., Müller, K.A., and Bednorz, J.G., Phys. Rev. B36, 4011 (1987).
3. Mota, A.C., Visani, P., and Pollini, A., Phys. Rev. B37, 9830 (1988).
4. Mota, A.C., Visani, P., Pollini, A., Juri, G., and Jerôme, D., Physica C 153-155 , 1153 (1988).

FLUX PINNING IN HIGH T_c SUPERCONDUCTORS

P.H. Kes*, J. van den Berg*, C.J. van der Beek*, J.A. Mydosh*, L.W. Roeland+, A.A. Menovsky+*, K. Kadowaki+, and F.R. de Boer+.

* Kamerlingh Onnes Laboratory, State University Leiden, P.O.Box 9506, 2300 RA Leiden, The Netherlands.

+Physics Laboratory University of Amsterdam, P.O.Box 20215, 100HE Amsterdam, The Netherlands.

Magnetization measurements on single crystals of $YBa_2Cu_3O_7$ have been carried out in magnetic fields up to 38T both parallel to the c-axis and the basal plane. The critical current density and the volume pinning force F_p at 20.4K showed the same characteristics as obtained for earlier experiments above 65K. The field at which the irreversibility disappears (F_p=0) depends on temperature in a quite unexpected way, supporting the view of an extended irreversibility line down to 20K. The consequences for the upper critical field and the coherence length are discussed.

Experimental

In succession of magnetization experiments on an $YBa_2Cu_3O_7$ single crystal in fields up to 5T using a vibrating sample magnetometer [1], pulsed-field measurements were carried out in the 40T facility of the university of Amsterdam on crystals of the same batch. Integration of the voltage of a compensated and well-callibrated pick-up coil system requires a stacking of 10 crystals of typical dimensions 1.0x0.9x0.3mm^3 with their c-axes (along the smaller dimension) parallel [2]. Typical field-sweep rates ranged between 40 and 60T/s. In order to suppress heating due to flux flow, the sample was directly immersed in the cooling liquid. The magnetization curves were so measured at 1.3, 4.2, 20.4 and 77K. Trapped flux was removed after each pulse by heating to room temperature. The irreversible properties have been investigated with the field directed parallel as well as perpendicular to the basal plane (henceforth denoted by $H_{\parallel}$ and $H\perp$).

The critical current densities J_c and the bulk pinning forces $F_p=BJ_c$ were determined using the Bean model. For this purpose the dimensions $(l\times b\times h)_i$ of each crystal was measured. From the difference in magnetic moment Δm between the field-up and field-down curves, J_c follows from $J_c=\Delta m/\sum_i d_iV_i$. It has been assumed here that all crystals have the same J_c. The d_i still depend on the field orientation. For H$\|$ it is related to the thickness: $d_i\|=h_i/2$. For H$\perp$ it stands for an effective radius: $d_i\perp=\sqrt{(4l_ib_i/9\pi)}$. Since the Ginzburg-Landau parameter κ is large and the lower critical field is small, demagnetization effects may be ignored above 1T. It is for the same reasons and the fact that J_c does not much depend on B, that the rather crude Bean model may be successfully used. The heating effect due to flux flow may be estimated from the Bean model as well. The energy dissipated during a pulse is given by $U=E^2V_i\tau/\rho_f$ with τ the pulse duration (≈1sec), $\rho_f\approx\rho_nB/B_{c2}(0)$ the flux flow resistivity, and E=vB, where the velocity v at which the flux front moves follows from conservation of flux. Substituting typical values: $B_{c2}(0)/B\approx10$, $\rho_n\approx10^{-7}\Omega m$ the generated energy is about 3×10^{-7}J. Using the specific heat data of Von Molnar et al.[3] on untreated single crystals, the adiabatic temperature rise is estimated to be at most 3K at 4.2K and less than 40mK at 20K. Usually, the super-conducting properties are not very sensitive to temperature changes, if they occur at very low reduced temperatures. We therefore conclude that the pulsed-field data on J_c and F_p can be readily compared with the earlier results obtained with the vibrating sample magnetometer in Leiden.

Results and Discussion

In Fig.1 the J_c results at 2 and 4T for both field orientations are shown as a function of T. The pulsed-field data in H$\perp$ has been scaled down with a factor 0.35, since the Amsterdam experiment yielded a $J_c\perp$ at 20.4K as high as 1.1×10^{10}A/m^2 at 2T. This deviation may be due to differences in the J_c's of the individual crystals related to a difference in the densities of twin planes [4]. The ratio $J_c\perp/J_c\|$ for the Amsterdam results is 4.6, somewhat larger than the previous value of about 3 obtained in Leiden. Again this may be caused by the

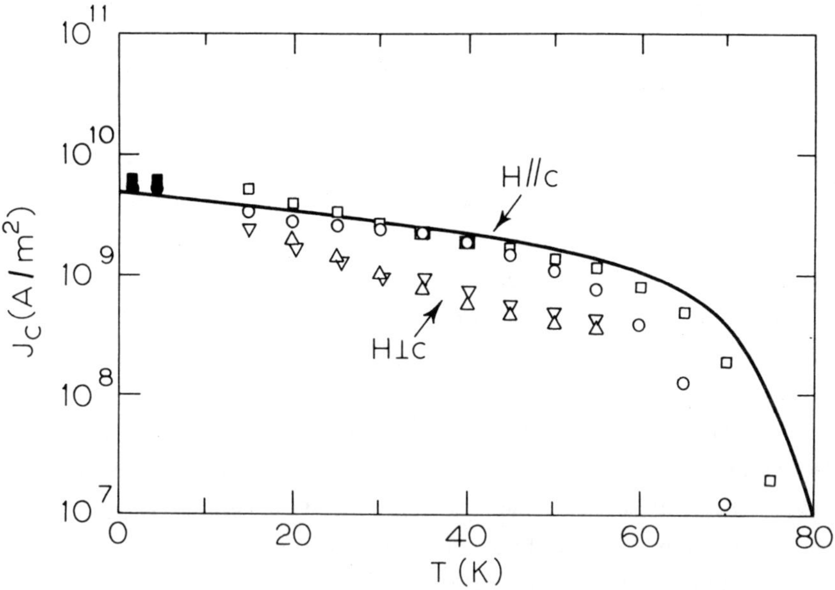

Fig.1. Critical current densities of YBaCuO single crystals vs. temperature at fields of 2T(□,∇) and 4T(○,Δ) directed ⊥ and ∥ to the c-axis. Solid symbols denote the pulsed-field data after scaling at 20K with the magnetometer results (open symbols).

difference in properties of each crystal. The drawn line in the figure displays a theoretical prediction for pinning by twin planes and H⊥ [4]. As stated in Ref.4 this line has been shifted down over a factor 3 for a better comparison with the Leiden data. As now follows from the Amsterdam results on the average of 10 crystals there would be a much closer agreement with the theory. However, this prediction depends on estimates of the scattering probability of electrons at a twin plane and some parameters of the superconductor that are still not well known, such as $B_c(0)$ and the BCS coherence length ξ_o.

All $F_p(B)$ curves at 1.3 and 4.2K and the H$\parallel$ result at 20.4K are almost linearly increasing functions of field up to 38T. The result for H$\perp$ at 20.4K is somewhat more interesting as it shows a maximum at bout 21T, see Fig.2. This behaviour resembles the characteristic field dependence observed above 65K on the Leiden single crystal [1]. For comparison the result of the 70K experiment is given in Fig.2 scaled in such a way that the maxima coincide. The behaviour above the maximum is quadratic in field. This suggests that the field at which the irreversibility disappears $B_o(F_p=0)$ for 20.4K can be about 63T. On the other hand $F_p(B)$ curves often decrease linearly with B as well. This extrapolation gives B_o=51T.

Generally, B_o is identified with the upper critical field B_{c2}. But this identification has recently been questioned by Malozemoff et al.[5]. They argue that due to "giant" flux creep effects, which especially occur in the high temperature superconductors [6], $B_o(T)$ rather indicates the "irreversibility line". Fig.3 displays all $B_o(T)$ data collected for the same batch of single crystals. The inset shows a

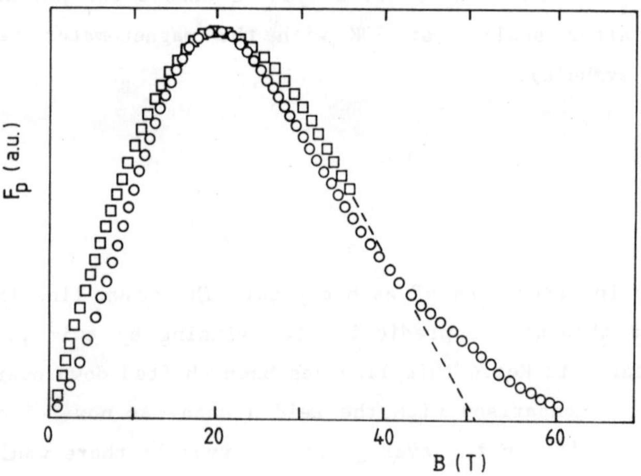

Fig.2. Comparison of $F_p(B\perp)$ results at 20.4K (□) and 70K (○). The latter are scaled up so that the maxima of both curves coincide.

blow-up of the same data close to T_c. Included are the results of resistance measurements by Wijngaarden et al.[6], both the zero-resistant points and the extrapolations $B_N(T)$ to the normal state resistivity line. The latter data remarkably coincides with H_{c2} values obtained from sensitive magnetization measurements on oriented crystals by the Ames group [7].

The two lines in the figure indicate the theoretical $H_{c2}(T)$ adapted to both B_o values at 20K [8]. The same lines are also drawn in the inset. The coherence length that would follow from these lines is either 1.8 or 2.0nm. It is clear, however, that either the conventional theory is not applicable to the high temperature superconductors, or that B_o is not the upper critical field, since all data above 65K lies

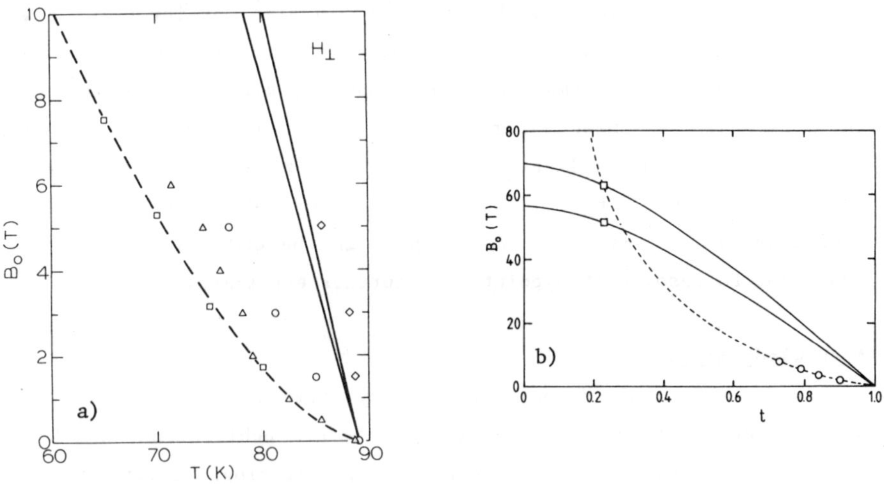

Fig.3. $B_o(T)$ data in H⊥ for YBaCuO single crystals of the same batch both just below T_c (a) and for all reduced temperatures t (b). Drawn lines: theory for conventional superconductors [8] adapted at 20K data. Dashed line: $B_{irr}(T)$ adapted at the high temperature data. Symbols refer to B_o(□), ac-susceptibility (100Hz) anomalies (Δ), and resistance vs. field data: (O) zero resistance, (◊) transition point determined by extrapolation to normal state resistance line.

well below the theoretical curves. This is especially worth noting for the zero resistance points.

The analysis can also be given in terms of the irreversibility line which would obey [5]: $B_{irr}(t) \propto (B_c^2(t)\xi(t)/t)\ln(f_o/f)$. The dashed line represents this behaviour with the proportionality constant determined at the four high-temperature data. It should be noted that the frequency f, typical for the pulsed-field measurements, is about 300 times larger than for the magnetometer. Taking this into account would shift the 20K data down by about 25 per cent to just under the irreversibility line. Nevertheless, the agreement with this interpretation of B_o is not too bad. On the other hand, one should keep in mind that a firm theoretical basis for it is still lacking.

Assuming that B_o rather represents B_{irr} than B_{c2}, means that we still don't know $B_{c2}(T)$ for H⊥ and probably also not for H∥. The wide resistance transitions in a field already indicated that flux-flow resistance due to thermal activated hopping might hamper the right interpretation of these experiments. A very tentative, but, if true, also very important implication would be that rather B_N represents B_{c2}. From the slope of $B_N(T)$ a coherence length of 0.8nm would follow indicative for a real-space Cooper-pair in the CuO planes. It is very important to resolve this point with further experiments.

Acknowledgements

We like to thank A.P. Malozemoff for very stimulating discussions, T.J. Wijngaarden for sending us his results before publication, and M.J.V. Menken for his assistance in the preparation of single crystals.

References

1. J. van den Berg, C.J. van der Beek, P.H. Kes, J.A. Mydosh, A.A. Menovsky and M.J.V. Menken, Proceedings MMS and HT_cS, Interlaken; Physica B., to be published.
2. L.W. Roeland, P.H. Kes, Li Jang-Ning, K. Kadowaki, A.A. Menovsky, Huang Ying-Kai, J. v.d. Berg, C.J. v.d. Beek, M.J.V. Menken, and F.R. de Boer, Contribution to the Symposium on High Field Magnetism, Leuven, July 1988.

3. S. von Molnar, A. Torressen, D. Kaiser, F. Holtzberg, and T. Penney, Phys. Rev. B37, 3762 (1988).

4. P.H. Kes, Ibidem Ref.1, to be published.

5. A.P. Malozemoff, T.K. Worthington, Y. Yeshurun, F. Holtzberg, P.H. Kes, Phys. Rev. B, Rapid Communications, to be published.

6. R.J. Wijngaarden, K. Heeck, R. Griessen, A.A. Menovsky, and M.J.V. Menken, Ibidem Ref.1, to be published.

7. M.M. Fang, V.G. Kogan, D.K. Finnemore, J.R. Clem, L.S. Chumbley, and D.E. Farrell, Phys. Rev. B37, 2334 (1988); L.A. Schwartzkopf, M.M. Fang, L.S. Shumbley, and D.K. Finnemore, Ibidem Ref.1, to be published.

8. N.R. Werthamer, E. Helfland, and P.C. Hohenberger, Phys. Rev.147 295 (1966).

MÖSSBAUER STUDIES OF ^{57}Fe SUBSTITUTION OF Cu IONS IN SUPERCONDUCTING OXIDES

E. Baggio Saitovitch
Centro Brasileiro de Pesquisas Físicas
Rua Dr. Xavier Sigaud, 150, Urca
22290 Rio de Janeiro, RJ, Brasil

Since the discovery of high-T_c superconductivity in $YBa_2Cu_3O_7$ oxides several studies of metal ions substitutions were reported. The observed depression on T_c without a systematic correlation with the charge and magnetic moment of Cu substituents claims for more detailed information about its local properties as can be revealed by ^{57}Fe Mössbauer spectroscopy. The results for different iron concentrations combined with the modifications of the superconducting transition are discussed concerning the presence of magnetic moments on the Fe ions and the preferential occupation of Cul sites, recently confirmed by neutron and electron diffraction experiments. The oxygen coordination for the different iron species are proposed on the basis of their dependence on Fe concentration, their behavior at high temperature as well as the electron diffraction and electron microscopy measurements reported for $Fe:YBa_2Cu_3O_7$ samples.

1. INTRODUCTION

Since the discovery of high T_c superconducting oxides intense efforts have been done to achieve a better understanding of the special properties of these oxides which could be responsible for superconducting materials. From the several theoretical models proposed until now it is still not possible to identify clearly which mechanism is the dominant interaction contributing to pairing formation.

From the phase diagrams of Y-Ba-Cu-O and La-B-Cu-O (B = Ca, Sr, Ba) the compositions of $YBa_2Cu_3O_7$ and $La_{1.85}B_{.15}CuO_4$ were established as responsible for the T_c values of 90 and 30-40K respectively. The effect of various substituents have been investigated mainly in the $YBa_2Cu_3O_7$ oxides in an attempt to enhance T_c further on. All the trivalent rare earths which substitute the Y ions do not induce modifications in the superconducting behavior(1). The same is not true for the Cu substituents and several papers(2) described the results obtained for various metals, which in principle should occupy the Cu sites with, in some cases, dramatic effect on T_c[2]. This depression of T_c has no clear systematic correlation with the charge and magnetic moments of Cu substituents. However, the Cu substitution must be analysed very carefully, since it can induce modification on the oxygen content of the $YBa_2Cu_3O_7$ samples which has been widely proved to be essential for the H-T_c values.

Additional information for a better understanding of how the Cu substituents perturb the superconducting behavior of H-T_c oxides can be achieved from hyperfine interactions methods like Mössbauer effect and perturbed angular correlations since they reflect the microscopic charge distribution at the substituent nucleus. However, this methods are restricted to some substituents which present appropriated isotopes. The combination of such experiments with usual macroscopic characterization methods can provide important pieces of information. Here we report the contribution obtained from Mössbauer spectroscopy to the understanding of Cu substitution by Fe in the superconducting oxides.

Since the first Mössbauer results were presented in the International Conference on Application of Mössbauer Effect (ICAME), in Melbourne, August 1988[3], different interpretations appeared. The main agreement corresponds to the magnetic patterns for the Mössbauer spectra of Fe:$YBa_2Cu_3O_7$ at 4.2K which indicate that iron has a magnetic moment. The presence of two crystallographic sites for copper, namely Cu1, with square planar oxygen coordination, and Cu2, with pyramidal coordination, together with the high sensitivity of ^{57}Fe to the oxygen content of the samples, induced many authors to analyse the Mössbauer spectra with two main asymmetric doublets[4] and propose, as wrong assignement, that the one with Quadrupole Splitting (QS)$\cong$2.0mm/s, is due to the occupation of Cu2 sites[5]. Different interpretations are now emerging from more carefull experiments and in the recent papers[6] the number of iron species has increased. A general agreement established QS $\cong$ 2.0mm/s to be characteristic of iron species in the oxygen deficient phase, converging to our previous results[7].

2. EXPERIMENTAL

The $YBa_2(Cu_{1-x}Fe_x)_3O_7$ samples preparation has been previously reported(7). Several treatments in O_2 flux at 500C were performed to achieve the best oxygen content for the samples. The X-ray analysis was made to determine the sample crystal structure.

The superconducting behavior was observed by DC magnetization measurements with vibrating sample magnetometer for sample cooled in zero magnetic field(7).

The Mössbauer spectra (MS) for several x values were obtained at 4.2K and 293K with $^{57}Co/Rh$ source, keeping the same temperature for source and absorber. The 293K MS were fitted assuming always doublets with symmetrical pairs of lines, while the 4.2K MS display complex magnetic splitted pattern which demand more elaborated fitting(8).

3. MÖSSBAUER RESULTS FOR DIFFERENT IRON CONCENTRATIONS

The MS obtained for the $YBa_2(Cu_{1-x}Fe_x)_3O_7$ samples at room temperature exhibit different results for x above and below some critical value x ≅.01. In the low concentration limit, x = .001, .005, .01, shown in Fig. 1, the MS present mainly three doublets (A, B, C) corresponding to iron species with isomer shifts (IS) of ≅ .18mm/s, .03mm/s, .20mm/s and quadrupole splittings (QS) of 1.56mm/s, 1.35mm/s, .30mm/s, respectively. A fourth iron species (D) with IS = .05mm/s and QS = 1.97mm/s, characteristic of the $Fe:YBa_2Cu_2O_6$ insulator oxide, may be present in small proportion as a signature of an improperly

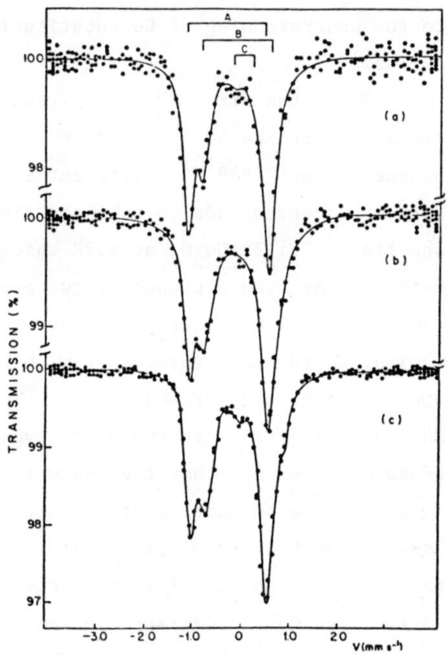

Fig.1: Mössbauer spectra of $YBa_2(Cu_{1-x}Fe_x)_3O_7$ at room temperature: a) x=.001, b) x=.005 and c) x=.01.

formed compound, even if the X-ray diffraction indicates a pure orthorhombic phase. The Mössbauer spectroscopy, in this case, is more sensitive than the X-ray analysis, since it can show the presence of oxygen deficient islands which are not big enough to be displayed as a crystallographic phase in the X-ray spectrum. The doublets A and B, which dominate the MS at low iron concentration, are sensitive to the sample oxygen content: they both transform reversible to D when the reduced oxide in the tetragonal structure is prepared either by quenching or annealing in Ar atmosphere[(7)]. This leads to the conclusion that Fe occupies preferentially a Cu site which is very sensitive to the O_2 content, namely Cul site[(7)]. This result has been recently confirmed by neutron and electron diffraction experiments[(9,10)]. The Fe species associated with A, B and D are associated with the occupation of Cul site albeit with different oxygen environments. However, the assignement of C is still to be established.

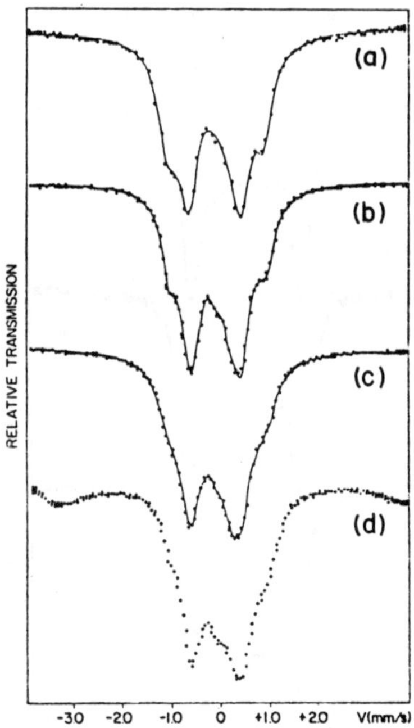

Fig. 2: Mössbauer spectra of $YBa\,(Cu_{1-x}Fe_x)_3O_{7+\delta}$ at room temperature: a) x=.10, b) x=.15, c) x=.20 and d) x=.25.

The increase of iron concentration produces modifications in the ^{57}Fe MS as can be seen in Fig. 2 compared with the Fig. 1. The spectra become more symmetrical due to the absence of the species corresponding to A related to the low iron concentration. Despite the fact that the samples with $.10 \leq x \leq .25$ exhibit similar MS spectra, the line width corresponding to B, with reduced QS ($\cong$ 1.1mm/s), increases as well as the relative intensity of C. The MS for x = .25 (Fig. 2d) indicate the precipitation of a magnetic phase in agreement with the X-ray analysis which show the presence of extra reflection lines for $x \geq .20$, establishing the limit of solubility of Fe in $YBa_2(Cu_{1-x}Fe_x)_3O_7$. The preferential occupation of Cu1 site by iron may be modified at higher concentration mainly if we consider that Cu2 site is two times more abundant than Cu1 site. The iron species corresponding to Cu2 site should have a different electronic configuration and consequently produce new absorption lines in the MS. However, they were not observed and consequently the possibility of such occupation needs to be discussed considering Fe species corresponding to C, or similar to B.

In order to shed some ligth on this question Mössbauer studies were performed on $YBa_2(Cu_{1-x}Fe_x)_3O_{6+\delta}$ samples, since the Cu2 site is less sensitive to the oxygen content of the sample. The MS at 458K obtained for several Fe concentrations are shown in Fig. 3. They indicate that besides D, two new doublets E and F, with QS of ≅.70mm/s and .50mm/s, are present. The species corresponding to B and C disappeared. Thus it becomes difficult to associate them to occupation of Cu2 site. F develops a magnetic spectrum at low temperature.

From these experimental results the doublets A, B and C are assumed to be due to Fe species having higher oxygen coordination present in the orthorhombic structure of $Fe:YBa_2Cu_3O_7$. In order to test their relative stability the MS were obtained in air at high temperatures. An upper limit of 300C was taken since, even in air, at this temperature D was formed, probably due to the loss of oxygen.

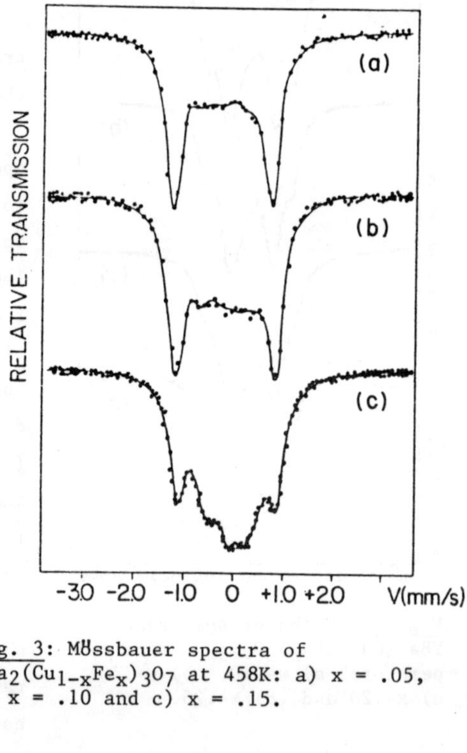

Fig. 3: Mössbauer spectra of $YBa_2(Cu_{1-x}Fe_x)_3O_7$ at 458K: a) x = .05, b) x = .10 and c) x = .15.

The MS obtained in air for sample with x = .005 at the indicated temperatures are shown in Fig.4. From these data it is possible to determine that

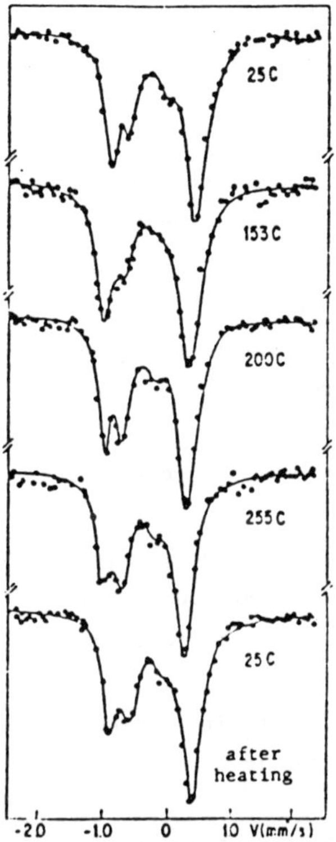

Fig. 4: Mössbauer spectra for $YBa_2(Cu_{.995}Fe_{.005})_3O_7$ at the indicated temperatures.

the relative intensity of B increases at the expenses of A and that this effect is partially reversible with temperature[12]. The behavior of A is similar to that observed for the twin boundary in pure $YBa_2Cu_3O_7$ by TEM[13]. The observation of twin boundary in several small single crystals from samples with x = .05 by electron microscopy[9] indicate that they are only present in the specimens with reduced iron concentration. This corresponds also to the concentration dependence of A which disappears at concentrations $0.01 \leqq x \leqq .05$, depending on preparation condition.

4. POSSIBLE MICROSCOPIC STRUCTURE FOR DIFFERENT IRON SPECIES IN $YBa_2(Cu_{1-x}Fe_x)_3O_7$ OXIDE

The understanding of the modification induced by iron substituents in the orthorhombic structure of $YBa_2Cu_3O_7$ as well as its different microscopic structures may be achieved by combining Mössbauer results with additional structural informations obtained by EXAFS, neutron X-ray and electron diffraction, electron microscopic and specific heat measurements. The critical Fe concentration, which stabilizes the tetragonal structure for $YBa_2(Cu_{1-x}Fe_x)_3O_7$, may have different values depending on the preparation condition.

The crystallographic studies reported for samples with x = .05[9] and x = .10[10] revealed the preferential occupation of Cu1 sites by iron, without an agreement concerning the occupation of Cu2 sites to be zero. The combination of X-ray, neutron and electron diffraction with electron microscopy measurements in several small single crystals revealed the presence of orthorhombic microdomains even when an overall tetragonal structure is observed[9]. Only when domains predominate in the single crystals, the twin boundaries are observed. Structural studies and carefull analysis of electron microscopy for more concentrated samples (x = .10) allow to propose different oxygen coordinations for Fe ions and the formation of linear clusters separated by 10 to 20 Å domains of the orthorhombic structure with reduced iron concentration[10]. Moreover, the increase of oxygen content with iron concentration is consistent with the EXAFS results which suggest that iron has mainly five oxygen neighbors[14].

Recently a very careful study was reported for samples with $0 \leqq x \leqq .04$ where specific heat measurements were combined with all the structural techniques already mentioned[15]. The critical iron concentration which induces tetragonal structure gives origin to a non homogeneous superconducting phase despite there is no iron segregation. The twinning observation depends also, as reported in references[9], from the iron concentration in the grains.

The ensemble of above results may lead to several conclusions, but the main one concerns with the homogeneity of iron distribution and the oxygen content. These two parameters can be responsible for the discrepancies obtained from the studies performed in samples with the same nominal concentration either by Mössbauer spectra or X-ray diffraction. Consequently it is worthwhile to study, in more detail, samples from independent batches with the same nominal concentration. The analysis with complementary techniques from microscopic to macroscopic level on the same sample is essential for the understanding of structural changes induced by iron.

In the following of this section we will propose microscopic assignements for some of the several iron species observed by Mössbauer spectroscopy, mainly based on the dependence of their relative intensities on iron concentration and oxygen content as well as the structural informations described above. Even for a fixed iron concentration several configurations may occur for Fe in Cu1 site mainly due to disorder in the occupation of oxygen sites in the Cu1 plane established at the preparation conditions; the Fe coordination can range from 2 to 6 (Fig. 5) leading to different Mössbauer doublets.

The increase of oxygen content with Fe concentration may be due to progressive occupation of the oxygen vacancies resulting in the transition to the tetragonal structure. The assignement which we proposed to A[12], namely the occupation of twin boundary, is based on the temperature behaviour and the x dependence of this iron species. Its relative intensity depends on the preparations as observed for the twinnings, and both disappear when the tetragonal structure is established. In this case we consider A to be the species most characteristic of the orthorhombic phase and which may have the configuration shown in Fig. 5d.

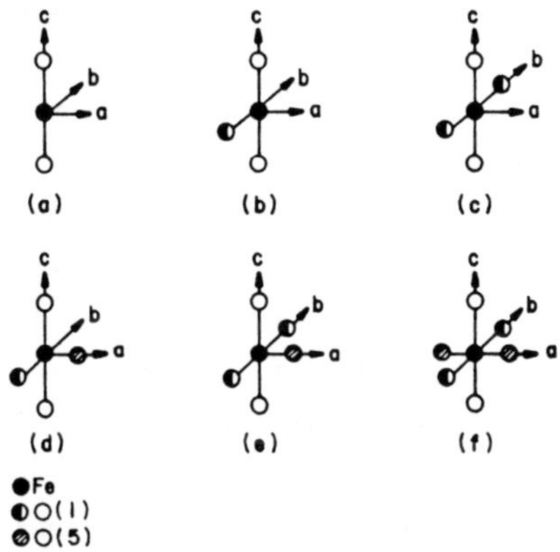

Fig. 5: Local environment of Fe at Cu1 site.

The doublet B dominates the MS for higher iron concentration, when the tetragonal structure is stabilized, and remains even for concentrations as high as x = .25. This species reflects a modification in the local structure of Cu1 site as compared with the pure $YBa_2Cu_3O_7$. It is compatible with the configuration shown in Fig. 5e with an extra oxygen in Cu1 plane resulting in the observed increase of oxygen content with iron Similar configuration has been reported from EXAFS measurements on $YBa_2(Cu_{1-x}Fe_x)_3O_7$ samples[14].

The configurations assigned to the Fe species associated with A and B need to be consistent with the Mössbauer results at high temperatures since, as can be seen in Fig. 4, the relative intensity of B increases at expenses of A. Indeed the configuration attributed to A can transform in to B if an oxygen atom from a neighboring ion hopps to the

O(1) site and completes the 5th coordination as is shown in Fig. 5e.

The electron diffraction results[10] indicate the formation of iron clusters rich in oxygen content which could include an octahedral coordination for Fe in Cu1 site as shown in Fig. 5f. This configuration has high local symmetry and its relative intensity should be small and increase with Fe concentration. It may be associated with C which displays similar features and, moreover, is absent in the MS of reduced samples.

The species corresponding to Fig. 5e, d and f are related to samples having high oxygen content. In the process of preparing the reduced sample disorder may occur in the occupation of oxygen sites in the Cu1 plane which will lead to several other configurations. In such reduced samples with low iron concentration the MS is mainly formed by the doublet D, which can be attributed to axial coordination exhibited in Fig. 5a, similar to what is observed for Cu in $YBa_2Cu_3O_6$, or to configurations shown in Fig. 5b, if Fe binds an extra oxygen. This seems to be consistent with EXAFS measurements reported for Fe: $EuBa_2Cu_3O_6$ samples[16].

There are still two iron species, E and F, which are present in reduced samples whose relative intensities increase with iron concentration (see Fig. 3). It is possible that the configuration proposed for E and F need to take into account not only Fe-oxygen coordination in Cu1 site but also the configuration of Fe and Cu as next neighbors. In this case more structural informations are necessary to understand them. However, the magnetic ordering temperature for F may indicate that it corresponds to occupation of Cu2 or iron clustering at Cu1 plane.

5. THE INFLUENCE OF IRON CONCENTRATION ON THE SUPERCONDUCTING TRANSITION OF $YBa_2(Cu_{1-x}Fe_x)_3O_7$

It is well established that Fe substituents depress T_c in $YBa_2Cu_3O_7$ compounds; initial studies compared 10% of Cu substitution by several transition metals Fe, Co, Ga and Zn. The latter induces the strongest effect[2] on T_c.

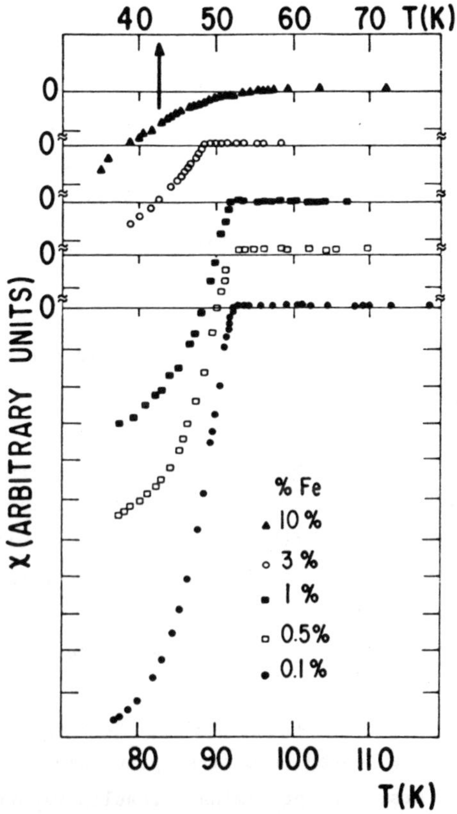

Fig. 6: Temperature dependence of the magnetic susceptibility χ for different Fe concentrations. The upper scale refers to the x = .1 sample.

Fig. 6 shows how the superconducting transition, as seen by DC magnetization measurements[(7)], depends on iron concentration. The T_c onset stays constant until x = .01, but the fraction of Meissner effect decreases. Above x = .01 this last effect is enhanced and the T_c onset decreases further down until being supressed for x=.15. These two effects should be analysed with the information obtained from Mössbauer spectroscopy and X-ray diffraction. Since the magnetization measurements give a bulk information this result is an integration of part of the sample where no iron is present and parts where it gradually depresses T_c, resulting in a broadening of the transition. The observation of superconductivity until x = .10 is in agreement with the X-ray analysis of our sample which still sees part of the orthorhombic structure; only with x = .15 the tetragonal structure is established and no superconducting transition is observed. Consequently, as long as considerable fraction of the sample has an orthorhombic structure, T_c may still be observed. The measurements of T_c for samples which have already pure tetragonal structure, induced by iron substitution, may be understood on the basis of crystallographic studies[(9)] which indicate that this structure contains orthorhombic microdomains responsible for finite T_c values.

Consequently, the iron substitution always perturbs the superconducting transition, but completely depresses T_c only when the concentration stabilizes the tetragonal structure.

6. CONCLUSION

The combination of Mössbauer results with several structural studies performed on $YBa_2(Cu_{1-x}Fe_x)_3O_y$ samples help to propose microscopic configurations for the several iron species observed in the Mössbauer spectra. This technique has proved to be very sensitive to the quality and oxygen content of the samples. Unhappily the complete structural studies were not performed on the same samples used for Mössbauer spectroscopy and their extreme dependence on preparation condition must be taken into account for the validity of our interpretation.

The observed preferential occupation of Cu1 site may depend on the condition of the sample preparation, and we believe that this question needs further studies, because either no Fe is present in Cu2 sites or it has similar hyperfine parameters as B since the line width of this doublet becomes broad with increasing Fe concentration. This, however, is inconsistent with the Fe species observed in the reduced samples[17].

The validity of the microscopic configuration for Fe in $YBa_2Cu_3O_y$ may be tested from first principles calculations of the hyperfine parameters, like preliminary results reported in this issue[18]. The fact that the calculated isomer shift is in good agreement with the experimental values is an indication of the importance of d electron delocalization, through interaction with oxygen p electrons. Consequently the ionic configuration model seems not to be appropriate for Fe in this compounds. The local d electron contribution to the electric field gradient (EFG) is too small compared with the observed values for the quadrupole interaction. This indicates that the lattice contribution to the EFG at Fe nucleus is important and may explain the good agreement reported between the experimental quadrupole interaction and the values obtained from point charge calculations[19].

It is still puzzling why the isomer shift of several iron species, except A and F, stay in a very narrow range. One possible explanation is that only A reflects the metallic character of $YBa_2Cu_3O_7$ while the others species (B,C,D,E) correspond already to the semiconducting phase induced by Fe or by the reduced oxygen content. In order to confirm this explanation detailed calculations for the several configurations are necessary.

The combination of all structural, specific heat and DC magnetization measurements confirmed the early Mössbauer results[9], which indicated that above x = .01 the Fe substitution perturbs the orthorhombic structure of $YBa_2Cu_3O_7$ and proper studies should be done below this value.

With respect to the superconductivity we may conclude that iron always perturbs the transition, probably due to the modification in the local structure associated with the occupation of the oxygen vacancies in O(5) sites. The long range order of the oxygen vacancies is necessary to establish high T_c values in $YBa_2Cu_3O_7$ oxide and the Cul substituition by iron supresses it with a dramatic effect on superconductivity.

Acknowledgements

We would like to thank Drs. I. Souza Azevedo, R.B. Scorzelli, C.A. dos Santos and D. Ellis for fruitfull discussions during the preparation of this paper. We gratefull acknowledge valuable discussion with Prof. H. Micklitz, during his stay with us, and his carefull reading of the manuscript.

References

(1) H.A. Borges, R. Kwok, J.D. Thompson, G.L. Wells, J.L. Smith, Z.Fisk, and D.E. Peterson, Phys. Rev. B36,2404(1987).

(2) Y. Maeno, T. Tomita, M. Kyogoku, S. Awaji, Y. Aoki, K. Hoshino, A. Minami and F. Fujita, Nature 328,512(1987); J.M. Tarascon, P.Barboux, P.F. Miceli, L.H. Greene, G.W. Hull, M. Eibschutz, and S.A.Sunshine, Phys. Rev. B37,5932(1988).

(3) In: Proceedings of the International Conference on the Applications of the Mössbauer Effect ICAME (Melbourne, August 1987), Hyperfine Interactions 42,1221-1258(1988).

(4) E.R. Bauminger, M. Kowitt, I. Felner, and I. Nowik, Solid State Comm. 65,123(1988).

(5) Z.Q. Qiu, Y.W. Du, H. Tang, J.C. Walker, W.A. Bryden, and K.Moorjani, J. Magn. Magn. Mat. 69,L221(1987); Q.A. Pankhurst, A.H. Morrish, and

X.Z. Zhou, Phys. Lett. A127,231(1988); T. Tamaki, T. Komai, A. Ito, Y. Maeno, and T. Fujita, Solid State Comm. 65,43(1988).

(6) C. Blue, K. Elgaid, I. Zitkovsky, P. Boolchand, D. McDaniel, W.C.H. Foiner, J. Oostens, and W. Huff, Phys. Rev. B37,5905(1988).

(7) E. Baggio-Saitovitch, I. Souza Azevedo, R.B. Scorzelli, H.Saitovitch, S.F. da Cunha, A.P. Guimarães, P.R. Silva, and A.Y. Takeuchi, Phys. Rev. B37,7967(1988).

(8) E. Baggio-Saitovitch, J.F. Litterst, I. Souza Azevedo, and R.B. Scorzelli, in this issue.

(9) G. Roth, G. Heger, B. Renker, J. Pannetier, V. Caignaert, M.Hervieu, and B. Raveau, Z. Phys. B - Cond. Matt. 71,43(1988).

(10) P. Bordet, J.L. Hodeau, P. Strobel, M. Marezio, and A. Santoro, Solid State Comm. 66,435(1988).

(11) E. Baggio-Saitovitch, I. Souza Azevedo, and R.B. Scorzelli, in this issue.

(12) E. Baggio-Saitovitch, I. Souza Azevedo, and R.B. Scorzelli, in Proceedings of the 5th JIMIS (Kyoto March 1988), to appear in Transactions of Japan Institute of Metals, 1988.

(13) N. Yamamoto, Y. Hirotsu, Y. Murata, O. Tomita, M. Takata, and S. Nagakura, in Proceedings of the 5th JIMIS (Kyoto March 1988), to appear in Transactions of Japan Institute of Metals, 1988.

(14) J.B. Boyce, F. Bridges, T. Claeson, R.S. Howland, T.H. Geballe, and M. Nygren, in Proceedings of the Conference on High T_c superconductors and Materials and Mechanisms of Superconductivity (Interlaken, March 1988), to be published in Physica C, 1988.

(15) A. Junod, A. Bezinge, D. Eckent, T. Graf and J. Muller, to be published in Physica C, 1988.

(16) K. Donnelly, J.M.D. Coey, S. Tomlinson and J.M. Greneche, to be published.

(17) The Fe species corresponding to F has been observed by many authors in the Fe:$YBa_2Cu_3O_7$ samples and attributed to the occupation of Cu2 site. However, F is only present in some of our reduced samples. In order to be consistent with that attribution Fe would have to migrate from Cu1 site to Cu2 site when the sample is annealed in Ar at 600C.

(18) D. Ellis, in this issue.

(19) E.R. Bauminger, D. Edery, I. Felner, M. Kowitt, Y. Lehavi and I. Nowik, in Proceedings of the Third Seeheim Workshop on Mössbauer Spectroscopy (Seeheim, May 1988).

SELF-CONSISTENT CHARGE AND SPIN DISTRIBUTIONS IN OXIDE SUPERCONDUCTORS

D. E. Ellis, Northwestern University, Evanston, IL 60208, and Centro Brasileiro de Pesquisas Fisicas, Rio de Janeiro

D. J. Lam and B. W. Veal, Argonne National Laboratory, Argonne IL 60439

Abstract

The self-consistent field local density theory provides a useful first principles theoretical basis for the description of electronic charge and spin distributions in complex materials. The ground state energy levels and charge and spin densities of copper-oxide based superconductors are discussed in terms of both band structure and molecular cluster models. X-ray photoemission and absorption data are considered, along with charge transport properties , to provide essential data necessary to understand basic bonding and excitation mechanisms. An embedded cluster approach is used to represent cluster-host interactions, with long-range terms included by Ewald summation. X-ray absorption near-edge spectra (XANES) are calculated for the Cu K-edge and compared with experiment in an effort to identify different valence states assumed to be involved in superconductive electron-pairing in certain models. The effects of probable defects and impurities on the electronic structure, especially those due to transition metals on substitutional sites, are calculated in the cluster scheme. Electric field gradients and isomer shifts are determined at Fe- substituted Cu sites for comparison with Mossbauer data and with the unperturbed host Cu sites.

I. INTRODUCTION

The oxide superconductors belong to a larger class of ceramic materials which display a variety of interesting electrical, magnetic, optical and mechanical properties. Although they have been intensively studied, the transition metal oxides have proved to be difficult to understand in detail, due in part to structural complexity and in part to the d-electron correlations which "interfere" with simple electronic structure descriptions. For example, the self-consistent field local density theory has evolved as a highly successful model of electron distributions with applications to atomic, molecular, crystalline, and amorphous materials.[1,2] However, the averaging over atomic configurations which is carried out to generate an effective potential can suppress the intraatomic Coulomb and exchange correlations which are critical for the excitations of the many electron system. The question of local correlations is critical in oxide superconductors: thus conductivity and thermopower measurements at elevated temperatures can be nicely explained by small polaron conductivity models which, by fitting to experiment yield definite predictions of the concentration of Cu^{+1}, Cu^{+2} and Cu^{+3} species as a function of temperature and oxygen stoichiometry[3] Spectroscopic measurements at room and low temperatures however have difficulty in verifying the presence of the trivalent species in any significant concentration. Indeed, both band structure[4-9] and cluster calculations[10-13] on ground state properties and low -lying excitations typically suggest a kind of average valence $d^{9+x}s^{y}p^{z}$ different from that obtained by traditional charge-balance arguments. This puzzle has been the subject of numerous experimental and theoretical studies over the past year, but is still not completely resolved.

Band structure calculations on the oxides have had a mixed success; in the semiconducting and insulating compounds a very acceptable ground state charge and spin density is generally found. On the other hand, the fundamental excitation gap tends to be seriously underestimated, for reasons which are now understood. Cluster calculations can be carried out in a variety of theories - semiempirical models such as the extended Huckel scheme[12], local density approaches[10,11], and still more rigorous valence bond and configuration interaction methods[13] have been used. There is naturally a relation between the complexity of the theory used and the size of the cluster treated, so boundary conditions used and treatment of the interaction between the finite cluster and infinite host become important issues. By construction, cluster models are well suited to description of localized interactions and excitations; in this brief report we will concentrate on the use of cluster models in the interpretation of local bonding and excitation mechanisms.

X-ray photoemisssion spectra (XPS) in the copper L-edge region are presented, and discussed in terms of the presence or absence of monovalent, divalent, and trivalent states of Cu invoked in models of superconductive electron pairing. The corresponding Cu K-edge absorption spectra (XANES) are discussed and compared with calculated cross-sections from the same perspective, and indications of the strong mixed ionic-covalent interaction between copper and its oxygen neighbors are extracted. The effects of probable defects and impurities on the electronic structure, especially those due to transition metals on substitutional sites are calculated in the cluster scheme, using local density theory. A preliminary report is given of electric field gradients and isomer shifts determined at Fe-substituted Cu sites for comparison with Mossbauer data and with the unperturbed host Cu sites.

II. EXPERIMENTAL PHOTOEMISSION SPECTRA

X-ray photoemission (XPS) measurements were taken with a Hewlett Packard 5950A photoelectron spectrometer. The superconducting oxides samples were cleaved or scraped in a vacuum of about 10^{-8} Torr and were then inserted directly into the measurement chamber where the vacuum was approximately 1×10^{-9} Torr. Cu 2p photoelectron spectra are shown, in Fig. 1, for the superconductor $La_{1.85}Sr_{0.15}CuO_4$, where they are compared with corresponding spectra for CuO and Cu_2O. Data for $YBa_2Cu_3O_{7-x}$ and $Bi_2Sr_2CaCu_2O_x$ are given in Fig.2. The strong similarity of the spectral features for the superconducting compounds indicates that, like the standard compound CuO, Cu is predominantly divalent in all of these oxide superconductors (with the $3d^9$ ground state configuration). This result has also been noted by several other investigators[14-19].

If it is assumed that oxygen occurs as doubly charged O^{2-} ions, then charge balance arguments would suggest that a substantial Cu^{3+} admixture should also be present. The inability of the spectroscopic techniques to reveal the presence of Cu^{3+} suggests that charge compensation might occur at the oxygen sites with formation of an appropriate concentration of single charge O^- ions.

The remarkable strong satellite structures that are observed in the oxide samples are quite characteristic of cations with unfilled shells that are constituents of insulating compounds. The satellites are generally considered to be attributable to final states of the core-ionized atoms that are sufficiently metastable to be observed in the photoemission process. Such stable final states of 3d elements are rarely observed in metallic systems. In Fig. 1 , for example, the Cu_2O spectra show weak satellites which vanish entirely for Cu metal.

The intense satellites in the oxides suggest that Cu ions show both the fully screened ($3d^{10}$) final state of the core-ionized Cu atom and the $3d^9$ configuration (also see Refs. 14 and 17). The inability of the system to select only the lower energy screened final state normally indicates that the local (Cu) environment is not metallic. Alternatively, if the 3d levels in the oxides are significantly more localized than in Cu metal, poor screening of the core hole could also occur, as is believed to be the case in Ce metal. Cluster calculations of the ionization states have been undertaken to investigate this possibility.

As the oxygen stoichiometry is reduced, the $YBa_2Cu_3O_{7-x}$ spectra show increasing resemblance to Cu_2O indicating the presence of monovalent copper. This reduction of copper with removal of oxygen is in agreement with general chemical arguments.

III. THEORETICAL APPROACH

The self-consistent field local density theory[1,2] is used in the framework of an embedded cluster model[20,21] of the solid. The effective Hamiltonian can be written as

$$h = T + V_C + V_{xc} \tag{1}$$

where T and V_C are the usual kinetic energy and Coulomb potential, and V_{xc} is some approximation to the exchange and correlation interactions among the electrons. In the work reported here, we use the simplest spin-polarized exchange of the Kohn-Sham-Slater type, with exchange parameter $a = 0.7$. The Coulomb and exchange potentials are formed from the charge and spin density of the infinite extended system, generated by summation of model atomic(ionic) densities which are themselves derived from the self-consistent iteration process. Details of this procedure are given in the literature[20-23]. Since the lattices treated here are ionic, and represented by extended rather than point-ion charge and spin, we have developed generalized Ewald summation procedures for the efficient calculation of V_C. In general, about three hundred terms in both direct and reciprocal space give adequate convergence for all points of the integration mesh used in forming matrix elements of the Hamiltonian. The embedding potential is completed by adding a repulsive pseudopotential designed to prevent cluster electrons from migrating to core levels of exterior (host) atom potentials. A simple parametrization gives satisfactory results, showing little sensitivity to choice of parameters.

Variational solutions of the Schrodinger equation are found by expanding the cluster wavefunctions in a linear-combination-of-atomic-orbitals (LCAO) basis set formed from numerical solutions of free atom(ion) potentials. Some effort is made to optimize the basis by iteratively inserting the cluster effective atomic configurations in the basis generation procedure. The single-site potentials are augmented by a potential well of adjustable depth and radius to induce additional basis states not bound in the isolated ion. All matrix elements are calculated by three dimensional numerical integration, sufficient precision being obtained with a few hundred points per atom arranged on either quasirandom or random-plus-core region meshes. The spin density is calculated as

$$\rho_s = \sum_i n_{si} |\psi_{si}|^2 \qquad (2)$$

where ψ_{si} and n_{si} are approximate eigenfunctions and Fermi-Dirac occupation numbers respectively. The ground state is determined by occupying the orbitals in order of increasing energy; certain excitations can be modelling by altering the occupation number vector as in the transition state procedure[1]. The charge density is given as

$$\rho_c = \rho_\uparrow + \rho_\downarrow .$$

The spectral densities, such as the charge- and spin-densities of states are generated by placing a Lorenztian dispersion curve at each energy level in order to present a continuous distribution representative of the solid. In order to generate partial densities of states, we also weight each Lorentzian by the amplitude of the charge contribution of interest to the level. Thus,for example, we may examine the 3d charge and spin character of the copper atom at site 1 as a function of energy.

IV. COMPUTATIONAL RESULTS

A. Charge and Spin Density

We present in Table 1 the self-consistent ionic configurations in the reference monovalent(Cu_2O), divalent (CuO), and trivalent ($KCuO_2$) compounds. It is evident that the full formal valency is never developed, as is seen in virtually all theoretical calculations on charge density in compounds. Although the 3d-population does decrease in the sequence $Cu^{I} > Cu^{II} > Cu^{III}$, far less than one electron is transferred with each oxidation state change. The d-electron count correlates well with the observed shift in the Cu X-ray K-edge for absorption; the observed shifts of 1.7 (I -> II) and 3.8 eV (I -> III) lead to a predicted shift of 6.3 eV per d-electron occupancy change.

Table 1. Self-consistent atomic orbital populations and ionic charges for reference compounds and superconductors as calculated by embedded cluster method.

	Cu_2O	CuO	$KCuO_2$	La_2CuO_4	$YBa_2Cu_3O_7$		
	Cu_1				Cu_1	Cu_2	
3d	9.54	9.24	8.98	9.18	9.03	9.23	
4s	0.69	0.16	0.13	0.05	0.30	0.23	
4p	0.11	0.03	0.04	0.06	0.19	0.21	
Q_t	0.66	1.57	1.85	1.72	1.48	1.33	
	O_1					O_2	O_3
2s	1.91	1.97	1.97	1.99	1.94	1.96	1.96
2p	5.42	5.60	5.45	5.85	5.35	5.54	5.61
Q_t	-1.34	-1.57	-1.42	-1.84	-1.30	-1.50	-1.56

A second factor worth noting is the sizable occupancy of Cu 4s and 4p orbitals in these systems. The absolute magnitude is not meaningful, since Mulliken populations of diffuse orbitals are notoriously sensitive to basis set changes; however, they do indicate a considerable mixing of diffuse Cu conduction states with the oxygen ligands. In fact, for these and a number of other Cu environments studied we are impressed by the relative stability of the d^9 configuration; most charge transfer driven by the environment occurs in the sp shells. When we compare the atomic configurations of Cu in the La- and YBa_2- based superconductors, we see a close resemblance to the divalent

CuO system. There are significant differences which can be seen in analysis of density maps, but there is no evidence for the trivalent species expected on charge balance arguments for the 1-2-3 compounds. This at least is consistent with the evidence presented above from XPS data. When the clusters (ranging from 9 to 39 atoms in size) representative of the superconductors are perturbed by forming vacancies or performing substitutions the Cu^{II} species proves remarkably resistant to change; most of the charge acceptor/donor behaviour is seen on the oxygen sites. This too is consistent with the idea that O^- species may be present, although such diffuse charge distributions can hardly be assigned to a single atomic site. When environments deficient in oxygen are generated by creating vacancies, the neighboring Cu ions are observed to accumulate charge, moving in the direction of Cu^{I} as generally expected.

In Fig. 3 we show the partial density of states for Cu(2), the planar site in $YBa_2Cu_3O_7$, computed from a 25 atom embedded cluster model. The relevant features seen here are

(1) the considerable width of the Cu-d distribution ~10 eV due to the strong covalent mixing with oxygen-p states

(2) the existence of considerable Cu-sp character in the occupied region $E < E_F$, in contrast to the classical oxidation state picture,

(3) the presence of well defined d- and sp-excitations immediately above E_F.

The Cu(1) chain sites show similarly well defined occupied and excited state features over the same interval. Thus on the basis of self-consistent-field cluster calculations one can point to the simultaneous importance of d- and sp-excitations just above E_F (which are of course mixed intimately with oxygen-p). Unfortunately we do not yet have the capability of calculating relative transition rates for specific (photon, phonon, electron-electron) valence excitation mechanisms, to separate contributions from plane and chain sites.

One way to probe the behaviour of different lattice sites is to carry out spectroscopic studies on substituted systems. In particular, data from Mossbauer[24,25], NMR, NQR[26], XANES[10,27-29] and other local-site measurements for transition metal substituents have promise of revealing the special nature of copper and its relation to crystal structure in controlling superconductive properties. In Fig. 4 we show the calculated valence charge density along a-, b-, and c-axes for the case of iron substituted at the Cu(1) chain site in $YBa_2Cu_3O_7$, taken from an embedded cluster. The strong anisotropy of ρ about the metal site and the build-up of covalent charge density along the Fe-O bond directions is evident. This anisotropy should lead to a noticeable angular dependence of X-ray absorption features, which may soon be available from partially oriented or thin film samples. The corresponding spin density $\rho_\uparrow - \rho_\downarrow$ is displayed in Fig. 5; in addition to the large moment (3.0 μB) associated with the Fe 3d shell, one can see the induced negative polarization in the neighboring O 2p shell and the (positive) valence part of the induced s contact densities at both sites. Contour maps of charge and spin densities in the (a,b) plane are given in Fig. 6 for this case. Here we can see an increased covalent charge in the Cu-O bond as compared to Fe-O and the $3d^{9.1}$ configuration of the Cu site. The spin density map shows a surprisingly large polarization of the second neighbor Cu on an adjacent chain, in addition to the smaller in-chain polarization discussed above. This neighbor-chain polarization without any intervening oxygen is probably overestimated due to cluster truncation effects, but certainly provides some food for thought about magnetic interactions in-chain and off-chain among Cu ions.

^{57}Fe Mossbauer spectra for Fe-substituted superconducting oxides have been reported by a number of groups[24,25]. The data on $La_2Cu_{1-x}Fe_xO_4$ are consistent with random replacement of Cu by Fe, while three- or four-site models have been invoked to describe the complex spectra of $YBa_2Cu_{3-x}Fe_xO_{7-y}$. From the Mossbauer data one can extract isomer shift (IS), related to $\rho_c(0)$, electric field gradient (EFG), related to $< \rho_c Y_{2m}/r^3 >$, and magnetic hyperfine fields (H_{hf}) connected with ρ_s (0). Nuclear magnetic resonance (NMR) and nuclear quadrupole resonance (NQR) have been used to extract EFG data at Cu sites[24-26].

The Mulliken atomic orbital populations given in Table 2 for Fe-substituted $YBa_2Cu_3O_7$ can be used to form estimates of IS and H_{hf}. The Fe 4s population differs by 0.1e from Cu(1) to Cu(2) sites, the former having the larger valence charge density. A detailed analysis of the Mossbauer spectra will be given elsewhere; here a simple estimate based only on the local 4s contribution gives

$$\Delta IS \quad = \quad A \Delta \rho_c(0) \quad = \quad 0.12 \text{ mm/s}$$

for the IS difference between the two sites. Here we have used the value A = -0.25 mm-a_0^3/s. The oxygen enriched Cu(1) site shows a predicted further shift of 0.12 mm/s. These values span the range of observed Δ-values. The total valence densitites ρ (0) given in Table 2 show that there are significant overlap contributions from ligand sites, and that the 4s populations do not really provide a quantitative measure of density differences. The predicted IS difference between Cu(1) and Cu(2) on the basis of $\rho_c(0)$ is 0.43 mm/s. The three-fold spectra reported by Baggio-Saitovitch et al. for dilute concentrations of Fe span a range of ΔIs =0.37 mm/s. It is noteworthy that the Cu(2) and Cu(1)-octahedral oxygen enriched sites are calculated to have very similar electronic configurations.

In the Cu(1) and Cu(2) sites, and also in the oxygen enriched Cu(1)+O_2 environment, iron appears to prefer the Fe^{3+} high spin configuration. In addition to the d^5s^0 free ion occupancy, Fe accumulates 0.5-0.8e in the 3d shell and ~0.5e in the 4sp shells through bonding interactions with oxygen ligands. The EFG is due primarily to anisotropic d-electron contributions; however we see that a significant 4p component is also present in this system. Because of the considerable change in electronic configuration on substitution, there is no particular reason for EFGs to be comparable for Cu and for Fe. In fact, the values calculated for Cu(1) and Cu(2), see Table 2, are substantially larger in magnitude than the values given for Fe. The experimental value of the ^{57}Fe (I=3/2) nuclear quadrupole moment is somewhat uncertain. If we use the currently favored value Q = 8.2 fm^2 [30,31], then we can estimate the Mossbauer quadrupolar splitting as

$$\Delta E = eQV_{zz}(1 + \eta^2/3)^{1/2}/2 = B\,V_{zz}(1 + \eta^2/3)^{1/2}$$

with B = 0.83 mm-a_0^3/s . With this value of the coupling constant B, the calculated valence contributions from Table 2 are only of the order of 25% of the observed values. This indicates the (obvious) importance of contributions of the remainder of the ionic lattice exterior to the cluster and the (likely) importance of Fe-2p,3p inner shell polarization. These additional terms have to be calculated in order to make firm contact with experiment. It is known that lattice sums in the ionic point charge model give correct order of magnitude results.

Further, the calculated ratio of the EFG for copper, Cu(1)/Cu(2)=2.1 stands in contradiction to interpretations of the origin of the NQR signals at 31.5 and 22.1 MHz[26]. The low frequency peak is observed to diminish and disappear with decreasing oxygen stoichiometry and is thus assigned to Cu(1). This suggests again that ionic lattice terms must be combined with local electronic contributions in order to obtain meaningful EFG predictions.

Table 2. Fe atomic orbital populations for charge and spin and valence electron contributions to electric field gradient (e/a_0^3) at Fe and Cu sites in $YBa_2Cu_{3-x}Fe_xO_7$.

Fe on site:	Cu(1)		Cu(2)		Cu(1)+O_2[a]	
	charge	spin				
3d	5.77	2.97	5.73	3.98	5.47	3.92
4s	0.32	0.05	0.20	0.08	0.08	0.04
4p	0.17	0.03	0.37	0.17	0.11	0.03
Q_t, Q_s	1.74	3.05	1.70	4.23	2.34	3.99
$\rho(0)$	7.21	1.66	5.51	0.72	5.76	0.76
$<(3z^2-r^2)/r^5>$	-0.24		0.12		0.18	
$<(x^2-y^2)/r^5>$	0.58		0.002		0.01	
Cu						
$<(3z^2-r^2)/r^5>$	-0.68		-0.32			
$<(x^2-y^2)/r^5>$	-0.02		+0.10			

---

a) Two additional oxygen ions placed at normally vacant a-axis sites; resulting six-fold coordination is distorted octahedral.

B. X-Ray Absorption

Further evidence can be obtained by comparing experimental XANES data and calculated Cu K-edge cross-sections. Experimental results for Cu_2O, CuO and $KCuO_2$ show, as mentioned above, a shift of the main "white line" peak to higher energy with increasing oxidation state. A subsidiary peak appears on the low energy side of the white line in the monovalent compound. Calculations have been made by several groups of the absorption cross-sections using the multiple scattering (MS-Xa) method to determine continuum final states/10,32-34/. Self-consistent potentials in the present work were generated by the discrete variational (DV-Xa) method in the LCAO basis described above, and then projected onto the muffin-tin approximation rquired by the MS scheme. When final state relaxation effects are taken into account by the transition state procedure, the edge position can be placed on an absolute energy scale. The overall agreement with experiment is found to be quite good, demonstrating both the adequacy of the single particle approximation and the local density theory for these core level excitations.

Early comparisons of XANES and calculations on the $La_{2-x}(Sr,Ba)_xCuO_4$ superconductor suggested that, in addition to the dominant Cu^{II} features, some evidence for the Cu^{III} species could be seen in a secondary peak in $\sigma(\omega)$ about 5 eV above the main absorption band[10]. Later it was realized that strong back-scattering from second neighbor oxygen ions contributes in the same region attributed to Cu^{III}, thus precluding a clean identification[10,32,33]. Measurements made on the $YBa_2Cu_3O_{7-y}$ compounds for varying oxygen stoichiometry "y" clearly reveal the reduction of Cu^{II} to Cu^{I} with oxygen depletion, but firm evidence for the presence of Cu^{III} appears to be lacking. Detailed cross-section calculations haave now also been made; the agreeement between theory and experiment is about as good as in the reference compounds. The use of a "trapping time" analysis of contributions to the cross-section produces a quantitative measure of relative size of oxygen s-, p-, and d-wave contributions in comparison to Cu l-wave contributions. For example, we can show in $YBa_2Cu_3O_7$ that the white line is composed approximately of 80% Cu 1s -> 4p single site transitions with the remainder dominated by nearest neighbor oxygen p-wave (10%), followed by second neighbor oxygen.

Does the absence of well defined Cu^{III} features in XPS and XANES measurements and calculations on these materials indicate the non-existence of low-lying Cu^{II} -> Cu^{III} excitations? Hardly; but it does suggest that such excitations are not very much involved in the relaxation processes involving core-hole screening, nor in the formation of dipole-allowed final states. This makes some sense in view of experience with transition metal compounds; the $d^n \rightarrow d^{n+1}$ locally screened final state is generally identified with the main line, and $d^n(sL)^1$ diffusely screened (here L = ligand) final states can be associated with satellite structures. Resonance emission experiments may possibly be used to observe two-electron processes involving $d^9 \rightarrow d^8$ transitions, but this takes us rather far from the original question about ground state populations and their relevance to superconductivity.

A more promising approach is to turn to excitation energies in the 10 meV - 1 eV range, with both band structure and cluster models show a considerable mixing of Cu d and O 2p character. We have calculated in La_2CuO_4 that charge transfer transitions (involving ~0.3 e Cu 3d -> O 2p) lie only ~20 meV above the occupied states and that motion of axial oxygen ions or substitution of La by Sr is sufficient to invert (Cu d,O p|$3z^2$-r^2), (Cu d,O p|x^2-y^2) level ordering and populations.

C. Lattice Distortions

In order to describe coupling between lattice distortion, vibrational modes and electronic states in a quantitative fashion it is necessary to carry out self-consistent calculations for different lattice displacements. This has been carried out for uniform "breathing modes" by band structure methods and for selected local displcements by cluster methods. For example, the lattice in La_2CuO_4 is found to be quite soft with respect to vertical displcements of the axial oxygen, $d^2 E_{tot}/ d z^2 = 40\ eV/a_0{}^2$, in comparison to in-plane breathing or rotational modes[34]. This is a reminder that although the in-plane Cu-O network may be critical in forming the ground state electronic band structure, the low-lying excitations may arise largely from out-of-plane transitions. A systematic effort to map out significant portions of the potential energy surface connected with lattice distortions is in progress.

Acknowledgements

Work supported in part by US DOE, BES-Materials Sciences, contract no. W-31-109-ENG-38. DEE thanks Centro Brasileiro de Pesquisas Fisicas for hospitality and use of their facilities during the course of this work. We also thank E. Baggio-Saitovitch, B.D. Dunlap and E.E. Alp for helpful discussions. Computations were carried out in part at the Laboratorio Nacional Para Computacão Cientifico, Rio de Janeiro.

References

1. Slater, J.C., The Self-Consistent Field for Molecules and Solids,(McGraw-Hill, New York, 1974).
2. Dahl, J.P. and J. Avery, eds., Local Density Approximations in Quantum Chemistry and Solid State Physics,(Plenum, New York, 1984).
3. Mason, T.O., to be published.
4. Mattheiss, L.F., Phys. Rev. Lett. 58, 1028 (1987).
5. Yu, J., A.J. Freeman and J.-H. Xu, Phys. Rev. Lett. 58, 1035 (1987);

Massidda, S., J. Yu, A.J. Freeman and D.D. Koelling, Phys. Lett. A122 198 (1987); Yu, J., S. Massidda, A.J. Freeman and D.D. Koelling, Phys. Lett. A122, 203(1987).

6. Xu, Y., G. Zhao, F. Zandiehnadem and W.Y. Ching (to be published)
7. Kasowski, R.V., W.Y. Hsu and F. Herman, Sol. Sta. Comm. 63, 1077 (1987).
8. Hatsugai, Y. and T. Fujiwara, Sol. Sta. Comm. 65, 1271 (1988).
9. Szpunar, B., V.H. Smith Jr. and R.W. Smith, Physica C 152, 91 (1988).
10. Alp, E.E., G.K. Shenoy, D.G. Hinks, D.W. Capone II, L. Soderholm, H.-B. Schuttler, J. Guo, D.E. Ellis, P.A. Montano and M. Ramanathan, Phys. Rev. B35, 7199 (1987); Alp, E.E., G.K. Shenoy, L. Soderholm, G.L. Goodman, D.G. Hinks, B.W. Veal and D.E. Ellis, to be published; Guo, J., D.E. Ellis and G.L. Goodman (submitted).
11. Chen, H., J. Callaway and P.K. Misra, Phys. Rev. B(to be published); and Chen, H., J. Callaway and P. K. Misra (submitted).
12. However, semiempirical Huckel calculations have been interpreted in terms of Cu^{II} and Cu^{III} charge states: Curtiss, L.A., T.O. Brun and D.M. Gruen, Inorg. Chem.(to be published).
13. Valence bond calculations predict that $Cu^{II} \rightarrow Cu^{III}$ transitions are unfavorable with respect to excitations of oxygen ligands; Guo, Y., J-M Langlois and W.A. Goddard III, Science 239 896 (1988).
14. Steiner, P., V. Kinsinger, I. Sander, B. Siegwart, S. Hufner, C. Politis, T. Hoppe and H.P. Muller, Z. Phys. B67, 497(1987).
15. Sasaki, T., Y. Baba, N. Masaki and I. Takano, Jpn. J. Appl. Phys. (submitted).
16. Shen, Z., J.W. Allen, J. J. Yeh, J.S. Kang, W. Ellis, W. Spicer, I. Lindau, M.B. Maple, Y.D. Dalichaouch, M.S. Torikachvili and J.Z. Sun, Phys. Rev. (submitted).
17. Fujimori, A., E. Takayama-Mulomachi, Y. Uchida and B. Okai (to be published).
18. Nocker, N., J. Fink, J.C. Fuggle, P.J. Durham and W.M. Temmerman, (to be published); Fink, J., N. Nocker, H. Romberg, J.C. Fuggle, P.J.W. Weijs, R. Schoorl, P.J. Durham, W.M. Temmerman and B. Gegenheimer (to be published).
19. Veal, B.W., A. P. Paulikas, D.J. Lam and J.W. Downey (unpublished).

20. Ellis, D.E., G.A. Benesh and E. Byrom, Phys. Rev. B20,1198(1979).

21. Ellis, D.E. and G.L. Goodman, Intl. J. Quant. Chem. 25, 185 (1984).

22. Baerends, E.J., D.E. Ellis and P. Ros, Chem. Phys. 2, 41 (1973).

23. Rosén, A., D. E. Ellis, H. Adachi and F.W. Averill, J. Chem. Phys. 65,3629(1976).

24. Baggio-Saitovitch, E., I. S. Azevedo, R.B. Scorzelli, H. Saitovitch, S.F. de Cunha, A.P. Guimaraes, P.R. Silva and A.Y. Takeuchi, Phys. Rev. B(to be published).; also see article in this volume, and references therein.

25. Dunlap, B.D., et al. (to be published)

26. Mihaly, L., this volume, and references therein.

27. Alp, E.E., L. Soderholm, G.K. Shenoy, D.G. Hinks, B.W. Veal and P.A. Montano, Physica (to be published).

28. Heald, S.M., J.M. Tranquada, A.R.Moodenbaugh, and Y. Xu (to be published); J.M. Tranquada, S.M. Heald and A.R. Moodenbaugh, Phys. Rev. B36, 5263 (1987); Tranquada, J.M., S.M. Heald, A.R. Moodenbaugh and M. Suenaga, Phys. Rev. B35, 7187 (1987).

29. Lytle, L.W., et al. , Phys. Rev (to be published).

30. Vajda, S., G.D. Sprouse, M.H. Rafailovich and J.W. Noe, Phys. Rev. Lett. 47, 1230 (1981).

31. Duff, K.J., K.C. Mishra and T.P. Das., Phys. Rev. Lett. 46, 1611 (1981).

32. Durham, P.J., W.M. Temmerman and G.M. Stocks, J. de Physique (to be published).

33 Garg, K., A. Bianconi, S. Della Longa, A. Clozza, M. DeSantis and A. Marcelli (submitted).

34. Khowash, P.K. and D. E. Ellis (unpublished)

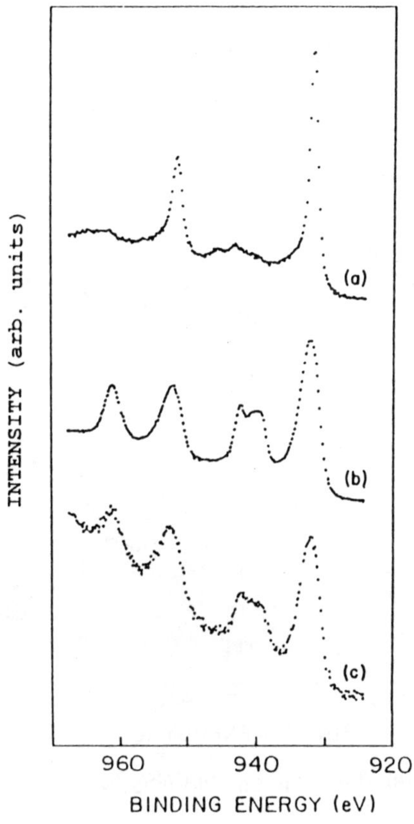

1. XPS of Cu 2p region: (a) Cu_2O , (b) CuO , (c) $La_{1.85}Sr_{0.15}Cu\ O_4$

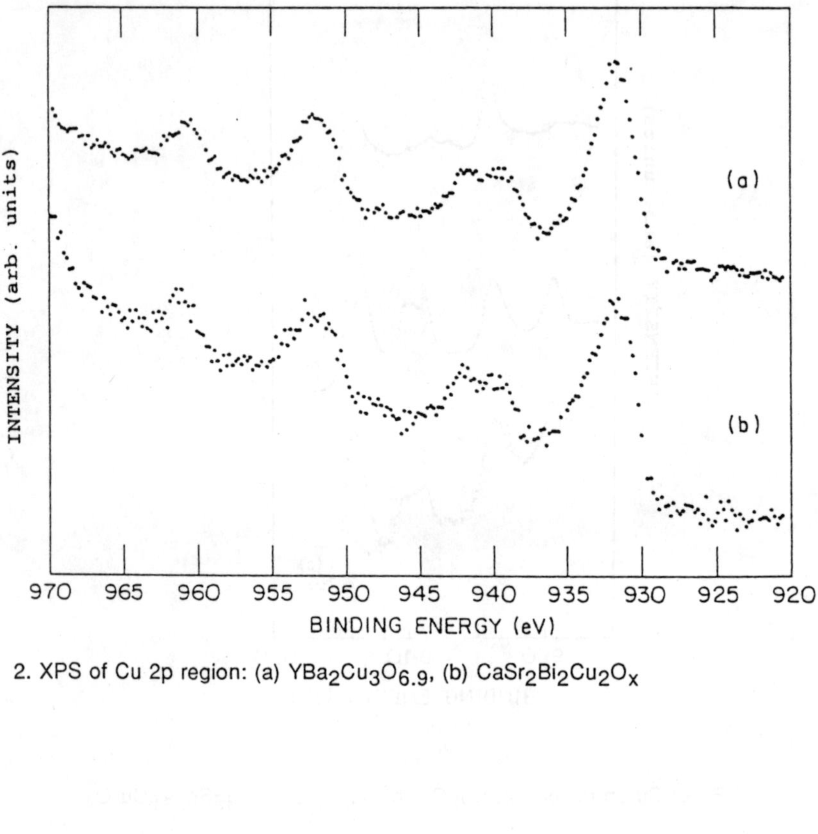

2. XPS of Cu 2p region: (a) $YBa_2Cu_3O_{6.9}$, (b) $CaSr_2Bi_2Cu_2O_x$

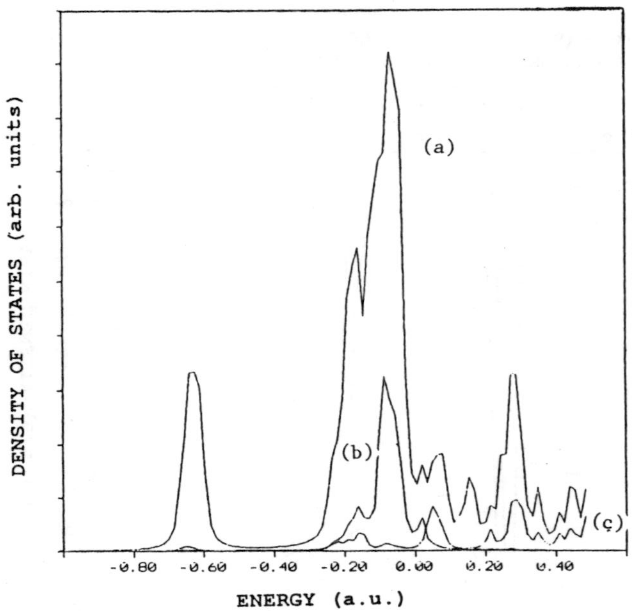

3. Charge density of states in $YBa_2Cu_3O_7$ from cluster calculations: (a) Total , (b) Cu(2) 3d partial DOS, (c) Cu(2) 4sp partial DOS. E=0 denotes Fermi energy.

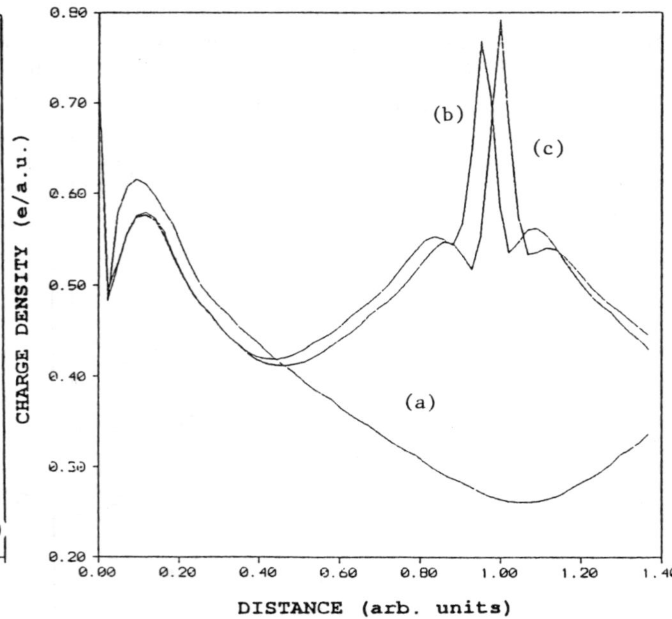

4. Valence charge density along crystallographic axes in Cu(1) -> Fe substituted $YBa_2Cu_3O_7$; (a) a-axis, (b) b-axis containing Fe-O-Cu- chains, (c) c-axis. Fe is at origin, secondary peaks are oxygen sites.

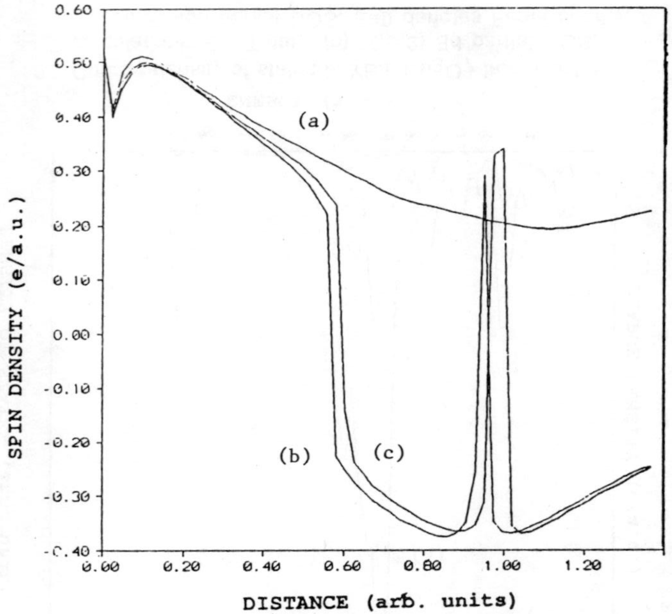

5. Valence spin density along crystallographic axes in Cu(1) -> Fe substituted $YBa_2Cu_3O_7$; (a) a-axis, (b) b-axis, (c) c-axis.

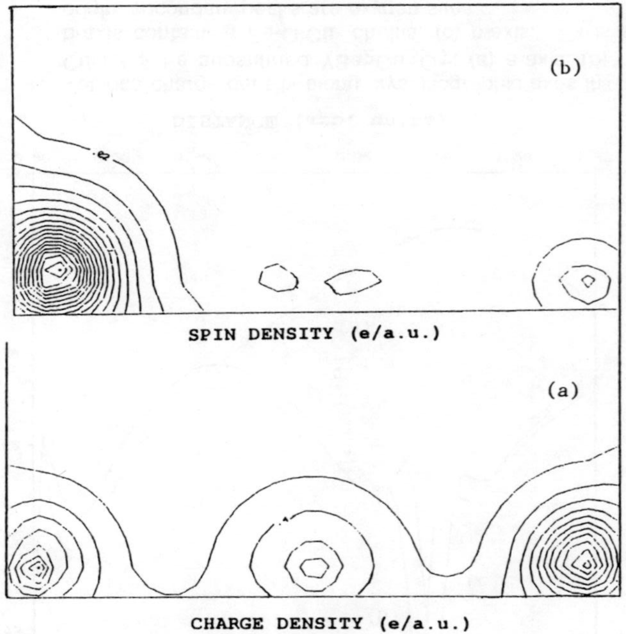

6. Contour maps of (a) charge and (b) spin density of Fe-substituted $YBa_2Cu_3O_7$ in (a,b) plane. Chains lie along the horizontal axis with Fe at the lower left corner.

CONTRIBUTED PAPERS

HIGH Tc SUPERCONDUCTIVITY IN THE Bi-Sr-Ca-Cu-O SYSTEM

E. Chavira, R. Escudero, D. Rios-Jara and L. M. Leon.
Instituto de Investigaciones en Materiales-UNAM.
Apdo. Postal 70-360, 04510 México D.F., México.

ABSTRACT

We have studied some structural and transport characteristics of Bi-based superconducting compounds. In order to analyse structural characteristics and mixtures of phases, X-ray powder diffraction patterns were obtained from the compounds and compared with computer generated ones. Zero resistance at 104K was measured in Pb-doped samples. On the basis of these results, we discuss the influence of $Cu\text{-}O_2$ planes on the superconducting transition temperature.

INTRODUCTION

Since the discovery of high Tc superconductivity in Bi-based compounds, first at 20K by the group of Raveau [1] and later at about 110K by the group of Maeda [2], much work has been done [3-5]. Something facinating in these new compounds is that they present a different crystallographic structure than that of the 1:2:3 compounds. In particular, they present only one type of Cu-O configuration, the $Cu\text{-}O_2$ planes, while in the 1:2:3 compounds two Cu-O configurations exist, the $Cu\text{-}O_2$ planes and the Cu-O chains. This subtle though important difference allows us to conclude about the relevance of these $Cu\text{-}O_2$ planes in high Tc superconductivity. Moreover, there are several experimental indications about the correlation between the number of $Cu\text{-}O_2$ planes and Tc. Another important difference between these two families of compounds is that, until now, it has become somewhat difficult to prepare the Tc=110K single phase of the Bi-based compounds. In this paper we report some experimental results concerning the preparation of Bi-based compounds and some improvements in Tc attained by the introduction of small quantities of Pb in the samples.

EXPERIMENTAL

In order to study the problem of how to attain the highest transition temperature, we sintered different compositions as shown in Table I. Powders of Bi_2O_3, $CaCO_3$, $SrCO_3$, CuO and $Pb(C_2H_3O_2)_2 2Pb(OH)$ were mixed, compressed into pellets, fired and sintered in air between 855°C and 880°C for different total times (see Table I). Parts of the pellets were used to measure resistance versus temperature curves, down to 10K, by using the

Table I. Sample preparation.

No	Compound	Tc(K)	Temp.(°C)	Time(Hr)	Phases*
1)	$Bi_4Sr_3Ca_3Cu_4O_y$	30	855	24	H+P
2)	$Bi_4Sr_3Ca_3Cu_5O_y$	41	855	264	H+P+C
3)	$BiSrCaCu_2O_y$	58	880	16	H+P
4)	$Bi_4Sr_3Ca_3Cu_6O_y$	66	855	48	H+P+C
5)	$Bi_{2.15}Sr_{1.65}Ca_{1.17}Cu_2O_y$	70	880	32	H+P
6)	$Bi_{1.8}Pb_{0.6}Sr_2Ca_2Cu_3O_y$	95	865	72	H+A

* H=80K phase, A=110K phase; P=Ca-Bi and C=CuO precipitates.

four point contact technique. Another part of the pellets was used to obtain X-ray powder diffraction patterns, with Cu-Kα radiation and a secondary monochromator. X-ray powder diffraction patterns were generated with the aid of the Lazy-Pulverix program [6]. Different diffractograms were obtained for several possible space groups and atom combinations within the 2212 (80K) and the 2223 (110K) structures, and were compared with the experimental ones. The best choice for the space group was Fmmm and it was used in the generated diffractograms presented in this work.

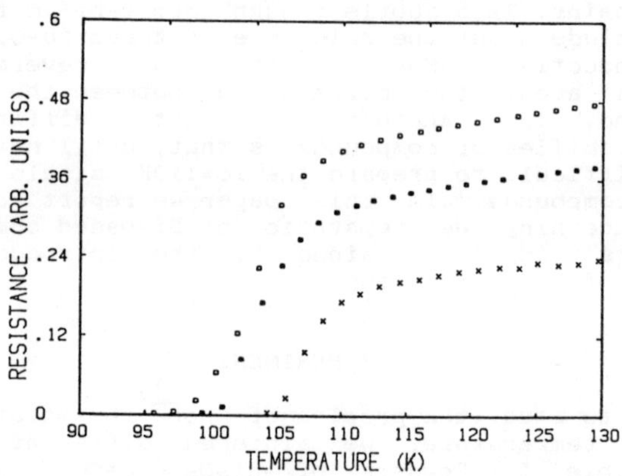

Fig.1. Resistance versus temperature curves for the sample containing Pb, after 3 days (Tc=95K), 6 days (Tc=99K) and 10 days (Tc=104K) of reaction treatment at 865°C.

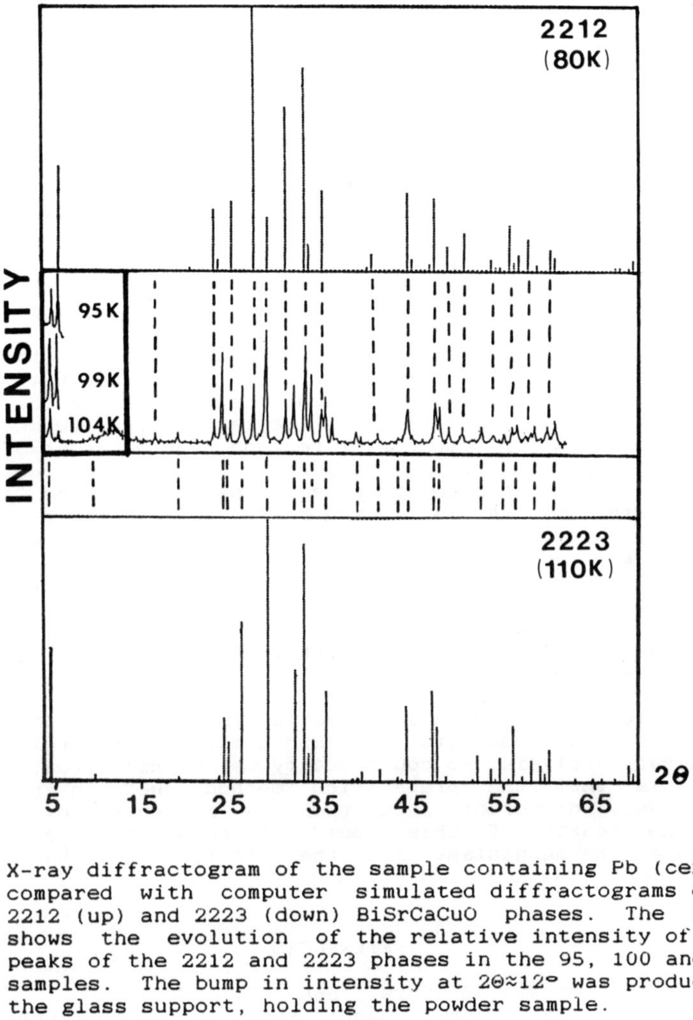

Fig. 2. X-ray diffractogram of the sample containing Pb (center), compared with computer simulated diffractograms of the 2212 (up) and 2223 (down) BiSrCaCuO phases. The insert shows the evolution of the relative intensity of (002) peaks of the 2212 and 2223 phases in the 95, 100 and 105K samples. The bump in intensity at 2Θ≈12° was produced by the glass support, holding the powder sample.

ACKNOWLEDGEMENTS

Financial support from the PUSATT-UNAM and CONACYT is greatfully acknowledged. We also thank F. Morales for general colaboration.

RESULTS

X-ray diffractograms from samples 1-5 in Table I were all alike, with the characteristic spectra of the 80K phase and some small peaks of Ca-Bi and CuO precipitates. Only sample 6 showed no precipitates, but extra peaks which, as we will show latter, were identified as belonging to the 110K phase. It is important to note the increase in Tc (up to 95K) measured for sample 6 containing Pb. When this latter sample was heated again at 865°C for a longer time (10 days), the transition temperature rose to 104K. Samples with the same stoichiometry, but without Pb addition, showed lower Tc after the same heat treatment. Resistance versus temperature curves for sample 6 showing zero resistance at 95K, 99K and 104K are presented in Fig. 1.

Fig. 2 shows the X-ray powder diffractogram of the sample containing Pb (center of the figure), together with computer simulated patterns of the orthorhombic structures of the 80K (up) and 110K (down) phases. The lattice parameters were measured as a=5.41Å, b=5.42Å for both structures and c=30.9Å and c=37.0Å for the 80K and 110K phases respectively. The unit cell of the 110K phase was constructed by considering extra Cu and Ca layers as suggested by several authors [5,7]. All the peaks in the diffractogram of Fig. 2 can be explained by peaks predicted by the simulated diffraction patterns. These results confirms the correct choice for the 110K structure. The best fittings were obtained by considering some substitution of Ca atoms by Bi or Pb atoms in both structures, as suggested by Sunshine et al for the 80K phase [8]. A more detailed description of these results will appear elsewhere.

To conclude, we observe a dramatic increase in Tc when Pb impurities are added to the Bi compounds. According to this preliminary analysis, Pb could occupy Ca (or Bi) sites. Important enough is that the preparation method used in this work is definitively different from the one presented by Kitazawa [5]. The reason of this could be related with some drastic changes of the mechanisms and the kinetics of the reaction, introduced by the addition of Pb.

REFERENCES

[1] C. Michel et al., Z. Phys., B68, 421 (1987).
[2] H. Maeda et al., Jpn. J. Appl. Phys., 27, L209 (1988).
[3] J. M. Tarascon et al., submitted to Phys. Rev. B.
[4] R. M. Hazen et al., Phys. Rev. Lett., 60, 1174 (1988).
[5] K. Kuwahara et al., this volume.
[6] K. Yvon et al., J. Appl. Cryst., 10, 73 (1977).
[7] P. Grant, to be published in Science.
[8] S. A. Sunshine et al., submitted to Phys. Rev. Lett.

SUPERCONDUCTIVITY AND STRUCTURE OF $YBa_2(Cu_{1-x}Mn_x)_3O_7$

R.F. Jardim, S. Gama, O.F. de Lima and I. Torriani
Instituto de Fisica Gleb Wataghin, UNICAMP
C.P. 6165 - CEP 13081 - Campinas - S.P. -Brazil

ABSTRACT

We have prepared samples of the following form $YBa_2(Cu_{1-x}Mn_x)_3O_7$, with x=0.0, 0.025, 0.05, 0.075, 0.10, 0.15 and 0.20. Superconductivity with T_c in the range of 84 and 92 K is reported by resistivity and magnetic susceptibility measurements. The substitution did not affect the onset in the resistive transition. X-ray diffraction analysis revealed additional phases increasing with the Mn content. A strong correlation between the width in the resistive transition and additional phases has been observed.

1. INTRODUCTION

Since the discovery of the high transition temperature superconductor $YBa_2Cu_3O_7$ by Wu et al.[1], there has been a great amount of work devoted towards understanding the nature of the superconducting mechanism[2,3]. It is know that when the oxygen content is changed from 7.0 to 6.5, the material gradually loses superconducting and becomes insulating. At the same time, the crystal structure undergoes a transition from an orthorhombic to a tetragonal phase due to the loss of oxygen vacancy ordering along the Cu-O chains[4]. Substitution of Cu by other 3d transition elements is probably the most effective means with which one can probe the role of Cu in the superconductivity of $YBa_2Cu_3O_7$[5]. Our aim in this work is to study how the substitution of Mn for Cu changes the superconducting properties of $YBa_2Cu_3O_7$ compound.

2. EXPERIMENTAL METHODS

We have prepared the mixture $YBa_2(Cu_{1-x}Mn_x)_3O_7$, where x=0.0, 0.025, 0.05, 0.075, 0.10, 0.15 and 0.20, by mixing appropriate proportions of Y_2O_3, $BaCO_3$, CuO and MnO_2, all with purities of 99.9%. The powders were cold pressed into disck-shaped pellets and calcined at 950°C. The resultant mixture was pulverized, pressed again into disk-shaped pellets, sintered at 950°C for 50 hours in flowing oxygen and then furnace cooled to room temperature.

Electrical resistivity was measured by a standard D.C. four terminal method and T_c was chosen to be the midpoint of the resistive transition. Magnetic susceptibility was measured in an A.C. inductance bridge with the modulation field about 3 Oe. The crystal structure of the samples was studied by X-ray diffraction using Cu-K_α radiation.

3. RESULTS AND DISCUSSION

Fig. 1 shows the dependence of T_c versus the Mn concentration. It is very interesting to note that the transition onset temperature (T_{onset}) and the zero resistance temperature (T_{co}) are constant with the Mn content in the range 0.0<x<0.10, suggesting that the substitution does not cause a measurable change in T_c. On the other hand for x>0.10 the Mn substitution provides a significant change in T_c, although not large enough to destroy the superconductivity. Indeed, T_{onset}=93 K was almost invariant for all samples. T_c presents a non linear decrease with the Mn content and the transition width gradually broadens probably due to the gradual occurrence of non-superconducting additional phases.

Fig.2 shows the diffractograms for all the samples. An analysis of the peak positions and intensities shows that the parent sample, x=0.0, has the orthorhombic structure with lattice parameters a=3.823 Å, b=3.886 Å and c=11.680 Å. The other samples preserve the basic orthorhombic structure, though some modifications of the diffraction pattern are observed. Up to x=0.025 Mn content, the samples appear as single phase with practically no changes in the spectrum. For the sample with x=0.05 new phases start appearing most likely Y_2BaCuO_5, which accounts for most extra peaks. It is probable, however, that the orthorhombic structure might have suffered some structural modifications, judging from the changes observed in several critical regions of the spectrum[6]. We note that the splitting of the most intense peaks corresponding to the (110), (103) and (013) reflections of $YBa_2Cu_3O_7$, lose resolution in a fashion similar to when the sample goes trough the orthorhombic-tetragonal phase transition[7].

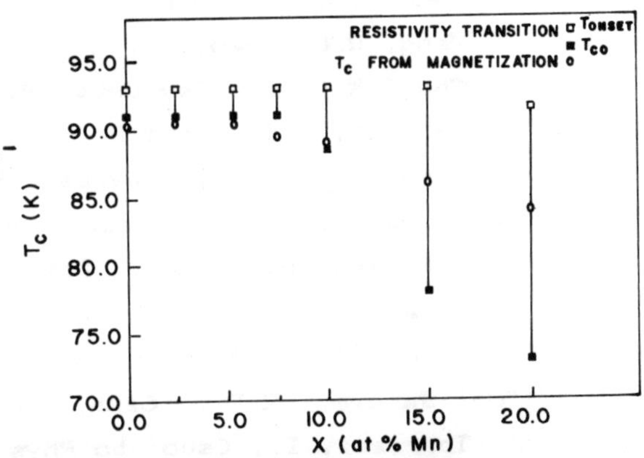

Fig.1 - The critical temperature plotted as a function of the Mn concentration.

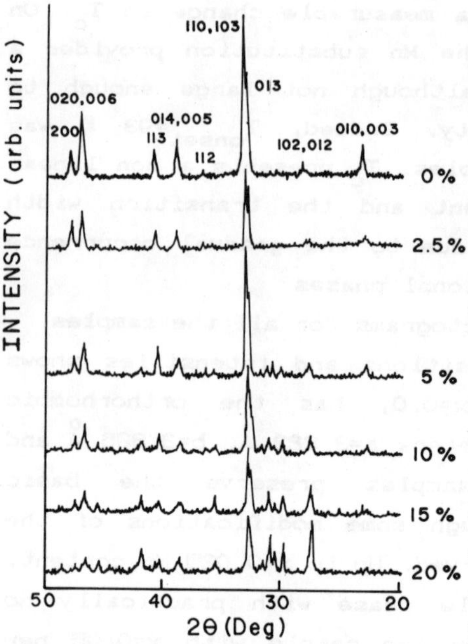

Fig. 2 - X-ray diffraction diagrams of the samples studied.

The (102) and (012) peaks seem to fuse to a single, quite intense reflection. The final peak positon for this reflection is not the same as that of the (012) reflection of the $YBa_2Cu_3O_{6.5}$ tetragonal phase. Nevertheless, the evolution of the (102) reflections as a function of the Mn content resembles the changes observed in de-oxigenation experiments in the 123 compound, which the phase transiton from orthorhombic to tetragonal is observed[7]. The authors acknowledge support from FAPESP.

4. REFERENCES

1- Wu, W.K., Asburn, J.R., Torng, C.J., Hor, P.H., Meng, R.L., Gao, L., Huang, Z.J., Wang, Y.Q. and Chu, C.W., Phys.Rev.Lett. 58, 908 (1987).

2- Hirsch, J.E., Phys.Rev.Lett. 59, 228 (1987).

3- Emery, V.J., Phys.Rev.Lett. 58, 2790 (1987).

4- Tarascom, J.M., McKinnon, W.R., Greene, L.H., Hull, G.W. and Vogel, E.M., Phys.Rev.B 36, 226 (1987).

5- Xiao, G., Streitz, F.H., Gaurin, A., Du, Y.W., and Chien, C.L., Phys.Rev.B 35, 8782 (1987).

6- Jardim, R.F., Gama, S., de Lima, O.F. and Torriani, I., (sub. to Phys.Rev.B Rapid Commun.).

7- de Lima, O.F., Jardim, R.F., Gama, S. and Torriani, I., (this conference).

SUPERCONDUCTIVITY IN Bi-Sr-Ca-Cu CERAMIC OXIDES

M.T.Causa[1],A.Fainstein[1],C.Fainstein[1],C.Puglisi[2],
L.B.Steren[1],M.Tovar[1],N.Veglio[1],and R.Zysler[1].

1. Centro Atómico Bariloche and Instituto Balseiro, 8400 Bariloche, RN, Argentina.
2. Instituto Nacional de Tecnología Industrial, Depto de Mecánica, Buenos Aires, Argentina.

We have been able to obtain a reproducible ceramic superconductor within the Bi-Sr-Ca-Cu-O system, with an onset temperature To=109K as measured by magnetic susceptibility. We describe the method of preparation, and show results from resistance vs temperature measurements, x-rays powder diffraction, and magnetic susceptibility. Results from EDAX analysis are discussed.

The interest in ceramic materials due to their physical properties has increased tremendously in the last two years with the discovery of ceramic superconductors with high critical temperatures. To study the physical properties of those materials it is necessary to develop methods of obtaining reproducible samples, that can be analysed trough increasingly sophisticated techniques. We report here a method for sample preparation within the superconducting Bi-Sr-Ca-Cu-O system, wich yields reproducible samples as defined by measurements of electrical resistance vs temperature, x-ray powder diffractograms, magnetic susceptibility, and EDAX analysis.

Reagent grade oxides (Bi_2O_3 , CuO) and carbonates ($CaCO_3$, $SrCO_3$) in quantities to satisfy the formula $Bi_4Sr_3Ca_3Cu_4O_{16+y}$ were thoroughly mixed in an agate mortar and pestle, lightly pressed (1000Kg/cm^2) in the shape of 12mm diameter pellets, and heated in air at 800°C for 18hs. The reacted material was quenched in air, grinded and pressed again into pellet shape with a load of 2500 Kg/cm2 , and kept for another 18hs at 875°C. The samples, quenched again in air , were of black color, with an uniform distribution of microcrystals. The temperature reached in this second heat treatment was high enough to distort the shape of the pellets without reaching the melting point.

X-ray diffractograms were fitted assuming a tetragonal cell with lattice parameters a = 3.82(1)Å and c = 30.68(1)Å. These values are in good agreement with results obtained by other authors(1,2), who give for the resulting compound the formula $Bi_4Sr_4Ca_2Cu_4O_{16+y}$. The electrical resistivity of the samples showed a metallic-like behavior

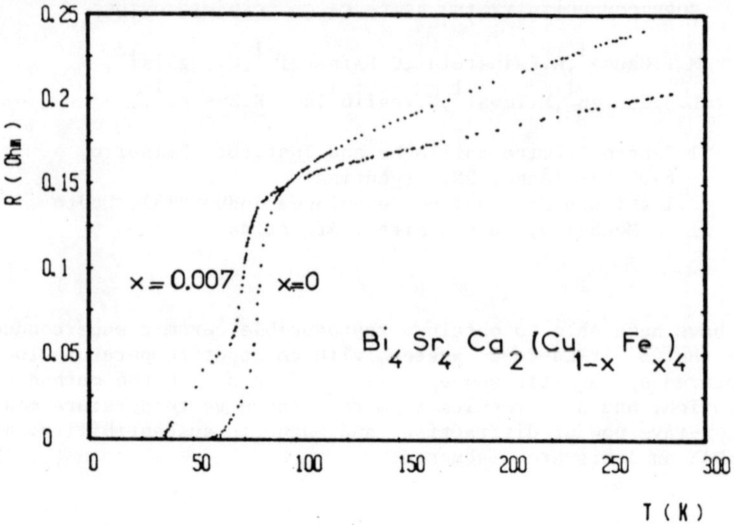

Figure 1. Electrical resistance vs temperature.

as shown in Fig.1. A superconducting transition was observed with a mid-point at 81K, reaching zero resistance around 60K. The onset of this transition is broad showing a small change of slope at about 110K and a more pronounced one at 85K.

Magnetization measurements were performed as a function of temperature using a Faraday magnetometer. The magnetic susceptibility measured at 5000Gauss was found to be Pauli-like above 120K, with $X_p = 3.04\ 10^{-4}$ emu/mol Cu . Figure 2 shows the diamagnetic behavior associated with the superconducting transition. As shown in the insert it starts at 109K and increases sharply around 85K. The Meissner effect measured at 50Gauss, reaches 16% at the lowest temperature measured of 63K. The behavior of both resistivity and magnetization indicates the coexistence of two superconducting phases. One with an onset at 110K and the other with an onset at 85K. This is in agreement with the results of Tarascon et al.[(3)] who show that the 110K material grows from the 85K material by annealing for several days at 885°C.

We have also prepared a sample with a starting nominal composition $Bi_4Sr_3Ca_3(Cu_{.993}Fe_{.007})_4O_{16+y}$. For this sample x-ray diffractograms were fitted assuming a tetragonal structure, with lattice parameters a = 3.82(1)Å and c = 30.64(1)Å . Fig.1 shows the electrical resistance vs temperature, where the onset of these transition is observed at 100K and a sharper drop in resistance at 76K.

The samples were also observed with an electron microscope with EDAX. Although the analysis was made difficult by the quality of the surface, it was possible to conclude that the average composition of the material was $Bi_4Sr_4Ca_3Cu_4O_{16+y}$.

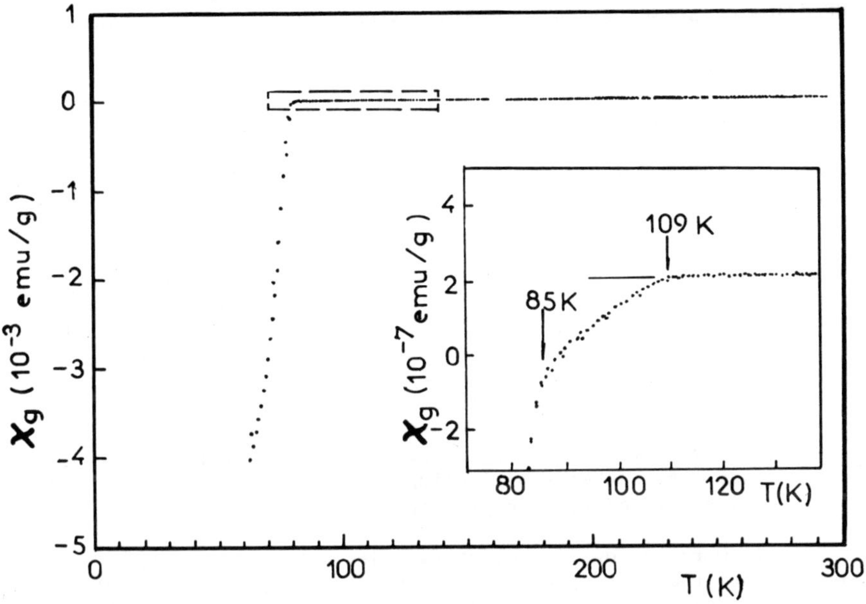

Figure 2. Magnetization vs temperature for $Bi_4Sr_4Ca_2Cu_4O_{16+y}$. The insert shows the onset of superconductivity and the changes in slope at 109K and at 85K.

In conclusion, we have described a method for obtaining the 85K ceramic superconductor. A small percentage of this material was observed to be superconducting at a higher temperature, near 110K. The addition of 0.07% Fe sustituting for an equal amount of copper results in a decrease of Tc of about 10K.

Partial support by CONICET (Argentina) is gratefully aknowledged.

References

1. J.M.Tarascon et al.,submitted to Phys Rev B., 02/11/88
2. J.B.Torrance et al.,submitted to Solid St Comm, 03/18/88
3. J.M.Tarascon et al., preprint, april 1988.

ELECTRONIC STRUCTURE OF COPPER OXIDE AND FLUORINE SUBSTITUTED CLUSTERS BY THE ITERATIVE EXTENDED HUCKEL METHOD

M.Weissmann, A.Saul, A.Levy Yeyati and A.M.Llois

Departamento de Física, Comisión Nacional de Energía Atómica, Av. del Libertador 8250, 1429 Buenos Aires, Argentina.

We study the electronic structure of Cu-O clusters that appear in the new high Tc superconductors by the iterative extended Hückel method. This allows us to estimate some electronic properties, such as the density of states, the charge transfers, the d orbital occupation, the degree of covalency, the intrasite d electron repulsion and the electric field gradient at the Cu site. These results are compared with other types of calculations and with experiments. The differences when F is incorporated in the samples, either substituting O or as an interstitial were also studied by this procedure.

In this work we consider clusters related to the two copper environments that are present in the high Tc compound $YBa_2Cu_3O_7$. As a first approximation we take into account only one Cu atom and its nearest neighbor oxygen atoms, in the conventional crystal structure for the superconducting phase (1). For these clusters and others with F added or substituting O, we calculate the energy levels and wave functions by the iterative extended Hückel method. Previous work on clusters, using different approximate methods of calculation, have already been reported by several authors (2) and used to study some electronic properties. We believe these results may be helpful for the interpretation of experiments that depend strongly on the local environment surrounding a Cu atom. Although these calculations will not be useful to explain the superconductivity they can also contribute to the theory through the choice of better parameters for the model Hamiltonians.

The method used for the calculation is one of the standard semi-empirical ones used in quantum chemistry. It is in essence similar to a Hartree-Fock solution of the Hubbard Hamiltonian, in a non-orthogonal basis. The LCAO wave function is given as:

$$\Psi_m = \sum_i c_{mi}\,\phi_i \qquad (1)$$

ϕ_i being the atomic orbitals 2s and 2p for O and F ; 3d, 4s and 4p for Cu. The secular equation:

$$\sum_i c_{mi}\,(\,H_{ij} - E_m\,S_{ij}\,) = 0 \qquad (2)$$

is solved in a self-consistent way, as the matrix elements H_{ij} are supposed to depend on the atomic charges, and therefore on the coefficients c_{mi}. The atomic charge on atom A is defined as:

$$Q_A = \sum_{i \in A} \sum_j S_{ij} P_{ij} - Z_A \qquad \text{with} \qquad P_{ij} = \sum_m n_m c_{mi} c_{mj} \tag{3}$$

where n_m is the occupation of the mth level, S_{ij} is the overlap between atomic orbitals ϕ_i and ϕ_j and Z_A is equal to the number of valence electrons. The Hamiltonian matrix elements are parametrized as follows:

$$H_{ii} = p + q\, Q_A + r\, Q_A^2 \qquad\qquad H_{ij} = K\, S_{ij}\, (H_{ii} + H_{jj}) \tag{4}$$

for orbital i belonging to atom A and i not equal to j. For our calculations we have used the standard parameters p,q,r,K for O,F,and Cu. We have also included the Madelung correction to the Hamiltonian matrix elements, that takes into account the electrostatic field produced by neighbor atoms, as they are not neutral due to charge transfers.

This method of calculation reproduces reasonably well the interesting features of an ab-initio calculation for the CuO molecule (3) ,for example the charge transfer (.7e) and the d orbital occupation (9.9). We therefore believe it may be adequate for our purpose.

To calculate the electric field gradient (EFG) at the copper sites we consider, as in our previous work (4), two contributions: one due to the departure from the spherical distribution of the valence electrons around the probe atom V^{loc} and another due to the contribution of the neighbor oxygen atoms V^{nn}. For Cu, the d orbitals are the most important, as the p orbitals are more diffuse and less occupied. Using the properties of spherical harmonics the different components of V^{loc} can be written as a linear combination of the P_{ij}, which result directly from our calculation, multiplied by the average of r^{-3} for the corresponding orbital (5). This was evaluated from the radial part of the atomic wave function in Hermann and Skillman (6) giving $\langle r^{-3} \rangle_{3d} = 57.1\ A^{-3}$. For $\langle r^{-3} \rangle_{4p}$ we used $30\ A^{-3}$, an estimation based on the Ga 4p orbital.

The neighbor oxygen atoms were taken as point charges, the Q_O given by expression (3). Therefore V^{nn} contains the position and charge of the oxygen atoms and the Sternheimer factor of the probe atom, taken from ref. 7 to be 18.37.

A first assumption to be made in a cluster calculation of this type concerns the total charge of the cluster. There is considerable uncertainty regarding the Cu valence and the ionicity or covalency of the Cu-O bond. Some authors have assumed totally ionic clusters, with components Cu^{2+} and O^{2-} (8). Others believe that this is not the case (2) and there are experimental reports of a Cu^+ state(9). If we assume for our calculations that all atoms are totally ionized, the clusters have large negative charges, such as -6 for the cluster CuO_4 around the Cu1 position and -8 for the cluster CuO_5 in the Cu2 position of $YBa_2Cu_3O_7$. However, the self-consistent results given in the table show substantial covalency, so that part of the negative charge goes into the Cu 4p orbitals. This makes the Cu atom negatively charged and therefore this picture does not seem appropriate. If partially ionic clusters are considered, such as $(CuO_4)^{-3}$ and $(CuO_5)^{-4}$, the results appear much more reasonable. This implies that Y and Ba are also partially ionized and donate only part of their valence electrons to the Cu-O planes and chains.

In the table we show the atomic charges and orbital occupations for the different clusters considered, that are shown in figure 1. In all cases the configuration is $3d^9$, so that the change in the copper charge is due to the occupation of 4s and 4p orbitals. Figure 2 shows the density of states, obtained by broadening the cluster levels with a Gaussian of width .74 eV, for two of the clusters. When compared with previous calculations on clusters (2) and with band theory (10) we find there is quite good agreement with the less charged cluster. The states near the Fermi level are a mixture of Cu 3d and O 2p states, the lower peak (4eV below Ef) being predominantly 3d while the higher peak, close to Ef, being 90% oxygen 2p. The shape of this density of states does not change significantly for small changes in the cluster charge or size, but in the highly charged clusters it is very different, with almost zero density of states at Ef. Due to the inconsistency between the assumed large negative charge and the covalency obtained we do not consider these clusters adequate.

Comparison with the experimental results (8,11) shows that for the less charged case the peaks near Ef are shifted by about 2eV above the experimental results, as for most other theoretical calculations. This is possibly due to the mean field approximation used for the electron-electron interactions. It is not clear why the highly charged clusters seem to give even better agreement with the experimental density of states.

To estimate the importance of electron repulsion we have calculated the Hubbard U by plotting the self-consistent matrix element H_{dd} against the orbital occupation n_d for CuO, CuO_2, CuO_4, CuO_5, Cu_2O_7,etc. As can be seen in the table these results do not depend strongly on the charge of the cluster, but n_d changes from 9.9 for CuO to 9.0 in CuO_5. There is an almost linear dependence and the slope U is approximately 8eV, in agreement with estimates from experiment (8,9). This value is possibly too large to be considered in a mean field approximation.

We have also performed calculations of the EFG at the Cu site, a property that can be used to interpret NQR and Mossbauer results. The ratio of our calculated EFG for Cu1 and Cu2 is in quite goood agreement with that of the NQR frequencies (12)(31.5 and 22 MHz respectively), considering the number of approximations involved. In a previous paper (4) we have also shown that this type of calculation can qualitatively explain the Mossbauer spectra obtained by introducing Fe in the samples.

Orbital occupancy and total charge on the Cu atom, local and nearest neighbor contributions to the EFG (in V/A^2) for the different clusters. V_{zz} indicates the direction of the principal axis of the EFG tensor.

Cluster	Charge	Orbital occupation 4s	4p	3d	Q_{Cu}	V^d_{zz}	V^p_{zz}	V^{nn}_{zz}	V^{tot}_{zz}
a	-1	0.50	0.38	9.14	0.98				
a	-3	0.59	0.63	9.11	0.67	-436	-73	143	-366
a	-6	0.68	1.79	8.97	-0.44				
b	-1	0.48	0.45	9.07	1.00				
b	-4	0.57	0.68	9.07	0.68	-239	-47	100	-186
b	-8	0.68	1.32	9.23	-0.23				
c	-4	0.56	0.71	9.10	0.63	-306	-52	96	-262
d	-5	0.39	0.72	8.87	1.02	49	23	70	142
e	-3	0.55	0.57	9.20	0.68	-360	-70	144	-286
f	-3	0.41	0.53	9.31	0.75	-286	-82	148	-220
g	-3	0.37	0.20	9.87	0.56	115	-15	138	238
h	-4	0.55	0.72	9.00	0.73	-476	-74	89	-461

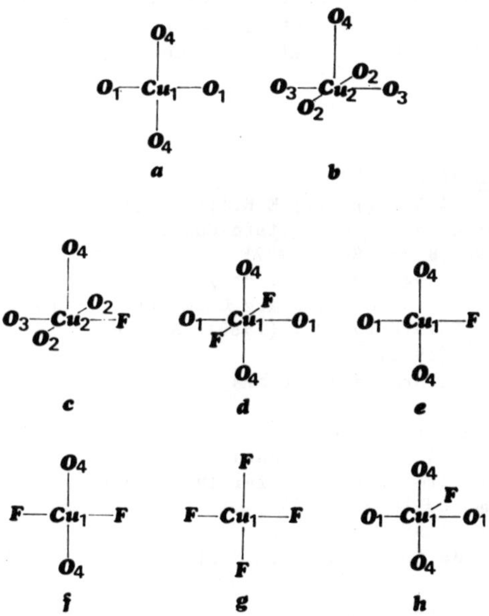

Figure 1. Clusters studied. Notation and distances from reference 1.

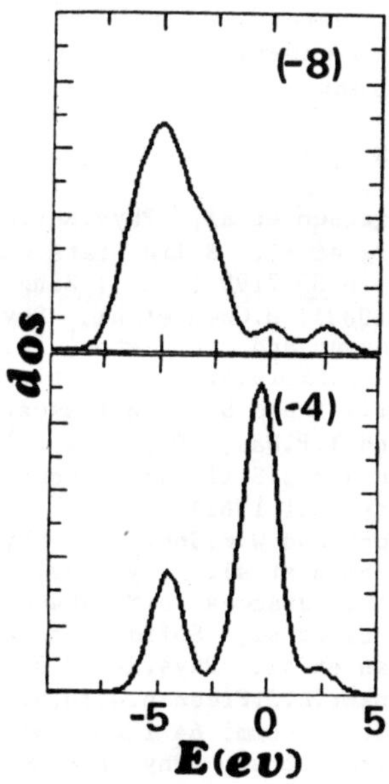

Figure 2. Calculated density of states of cluster b with charges -4 and -8.

As there have been reports of F incorporation in the high Tc superconductors, that can even produce an increase in Tc (13),we believe that NQR experiments on the fluorinated samples could be useful to establish the position of F in the crystal. It could either be replacing O or occupying the vacant sites, and this information could be of great use to understand its effect on the superconductivity. We have therefore calculated the EFG for a few clusters with F in different positions,shown in figure 1, to see if there is any appreciable difference between them. The results are given in the table. In all cases we have assumed partial ionization. As there are no vacant sites for F around a Cu2 site, only one cluster with a substitutional replacement was considered in that case. The EFG is appreciably increased by this replacement and would therefore lead to a weakening of the corresponding low frequency line and the appearance of a new intermediate line between those of Cu1 and Cu2. In the Cu1 site there may be substitution of O by F or addition of F to the original cluster, as suggested in ref.14. The EFG can either decrease if the cluster is more octahedral-like, as is the case of (CuO_4F_2), or increase in the less symmetric groups. The differences are large and could certainly be detected experimentally.

To summarize, we have shown that these calculations are consistent with quite different experimental resuls and we therefore believe they give the correct qualitative features of the electron distribution in the Cu-O clusters. We hope they may be used to interpret or even predict new experimental results, related to local electron distributions.

REFERENCES

1.J.D.Jorgensen et al, Phys.Rev.B 36 3608 (1987)
2.H.Eschrig et al, Solid State Comm. 64 521 (1987); E.E.Alp et al, Phys.Rev.B 35 7199 (1987); Zeng Zhi et al, Solid State Comm. 65 39 (1988); H.Chen et al, Phys.Rev.B 36 8863 (1987)
3.P.V.Madhavan and M.D.Newton, J.Chem.Phys. 83 2337 (1985)
4.A.Saul,A.M.Llois,A.Levy Yeyati,M.Weissmann Solid State Comm. 66 491 (1988)
5.H.M.Petrilli and S.Frota Pessoa, J.Phys.F 15 2307 (1985); A.Coker, Y.Lee and T.P.Das, Phys.Rev.B 13 55 (1976)
6.F.Herman and S.Skillman, Atomic Structure Calculations (Prentice Hall 1962)
7.F.D.Feiock and W.R.Johnson, Phys.Rev. 187 39 (1969)
8.Zhi-xun Shen et al, Phys.Rev.B 36 8414 (1987);V.R.Marathe and A.Trautwein, Advances in Mossbauer Spectroscopy (Elsevier 1983 p.398)
9.A Bianconi et al, Solid State Comm. 63 1135 (1987)
10.F.Herman et al, Phys.Rev.B 36 6904 (1987)
11.I Redinger,A.J.Freeman.J.Yu,S.Massida (preprint);J.M.Mariot et al Solid State Comm. 64 1203 (1987)
12.W.W.Warren et al, Phys.Rev.Lett. 59 1860 (1987)
13.B.M.Tissue et al, Solid State Comm. 65 51(1988); S.R.Ovshinsky et al, Phys.Rev.Lett. 58 2579 (1987)
14.T.Kistenmacher, Phys.Rev.B 36 7197 (1987)

A THEORY FOR THE ANTIFERROMAGNETISM OF La_2CuO_4

G.G. Cabrera and Djalma Medeiros
Instituto de Física "Gleb Wataghin", Universidade Estadual de Campinas, Caixa Postal 6165, 13.081 Campinas, São Paulo, Brazil,

and

Miguel Lagos
Facultad de Física, Universidad Católica de Chile, Casilla 6177, Santiago 22, Chile

ABSTRACT

New solutions recently proposed by two of us for the antiferromagnetic Heisenberg Hamiltonian in the quasi-Ising asymptotic regime[1], also apply to the equivalent Hubbard model with a half-filled band in the large on-site repulsion limit. The ground state of the above models turns out to be a quantum mechanical coherent state, where the dominant components are of the Néel type. However quantum fluctuations are present and lead to the reduction of the staggered magnetization. This effect is similar to the one observed in $La_2CuO_{4-\delta}$ in recent experiments, where a considerable reduction of the effective magnetic moment ascribed to Cu^{++} ions has been reported[2]. Our solution for the ground state is neither variational nor perturbative and can be envisaged as a resonant quantum superposition of many electronic configurations displaying long-range order. We compare our solution with exact results and large size ($N \sim 20$) numerical simulations for the linear chain. We also discuss the extension of our approach to higher dimensions.

As a prototype copper oxide, $La_2CuO_{4-\delta}$ appears as a likely candidate for a two-dimensional antiferromagnetic system, given its lamellar structure and the highly directional bonding in the CuO_2 planes. It does show three-dimensional antiferromagnetic ordering with a Néel temperature in the range of 50-300 K, depending on oxygen concentration[2]. For different lanthanum and oxygen concentrations the same compound may be superconducting[3], thus giving strong support to the idea that antiferromagnetism may play a central role in the mechanisms involved in the recently discovered high-T_c superconductivity[4]. There is also evidence that a two-dimensional antiferromagnetic correlation phase without long-range order develops over distances ~ 200 Å[5], with unusual properties which resemble the behavior of the one-dimensional Heisenberg model. At this stage, a reliable description of the highly correlated antiferromagnetic insulating phase in these low dimensional materials is desirable. This implies an improved understanding of the ground state and its excitations for spin - 1/2 Heisenberg systems. We note that low dimensions and low spin values are particularly susceptible to quantum fluctuations, which in some cases may suppress any phase transition.

In this paper we would like to present a general analytical solution of the anisotropic spin - 1/2 Heisenberg model with antiferromagnetic coupling, which is asymptotically exact in the high correlation limit.

We write the Heisenberg Hamiltonian for the linear chain as

$$\hat{H}=J\sum_{m=1}^{N}\left[S_z(m+1)S_z(m)+\frac{\alpha}{2}\{S_+(m+1)S_-(m)+S_+(m)S_-(m+1)\}\right] \quad , \tag{1}$$

where $J > 0$ is the antiferromagnetic exchange constant, and $S_z(m)$, $S_\pm(m)$ are the usual spin components for $S = 1/2$ at the m-th site. The parameter α takes into account anisotropy effects in the spin coupling, being $\alpha = 0$ the Ising limit and $\alpha = 1$ the isotropic case. We are mainly interested in the range $0 \leq \alpha \leq 1$; indeed our analytical solution is proposed in the regime of small values of α (quasi-Ising region). As long as α remains smaller than 1, the ground state is a doublet and presents long range order due to the dominant character of the Néel components. Next contributions to the ground state come from configurations which are obtained from the Néel state by flipping pairs of neighboring spins and are contained in the manifold of total $S_z = 0$. The relative weight of the latter increases with α, and eventually at $\alpha = 1$ the system gets disordered. This fact has been systematically observed in finite-size calculations for Heisenberg chains[6]. Numerical results for $N = 16$ as a function of α are shows in Table 1.

The above considerations heuristically lead to the definition of the following even and odd operators:

$$\phi_e^\dagger \equiv \sqrt{\frac{2}{N}}\sum_{m\ \text{even}} S_+(m+1)S_-(m) + \alpha\sqrt{\frac{N}{8}} \quad ,$$

$$\phi_o^\dagger \equiv \sqrt{\frac{2}{N}}\sum_{m\ \text{odd}} S_+(m)S_-(m+1) + \alpha\sqrt{\frac{N}{8}} \quad , \tag{2}$$

where N is the total number of sites in the chain. The above operators reverse spin pairs in the Néel state defined through $S_z(m) = \frac{1}{2}(-1)^m$.

Table 1. Ground state structure for a system with N = 16 spins as a function of α. P_N is the probability of finding the system in one of the Néel state, while P_1 is the total weight assigned to configurations coming from the Néel states with single flips of neighboring spins. G(8) is the correlation function $\langle S_z(m)S_z(m+8)\rangle$, which would extrapolate to the square of the staggered magnetization in the infinite size limit, $N \to \infty$.

α	0.2	0.4	0.6	0.8	1.0
G(8)	.23033	.16935	.09278	.04671	.02825
P_N	.85327	.52115	.23957	.12066	.07454
P_1	.06684	.15631	.14927	.11495	.09198

In the quasi-Ising limit, we replace the $S_z(m)$ operator by its eigenvalue for the Néel state. The commutator algebra for ϕ_e and ϕ_o is then simplified to a bosonic algebra. Within the same approximation, our model Hamiltonian (1) can be partially diagonalzied in the subspace $S_z = 0$,

$$\hat{H} = J(\phi_e^\dagger\phi_e + \phi_o^\dagger\phi_o) + E_g(\alpha) + H'(\alpha) \quad , \tag{3}$$

where $E_g(\alpha)$ is the ground state energy (absence of quasiparticles), and $H'(\alpha)$ represents additional degrees of freedom which are not contained in the manifold of total $S_z = 0$. The ground state is constructed as the vacuum of the ϕ-excitations, i.e. as solution of $\phi|g\rangle = 0$, while the Néel state $|N\rangle = |\downarrow\uparrow\downarrow\uparrow\ldots\rangle$ satisfies $\phi|N\rangle = \alpha\sqrt{\frac{N}{8}}|N\rangle$. The solution of the above problem can be given in the form of a <u>quantum mechanical coherent state</u>[1)]

$$|g\rangle = \exp\{-\alpha\sqrt{\frac{N}{8}}(\phi_e^\dagger-\phi_e)\}\exp\{-\alpha\sqrt{\frac{N}{8}}(\phi_o^\dagger-\phi_o)\}|N\rangle \quad , \tag{4}$$

with an extremely interesting structure. Fluctuations over the Néel state $|N\rangle$ are induced by the boson fields ϕ_e and ϕ_o, but the weight (and phase) of a given configuration is not arbitrary: the whole superposition in (4) describes a highly ordered structure displaying long-range order, the α parameter being related to the half-width of the wave packet. This result means that the usual picture of sublattices in antiferromagnets can be saved, in spite of the complications introduced by the ground state structure.

The same approach described above can be given in higher dimensions[6),7)], assuming that the lattice has no geometric frustration to the antiferromagnetic ordering, i.e. it can be separated into two interpenetrating equivalent sublattices. In this case the boson field ϕ has as many components as the number of nearest neighbors in the lattice (coordination number z). In the quasi-Ising limit, the ϕ's satisfy a boson commutator algebra and the ground state can be obtained as a generalized coherent state for a many modes system[7]. The ground state energy to second order in α turns out to be

$$E_g(\alpha) = E_N\left(1 + \frac{\alpha^2}{z-1}\right) \qquad , \tag{5}$$

where $E_N = -\frac{zJN}{8}$ is the energy associated to the Néel state. Eq. (5) reduces to the linear chain result for z = 2. Up to second order in α, the ground state energy depends on the lattice structure only through the coordination number z. Eq. (5) also shows that the energy correction in relation to the Néel state energy diminishes with the coordination number, and hence with the dimensionality, meaning that our approach has a more extended range of validity in higher dimensional systems. However from numerical simulations for very large chains $(N \sim 20)$[8], it is concluded that our asymptotic solution is accurate to better than 1% for the range $0 \leq \alpha \leq .5$, strongly suggesting that the proposed structure of wave function (4) is essentially correct. Breakdown of our theory seems to be related to the fact that in our approach spin wave bands collapse into discrete levels, the gap of excitations being independent of the anisotropy α[1]. The coherent state given in eq. (4) has a number of interesting properties which allows the analytical calculation of any matrix element in a compact and neat way. Details are given in Ref. 1, 6 and 8, as well as a derivation and construction of the excited states.

1. Miguel Lagos and G.G. Cabrera, Solid State Commun. in press, Phys. Rev. B, to be published.
2. D. Vaknin et al, Phys. Rev. Letters 58, 2802 (1987); S. Mitsuda et al, Phys. Rev. B36, 822 (1987).
3. A. Khurana, Physics Today, September 1987, p. 17 and references therein.
4. P.W. Anderson, Science 235, 1196 (1987); G. Baskaran, Z. Zou and P.W. Anderson, Solid State Commun. 63, 973 (1987); V.J. Emery, Phys. Rev. Letters 58, 2794 (1987).
5. G. Shirane et al, Phys. Rev. Lett. 59, 1613 (1987).
6. D. Medeiros, G.G. Cabrera and M. Lagos, unpublished.
7. G.G. Cabrera and M. Lagos, unpublished.
8. M. Lagos, M. Kiwi, E.R. Gagliano and G.G. Cabrera, Solid State Commun., in press.

ELECTRONIC PROPERTIES AND SUPERCONDUCTIVITY IN $YBa_2Cu_3O_{6+x}$

Enrique V.Anda[+] and Luis Moran Lopez[++]
+ Universidade Federal Fluminense; Instituto de Física
Outeiro de São João Batista s/n, Centro, Niterói, RJ, Brasil
++ Universidad Autonoma San Luis Potosi;
Instituto de Fisica - San Luis Potosi - Mexico

ABSTRACT

The Superconductivity critical temperature of $YBa_2Cu_3O_{6+x}$ reduces rapidly with x and it looses its superconducting properties in the tetragonal phase. To explain this behavior the density of states, $\rho(w)$, is calculated for the planes CuO_2 and CuO_{2-x}. The value of $\rho(E_f)$ as a function of x provides a qualitative interpretation of the dependence of Tc as a function of x and emphasizes the role played by the basal plane CuO_{2-x} in the superconductivity of this material.

INTRODUCTION

The study of the crystolographic and electronic structure of high Tc superconducting materials is an essential step to understand superconductivity in these systems. Neutron powder diffraction experiments have revealed the crystalographic structure of some of these compounds(1,2). It is now well established that reducing the temperature $YBa_2Cu_3O_{6+x}$ goes through a tetragonal orthorombic phase transition. The orthorombic phase consists of nearly square planar CO_2 layers weakly bonded through Cu-O-Cu bonds to two CO_3 chains (x=1). The transition from the orthorombic to the tetragonal phase is an order-disorder transition in which the disordering of the oxygen atoms in the central plane destroys the linear chains present in the orthorombic structure. The experimental results show that the disordering of the oxygen atoms between the (1/2,0,0) and the (0,1/2,0) sites occurs in a continuous way from the orthorombic to the tetragonal phase, while the oxygen composition decreases monotonically with increasing temperature. This transition has been studied theoretically(3) using the cluster variational method which was able to explain most of the features observed in neutron diffraction.

It has been shown that the samples produced by rapid cooling exhibit reduce superconducting transition temperatures and in particular those cooled from above the orthorombic tetragonal transition temperature are not superconducting[2]. The understanding of the correlation of these structural properties with the superconducting properties will require a complete band structure calculation including the dependance with the variable x and the order parameter η which governs the order-disorder configuration of the basal plane. This study might give some crutial clues in relation with a more profound knowledge of superconductivity in $YBa_2Cu_3O_{6+x}$.

This paper, using a rather simplified model, represents an effort in this direction.

THE BAND STRUCTURE OF $YBA_2Cu_3O_{6+x}$

In the last year several band structure calculations of $YBa_2Cu_3O_{6+x}$ [4,5,6] have appeared in the literature which use different local-density functional approximation schemes. They coincide in establishing that the main features of the band structure at the Fermi level can be obtained by considering only two bands which arise from strong nearest neigbour (pdσ) interaction between Cu 3d orbitals of ($x^2 - y^2$) symmetry and neigboring O_{2p} orbitals directed along the CuO bonds in the plane. As suggested by Matheiss et al [5], the dispersion of these two bands can be well reproduced by a simple tight binding model with two parameters ($\epsilon_{Cu} = \epsilon_o = -3.2$ eV; $V_{pd} = -1.85$ eV, energies taken with respect to the Fermi level).

These calculations were done for a situation in which x=1. Most of the speculation in relation with a change of x is based upon the idea that the only effect of reducing the oxygen content is to increase the number of eletrons, as oxygen behaves like an acceptor. However, it will be shown below that the energy band corresponding to the basal plane suffers a very important modification as a function of x, ruling out the possibility of assuming a rigid band model.

In this paper we calculate the band structure of $YBa_2Cu_3O_{6+x}$ as a function of x and η, where η is the order-disorder parameter which controls the tetragonal orthorombic transition. In order to do so a tight binding Hamiltonian is assumed and according to the local density functional calculation, only two bands are included.

We restrict ourselves to the region of energy near the Fermi level which controls all the relevant physical properties. We propose a Hamiltonian for holes in the Cu planes given by

$$H = \sum \epsilon_o^i \, p_{i\sigma}^+ \, p_{i\sigma} + \sum \epsilon_{Cu}^j \, d_{j\sigma}^+ \, d_{j\sigma} + \sum_{\langle ij \rangle} V_{ij}^{pd} \, (p_{i\sigma}^+ d_{j\sigma} + d_{j\sigma}^+ p_{i\sigma}) \quad (1)$$

where $p_{i\sigma}$ and $d_{j\sigma}^+$ creates a p hole on the oxygen site and a d hole on the copper site and the V^{pd} is the hybridization matrix element. The effective diagonal elements ϵ_o^i and ϵ_{Cu}^j are taken to be equal, as suggest by Matheiss et al [5] and they include in a mean field approximation all the interatomic and the intraatomic electronic interaction.

For the basal plane the diagonal element ϵ_o^i is a random variable with a probability distribution given by:

$$P(\epsilon_o^i) = \frac{1}{2}(x \pm \eta)\,\delta(\epsilon_o^i) + \left(1 - \frac{(x \pm \eta)}{2}\right)\delta(\epsilon_o^i - \alpha) \qquad (2)$$

We have taken advantage of the equivalence between the absence of an atom on site i and an atom with an infinite diagonal element at that site.

The factors $\frac{x+\eta}{2}$ and $\frac{x-\eta}{2}$ correspond to the probability of occupation of the oxygens along the direction of the chains and perpendicular to them.

The tetragonal orthorombic phase transition is driven by the disordering of the oxygen atoms in the Cu basal plane. This simply means that $\eta=0$ corresponds to the tetragonal phase and $\eta\neq0$ to the orthorombic phase.

It has been established experimentally [1,2] that the oxygen content of the non basal plane maintains the composition CuO_2 independent of x. As a consequence these planes can be considered to be identical to the basal plane for a situation in which $\eta=0$ and $x=2$.

We have supposed that the CuO planes do not interact between them. This seems to be a good approximation because the oxygen that is in a bridge position between the planes is nevertheless much nearer to the basal plane than to the upper or lower planes. Morover it binds to the dz^2 orbital of the copper which is orthogonal to the dx^2-y^2 orbital, responsible for the dispersion in the planes. All the p and d orbitals that do not contribute to the pd interaction, are supposed to be below the Fermi level as it can be infered from the local density functional calculations. In the model proposed, the only effect of Y and Ba is to provide 7 electrons to the CuO planes.

For the whole physical region in which $0<x<1$ the disorder of the oxygens has fundamental consequences on the nature of the eigenstates as it introduces localization, and on the density of states (DOS) because it originates peaks and gaps due to configurational fluctuations. These effects are particularly important in this case because of the low value of the mean coordination number. ($\langle Z\rangle \langle 3$) . As we will expect that the value of the DOS at the Fermi level will control the superconducting transition temperature Tc, the disorder has to be treated adequatelly in order to establish a correlation between x and Tc.

The basal plane, which is the only disordered part of the system, will be represented by a ringless Bethe lattice. We consider this to be a very good approximation because in the orthorombic phase the basal plane is almost constituted by one dimensional structures while in the tetragonal phase as $x < 0.5$ the probability of having a closed loop (absent in the Bethe lattice) is almost negligable $p_{c\ell} \simeq \left(\frac{x}{2}\right)^4 < 0.0039$.

The DOS of the other two planes corresponds to a standard square lattice decorated by the oxygen atoms.

To treat the disorder introduced by the basal plane we use a simple extension of the real space renormalization group formalism applied to a Bethe lattice to incorporate the existance of two species: Cu with z=4 and O with z=2. The advantage of applying real space renormalization techniques, in comparison with any other mean field approximation is based on the fact that in a renormalization formalism spatial fluctua - tions of all orders are considered, which is essential for systems with low coordination number.

RESULTS

We have calculated the DOS of the basal plane as a function of the oxygen content x and of the order-disorder parameter η.

The position of the Fermi level have been calculated assuming that

$$2\int_{-\alpha}^{\epsilon_f}\Big[2\,\rho_{Cu(2)}^{(w)} + \rho_{Cu(1)}^{(w)} + 4\,\rho_{O(2)}^{(w)} + 2\,\rho_{O(4)}^{(w)} + \frac{1}{2}\,(x+\eta)\,\rho_{O(1)}^{(w)} + \frac{1}{2}\,(x-\eta)\,\rho_{O(5)}^{(w)}\Big]\,dw = 16$$

where the $\rho(w)$'s correspond to the DOS of a cooper and oxygen belonging to the two types of planes of the unit cell. In particular O(4) represents the oxygen bounded perpendicular to the basal plane and those on the plane O(1) along the linear chains and O(5) perpendicular to it. The integral is equal to 16 because each cooper contributes with 3 electrons and YBa_2 with 7 electrons. The number 2 at the front of the integral results from the spin degeneracy. In fig.1 it is shown the dependance of the Fermi level with x. As the concentration of oxygen reduces from x=7, even if the DOS at the basal plane shrinks to smaller frequency values, the Fermi level is displaced to greater values of w because the number of holes per site reduces.The dependance is almost linear for the range $0< x < 1$. We have inclused in fig. 1 two experimental points obtained from photoemission studies. A more sistematic experimental measurement of the Fermi level as a function of x could permit a more complete comparison with our theoretical results.

We have represented in fig. 2 and 3 the DOS of cooper and oxygen on the basal plane for x=0.8, and η=0.8. This corresponds to a situation for which the system is a superconductor. The DOS at the Fermi level per atom is almost the same for cooper and oxygen along the linear chain. As shown in fig. 3 the DOS of O(1) is big in most of the frequency range for which the DOS of O(5) is small. For the complementary part of the support of the DOS the situation is opposite. This implies that ordering produces an important concentration of the DOS over the chain for most of the frequency range, which emphasis the importance of the chain for superconductivity in this compound.

In order to compare the effect that the crystalographic ordering has on the DOS we have calculated the completely disordered case, η=0. It is clear from the figures that the DOS is much smaller for the tetragonal phase than for the orthorombic phase. A complete analysis of the DOS as a function of x and η rules out the possibility of supposing that the only effect of reducing the oxygen content is to increase the number of electrons per site. This can be concluded from fig.2 where we have represented as well the DOS of cooper for x=2 and η=0.

In order to have a understanding of the behavior of the total DOS, as a function of x and η we show in fig.4 the dependance of $\rho_T(\epsilon_f)$ with x. The DOS at the Fermi level increases monotonically as a function of x, in the physical interval $0 < x < 1$. This behavior is greatly enhanced for the orthorombic phase at the cooper and oxygen site of the basal plane as it is shown in the figure. As it is plausible to admit a similar dependance of the critical temperature with $\rho_T(\epsilon_f)$ as in the case of B.C.S. theory, the results could explain why the critical temperature increases with x in the range $.5 < x < .9$ and why the orthorombic phase is a better candidate for superconductivity than the tetragonal phase. In fig.4 we show as well the dependance of the DOS at the Fermi level for the cooper and oxygen atoms of the other two planes. The DOS is almost independent of x in this case reducing a little when x increases. These results will emphasize the importance of the basal plane for superconductivity for which the DOS is extremely sensible to the oxygen concentration. The diferential behavior is of course a consequence of the fact that the change in the oxygen composition affects only the basal plane, situation which seems to be very well established experimentally( ).

Even if we believe that these results are indicating a plausible situation, the knowledge of the microscopic mechanism which produces superconductivity is essential to establish in a realiable way that the density of states is one of the crutial parameters in determining the superconductive temperature.

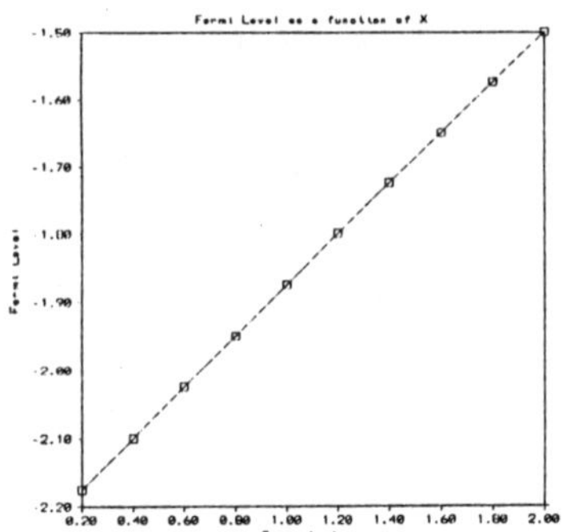

Figure 1

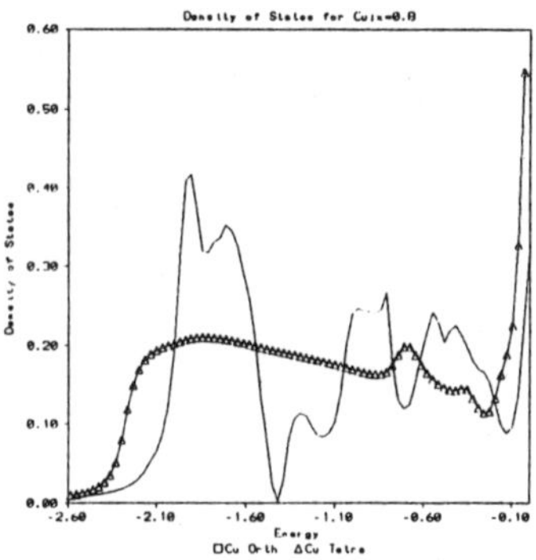

Figure 2

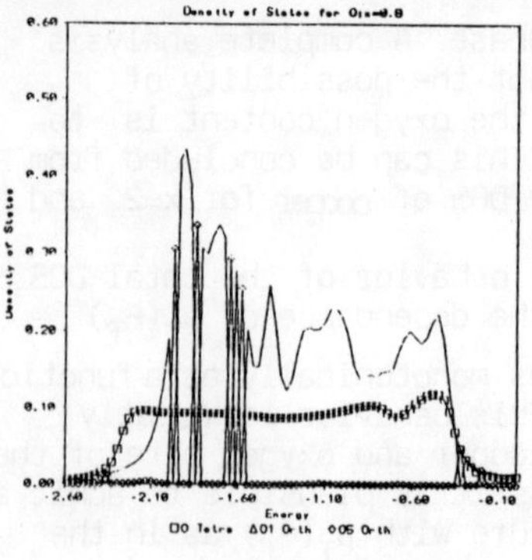

Figure 3

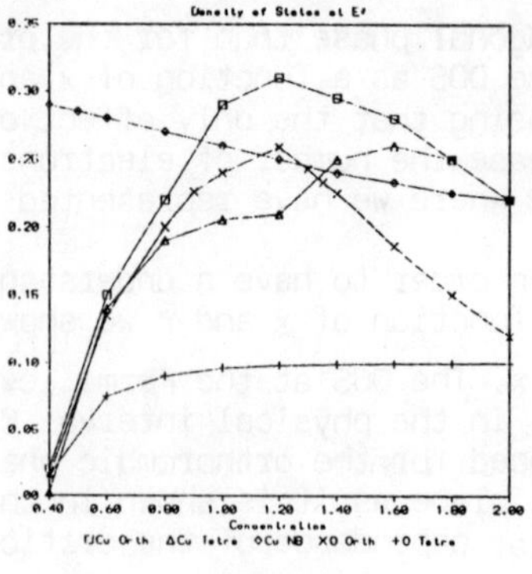

Figure 4

REFERENCES

1) J.D.Jorgensen, M.A.Beno, D.G,Hinks, L.Soderholm, K.L. Volin, R.L.Hitterman, J.D.Grace, I.Schuller, C.V.Sogre, K.Zhang and M.S.Kleefisch, Phys.Rev. B, 36,7,3608, 1987.

2) J.D.Jorgensen, B.W.Leal. W.K.Kwog, G.Crabtree, U.Umezawa, L.J.Nowicki and A.P.Paulikas, Phys.Rev. B, 36,10,5731, 1987

3) J.M.Sanchez. F.Mejía-Lira and J.L.Morán Lopez - To be published.

4) L.F.Matheiss and D.R.Hamann, Soli.St.Com., 63,5,395,1987.

5) N.Y.Ching, Yongnian Xu, Gnang-Lin, K.W.Wong, F.Zandichnadom, Phys.Rev.Lett. 59,12,1333, 1987.

6) S.Massida, Jaejun Yu, A.J.Freeman, Phys.Lett.A, 122,3,198, 1987.

A MAGNETIC FRÖLICH MODEL FOR THE OXIDE SUPERCONDUCTORS

Mauro M. Doria
CNLS MS-B258, Los Alamos National Laboratory,
Los Alamos NM 87545, USA
Hortencio A. Borges
Depto. de Física, Pontifícia Universidade Católica do Rio de Janeiro,
Rua Marquês de São Vicente, 225, Rio de Janeiro RJ 22453, Brasil

ABSTRACT

The families of Oxide Superconductors, represented by $La_{2-x}(Sr,Ba)_xCuO_4$ and $RBa_2Cu_3O_{6+x}$ (R a Rare Earth element), have a temperature-concentration phase diagram which show for increasing values of x, antiferromagnetic order with decreasing Néel temperature and the eventual onset of superconductivity. We discuss a microscopic model for the Cu-O planes that is in qualitative agreement with this phase diagram. It differs from the theories of Aharony et al. and Goddard et al. by the fact that the superconducting condensate is purely of the singlet type here.

The common feature of all the high temperature oxide superconductors with Tc>30K, including the Bi and Tl based compounds, is the existence of planes formed by the copper and oxygen atoms (CuO_2). Another striking common feature of the oxide superconductors, so far only studied in the so-called 214 (Tc=40K) and 123 (Tc=90K) compounds, is their diagram critical temperature versus concentration x. Both superconducting materials are connected to "parent" insulating compounds (x=0) displaying long range antiferromagnetic order with copper from the CuO_2 in a $3d^9$ configuration (Cu^{++})[1,2]. As the concentration x increases the Néel temperature decreases and eventually a superconducting critical temperature becomes observable at a critical doping x_2, reaching a maximum at x_3 and finally disappearing at x_4[3-7]. The region between the antiferromagnetic order and the superconducting state is presently not well understood and experiments there may play a crucial role in determining the mechanism responsible for high Tc superconductivity.

Here we report on a theory recently proposed[8] to explain the phase diagram temperature versus concentration. In this theory the localized magnetic moments responsible for the antiferromagentic order at small values of x also lead to the superconducting phase at larger x. As argued by Emery[9], the most likely scenario in this case is to have the holes in the oxygen sites and the pairing mechanism should involve the local spin configurations on the Cu sites. Theories within this scenario have been proposed by Goddard et al.[10] and Aharony et al.[11]. The present Hamiltonian is similar to the one discussed by these authors but with a quite important distinction. This paper's Hamiltonian favors a pure singlet state whereas those authors' theories imply an admixture of triplet and singlet pairing between carriers. Their theories lead to a macroscopic superconducting wave function without a well defined parity symmetry and the present one doesn't. I write the corresponding Hamiltonian as follows:

$$H_{eff} = H_p + H_d + H_{pd} \tag{1}$$

where

$$H_p = -t \sum_{\langle i,j\rangle \sigma} (A^{+}_{i\sigma} A_{j\sigma} + c.c.) + U \sum_i n_{i\uparrow} n_{i\downarrow} \tag{2}$$

$$H_d = J_{dd} \sum_{\langle d_i d_k \rangle} S_{di} \cdot S_{dk} \tag{3}$$

$$H_{pd} = -G \sum_{\langle d_i\, j\, d_k \rangle} n_j\, S_{di} \cdot S_{dk} \tag{4}$$

where $G > J_{dd} > 0$, $n_{j\sigma} = A^{+}_{j\sigma} A_{j\sigma}$ and $n_j = n_{j\uparrow} + n_{j\downarrow}$, describes the total hole density on the linking oxygen bond between the nearest copper spins $\mathbf{S}_{di}$ and $\mathbf{S}_{dk}$. The vacuum for the operators representing the oxygen holes corresponds to the O^{--} state.

The qualitative agreement of the present theory with the temperature versus concentration diagrams can be seen as follows. As the concentration x increases the antiferromagnetic state tends to loose its long range order because of the increasing number of ferromagnetic pockets created by the carriers and T_N is expected to decrease. Finally one reaches a threshold where the number of carriers is large enough that two holes are sufficiently close to search together their energy minima; pairing begins to take place in the antiferromagnetic background and the superconducting state appears. For very large carrier concentration the antiferromagnetic spin background gets so saturated with carriers that it cannot provide the spin flipping mechanism of attraction and superconductivity ceases to existent at this moment. Aharony et al.[(11)] claimed a spin-glass phase between the antiferromagnetic and superconducting regions in the temperature-concentration diagram. Similar spin-glass picture follows from this theory too.

To understand the attraction mechanism of the above theory and its differences from Refs. (10) and (11) , first consider the situation of a single hole. The hole in the bonding oxygen induces a ferromagnetic interaction between the nearest copper atoms and therefore one of the copper spins is forced to flip its spin ferromagnetically. As a result one obtains 4 broken bonds that is, 4 bonds having ferromagnetic interaction. Introducing a second hole in one of the broken bonds will cause no further spins to flip from their original antiferromagnetic configuration whereas if introduced far away from the first one another set of 4 bonds must be broken. Thus it is energetically favorable for two holes to stay together. Refs.(10) and (11) give triplet pairing in this situation and in the present one pairing is of singlet type. A net gain of energy also exists if the holes are separated by two non-consecutive bonds. Such configurations can be achieved with just 6 broken bonds as one can easily check. For this situation Refs. (10) and (11) allow singlet pairing like the present one.

The attraction mechanism described here is based on similar energetic considerations of the so-called Frölich model[12]. The elastic energy whose competition with the electronic leads to fermion-fermion attraction in the Frölich model is replaced here by a half-filled band forming an AF Heisenberg ground state (Cu sites). Therefore superconductivity is due to the competion between different exchange interactions produced by the presence of local conducting charges (Oxygen holes).

This research is supported bu the U.S. Deartment of Energy under Contract N0. W-7405-ENG-36.

REFERENCES

(1) J.M. Tranquada et al., Phys. Rev. Lett. 60, 156 (1988).

(2) D.Petitgrand and G. Collin, Proc. of the Int. Conf. on High-Temp. Superc. and Mat. and Mech. of Superc., edited by J. Müller and J.L. Olsen, to be published in Physica C.

(3) J.H. Brewer et al.,Phys. Rev. Lett. 60, 1073 (1988).

(4) J.M. Tarascon et al. in Novel Superconductivity, edited by S.A. Wolf and V. Kresin, Plenum Press (1987).

(5) D.C. Johnson et al., Phys. Rev. B36, 4007 (1987)

(6) R.J. Cava et al., Nature 329, 423 (1987).

(7) Features like the plateau at x=0.5 in the Y-123 compound (see Ref.(6)) are being ignored here.

(8) M. Doria and H. Borges, submitted to Physical Review B.

(9) V.J. Emery, Phys. Rev. Lett. 58,2794 (1987).

(10) Y. Guo et al., Science 239, 896 (1988); G. Chen and W.A. Goddard, Science 239, 899 (1988).

(11).A. Aharony et al., Phys. Rev. Lett. 60, 1330 (1988); A. Aharony et al., Proc. of the Int. Conf. on High-Temp. Superc. and Mat. and Mech. of Superc., edited by J. Müller and J.L. Olsen, to be published in Physica C.

(12) H. Frölich, Phys. Rev. 79, 845 (1950).

ON THE ROLE OF INTERBAND PAIRS IN SUPERCONDUCTORS

O.L.T. de Menezes
Centro Brasileiro de Pesquisas Físicas
Rua Dr. Xavier Sigaud, 150 - CEP 22290 - Rio de Janeiro

ABSTRACT
The stabilization of interband Cooper pairs related to systems like heavy fermions and high temperature superconductors, could be of fundamental importance in the understanding of the superconductivity ground state.

INTRODUCTION

Since the formulation of the BCS theory of superconductivity [1] experimental results with deviations from the predictions have been found in many systems. The studies, to deal with these "anomalies", have either suggested other physical mechanisms or considered new interactions added to the original BCS Hamiltonian, which could be relevant for particular systems.

The discussion on the validity of BCS theory was revived, in the last years, after the discovery of the heavy fermion superconductors (HFS) [2]. Due to the anomalies on several properties, theoretical models have been proposed. Many of these try to explain the unusual behaviours suggesting theoretical explanations quite distinct from the original BCS framework. Recently, the new high temperature superconductors (HTS) systems [3] have been responsible for a large number of models and interpretations, as part of a research activity without precedent in the history of science.

We intend to consider the case of two band states contributing to the density of states at the Fermi energy in the normal state. This case, which could include both HTS and HFS systems, open the question of the stability of interband pairs constructed with electrons from each band state. Besides been called interband pairs, these Cooper pairs, have been also called exotic or hybrid, in order to distinguish them from the usual interband pairs. The presence of interband transitions and superconducting pairs can be responsible for very anomalous behaviours. Even in the case where phonon mediate the attractive interactions, strong deviations can be obtained. If hybrid pairs occur, something so far difficult to know, we claim that this extended BCS framework is able to describe a broad spectrum of behaviours, which have been, at first sight, interpreted as evidence hybridization is responsible for the mixed solution, in terms of the simultaneous superconductivity of the normal and heavy intraband and interband states.

Very interesting results for the interband pairs were obtained by Shuh [10] reaching important conclusions for the interband pairs framework. Using a non- phononic mechanism, similar to that of Fedro and Sinha [6], based on the d-f exchange interaction, they have obtained an initial critical field slope strongly enhanced with respect to the BCS value, a specific heat jump larger than that predicted by the BCS theory, and the same value of $2\Delta/kT_c \sim 3.45$, as expected in the BCS model. These results are in agreement with the features observed in $CeCu_2Si_2$ and show that the interband pairs approach gives very reasonable values.

On the role of the hybridization, it was shown [11] that the interband pairs of the d-f type are always present and should not be disregarded. This remains valid even in the case where the hybridization is large enough and even if the Fermi energy lies on one of the branches of the orthogonalized states. Very recently, this point was further discussed with particular emphasis on the HTS [12].

In 1984 the first measurements using ultrasound in HFS were done [13]. The large deviation of the attenuation from the BCS exponential and the interpolation of power laws give rise to the generalized idea that this could be understood as a final evidence of the triplet superconductivity in UPt_3. A disagreement between different fittings of the power law can be found in the literature [14]. However, if we look at the BCS model under a more generalized framework and take into account the presence of the two band state around the Fermi surface, the possibility of f-d transitions induced by the ultrasound, give us very interesting and unconventional results [15]. The strong absorption, experimentally observed below the critical temperature, could be explained within the BCS framework. This absorption could be strong enough to shadow the low temperature behaviour so that we could try to look for power laws instead of the exponential behaviour, which could remain at very low temperatures. It is very interesting to note that recent ultrasound attenuation measurements have been done in $YBa_2Cu_3O_7$ exhibiting the same strong absorption in the HFS [16]. Also, strong absorption was observed in La based HTS systems, with the frequency dependence of the peak position [17] in agreement with the theoretical predictions for the interband transitions [15]. In our understanding these are direct evidences showing that interband effects are very important in HTS systems as discussed in the case of HFS.

Other interesting features can be pointed out. To obtain the BCS classical result, it is necessary to suppose that the effective electron-electron interaction remains constant, within a range of $2\hbar\omega_D$, around the Fermi energy. However, in the case where two band states are present, this approximation is not straightforward. For the hybridized two band case, the assumption of a constant attractive interaction is not an adequate approximation. The dependence of the energy gap can be strongly affected, being responsible for deviations of the conventional BCS behaviour [18].

The theoretical approaches which treat the two band problem have neglected an important fact: the invariance of the Hamiltonian under inversion [19]. The correct inclusion of this is responsible for other of a non- BCS behaviour.

THEORETICAL SURVEY AND DISCUSSION

The theoretical questions which arise when one considers the extension of the BCS theory, for the case of overlapping bands, were first treated in the work of Suhl et al. [4]. They consider only interactions for the intraband s-s and d-d pairs. The "interband term" corresponds to the destruction, in one band, and creation, in the other, of a Cooper pair.

The possibility of interband pairs stabilization was first discussed in 1968 by Chow [5] studying the effects of non-magnetic impurities in the superconductivity of systems with overlapping bands. Due to the structure of the interband term the possibility of formation of Cooper pairs with one electron of each band state was considered.

The relevance of the intraband and interband electron-phonon (el-ph) interaction, as an important mechanism to describe anomalies, was many times used in the theoretical study of intermediate valence problems. After the interesting discover of the heavy fermion systems [2], with their anomalous properties, theoretical activity involving the HFS follows and several models have been proposed. *New* or *old* mechanisms with *new* or *old* suggestions on the formation of pairs have been studied. A typical and important aspect of these HFS systems is the hybridization between the two band states. In spite of the large number of theoretical models and experimental results, a final answer to the question of the relevant mechanisms underlying the unusual properties of the HFS is still lacking.

The possibility of superconducting singlet hybrid pairs, involving d- and f- electrons, related to the intermediate valence problem, was treated in 1982 [6]. The authors argued for the presence of a highly coherent state, driven by correlations and stabilized by the negative effective exchange interaction between the f- and d- conduction states. They obtained a superconducting behaviour and indicated values for T_c of a few milikelvins. They showed [7] that no pair solution appears when the chemical potential lies into the lower branches of the orthogonalized states, for the case of one electron per atom. The value of T_c decreases when the effective energy of the f- electrons goes down, exhibiting its highest values in the intermediate valence regime.

The general treatment of the el-ph interaction, taking into account the possible intraband and interband states, was developed by Entel et al. [8]. The authors concluded that the hybrid pairs should have a less important role in a Kondo lattice. Also, due to the presence of interaction between the possible intraband pairs, f-f and d-d superconductivity should occur simultaneously. In the case of strongly correlated f- electrons, the authors obtained a crossover from the HFS regime to the normal conduction electron superconductivity.

The importance of the hybridization and the generalized treatment of the el-ph scatterings was stressed in Ref.[9], under the assumption that a singlet ground state is to be expected. It was shown that the ground state should be built by a linear combination of intraband and interband states. In a very simple picture, it is possible to show that the Fermi sphere becomes unstable for formation of interband Cooper pairs. An energy gap - like equation can be obtained and the features which seem to be important for the description of HFS and HTS [15,20,21]. One straightforward consequence is the possibility of distinguishing, in a comprehensive way, between different interband pairs. On the other hand the superconducting temperatures can be enhanced by a factor close to four, with respect to the intraband BCS value. Another result is connected to the possibility that more than one mechanism can contribute to the stabilization of the Cooper interband pairs. E.g., exchange-like mechanisms can be added to the normal el-ph coupling and then an additional enhancement can be expected with respect to the pure phonon mediated interactions. Moreover, depending on the ratio between the interactions, singlet or triplet interband pairs can be the superconducting ground state [20].

FINAL COMMENT

The aim of the present discussion is to emphasize that the BCS treatment of superconductivity can display behaviours in a much richer picture than those considered to characterize "BCS systems". If two band states around the Fermi energy, are involved, it is possible to show that superconducting properties could be far from those observed in conventional systems being, however, understood within a BCS generalized framework.

REFERENCES

[1] J.Bardeen, L.N.Cooper, and J.R.Schrieffer, Phys.Rev.108,1175(1957).

[2] F.Steglich, J.Aarts, C.D.Bredl, W.Lieke, D.Meschele, W.Franz and H.Shäfer, Phys.Rev.Lett.43,1892(1979).

[3] J.G.Berdnorz and K.A.Muller, Z.Phys.B64,189(1986).

[4] H.Suhl, B.T.Matthias, and L.R.Walker, Phys.Rev.Lett.3,552(1959).

[5] W.S.Chow, Phys.Rev.172,467(1968).

[6] A.J.Fedro and S.K.Sinha, in Valence Instabilities, (Edited by P.Wachter and H.Boppart) p.371, North-Holland(1982).

[7] A.J.Fedro and S.K.Sinha, in Moment Formation in Solids, (Edited by W.J.L.Buyers) p.61, Plenum(1984).

[8] P.Entel, J.Zielinski, and M.Matschke, Z.Phys.B60,243(1985).

[9] O.L.T. de Menezes, Solid State Commun.56,799(1985).

[10] B.Schuh, Phys.Stat.Sol.(b)131,243(1985).

[11] O.L.T. de Menezes, Solid State Commun.57,825(1986).

[12] A.A.Aligia, preprint.

[13] D.J.Bishop, C.M.Varma, B.Batlogg, E.Bucher, Z.Fisk, and J.L.Smith, Phys.Rev.Lett.53,1009(1984).

[14] V.Muller, D.Mauer, E.W.Scheidt, Ch.Roth, K.Luders, E.Bucher, and H.E.Bommel, Solid State Commun.57,319(1986); B.S.Shivaram, Y.H. Jeong, T.F.Rosenbaum, and D.G.Hinks, Phys.Rev.Lett.56,1078(1986); I.J.Qian, M-F.Xu, A.Schenstrom, H-P.Baum, J.B.Keterson, D.Hinks, M.Levy, and B.K.Sarma, Solid State Commun.63,599(1987).

[15] A.A.Aligia and O.L.T. de Menezes, J.Magn.Magn.Mat.63&64,423(1987).

[16] M-Xu, H-P.Baum, A.Schenstrom, Bimal K.Sarma, Moises Levy, K.J.Sun, L.E.Toth, S.A.Wolf, and D.U.Gubser, Phys.Rev.B37,3675(1988).

[17] S.K.Sinha, private communication.

[18] O.L.T. de Menezes and A.A.Aligia, Jpn.J.App.Phys.26,1231(1987)..

[19] G.Czycholl and H.J.Leder, Z.Phys.B44,59(1981).

[20] O.L.T. de Menezes, J.Magn.Magn.Mat., in press (1988).

[21] O.L.T. de Menezes, Solid State Commun.65,1559(1988).

HIGH T_c SUPERCONDUCTING SHIELDS

O.G. Symko, W.J. Yeh, and D.J. Zheng, Dept. of Physics, University of Utah, Salt Lake City, Utah 84112, and S. Kulkarni, Ceramatec, Inc., Salt Lake City, Utah 84115.

ABSTRACT.

Superconducting shields made out of $Y Ba_2Cu_3O_7$ material have been developed for shielding a SQUID sensor at liquid nitrogen temperature. The shields consist of tubes of current density of 63 amperes/cm^2. The performance of the shields shows excellent characteristics at 77K, external fields being attenuated by factors larger than 10^6 when the external fields are less than 9.5 gauss. Larger fields show flux penetration due to insufficient current density in the shields.

INTRODUCTION AND RESULTS: We have developed r.f. biased SQUIDS made from $Y Ba_2 Cu_3 O_7$; they are simple to make and they show high sensitivity. For operation at 77K, a high degree of shielding is needed and hence, we present results on high T_c superconducting shields. The geometry is a long hollow cylinder; external magnetic fields applied to this shield are attenuated by induced currents in the shield which keep the flux inside the tube constant, as it is a resistanceless circuit.

The dimensions of the tubes are: 6.25 cm long, 1.20 cm o.d., and 0.96 cm i.d.. The tubes were made using a solid-state reaction method and conventional ceramic fabrication techniques. High purity powders (puratonic grade) were used to fabricate the shields. The calcinated powder of the ceramic was mixed with a binder and dried. Green ceramic tubes were made using this powder and isostatic pressing techniques. The tubes were fired at 960°C for 12 hours in air and slowly cooled in the furnace. Subsequently the tubes were annealed in an oxygen atmosphere for 18 hours at 850°C.

The characteristics of the tubes were studied at 77K using a Hewlett-Packard model 4288R Fluxgate Magnetometer. The field inside a tube, at its center, was measured when an external magnetic field produced by a solenoid was increased from zero. Figure 1 shows the details of the testing arrangement.

As the external magnetic field was increased, excellent shielding behavior was observed up to some characteristic field value and then flux started to enter the tube. This is shown in Figure 2. The higher the current density of the tube, the higher the value of this critical magnetic field. As a comparison, the behavior is shown without shield and with a μ-metal shield. The critical field of 9.5 gauss corresponds to a current density of approximately 63 amperes/cm^2 which is so far our best value for this geometry of tube. From our sensitivity, the attenuation is better than 10^6 at low fields. With the characteristics here presented, the shield is adequate for SQUID operation.

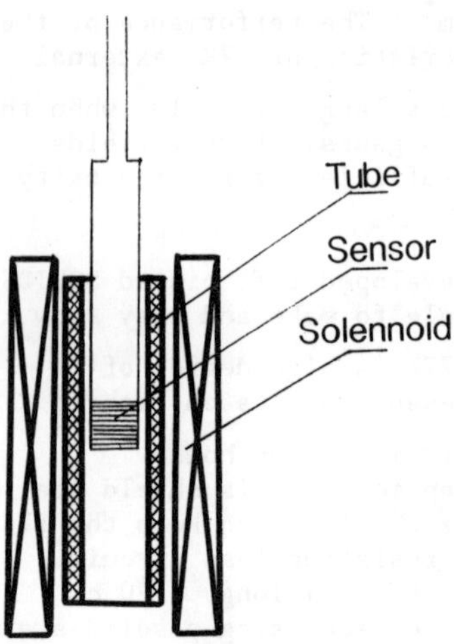

Figure 1.
Test Set-up

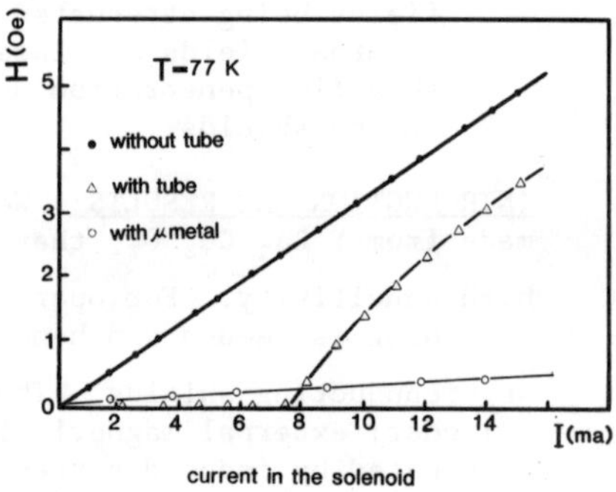

Figure 2.
Field inside shield as a function of external field

ANALYSIS OF LOW VELOCITY MÖSSBAUER SPECTRA OF A $YBa_2(Cu_{1.9583}Fe_{0.0417})_3O_\delta$ SUPERCONDUCTOR AS A FUNCTION OF TEMPERATURE

R.W. Gómez, S. Aburto, M.L. Marquina, V. Marquina, M.J. Jiménez and C. Quintanar. Facultad de Ciencias, UNAM. México D.F., 04510. México, and T. Akachi, R. Escudero, R.A. Barrio and D. Rios-Jara. Instituto de Investigaciones en Materiales, UNAM. México D.F., 04510. México.

ABSTRACT

The behaviour of the Mössbauer parameters as a function of temperature at a low velocity scan (~ ±1.7 mm/s) for a $YBa_2(Cu_{1.9583}Fe_{0.0417})_3O_\delta$ superconducting sample are presented. Special attention is paid to the asymmetry observed in the components of each of the three quadrupole doublets obtained for this sample and, in order to explain it, we propose the existence of a weak magnetic fields in the iron atom positions.

The substitution of Cu atoms with Fe atoms in the 1-2-3 superconductors has proven to be useful in the study of some characteristics of this system. In first place, this substitution has an effect on the critical temperature of the sample; in second place the iron atoms can be used as probes of the local suroundings of the site where they go to by means of Mössbauer spectroscopy [1-7]. It seems to be now a general agreemment that, at least for low iron concentrations, the Fe atoms substitute the cooper atoms in the Cu(1) sites of the well known structure of this system. In a previous work [8], we reported an anomalous behaviour of the Mössbauer parameters of a $YBa_2(Cu_{1.9583}Fe_{0.0417})_3O_\delta$ superconducting sample around 110 K.

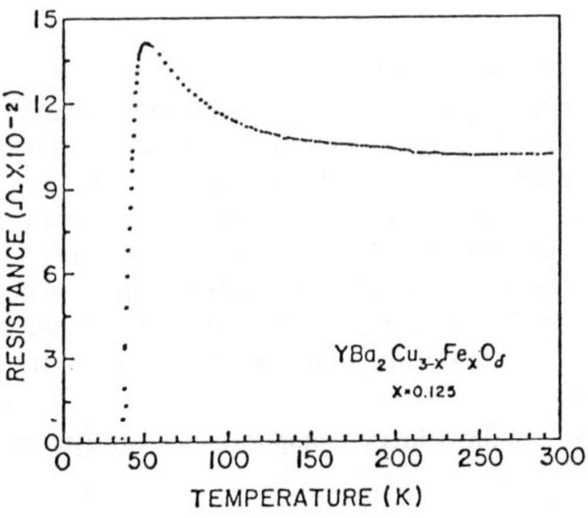

Figure 1. Resistance as a function of temperature.

In this work we present the results obtained for the same sample using a low velocity scan (~ ±1.7 mm/s) in a temperature interval from 11 to 120 K. The superconducting sample was prepared with the now standard method and the Mössbauer spectra were obtained in a transmition geometry with a Co in Pd source at room temperature and the sample in a closed cycle refrigerator.

The resistance of this sample (Fig. 1) shows a change from semimetalic behaviour above 240 K to semiconductivity like behaviour below 240 K. The superconductivity onset is at 50 K and zero resistance is reached at 35 K. The x-ray spectra showed that the sample is tetragonal, although this structure does not correspond to the high temperature tetragonal structure [9].

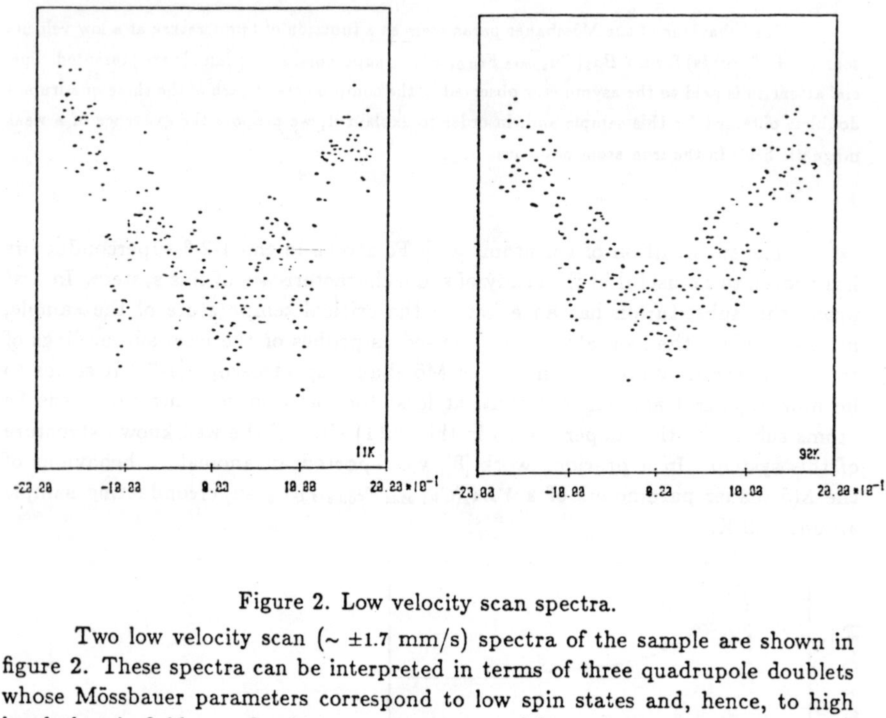

Figure 2. Low velocity scan spectra.

Two low velocity scan (~ ±1.7 mm/s) spectra of the sample are shown in figure 2. These spectra can be interpreted in terms of three quadrupole doublets whose Mössbauer parameters correspond to low spin states and, hence, to high local electric fields, confirming a previous result [8]. It is important to note the asymmetry in the absorption of the external doublet, and the difference in the widths of the two components of this doublet. This effect suggests the presence of a weak magnetic field (~ 6 koersted) in the position of the iron atoms that would be responsible for these asymmetries (here low field means low with respect to the internal fields in iron oxides, which are ~ 500 koersted).

Figure 3 shows the differences obtained in the computational adjustment assuming only three quadrupole doublets (Fig. 3a) and the ones resulting using a hamiltonian in which a combined electric and magnetic effects are taken into account (Fig. 3b):

$$H = [eQV_{zz}/\{4I(2I-1)\}]\left[3I_z^2 - I(I+1) + \eta(I_x^2 - I_y^2)\right]$$

$$-g\mu_N H(0)\left[I_z cos\theta + (I_x cos\phi + I_y sin\phi) sin\theta\right],$$

where

Q = electric quadrupole moment, e = charge of the proton,
g = nuclear g-factor, I = spin of nucleus,
μ_N = nuclear magneton, $H(0)$ = magnetic field
V_{xx}, V_{yy}, V_{zz} = components of the electric field gradient tensor,
$\eta = (V_{xx} - V_{yy})/V_{zz}$ = asymmetry parameter,
θ,ϕ = angles specifying the direction of $H(0)$ with respect to the electric fiel gradient,
I_x, I_y, I_z = components of the total angular momentum.

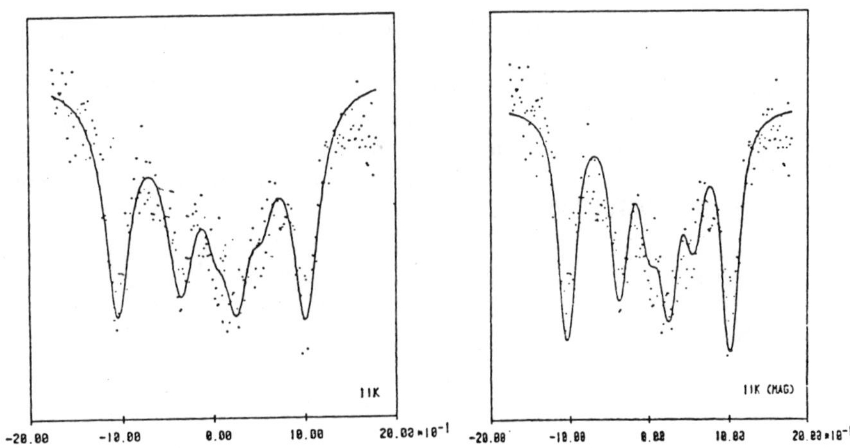

Fig. 3a Fig. 3b

Figure 3. Computational adjustment of the spectra.

In order to get good fits, it is necessary to allow for changes in the orientation (θ or ϕ or both) of the magnetic field with respect to V_{zz}. Our assumption is supported by other evidence of the existance of magnetic effects in the normal and superconducting states previously reported [10].

In conclusion, our results seem to indicate the existence of low magnetic field in the Cu(1) sites of the structure of the 1-2-3 superconducting system occupied by the substituting Fe atoms.

REFERENCES

[1] Z.Q. Qiu, Y.W. Du, H. Tang, J.C. Walker, W.A. Bryden and K. Moorjani. J. Mag. Mag. Mater., **69**, L221, (1987).

[2] M. Takano and Y. Takeda. Jpn. J. Appl. Phys., **26**, L1862, (1987).

[3] H. Tang, Z.Q. Qiu, Y.W. Du, G. Xiao, C.L. Chien and J.C. Walker. Phys. Rev. **B36**, 4018, (1987).

[4] X.Z. Zhou, M. Raudsepp, Q.A. Pankhurst, A.H. Morrish, Y.L. Luo and I. Maartense. Phys. Rev. **B36**, 7230, (1987).

[5] Q.A. Panhurst, A.H. Morrish, M. Rauttsep and X.Z. Zhou. J. Phys. C: Solid State Phys. **21**, L7, (1988).

[6] C. Saragovi-Badler, F. Labenski de Kanter, M.T. Causa, S.M. Dutrús, C. Fainstein, L.B. Steren, M. Tovar and R. Zysler. Submitted to Solid State Commun.

[7] T. Tamaka, T. Komai, A. Ito, Y. Maeno and T. Fujita. Solid State Commun. **65**, 43, (1988).

[8] R. Gómez, S.Aburto, M.L. Marquina, M. Jiménez, C. Quintanar, T. Akachi, R. Escudero, R.A. Barrio and D. Rios-Jara. Phys. Rev. **B36**, 7226, (1987).

[9] Y. Maeno and T. Fujita. Physica C, 153, (1988).

[10] S. Noguchi, J. Inoue, K. Okuda, T. Maeno and Y. Fujita. Jpn. J. Appl. Phys. **27**, L2087, (1988).

ESR STUDIES OF $YBa_2Cu_3O_{7-x}$, Y_2BaCuO_5 and $BaCuO_{2+w}$*

R.N.de Mesquita, J.H. Castilho, G.E. Barberis, C. Rettori
I. Torriani, O.F.de Lima, S. Gama and R. Jardim
Instituto de Física, Unicamp, 13081 Campinas (S.P.) Brasil

and

M.C. Terrile, H. Basso and O.R. Nascimento
Instituto de Física e Química, USP, 13560 São Carlos (S.P.) Brasil

Abstract

X and Q band ESR spectra of the high T_c superconductor $YBa_2Cu_3O_{7-x}$ and the related compounds Y_2BaCuO_5 and $BaCuO_{2+w}$ are reported. We study the room temperature spectra as a function of the oxygen concentrations x and w in the samples, and we fit the powder spectra with a model Hamiltonian for exchange narrowed coupled Cu^{2+} ions. We conclude that the Cu atoms in the superconductor are not ESR active, and that the observed resonances come from other residual phases with concentrations smaller than 5000 ppm, which cannot be observed by X-ray diffraction.

Several reports on $YBa_2Cu_3O_{7-x}$, Y_2BaCuO_5 and $BaCuO_{2+w}$ ESR spectra were published recently [1-6]. Although the general features of the different spectra reported are similar, the interpretations of the experiments disagree. The origin of the resonances in these compounds is understood by different authors in very different manner, some of them assuming that the resonances come from impurity phases. The main objective of this work is to clarify these divergencies, fitting a model for coupled Cu^{2+} ions to the experimental ESR data; we also study the ESR spectra as function of the oxygen concentration of the samples in order to better understand our results.

The samples were prepared according to the solid diffusion method which was reported previously [4]. The only difference was that for the present batch of samples the process of grinding and annealing during 20 hours at 950º C in an oxygen atmosphere were repeated 3 to 6 times in order to get as much as possible pure single phase samples.

The deoxygenated samples were obtained after heating under a vacuum of 10^{-3} Torr during 20 hours at various temperatures between 350º C and 450º C. The reoxygenated samples were obtained after a heat treatment at 450º C during 12 hours in an atmosphere of circulating oxygen, although only 3 hours were enough to recover almost the same original oxygen concentration, except for the samples of Y_2BaCuO_5 (see below).

The ESR experiments were carried out at room temperature in conventional ESR Varian E-Line X and Q-Band spectrometers. The samples were always run in form of fine powder (10 - 30 μm) and were in all cases kept in dry atmosphere.

The distances between Cu^{2+} ($3d^9$) ions in all these compounds is small enough to make their exchange (or superexchange) interaction important. Therefore, we assumed a system of coupled Cu^{2+}, with complete exchange narrowed hyperfine and Zeeman interactions as a model to fit the ESR resonances [7,8]. We fitted the corresponding Hamiltonian for S = 1/2 to the experimental spectra assuming also random orientation of the crystallites in the samples. More detailed analysis will be published elsewhere.

Table I - Fitting parameters for the ESR powder spectra

Sample	Microwave Band	g_x	g_y	g_z	ΔH(Gauss)
Y_2BaCuO_5	Q	2.08±0.01	2.12±0.02	2.25±0.01	70±10
	X	2.08±0.01	2.12±0.02	2.25±0.01	25±5
$BaCuO_{2.02}$(Fresh)	Q	2.055±0.005	2.055±0.005	2.24±0.01	100±10
	X	2.055±0.005	2.055±0.005	2.24±0.01	40±5
$BaCuO_{2.07}$(Old)	Q	2.05±0.02	2.12±0.02	2.22±0.02	202±20
	X	2.05±0.02	2.12±0.02	2.22±0.02	68±5
$YBa_2Cu_3O_7$	Q	2.05±0.02	2.09±0.03	2.24±0.02	188±20
	X	2.05±0.02	2.09±0.03	2.24±0.02	65±8

In the Y_2BaCuO_5 samples, no significant changes could be observed in the spectra after each deoxygenation and reoxygenation treatment. We could not attribute the mass losses of the samples to oxygen deficiencies because also during the reoxygenation the mass of the samples was still decreasing, probably due to the presence of small amounts of non-reacted carbonates. The spectra could be fitted with an orthorhombic Hamiltonian and the g-values obtained can be seen in Table I.

Freshly reacted $BaCuO_{2+w}$ samples present a tetragonal ESR spectrum (see Table I), whose intensity strongly depends on the samples oxygen concentration. This intensity decreases three orders of magnitude from the less (w = 0.0) to the most oxygenated (w = 0.07) sample. The ESR spectrum after one complete de and reoxygenation cycle completely reproduces the former spectrum, showing that oxygen comes out and enters in the samples in a reversible manner. We studied one of the $BaCuO_{2+w}$ samples ESR spectra as a function of time. We stored the more oxygenated of our samples in a dry atmosphere and we measured it in periods of one week during two months. We observed a strong change in the spectra, going from tetragonal to orthorhombic, with the g-values shown in Table I. This change was also observed by X-rays, presenting the appearance of unidentified spurious phases.

Two $YBa_2Cu_3O_{7-x}$ samples were studied by ESR as a function of oxygen concentration. One of them was grinded and annealed three times, and the other six times. The last one presents one order of magnitude smaller ESR intensity than the former. It is interesting to note that the resonances correspond to an orthorhombic phase with an increasing linewidth as the de and reoxygenation treatment proceeds. At the same time, X-ray analysis show that the change from orthorhombic to tetragonal phase occurs in the sample. The g-values of the fitting are comparable with those corresponding to the spurious phase in $BaCuO_{2+w}$ (see Table I).

Table II summarizes the relative intensities in our samples, which are related to the intensity of the same mass of the Y_2BaCuO_5 green phase in the same experimental conditions, which is stable. It can be observed that the more the thermal annealing, the smaller the

$YBa_2Cu_3O_{7-x}$ ESR signal.

Table II - ESR relative intensities

$YBa_2Cu_3O_{7-x}$ (1)		$YBa_2Cu_3O_{7-y}$ (2)		$BaCuO_{2+w}$	
x	Intensity	y	Intensity	w	Intensity
0.0	$4x10^{-2}$	0.0	10^{-3}	0.07	10^{-3}
0.2	$4x10^{-2}$	0.1	10^{-3}	0.05	10^{-1}
0.3	$4x10^{-2}$	0.3	10^{-3}	0.02	10^{-1}
0.0*	$2x10^{-2}$	0.0*	10^{-3}	0.08*	$2x10^{-2}$

(1) annealed 60 hs., (2) annealed 120 hs., (*) reoxygenated.

REFERENCES

*Partially supported by FAPESP and CNPq, Brazil.

1. C. Rettori, D. Davidov, I. Belaish and J. Felner, Phys. Rev. B**36**, 4028 (1987)
2. D. Shaltiel, J. Genossar, A. Grayevsky, Z.H. Kalman, B. Fisher and N. Kaplan, Solid State Commun. **63**, 987 (1987)
3. F. Mehran, S.E. Barnes, T.R. McGuire, W.J. Gallagher, R.L. Sundstrom, T.R. Dinger and D.A. Chance, Phys. Rev. B**36**, 740 (1987)
4. J.H. Castilho, P.A. Venegas, G.E. Barberis, C. Rettori, R.F. Jardim, S. Gama, D. Davidov and I. Felner, Solid State Commun. **64**, 1043 (1987)
5. D.C. Vier, S.B. Oseroff, C.T. Salling, J.F. Smyth, S. Schultz, Y. Dalichaouch, B.W. Lee, M.B. Maple, Z. Fisk and J.D. Thompson, Phys. Rev. B**36**, 8888 (1987)
6. R. Bartucci, E. Colavita, L. Sporetelli, G. Balestrino and S. Barbanera, Phys. Rev. B**37**, 2313 (1988)
7. P.W. Anderson, Jour. of the Phys. Soc. Japan **9**, 307 (1954)
8. R. Calvo and M.A. Mesa, Phys. Rev. B**28**, 1244 (1983)
9. I.K. Schuller, D.G. Hinks, M.A. Beno, D.W. Capone II, L. Soderholm, J.P. Locquet, Y. Broynseraede, C.U. Segre and K. Zhang, Solid State Commun. **63**, 385 (1987)

MOSSBAUER SPECTROSCOPIC STUDIES OF $YBa_2Cu_{3-x}Sn_xO_y$

R. C. Mercader* and G. Punte*
Departamento de Física, Facultad de Ciencias Exactas, UNLP, 1900 La Plata, Argentina.

P. Aja, M. Mondani, G. Quintana and H. Sirkin
Departamento de Física, Facultad de Ingeniería, UBA, Paseo Colón 850, 1063 Buenos Aires, Argentina.

Abstract

Resistivity measurements (77 - 300K), X-ray powder diffraction (300 K) and Mössbauer Spectra (77 - 300K) have been performed on samples of Y-Ba-Cu-O system doped with different Sn concentrations. The material obtained shows no significant drop in the critical temperature and a transition width as sharp as the undoped one for up to x=0.15. Mössbauer data indicate a +4 valence state for the Sn that is likely substituting the Cu site. Possible substitutions of Sn in other lattice sites are analized. The results obtained are correlated with other experimental studies.

Introduction

In spite of the wealth of both theoretical and experimental studies performed on the high temperature superconducting system $YBa_2Cu_3O_y$, it is not clear yet what the mechanisms responsible for superconductivity are. With

* Member of Carrera del Investigador Científico, CONICET.

the purpose of aiding in the understanding of microscopic aspects of superconductivity in this material, samples doped with tin were prepared to be investigated through Mössbauer spectroscopy (MS) of ^{119}Sn. MS is especially indicated to probe local properties like the type of bonds, vibrational behavior or electronic densities. If Sn replaces Cu in the sites of the perovskite type structure of the Y-Ba-Cu-O system, it is to be expected to bring forth information on the sites where the superconductivity mechanism is more likely to be sensitive. In addition, the effect of tin doping on the crystal structure and transport properties is also worth investigating.

Experimental

The samples were prepared through solid-state reactions of appropriate amounts of 99.9% pure Y_2O_3, CuO, $BaCO_3$ and SnO_2 by a process with similar conditions to those described elsewhere [1]. Several batches were obtained of nominal compositions $YBa_2Cu_{3-x}Sn_xO_y$ with x = 0.05, 0.10, and 0.15.

Powder X-ray diffraction patterns at room temperature, showed the orthorhombic $YBa_2Cu_3O_y$ structure [2,3] for all samples. No other phases were present within the limit of sensitivity (see fig.1).

A standard four-probe method was used to measure the resistance temperature dependence. Fig. 2 shows that the resistance goes to zero at 91K with a temperature transition width (10%-90%) of 5K for the x=0.15 tin substituted sample.

Mössbauer spectra were taken in a conventional constant velocity spectrometer in transmission geometry at several temperatures, from 77 to 300K, inside a liquid nitrogen cryostat with a 5 mCi $Ba^{119m}SnO_3$ source kept at room temperature. A 50-μm thick Pd filter was used. Spectra were fitted with a least-squares refinement program to Lorentzian line shapes with constraints. Isomer shifts are referred to $BaSnO_3$ at 300K. All spectra displayed one broad and slightly asymmetric pattern probably due to at least two unresolved Lorentzian lines. For reasons discussed below, the spectra have been fitted with two quadrupole doublets. Fig.3 shows the spectra of the x=0.15 sample at 77 and 300K. The parameters (in mm/s) for 77 and 300K are respectively:

ΔQ_1=0.46±0.04; 0.51±0.03, δ_1= 0.52±0.05; 0.40±0.03,
Γ_1= 0.87±0.05; 0.78±0.05, A_1= 37±5%; 31±4%
ΔQ_2=0.39±0.02; 0.34±0.03, δ_2=-0.09±0.05; -0.10±0.04,
Γ_2= 0.85±0.03; 0.89±0.04, A_2= 61±6%; 67±6%.

Discussion

The X-ray analyses of the samples show only the presence of single-phase orthorhombic $YBa_2Cu_3O_y$ structure without any change in the cell parameters for doping concentrations x=0.015, 0.10 and 0.15. A sample prepared with x=0.30 exhibited $BaSnO_3$ segregation. Good single-phase orthorhombic superconductors were obtained when undoped samples were prepared just by mixing and pulverizing. However, pelletizing under pressure and sintering were essential for the introduction of Sn in the structure even for the smallest tin concentrations. This finding together with the results of Pankhurst et al.[4], where the same type of compounds prepared under apparently the same conditions yielded dramatic changes in T_c when enriched ^{57}Fe was used instead of natural iron, show that in these very complex and multiphased systems the substitution is extremely sensitive to the preparation conditions.

In our case, the doping with tin did not change either Tc or the wi dth of the superconducting resistivity transition. This has also been found by other authors [5] who performed tin doping up to a level of 30%.

The Mössbauer results show a 4+ state for Sn. Based on the ionic radii of Sn^{4+}, 0.71A, and of Cu^{2+}, 0.69A, the assumption that Sn is substituting Cu in the structure seems plausible (Ba_2^+ radius is 1.35A and Y^{3+} is 0.93A). Additional evidence comes from two recent papers published during the course of the present investigation. Indeed, recoil-free fraction measurements of Sn sites have shown [6] that the behavior is different in the orthorhombic than in the tetragonal phase, and [7] that in the $EuBa_2u_{2.98}Sn_{0.02}O_y$ the f(T) of the Sn site is not like that of the Eu site, giving further support to the assumption of the substitution of Cu for Sn.

Regarding the substitution of Sn in one or both of the inequivalent sites of the Y-Ba-Cu-O structure, some authors indicate that there is a preferential substitution for the Cu(1) [7,8]. However, all the data obtained in this work could be fitted assuming substitutions for both copper sites, since the relative intensities of the MS lines, the quadrupole splittings and the isomer shifts are compatible with this picture. This has also been found by Yuen et al. [6]. These results stress even more the importance that the preparation has in the compound finally obtained.

Raman studies performed on Sn doped compound [5] are also consistent with Cu substitution, as the $\approx 500 cm^{-1}$ line corresponding to the Cu(1)-O(4)-Cu(2) stretching mode of the bridging O(4) oxygen, broadens anomalously and finally vanishes with increasing tin concentration.

The numerous papers on iron substitution [4,9-14] also give evidence that the probes in small percentages of doping replace the Cu in the structure rather than the heavier ions. This claim is based on the fact that different oxygen treatments influence the probe environment as revealed by the Mössbauer spectra. Indeed, besides the 2:1 relative intensities of the lines, our results for the quadrupole splittings for the two sites at room temperature: $\Delta Q_1/\Delta Q_2 = 1.5 \pm 0.22$ is coherent with the ratio of the similar assigned substitutions of Fe in the two Cu sites [9].

The value of $\delta \approx -0.10$ mm/s for the Cu(2) site means that its character is more ionic than the Cu(1) site according to the classification of Flinn [15]. This higher covalent character for Cu(1) site is consistent with the structural results that give the shortest Cu-O distance for the Cu(1)-O(4) (1.85Å).

No significant variation was found for the MS parameters at different temperatures between 77 and 300K, except for the isomer shift and the normalized area. We conclude that the local environment of Sn does not change through the superconducting transition, and, since the isomer shift variation can be accounted for by the variation of the second order Doppler shift on its own, neither does the electron density at the Sn site.

References

1. K. Kadowaki, Y. K. Huang, M. Van Sprang and A. A. Menovsky, Physica, 145B (1987) 1.
2. M. A. Beno, L. Soderholm, D. W. Capone II, D. G. Honks, J. D. Jorgensen, I. K. Schuller, C. U. Segre, K. Zhang and J. D. Grace, Appl. Phys. Lett. 51 (1987) 57.
3. J. M. Tarascon, W. R. McKinnon, L. H. Greene, G. W. Hull and E. M. Voyel, Phys. Rev. B 36 (1987) 226.
4. Q. A. Pankhurst, A. H. Morrish, and X. Z. Zhou, Phys. Lett. A 127 (1988) 231.
5. Zhang Ming-shen, C. Ziang, H. Gongjian, J. Xin, S. Huimin, W. Guigin, Z. Jiwan, J. F. Scott and P. D. Beale, Submitted to Solid State Comm.
6. T. Yuen, C. L. Lin, J. E. Crow, G. N. Myer, R. E. Salomon, and P. Schlottmann, N. Bykovetz, and W. N. Herman, Phys. Rev. B 37 (1988) 3770.
7. P. Boolchand, R. N. Enzweiler, I. Zitkovsky, J. Wells, W. Bresser, D. McDaniel, R. L. Meng, P. H. Hor, C. W. Chu and C. Y. Huang, Phys. Rev. B 37 (1988) 3766, and references therein.
8. T. M. Uen, T. J. Yang, Y. S. Gou, T. M. Chen, C. Y. Wang, L. T. Chen, H. L. Chang, J. J. Wu and F. W. Hsu, Symposium on Low Temperature Physics, September 7-8, 1987, Hsinchu, Taiwan.
9. E. Baggio Saitovitch, I. Souza Azevedo and R. B. Scorzelli, this Conference Proceedings and references therein.
10. C. Saragovi-Badler, F. Labenski-Kanter, M. T. Causa, S. M. Dutrús, C. Fainstein, L. B. Steren, M. Tovar and R. Zysler, this Conference Proceedings and references therein.
11. R. V. Gómez, S. Aburto, M. L. Marquina, M Jiménez, C. Quintanar, T. Akachi, R. Escudero, R. A. Barrio

and R. Ríos Jara, this Conference Proceedings and references therein.

12. X. Z. Zhou, M. Raudsepp, Q. A. Pankhurst, A. H. Morrish, Y. L .Luo, and I. Maartense, Phys. Rev. B 36 (1987) 7230.

13. H. Tang, Z. Q. Qiu, Y.-w. Du, G. Xiao, C. L. Chien and J. C. Walker, Phys. Rev. B 36 (1987) 4018.

14. J. M. D. Coey and K. Donnelly, Z. Phys. B 67 (1987).

15. P. A. Flinn, in "Mössbauer Isomer Shifts", ed. G. K. Shenoy and F. E. Wagner, North Holland, Amsterdam, 1978, 595.

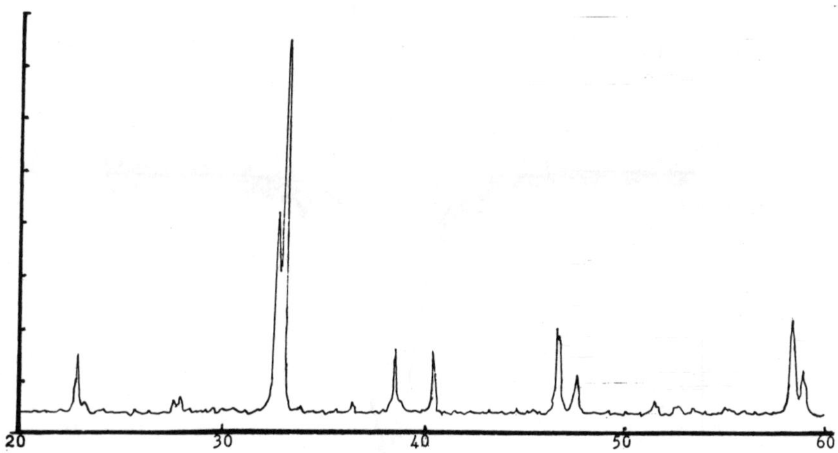

Fig. 1. X-ray diffraction pattern (x=0.15).

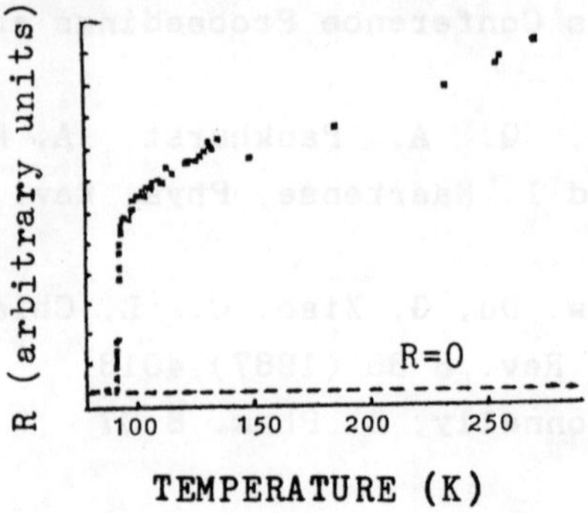

Fig. 2. Temperature dependence of resistance (x=0.15).

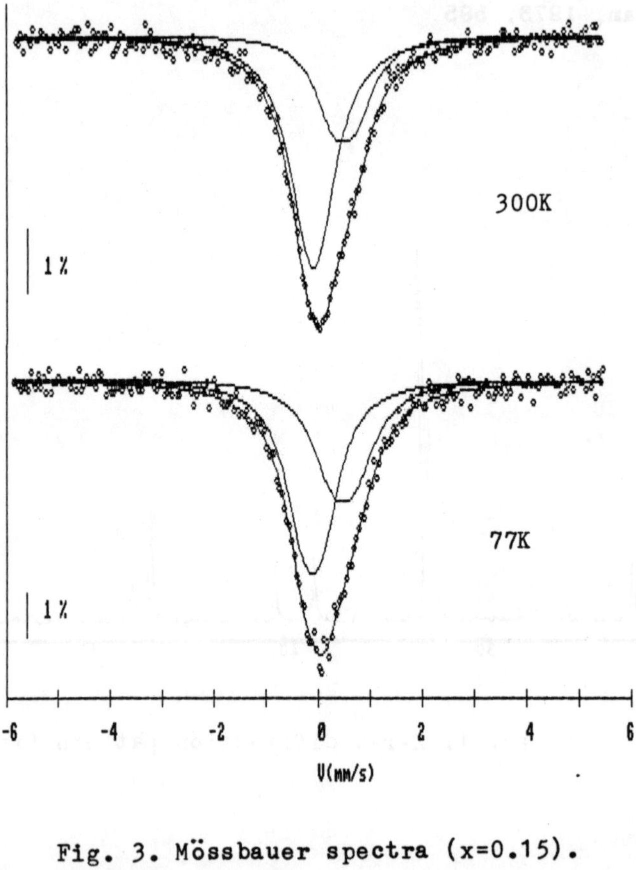

Fig. 3. Mössbauer spectra (x=0.15).

PHOTOACOUSTIC CHARACTERIZATION OF SUPERCONDUCTING OXIDES: THERMAL DIFFUSIVITY

E.C.SILVA, H.VARGAS, A.C.BENTO, R.F.JARDIM, S.GAMA
Instituto de Física, UNICAMP, C.P. 6165
13081 - Campinas - São Paulo - Brasil

E.A.PINHEIRO, F. GALEMBECK
Instituto de Química, UNICAMP, C.P.6154
13081 - Campinas - São Paulo - Brasil

1. ABSTRACT

The photoacoustic thermal characterization of some super conducting oxides is discussed. The physical quantity used as a thermal monitoring parameter is the thermal diffusivity, which is measured using the photoacoustic phase data. We demonstrated that thermal diffusivity is very sensitive to the composition and porosity variations. We discuss some correlations between the thermal diffusivity behavior and the superconducting oxides properties.

2. DIFFUSIVITY MEASUREMENTS BY PHOTOACOUSTIC EFFECT

Various methods have been developed to measure thermal diffusivity by means of the photoacoustic effect(1,2,3). We have used here the single modulation method(4,5). The method consists in measuring the relative phase lag $\Delta\phi = \phi_F - \phi_R$, at a single modulation frequency w, between the rear surface illumination (R) and the front surface illumination (F) (see figure 1). Using the thermal diffusion model of Rosencwaig and Gersho(6) for the production of the photoacoustic signal this phase lag is given by

$$\tan(\Delta\phi) = \tanh(\ell a_s)\tan(\ell a_s) \tag{1}$$

where ℓ is the sample thickness and $a_s = (w/2\alpha)^{1/2}$, α the thermal diffusivity.

When thermoelastic bending is present we can no longer use

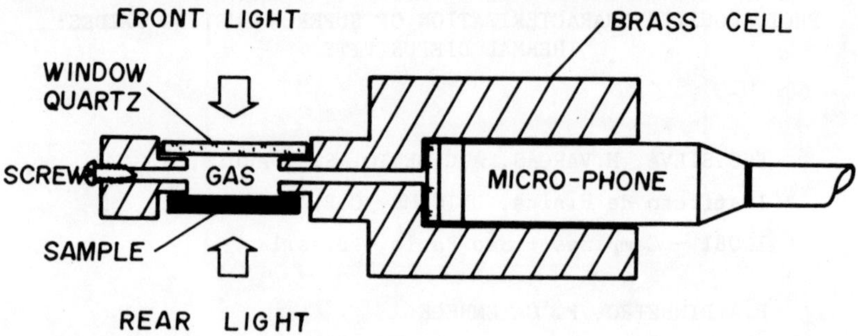

Figure 1 - Photoacoustic cell for α measurements

the phase lag method and one has to resort the front signal phase data. The thermoelastic coupling theory[7] predicts that the front phase $\emptyset_F$ varies as

$$\emptyset_F = -\frac{\pi}{z} + \arctan\left(\frac{1}{z+1}\right), \quad z = \ell a_s \qquad (2)$$

According to the behavior of the rear signal with the modulation frequency[8] we fit the equation (1) or (2) to calculate α.

3. RESULTS AND DISCUSSION

It is well known that thermal diffusivity is very sensitive to composition and microestructure variations of solid materials[9]. For the high-Tc superconducting oxides is the case. We have measured α for the superconductor $La_2\ Ba_4\ Cu_6\ O_{14+y}$. Several thin disks of 8mm diameter were prepared by coprecipitation and by applying an axial pressure of 200 MPa. After, to simulate different porosities,we applied different isostatic pressures to each disk. The results can be seen in figure 2. As expected α increase with decreasing porosity.

As an expressive example we measured α for the $Y\ Ba_2\ (Cu_{1-x}\ Mn_x)_3\ O_7$ series. These samples were prepared by mixing the oxides and with sinterization at 970°C by 100h. The structure, the superconducting and the magnetic properties of these samples were measured (X-ray, resistivity and susceptibility) and presented by R.F.Jardim and others (this conference). All the samples (0.0,5.0,

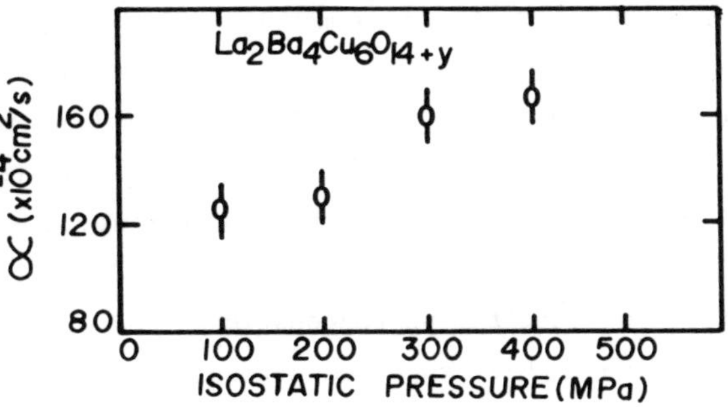

Figure 2 - Thermal diffusivity as a function of porosity

7,5, 10.0, 15.0 and 20.0 at % Mn) are superconductors with the same T_c (~92K) and with approximately same transition width.

Figure 3 shows that α remains constant up to 5.0% manganese, then decreases steeply and, after 15.0% manganese, a new and lower steady value is reached. These results are consistent with the fact that after 5.0% Mn new phase, not superconducting, are precipitated, like was well observed by x-Ray and optical metallography (R.F.Jardim and others, this Conference). Figure 4 shows the electrical resistivity measurements, at normal state (conducting), for samples, with different manganese concentrations. It's possible to infer that we

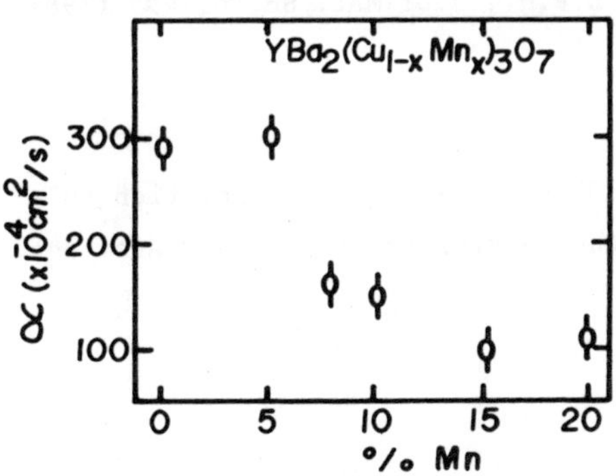

Figure 3- Thermal diffusivity as a function of concentration

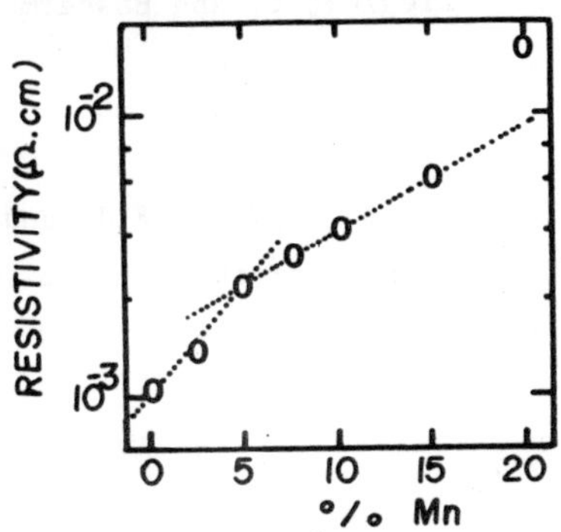

Figure 4 - Electrical resistivity as 150K as a function of concentration

observe two regions separated at approximately 5.0% Mn concentration. This result is also consistent with obtained for α. One possible interpretation for these results is that about 5.0% - 7,5% Mn a solubility limit at 970°C is attained for the manganese in the superconducting phase. So, other phases, not superconducting, are precipitated.

The results show that the photoacoustic method may become a useful technique for the high-Tc superconducting oxides characterization.

4. REFERENCES

1. Adamns, M.J. and Kirkbright, G.F., Analyst 102, 281 (1977).
2. Sugitami, Y. and Fujinami, M., Bull. Chem.Soc.Jpn.54, 722(1981)
3. Charpentier, P., Lepoutre F. and Bertrand, L.J.Appl.Phys.53 608 (1982).
4. Pessoa Jr., O.F. et allii, J.Appl. Phys. 59 (4), 1316 (1986)
5. Bento, A.C., Vargas, H., Aguiar, M.M.F. and Miranda, L.C.M., Phys. and Chem. of glasses, 28 (3), 127 (1987)
6. Rosencwaig, A. and Gersho, A.J.Appl. Phys. 47, 64 (1975)
7. Rousset, G. et allii, J.Appl. Phys. 47, 64 (1975)
8. Leite, N.F., Cella, N., Vargas, H. and Miranda, L.C.M., J.Appl. Phys., 61, 3023 (1987)
9. Ziegler, G. and Hasselman, D.P.H., J.of Mat. Sc.16, 495 (1981)

ACKNOWLEDGEMENT

E.C.S., H.V., R.F.J. and S.G. express their appreciation to FAPESP for the support in the participation of the Conference.

I-V CHARACTERISTICS IN SUPERCONDUCTING $YBa_2Cu_3O_{7-\delta}$ FILMS

M. AKINAGA*, D. ABUKAY, V. SOARES and L. RINDERER

Institute of Experimental Physics, University of Lausanne, CH-1015 Lausanne-Dorigny, Switzerland.

The temperature dependences of the electrical resistivity and I-V characteristics of the high-T_c $YBa_2Cu_3O_{7-\delta}$ films prepared by RF magnetron sputtering have been studied. We have clearly observed the I-V characteristics due to superconducting weak link where the grain boundaries play a dominant role.

PACS number: 74.75.+t, 74.50.+r

1. INTRODUCTION

The impact of new class of high-T_c superconductors[1,2] on the Josephson superconductive electronics opens new stimulating perspectives. However, the difficulties connected with the preparation of good quality films of these materials and the micro-fabrication of these films with very short coherence length have not eliminated yet. In order to investigate the electrical properties of thin film with small grain size, we tried to get such a film by rather special method. In this paper we would like to report the synthesis of the thin film of $YBa_2Cu_3O_{7-\delta}$ and the results due to the Josephson weak link by grain boundary.

2. SAMPLE PREPARATION

Thin films of $YBa_2Cu_3O_{7-\delta}$ were deposited onto (100) $SrTiO_3$ substrates water-cooled. Target were made by sintering a disk of 100 mm diameter and 5 mm thick with the composition (Y: Ba: Cu) = 1: 2: 4.5. Sputtering was carried out under the conditions: substrate temperature 20 $^{\circ}$C (water-cooled); Ar gas pressure 1.5 x 10^{-2} Torr; RF power 100 w; deposition rate 14 nm/min. The deposited films were annealed in a pure oxygen atmosphere at 800 ~ 820 $^{\circ}$C. In Fig. 1, the SEM image of annealed sample is shown. The grain sizes are almost in a order of 0.1 μm and this values are much less than those in bulk sample or in thin film deposited on usual hot substrate. We think this small size is due to rather low reaction energy on a water-cooled substrate during sputtering.

---

* On leave from Department of Physics, Fukuoka University of Education 811-41, Japan.

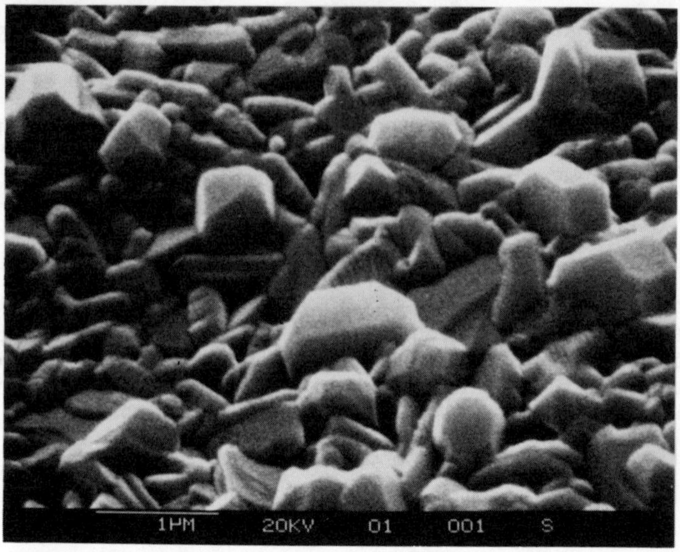

Fig. 1. SEM picture of $YBa_2Cu_3O_{7-\delta}$ film on (100) $SrTiO_3$ substrate.

3. EXPERIMENTAL RESULTS AND DISCUSSION

The electrical resistance of the samples were measured by usual four-probe method in detail. In Fig. 2, a typical I-V plots are shown together with the figure of resistance behavior in the region of the transition. These plots were acquired point by point in the stationary state for both directions of the current at T = 40 K just below the off-set temperature. This characteristics are conventional ones for superconducting weak link.[3] In this case, the excess current which is determined by the straight line (solid line in Fig. 2) from the normal resistive part holds about 10 mA.

Figure 3 shows each characteristics at different three temperatures, T = 20 K much below the transition temperature, T = 60 K midle point and T = 80 K just above usual on-set temperature 78 K. With increasing temperature, the characteristics approach to a normal ohmic ones and the excess current decreases. At T = 60 K, its characteristics are changed to modified ones where critical current seem to disappear but a " bump " on the I-V curve exists. Even at 80 K we can find such modified characteristics but it becomes rather faint.

We did not artificially make any fabricatrion on junction in the film where Josephson effect can take place. Then this weak link is " natural " one. It is easy to understand that the grain boundary in the film shown in Fig. 1 works as a superconducting weak link, namely, a grain boundary junction as reported.[4] We can image that very many grains contact each other randomly and they are forming a complicated network of Josephson junctions.

As shown in Fig. 1, the grain size is very small and it may result to rather high electrical resistivity in the film. Actually, this sample has a large value of resistivity 4×10^{-2} Ω cm. However, recalling the very short coherence length of this material, 10 ~ 20 A, it should be remarkable that the film deposited on a cooled substrate easily shows the properties due to superconducting weak link.

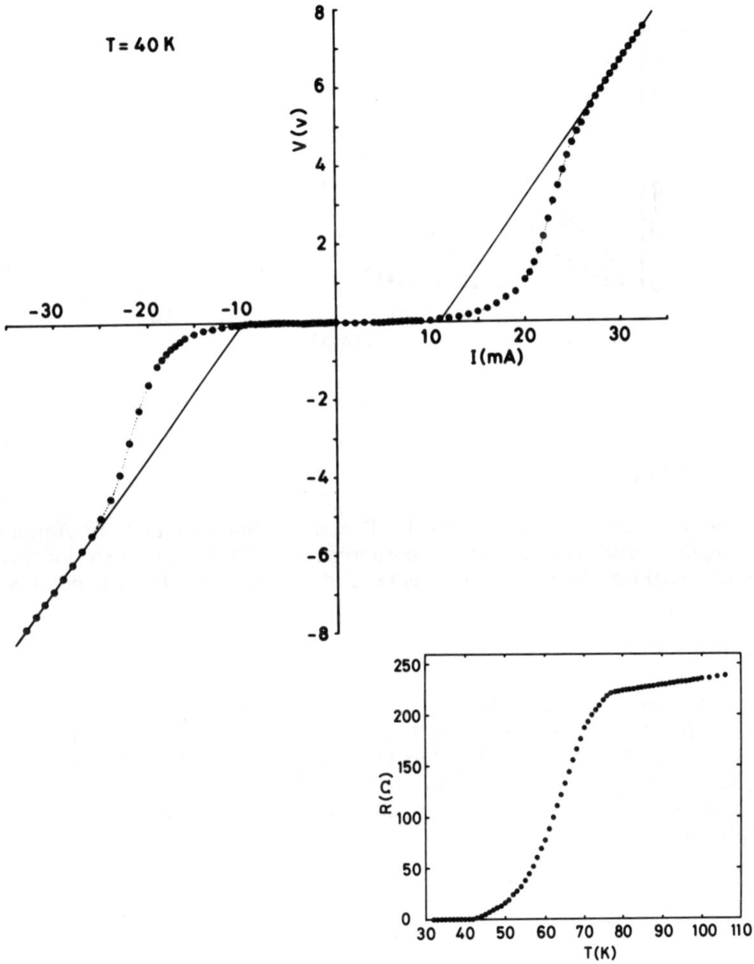

Fig. 2. I-V plots of $YBa_2Cu_3O_{7-\delta}$ film at T = 40 K. Inserted figures show the resistance behavior in the region of transition.

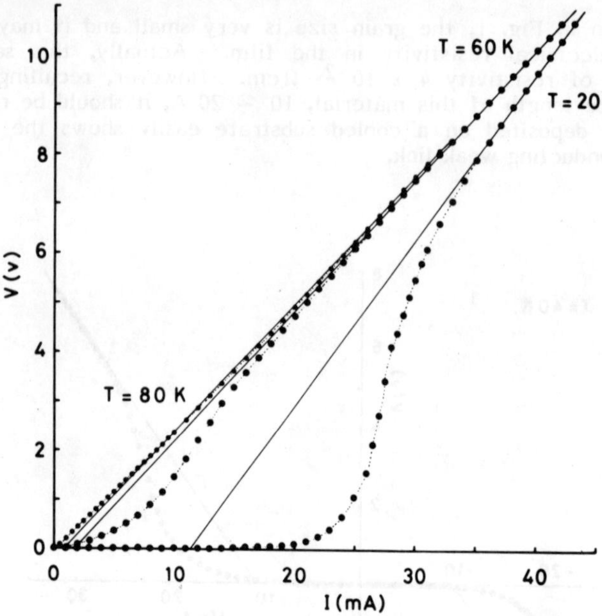

Fig. 3. I-V plots of $YBa_2Cu_3O_{7-\delta}$ film at T = 20, 60 and 80 K.

ACKNOWLEDGEMENT

The authors express thanks to Profs. Y. Shibuya and T. Aomine for their valuable discussions and encouragement. They are also indebted to the Fonds National Suisse de la Recherche Scientifique for financial suport.

REFERENCES

1) J. G. Bednorz and K. A. Müller: Z. Phys. 64 (1986) 189.
2) M. K. Wu, J. R. Ashburn, C. J. Torng, P. H. Hor, R. L. Meng, L. Gao, Z. J. Hung, Y. Q. Zang and C. W. Chu: Phys. Rev. Lett., 68 (1987) 908.
3) K. K. Likharev: Rev. Mod. Phys. 51 (1979) 101.
4) H. Tanabe, S. Kita, Y. Yoshizako, M. Tanouchi and T. Kobayashi: Jpn. J. Appl. Phys. 26 (1987) L1961.

CRITICAL CURRENTS IN THE HIGH T_c SUPERCONDUCTORS $Dy_1Ba_2Cu_3O_{7-y}$, $Y_1Ba_2Cu_3O_{7-y}$ AND $Bi_2Ca_2Sr_2Cu_6O_y$.

J. Aponte, H.C. Abache and M. Octavio
Centro de Física, Instituto Venezolano de Investigaciones Científicas, Apartado 21827, Caracas 1020-A. Venezuela

Abstract

We report on the temperature dependence of the critical current of the high T_c superconductors $Dy_1Ba_2Cu_3O_{7-y}$, $Y_1Ba_2Cu_3O_{7-y}$ and $Bi_2Ca_2Sr_2Cu_6O_y$. We find the critical current to vary linearly with (T_c-T). This is consistent with a network of Josephson-coupled superconducting grains, where the critical current is determined by the breakdown of the coupling. Close to T_c breakdown in the grains becomes significant resulting in a non-linear tail close to T_c. Using this model we obtain good quantitative agreement between our results and predictions of a Ginzburg-Landau model for this system.

Introduction

The new high T_c superconductors[1,2], are known to have a granular structure with the grains and their domains weakly-coupled to each other. However, there is yet little understanding about how this granularity affects the transport properties of these materials. In particular, if the critical current I_c of bulk materials is determined by the coupling between grains then, one expects I_c to vary linearly with temperature. In this paper, we present detailed experimental studies of the temperature dependence of I_c in $Dy_1Ba_2Cu_3O_{7-y}$, $Y_1Ba_2Cu_3O_{7-y}$ and $Bi_2Ca_2Sr_2Cu_6O_y$. We find that $I_c(T)$ can be described by a Ginzburg-Landau approach for single-phase materials. If more than one phase is present, then different T_c's need to be invoked to understand I_c (T).

Experiments and Discussion

Our samples were made by solid state reaction of the rare earth or bismuth and copper oxides with appropriate carbonates. The powders were thoroughly mixed and heated to 600 to 900°C for up to 18 hours. The powder was again mixed and introduced in the oven for 18 hours at temperatures between 750 and 900°C. Finally the samples were slowly cooled by decreasing the temperature in steps of 200°C for two hours in flowing oxygen.

Our best samples of $Y_1Ba_2Cu_3O_{7-y}$ were single-phase and had resistivity ratios $\rho_{300}/\rho_{100} \sim$ 2-3 as shown in Fig. 1 for a good sample. The resistivity curve or Fig. 1 can be interpreted as done by Halbritter et al [3], in which ρ is modelled by $\rho=\Sigma\rho_b+ p(\rho_o+\alpha T)$, where ρ_b is the barrier resistivity between the grains, ρ_o is the intrinsic residual resistivity of a grain and p takes into account the conduction path by the coupling

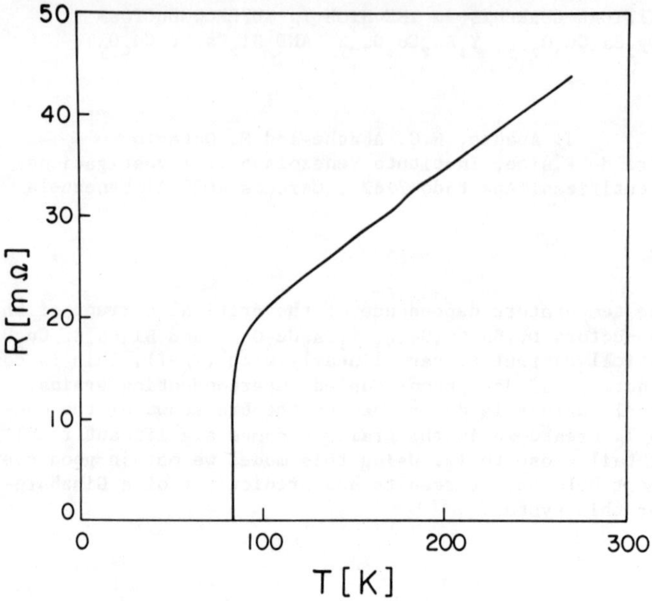

Figure 1. R vs. T curve for a single-phase $Y_1Ba_2Cu_3O_{7-y}$ sample.

between grains. For our best samples, ρ_b is quite small indicating that our samples are homogeneous, with $d\rho/dT$= 1.5 μΩ-cm/K.

For the critical currents we obtain two types of behavior depending on the homogeneity of the samples. If the sample is multiphase, then the temperature dependence of the critical current is like that shown in Fig. 2. There are two regimes to this curve, the resistance first goes to zero at 82K and I_c varies as (80-T), down to 75K,below this value the critical current grows linearly with (75- T). This is shown in detail in the insets in Fig. 2. Similar behavior can be found in multiphase $Bi_2Ca_2Sr_2Cu_6O_y$.

In single phase samples, the behavior is quite different as shown in Fig. 3 for a single phase $Y_1Ba_2Cu_3O_{7-y}$ sample, a behavior typical of all our Y and Dy "123" samples. The data is a logarithmic plot of I_c versus $(1-T/T_c)$, with T_c defined as the value that gave the best power law fit to the data. Except for a very small region near T_c, the variation is clearly linear down to the lowest temperatures measured.

A linear dependence of I_c is characteristic of Josephson-coupled weak-links, I_c is determined by the weakest of the barriers between grains. Given the percolative nature of the samples, this temperature dependence can be understood in more quantitative terms. The weak-coupling between grains have a wide distribution of strengths. If the tempera-

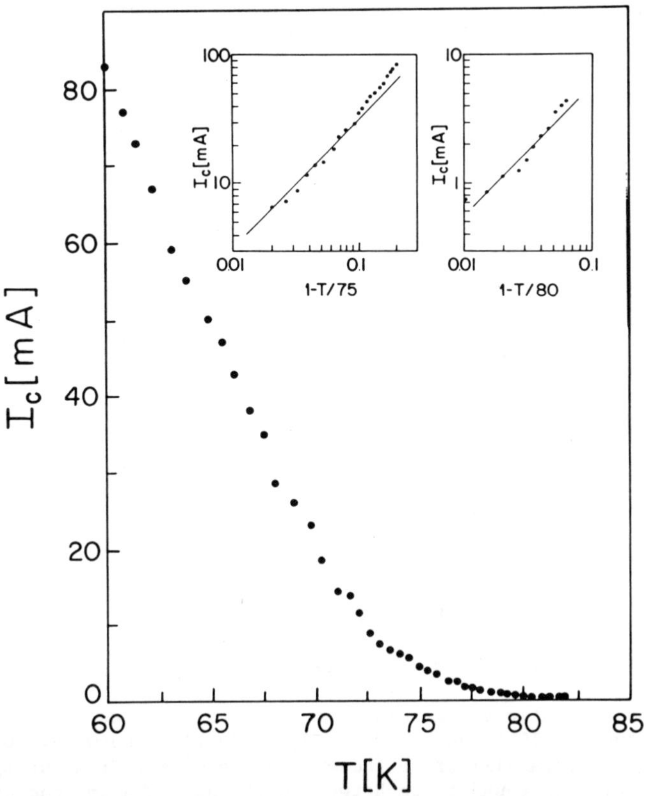

Figure 2. I_c vs. T plot for a multiphase sample.

ture dependence of I_c of a pair of grains varies as F(T), the variation of coupling stengths will effect the magnitude but not the temperature dependence of I_c, as long as the distributions of T_c's is not too wide. The form of $I_c(T)$ will be determined as discussed by Clem et al [4], by the dominant contribution of the free energy in the sample. If the condensation energy E_c of the grains is comparable to the coupling energy E_j, then it is possible to have a reduction of the gap in the grains. If however, this energy is greater than the coupling energy, the coupling energy is dominant and I_c follows the linear temperature dependence of weak-links. The ratio $\varepsilon(T)=E_j/2E_c$, can be parametrized as $\varepsilon(T)=\varepsilon_o G(T)$ where G(T) varies with T and $\varepsilon_o=2.93\hbar k_B/e^2\gamma T_{co}\rho_n a_o^2$, where a_o is the grain size, ρ_n is the normal state resistivity and γ is the Sommerfeld constant. The crossover temperature can be found to be[4] $T_x=T_c(1-0.882\varepsilon_o)$.

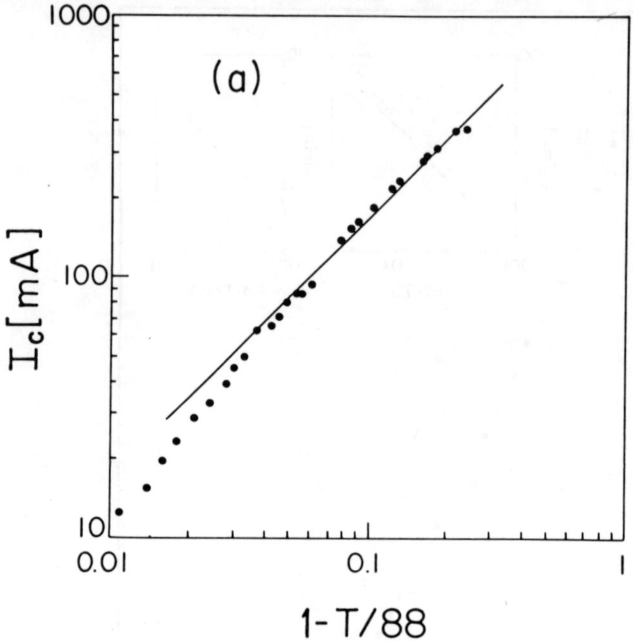

Figure 3. Critical current for a single-phase $Y_1Ba_2Cu_3O_{7-y}$ sample.

Using the characteristic parameters of these materials, we obtain that ε_o varies between $2x10^{-2}$ and $2x10^{-6}$ for grain sizes between 0.01 μm. and 1μm, parameters similar to those in our samples. From our I_c vs. T curves $(1-T_x/T_c)$ is found to be between 0.06 and 0.006 in good agreement with the range of domain sizes known to exist in these materials. Unfortunatelly this narrow range has not allowed us to determine in detail the temperature dependence of I_c a task complicated further by the possible effects of fluctutations[5].

In conclusion, we have shown that applying phenomenological concepts based on our understanding of the usual high T_c superconductors, it appears possible to understand the measured temperature dependence of the critical current. The magnitude of the crossover temperature where different behaviors dominate are known to be in very good agreement with models.

This work supported by CONICIT Grant S1-1828 and by the Third World Adacemy of Sciences.

References

1. J. Bednorz and A.K. Muller, Z. Phys. B64 189 (5986)
2. M.K. Wu, J.R. Ashburn, C.J. Torng, P.H. Hor, R.L. Menmg, L. Gao, Z.J. Huang, Y.Q. Wang and C.W. Chu, Phys. Rev. Lett. 58, 908 (1987).
3. J. Halbritter, H. Kupfer, B. Runtsch and H. Wuhl preprint (1988)
4. John R. Clem, B. Bumble, S. I. Raider, W.J. Gallagher, and Y.C. Shih, Phys. Rev. B35, 6637 (1987)
5. C.J. Lobb, Phys. Rev. B36, 3930 (1987)

MICROWAVE ABSORPTION AND RECTIFICATON IN SUPERCONDUCTING Y $Ba_2Cu_3O_7$

R. Durny, J. Hautala, S. Ducharme, D.J. Zheng, O.G. Symko, P.C. Taylor, Dept. of Physics, University of Utah, Salt Lake City, Utah 84112, and S. Kulkarni, Ceramatec, Inc., Salt Lake City, Utah 84115.

ABSTRACT

The magnetic field dependent non-resonant microwave absorption in Y $Ba_2Cu_3O_7$ provides an extremely sensitive and convenient method of characterizing the new high T_c superconductors. We also observe that the incident microwave radiation induces a d.c. voltage across samples with a strong field dependence at low magnetic fields. The results are explained in terms of dissipative fluxon motion along naturally occuring barriers inside the superconducting crystallites.

INTRODUCTION AND RESULTS: When we first showed[1)] that microwave absorption can characterize the new high T_c superconductor Y $Ba_2Cu_3O_7$, it became clear that this method offered a very large sensitivity for studies of ceramic superconductors. These systems exhibit granularity, grain boundaries, twins, and other defects; such defects can lead to Josephson barriers[2)] and thin bridges which can be studied with microwave absorption. We are presenting such studies and the effects of small magnetic fields on microwave induced voltage across the samples.

Samples of Y $Ba_2Cu_3O_7$ in the form of bars, crystals, and powders, with $T_c \sim 92K$, were studied in a Bruker ER 200D-SRC Spectrometer at X-band in a TE_{103} cavity. The samples were cooled down to 4K in an Air Products Helium Flow System.

Figure 1 shows the very low field modulated absorption signal at 30K with modulation amplitude being 0.0125 Oe for (a), 0.25 Oe for (H), and 2.5 Oe for (c). The measured signal shows a change of slope at around 10 Oe, and then it continues up to at least 12k Oe. We attribute the absorption to dissipation by fluxon motion in long Josephson junctions in the sample. When a junction dimension L is larger than λ_J, the Josephson penetration depth, Josephson fluxons can be induced in the junctions by the external magnetic field. Microwave induced currents exert a Lorentz force on these fluxons causing them to move in the barriers and dissipate energy due to surface losses and normal electrons. At higher fields, the losses probably are due to Abrikosov vortices set in motion in the bulk sample by the microwaves as proposed by reference 3.

To gain insight in this model, we looked at the d.c. voltage across a small bar of the sample, 10 x 1 x 1 mm, induced by X-band microwaves. The results are shown in Figure 2; a small magnetic field reduces the induced voltage. We attribute the induced voltage to the inverse a.c. Josephson effect; as the external magnetic field is increased, phase coherence is reduced thereby decreasing the induced signal.

References

(1) R. Durny, J. Hautala, S. Ducharme, B. Lee, O.G. Symko, P.C. Taylor, D.J. Zheng, and J.A. Xu, Phys. Rev. B 36, 2361 (1987).

(2) G. Deutscher and K.A. Muller, Phys. Rev. Lett. 59, 1745 (1987).

(3) K.W. Blazey, A.M. Portis, and J.G. Bednorz, Solid State Comm. 65, 1153 (1988).

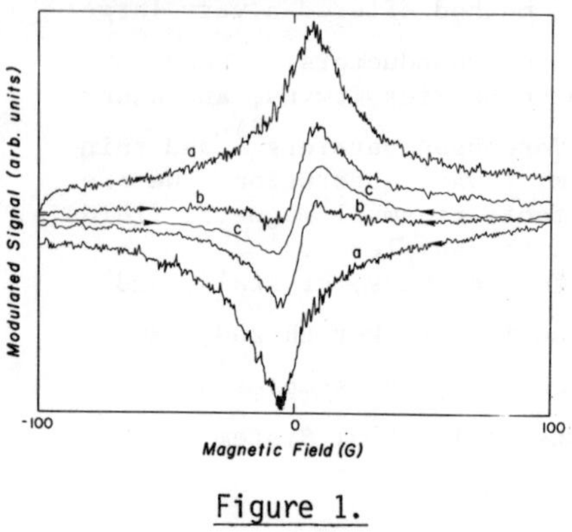

Figure 1.

Modulated absorption signal in a magnetic field

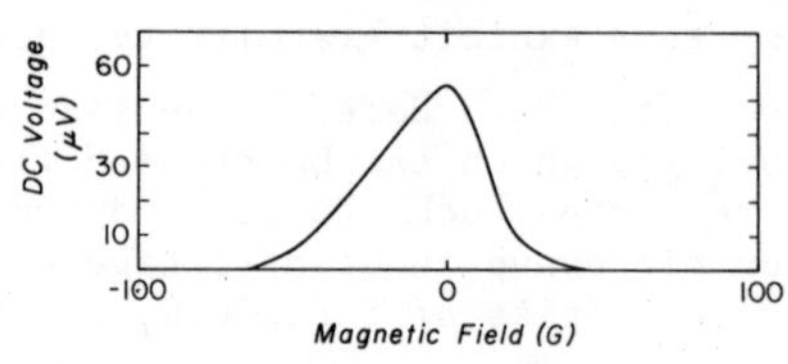

Figure 2.

Microwave induced voltage as a function of magnetic field.

ELECTRON SPIN RESONANCE OF THE HIGH T_c SUPERCONDUCTING MATERIAL $DyBa_2Cu_3O_{7-x}$

H.C. Abache, M. Octavio, J. Aponte and P. Silva

Instituto Venezolano de Investigaciones Científicas,
Centro de Física, Apartado 21827,
Caracas 1020A, Venezuela.

Electron Spin Resonance experiments, in the temperature range 76-350 K, are reported for polycrystalline sintered samples of $DyBa_2Cu_3O_{7-x}$. Our studies show that Cu is present only in the Cu(II) states. In addition we observed inhomogeneous broadening of the line shape and at low temperatures, the g-value increases gradually towards 2.3. Above T_c, the g-value does not change appreciably. Some interpretation regarding the EPR signal behaviour in the superconducting state is discussed.

Introduction

Intense interest in oxigen deficient perovskite superconductors has been generated by the discovery of superconductivity above 30 K in the La-Ba-CuO system by Bednorz et. al. (1) and above 90 K in the Y-Ba-CuO system by Wu et. al. (2). The role of the copper in the mechanism of superconductivity is a central issue. A proposal has been made (3,4) that in Perovskite - related oxides, copper may exist in any of the formal valence states Cu(+1), Cu(+2) or Cu(+3). Santoro et. al. (5) in the case of $YBa_2Cu_3O_{7-x}$ suggested that the structure of $YBa_2Cu_3O_6$ is best described as containing localized Cu(1+) and Cu(3+) ions. Mehran et. al. (6) observed a strong paramagnetic resonance signal characteristics of Cu(2+) ions in the high T_c superconductor Y-Ba-CuO. Another proposition has been made (4), that one of the interesting features of $YBa_2Cu_3O_{7-x}$ and related oxides is that the Cu(3+)/Cu(2+) ratios play a crucial role in the superconducting transition temperature. Such a situation tell us that there are several intriguing features associated with Cu(2+)-ions in such oxides which are not readily understood. We have undertaken a study of Electron Spin Resonance (ESR) of the Dy-Ba-CuO system above and below the superconducting transition temperature T_c = 88 K.

Experimental

The material used in this study had the nominal stoichiometry corresponding to $DyBa_2Cu_3O_7$ which was identified to be superconducting in the Dy-Ba-CuO system. The samples were prepared by the conventional solid-state reaction. We used Dy_2O_3, $BaCO_3$ and CuO as the starting materials. They were weighed stoichiometrically, mixed and ground. The mixture was heated in oxigen atmosphere at 900 °C for more than 15 hours and the reacted mixture was a ground sufficiently. After several repetitions of the presintering process, the sample was pressed into disks for final sintcring at 900 °C for about 7 hours in oxigen and slowly cooled from 900 °C to 50 °C over six hours.

The ESR measurements that we report here were all made on powders using a standard spectrometer of E-line Varian. These measurements were carried out at 9.3 GHz between 76 and 350 K. We have, additionaly investigated the ESR as function of temperature on samples which were made only from CuO, DyCuO and also on DyBaCuO irradiated by gamma ray (source Co-60).

Results and Discussion

In Figure 1. we show the experimental 9.3 GHz spectra for $DyBa_2Cu_3O_7$ at 118 K. The polycrystalline ESR spectrum of Dy-Ba-CuO, shows a small change with temperature, consistent with the small change in structure on reducing the temperature from 298 to 77 K, but gives no information on the change in stereochemistry involved. On cooling gradually a narrover line appears. Just below T_c, the ESR line was observed to weaken and apparently dissapear below 77 K. We plot the g-value (Figure 2.) as function of temperature. On cooling down from RT to about 100 K g does not change appreciably. Below 90 K there is a gradual increase in the g-values. The line-width as a function of temperature shows a sharp drop at the transition temperature $T_c \cong 100$ K. The analysis of the line-width either above or below T_c is not easy. Below T_c, the observed line-width increases dramatically by several orders of magnitude.

The samples made only with CuO, DyCuO materials exhibited and ESR signal of low intensity. We also investigate a sample of Dy-Ba-CuO irradiated (gamma-ray). The ESR spectrum of this sample at 119 K shows a drastically change in the slope of the base line and the intensity of the Cu line decreased gradually and vanished at about 155 K.

It can be conclude at present that, there exists a high T_c superconducting phase in Dy-Ba-CuO system. We believe that the ESR signal observed in the sample is characteristic of the ionic Cu(2+) in the superconducting phase. The rise of g towards 2.3 on further cooling gives further support to this proposal (the g-value reported for Cu(2+) fall within range 2.1-2.4 (7).

Several alternative explanations for the observed satellite line (Fig. 1) above T_c until approximately 200 K can be postulated. For example, one possibility is that the satellite might be essentially Cu(2p) lines. There is now much evidence from Low Energy Electron Diffraction (LEED) and X-Ray Photoelectron Spectroscopy (XPS) studies that such satellites are characteristic of cooper in the 2+ oxidation state.

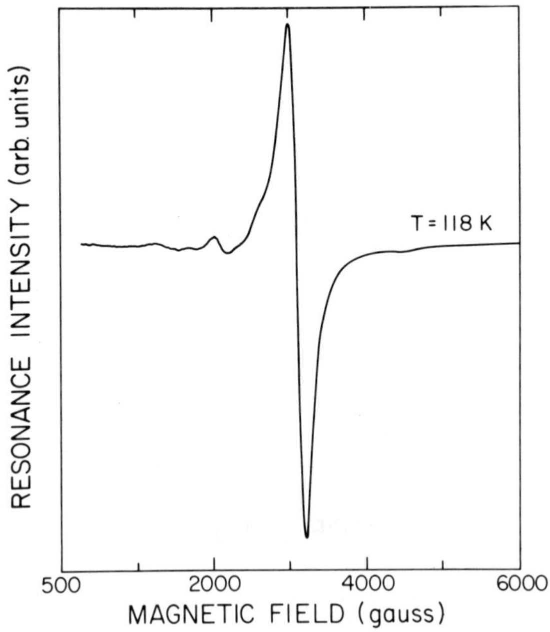

Figure 1. ESR spectrum of polycrystalline $DyBa_2Cu_3O_7$ taken at 9.3 GHz and 118 K.

The stron change in the slope of the base line of Cu ESR line (irradiated sample) is possibly suggesting the presence of a distribution of O....O bonds. This problem requires detailed investigation in powder samples. Work is in progress for Cu(2+) in the same compound and the role played for the oxygens, this will be published shortly.

Further experiments are necessary especially of single crystals at lower temperature and over a wide range of microwave frequencies.

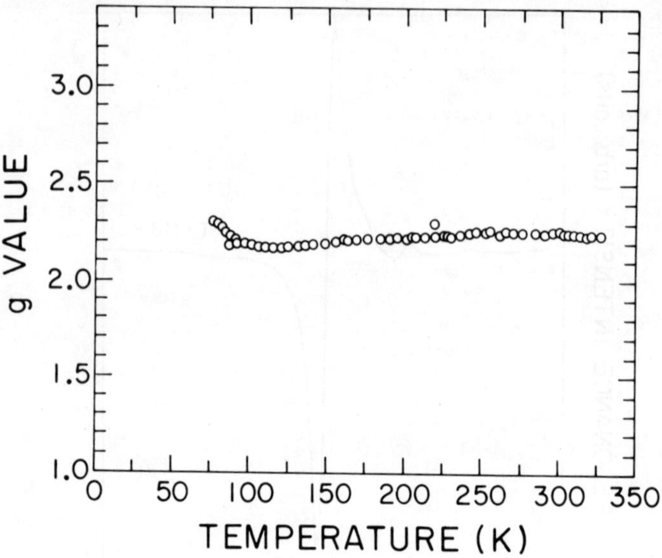

Figure 2. Temperature dependence of the g factor in the Dy-Ba-CuO system.

Acknowledgment

We are deeply grateful to S. Camero at Institute of Engineering Foundation for the technical support. This work was supported by CONICIT Grant S1-1828.

References

1. J.G. Bednorz and K.A. Muller, Z. Phys. B64, 189 (1986).
2. M. Wu, J.R. Ashburn, C.J. Torng, P.H. Hor, R.L. Meng, L. Gao, Z. Huang, Y. Wang and C.W. Chu, Phys. Rev. Lett. 58, 908 (1987).
3. J.B. Goodenough, Mat. Res. Bull. 8, 423 (1973).
4. C.N. Rao, P. Ganguly, K. Sreedhar, R.A. Mohan Ram and P.R. Sarode, Mat. Res. Bull. 22, 849 (1987).
5. A. Santoro, S. Miraglia, F. Beech, S. Sunshine, D. Murphy, L. Schneemeyer and J. Waszczak, Mat. Res. Bull. 22, 1007 (1987).
6. F. Mheran, S. Barnes, T. Mac Guire, W. Gallagher, R. Sandstrom, T. Dinger and D. Chance, Phys. Rev. B36, 740 (1987).
7. W. Low and E. Offenbacher, Solid State Phys. 17, 135 (1965).
8. D. Paul, G. Balakrishnan, N. Bernhoeft, W. David and W. Harrison, Phys. Rev. Lett. 58, 1976 (1987).

ESR IN THE Y-Ba-Cu-O SYSTEM

R.S. de Biasi, A.A.R. Fernandes, C.L. Ferreira,
R. Pascual and A. Saavedra

Seção de Engenharia de Materiais, Instituto Militar de Engenharia, 22290 Rio de Janeiro, RJ, Brazil

ABSTRACT

ESR measurements have been performed on three different phases of the Y-Ba-Cu-O system: a black orthorhombic phase, a black tetragonal phase and a green phase. Paramagnetic resonance lines with $g \simeq 2$ were observed in all phases, but only the superconducting phase (the black orthorhombic phase) showed a strong microwave absorption at low magnetic fields. This low-field absorption line is only observed below the superconducting transition temperature, as determined from magnetic susceptibility measurements, and thus seems to be characteristic of the superconducting state.

Superconductivity above the boiling point of liquid nitrogen in the Y-Ba-Cu-O system has been reported by Wu et al.[1] The superconducting phase was identified[2,3] as a layered, opaque, oxygen-deficient orthorhombic perovskite with composition $YBa_2Cu_3O_{7-y}$, i.e., a "black orthorhombic" phase. Depending on the initial composition and oxygen content of the samples, the Y-Ba-Cu-O system can exhibit two other phases, both non-superconducting: an opaque tetragonal perovskite with composition $YBa_2Cu_3O_{6+y}$, which may be called a "black tetragonal" phase, and a green orthorhombic phase[4] with composition Y_2BaCuO_5 (a "green" phase). In this work, ESR measurements on samples of the three phases above are reported in the temperature range 77-300 K.

Three different kinds of samples were prepared from Y_2O_3, $BaCO_3$ and CuO powders. Samples type I ("black orthorhombic") and type II ("black tetragonal") had the same starting composition, corresponding to the formula $YBa_2Cu_3O_x$, while in the case of type III ("green") samples the starting composition corresponded to the formula Y_2BaCuO_x. Finely ground powders were pressed into disks and fired at 950 ^oC for 24 hours in flowing oxygen. Type I samples were reground, sintered at 950 ^oC for 24 hours in flowing oxygen and cooled slowly to room temperature. Type II and type III samples were reground, sintered in air at 950 ^oC for 24 hours and quenched in air. Room temperature X-ray diffraction patterns of type I, type II and type III samples matched, respectively, the reported spectra[3,5,6] for $YBa_2Cu_3O_{7-y}$, $YBa_2Cu_3O_{6+y}$ and Y_2BaCuO_5. The superconducting transition temperature of type I samples, as determined from mutual inductance measurements,[7] was found to be about 90 K. Type II and type III samples did not show any superconductivity down to 77 K.

Small pieces (typical dimensions 2 mm x 2 mm x 1 mm) were cut from the sintered disks and mounted at the tip of a cold finger cryostat. ESR spectra were recorded in the temperature range 77-300 K using a Varian E-12 X-band spectrometer. The magnetic field was calibrated with a proton resonance gaussmeter.

The room temperature ESR spectra of the three phases are shown in Figs. 1-3. The spectra at 77 K are shown in Figs. 4-6. The spectrum of the "black orthorhombic" phase is broader at all temperatures between 77 and 300 K; in addition, below 90 K, it shows a strong low-field microwave absorption (see Fig. 4) which is absent in the other phases.

The temperature dependence of the g-factor and of the ESR linewidth for the three phases is shown in Figs. 7 and 8. The g-factor of the "black" phases increases with increasing temperature, while the g-factor of the "green" phase decreases slightly as the temperature is increased. Below 100 K, the linewidth of the "black orthorhombic" phase increases sharply. The room-temperature values are: for $YBa_2Cu_3O_{7-y}$, $g = 2.0902$, $\Delta H_{pp} = 34$ mT; for $YBa_2Cu_3O_{6+y}$, $g = 2.0840$, $\Delta H_{pp} = 23$ mT; for Y_2BaCuO_5, $g = 2.0920$, $\Delta H_{pp} = 25$ mT.

The results reported here seem to rule out the possibility, suggested by some authors,[8,9] that the ESR line which is observed at $g \simeq 2$ in the "black orthorhombic" (superconducting) phase is due to a small amount of the "green" compound Y_2BaCuO_5. However, it could be due to some other residual phase, such as[10] $BaCuO_2$. On the other hand, the low-field absorption that is observed at low temperatures is clearly related to the superconducting state.

The authors thank Prof. R.F. Ferreira, who performed mutual inductance measurements on selected samples. This work was carried out with financial support from FINEP, CNPq and Ministério do Exército.

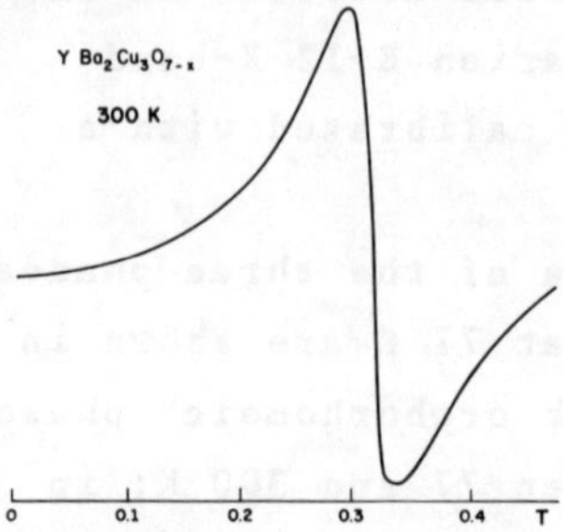

Fig. 1

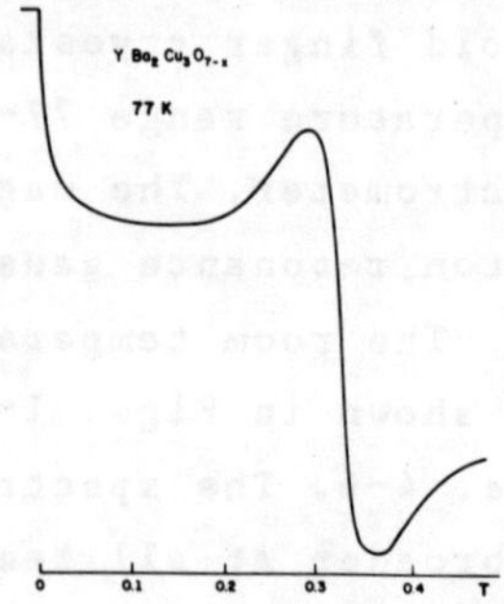

Fig. 4

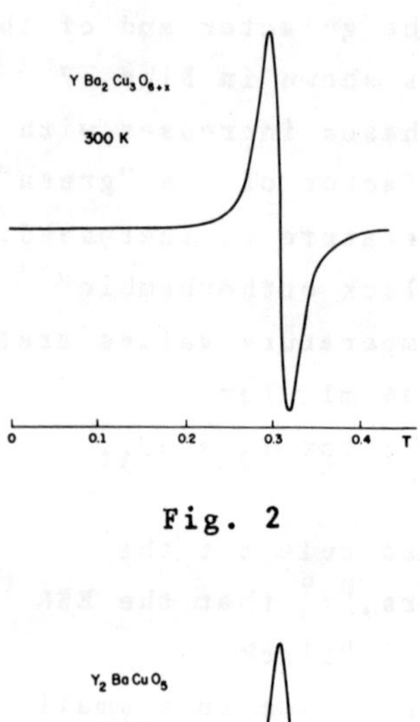

Fig. 2

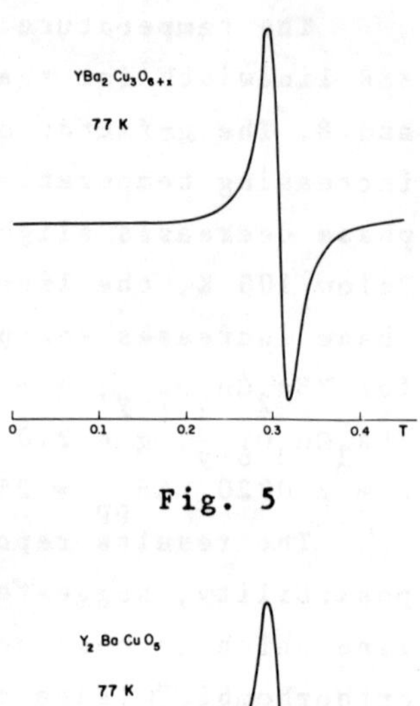

Fig. 5

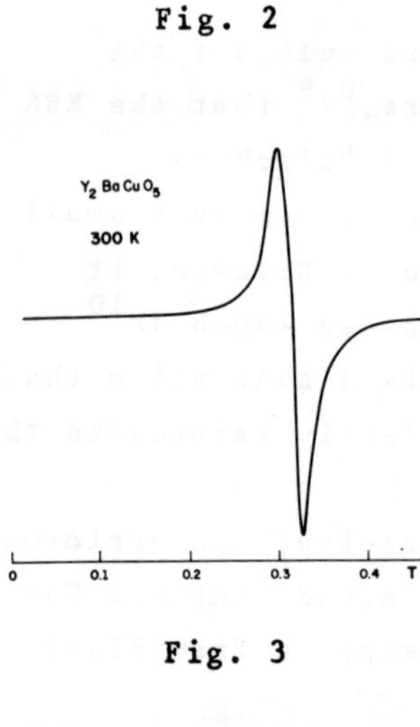

Fig. 3

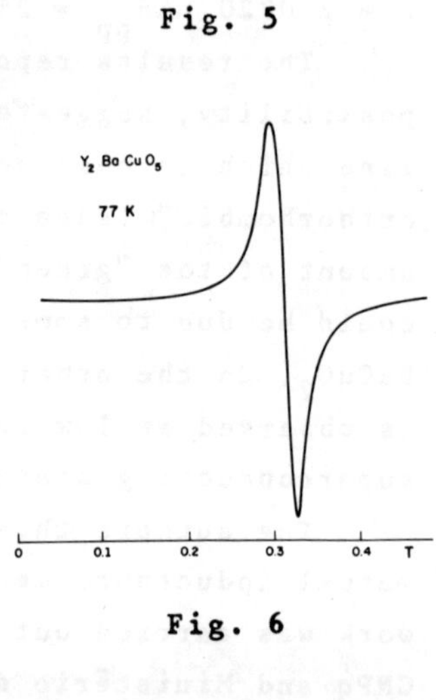

Fig. 6

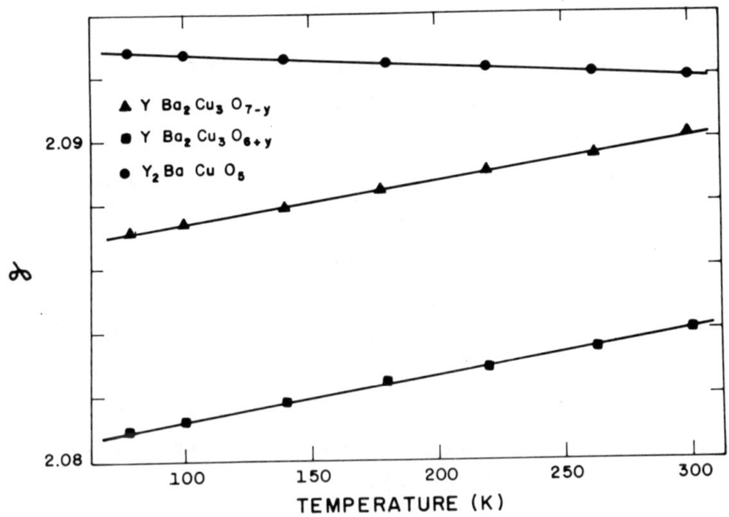
Y Ba2 Cu3 O7-y
Y Ba2 Cu3 O6+y
Y2 Ba Cu O5
2.09
2.08
g
100
150
200
250
300
TEMPERATURE (K)

Fig. 7

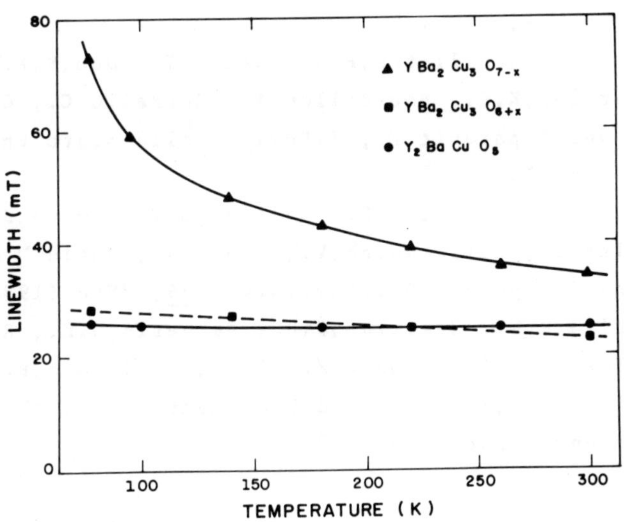
80
60
40
20
0
LINEWIDTH (mT)
Y Ba2 Cu3 O7-x
Y Ba2 Cu3 O6+x
Y2 Ba Cu O5
100
150
200
250
300
TEMPERATURE (K)

Fig. 8

REFERENCES

1. Wu,M.K., Ashburn,J.R., Torng,C.J., Hor,P.H., Meng,R.L., Gao,L., Huang,Z.J., Wang,Y.Q. and Chu,C.W., Phys.Rev. Lett. 58, 908 (1987).
2. Kadowaki,K., Huang,Y.K., Van Sprang,M. and Menovski, A.A., Physica 145B, 1 (1987).
3. Gallagher,P.K., O'Bryan,H.M., Sunshine,S.A. and Murphy, D.W., Mat.Res.Bull. 22, 995 (1987).
4. Michel,C. and Raveau,B., J.Solid State Chem. 43, 73 (1981).
5. Santoro,A., Miraglia,S., Beech,F., Sunshine,S.A., Murphy,D.W., Schneemeyer,L.F. and Waszczak,J.V., Mat. Res.Bull. 22, 1007 (1987).
6. Hazen,R.M., Finger,L.W., Angel,R.J., Prewitt,C.T., Ross,N.L., Mao,H.K., Hadidiacos,C.G., Hor,P.H., Meng, R.L. and Chu,C.W., Phys.Rev. B35, 7238 (1987).
7. Pereira,R.F.R., Meyer,E. and Silveira,M.F., Rev.Sci. Instrum. 54, 899 (1983).
8. Bowden,G.J., Elliston,P.R., Wan,K.T., Dou,S.X., Easterling,K.E., Bourdillon,A., Sorrel,C.C., Cornell, B.A. and Separovic,F., J.Phys.C:Solid State Phys. 20, L545 (1987).
9. Vier,D.C., Oseroff,S.B., Salling,C.T., Smyth,J.F., Schultz,S., Dalichaouch,Y., Lee,B.W., Maple,M.B., Fisk, Z. and Thompson,J.D., Phys.Rev. B36, 8888 (1987).
10. Mesquita,R.N., Castilho,J.H., Barberis,G.E., Rettori, C., Torriani,I., Lima,O.F., Gama,S., Jardim,R.F., Terrile,M.C., Basso,H and Nascimento,O.R., this conference, paper no. C82.

EPR ANALYSIS IN M-Ba-Cu-O COMPOUNDS

C. LOPEZ, A. MARIÑO, M. ESGUERRA* AND G. FRANCO.

Dpto. Física. Universidad Nacional de Colombia . Bogotá, Colombia.

* Centro Internacional de Física, Bogotá , Colombia

ABSTRACT

We report EPR measurements in the systems M-Ba-Cu- O (M= Y, La). An absorption line attribute to Cu^{++} was observed in both Y_2BaCuO_5 and $Y_1Ba_2Cu_3O_{7-y}$ compounds and their temperature dependence (100-300K) studied. The spectra of $La_1Ba_2Cu_3O_{7-y}$ (nominal composition) annealed in an oxygen flow showed a similar signal as the one observed in the Y compounds . As prepared La samples showed, however, an additional broad line (0.05 - 0.17 T) which decreases strongly by heat treatement in oxygen flow. Additionaly the behaviour of this line as a function of temperature showed two interesting features : The linewidth varies as $\Delta Hpp \propto e^{-T/To}$ and the line intensity as $I \propto e^{a/T}$.

Electron spin resonance (EPR) measurements in YBaCuO and LaBaCuO compounds with nominal composition $(La-Y)_1$, $Ba_2Cu_3O_{7-y}$ and Y_2BaCuO_5 were carried out on an X-band Varian spectrometer 4500 . The temperatures were measured with a calibrated Pt(100) resistance with the sample in a nitrogen gas flow system , in orden to compare intensities aproximately the same quantity of powder samples inside a quartz tube was used. Different samples to be denoted by $Y_1Ba_2Cu_3O_{7-y}$ (annealed in oxygen flow Tc = 90K, ΔTc=1-2 K) $Y_1Ba_2Cu_3O_{7-y}$ (not annealed) , Y_2BaCuO_5 (insulating) and $La_1Ba_2Cu_3O_{7-y}$ (annealed in oxigen flow and not annealed) were measured . The method of preparation was described previously [1, 2]. The $La_1Ba_2Cu_3O_{7-y}$ samples showed hopping conduction mechanism in the temperature range between 80 - 300K. No evidence for superconductivity was observed down to 50K in the annealed $La_1Ba_2Cu_3O_{7-y}$ sample. All the Y compounds annealed in oxygen flow and cooled slowly (aprox. 2°C/min) had mainly the orthorombic 1-2-3 phase; the La-compounds showed preferentially thetragonal structure.

An anisotropic EPR signal attributed to Cu^{++} [(3,4,5)] was observed both in annealed and not annealed $Y_1Ba_2Cu_3O_{7-y}$ samples. Our results showed that there is no appreciable change in g-values ; $g_{\parallel}$= 2.2033 , $g_{\perp}$= 2.0446 , Δg = 0.16, however, the intensity is no the same as can be seen in Figure 1a. Small lineshape differences can be attributed to variation in the lattice parameters by annealing treatment. On the other hand the significant changes in the intensity from sample to sample being the largest for $Y_1Ba_2Cu_3O_{7-Y}$ (not annealed) which is not superconducting have been associated with the appearance of the superconducting transition [6]. The weaker the EPR intensity, the stronger the superconducting signal and viceversa. Recent investigation report no EPR signal in $Y_1Ba_2Cu_3O_{7-\delta}$ single crystals[7]. In this context we agree that the residual EPR signal in the superconducting samples comes from impurity phases, like Y_2BaCuO_5 (green phase) and others. See Fig. 1b.

The resonance spectra for annealed and not annealed $La_1Ba_2Cu_3O_{7-Y}$ samples at room temperature are displayed in Fig.2a, the spectrum of the not annealed sample consists of two different signals. The narrower one shows the same characteristics as observed in Y-compounds. No appreciable change is observed in g-values ; $g_{\parallel}$= 2.2060, $g_{\perp}$= 2.0460, Δg= 0.16. The intensity of the Cu^{++} line, both in Y and La compounds (annealed or not annealed) exhibited a Curie-Weiss like behaviour in the temperature range 80-300K, without discernible change in shape or linewidth. The intensity decreases also markedly, as in Y-compounds, by heat treatment (see Fig.2a). The broader line on the other hand is an asymetric wide line, ΔH=0.1T and g=2.1 which decreases strongly by annealing in oxygen flow. Both, linewidth and intensity ($I=\Delta H_{pp}{}^2 Y_{pp}$) showed a strong temperature dependence. Resonance spectra at various temperatures for not annealed $La_1Ba_2Cu_3O_{7-Y}$ samples are showed in Fig.2b. No changes in g-values were observed. The linewidth and the intensity variation as a function of temperature are displayed in Fig.3 . An exponential behaviour of the form $\Delta H_{pp} \alpha\, e^{-T/T_0}$ with T_0 aprox. 400-500K for different samples was observed. Fig. 4a. The intensity however can be fitted to the empirical relation $I \alpha\, e^{a/T}$; a=230-260K . Fig. 4b.

The observed linewidth variation as a function of temperatura , unlike the Cu^{++} line can be associated with the existence of collective magnetic coupling, probably an antiferromagnetic type of interaction.That this collective magnetismus is eliminated by heat treatement is not completely clear, it is possible that this effect also exist in both annealed and superconducting samples and could be responsable for the inestability of the ground line, which is observed in all EPR spectra. The markedly existence of this interaction in not annealed La-compounds could be a consequence of the greater substitutional disorder in these samples[2]. Similar behaviour of ΔH vs T was observed specially in $Y_1Ba_2Cu_3O_7$ (quenched) no superconducting samples at lower temperatures (10K-40K) [6].

The existence of a broad line with a similar temperatura variation of the EPR line intensity was also reported in Gd-$Ba_2Cu_3O_X$ samples [8]. We consider however that more research is necessary to assign wich type of interaction is resposible for this behavior.

We wish to thank the colleges of the low temperature and magnetic resonance groups for fruitfully discussion.

REFERENCES

1- E. Posada, G. Franco, G. Holguin, A. Jiménez, A. Mariño, V. Niño and D. Rodriguez. Rev. Colombiana de Física , to be published.

2- M. Esguerra, A. Mariño, G. Franco, A. Jiménez, G. Holguin and E. Posada To be published in J. Phys. C. Solid State physics.

3- G. J. Bowdem, P.R. Elliston, K.T, Wan, S. X. Dou , K.E, Easterling, A. Baurdillon, C.C. Sorrell, B.A Cornell and F. Separovic, J. Phys. C. Solid State Phys **20**, 1545-552, 1987.

4- J. Stankowski, P.K. Kahal and N.S. Dalal, J.S. Moodera. Phys. Rev. B. Vol. **36**, 7126, 1987.

5- K. Kojima, K. Ohbayashi, M. Udagawa. Jpn. J .Appl. Phys. Vol. **26**, 5 , L766-L767 , May. 1987.

6- C. Rettori, D. Davidov, I. Belaish and I . Felner. Physical Rev. B. Vol.**35**, 7, Sep. 1987

7- J. Albino , O de Aguiar, A.A. Menovsky, J. van den Berg and H. B. Brom . J. Phys. C. Solid State physics. to be published .

8- F. Mehran, S. E. Barnes, C.C. Tsuei and T. R. McGuire , Phys. Rev. B, **3**, 7266 ,1987.

FIGURE CAPTION

Fig. 1 . 1a. The EPR spectra of $Y_1Ba_2Cu_3O_{7-Y}$ samples at room temperature. 1) Annealed in oxygen flow. 2) Not annealed. 1b. EPR spectrum of Y_2BaCuO_5(green phase) at room temperature.

Fig. 2. 2a EPR spectra of $La_1Ba_2Cu_3O_{7-Y}$ samples at room temperature. 1) Annealed in oxygen flow. 2) Not annealed. 2b. EPR spectra of not annealed $La_1Ba_2Cu_3O_{7-Y}$ at various temperatures.

Fig. 3. Temperature variation of the EPR linewidth ΔH and Intensity I of the broad line in not annealed $La_1Ba_2Cu_3O_{7-Y}$ samples.

Fig. 4. 4a Logarithmic variation of the EPR linewidth as a function of temperature of the broad line in not annealed $La_1Ba_2Cu_3O_{7-Y}$ samples and 4b. Logarithmic variation of the EPR intensity as a function of inverse temperature. In both cases the dashed lines are the best fit.

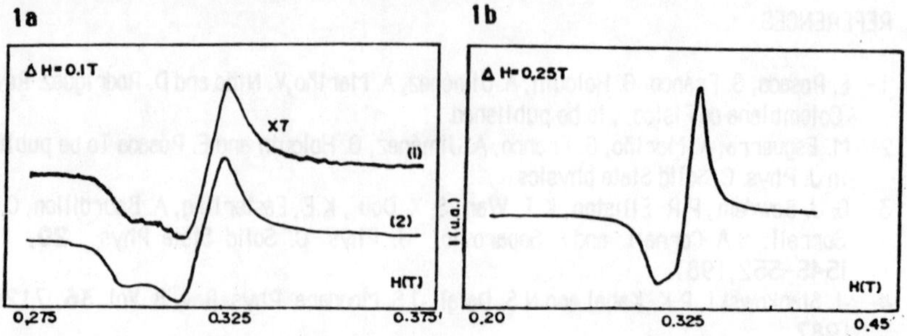
1a
ΔH=0.1T
X7
(1)
(2)
I(u.a)
H(T)
0.275
0.325
0.375
1b
ΔH=0.25T
I(u.a.)
H(T)
0.20
0.325
0.45

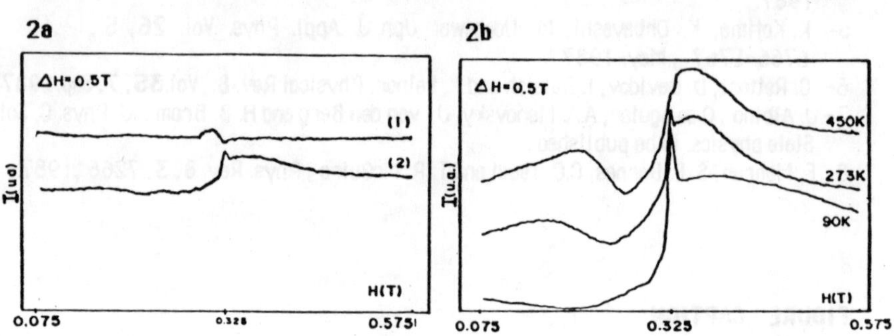
2a
ΔH= 0.5T
(1)
(2)
I(u.a)
H(T)
0.075
0.325
0.575
2b
ΔH=0.5T
450K
273K
90K
I(u.a)
H(T)
0.075
0.325
0.575

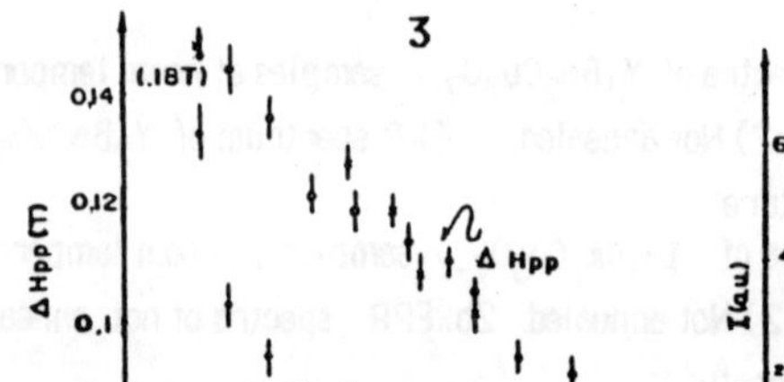
3
(.18T)
0.14
0.12
0.1
0.08
ΔHpp(T)
ΔHpp
I
60
40
20
I(a.u.)
100
200
300
400
T
(K)

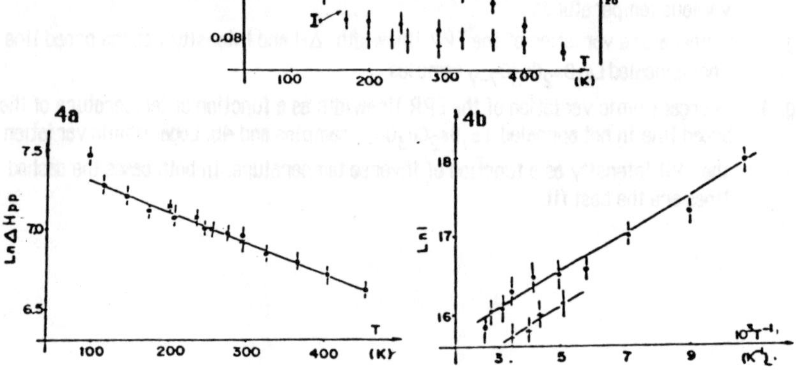
4a
7.5
7.0
6.5
LnΔHpp
100
200
300
400
T
(K)
4b
18
17
16
Ln I
3
5
7
9
$10^3 T^{-1}$
(K^{-1})

MAGNETIC PHASE PRESENT IN $YBa_2(Cu_{1-x}Fe_x)_3O_6$ OXIDES

E. Baggio Saitovitch, I. Souza Azevedo and R.B. Scorzelli

Centro Brasileiro de Pesquisas Físicas
Rua Dr. Xavier Sigaud, 150, Urca
22290 Rio de Janeiro, RJ, Brasil

The Mössbauer study of oxygen deficient ^{57}Fe: $YBa_2Cu_3O_6$ oxides was very important to establish the preferential occupation of Cu1 site by Fe at very low concentrations. The iron species characteristic of this phase (δ = 0.06mm/s, QS = 2.0mm/s) may be present in some ^{57}Fe: $YBa_2Cu_3O_7$ samples and induced many authors to propose as wrong assignement the occupation of Cu2 sites. Recent determination of antiferromagnetic ordering for Cu moments in Cu2 sites (T $\sim$ 450K) and our early observation of small proportion of a magnetic phase at room temperature for ^{57}Fe: $YBa_2Cu_3O_6$ lead us to perform systematic studies of $YBa_2(Cu_{1-x}Fe_x)_3O_6$ with x = 0.005, 0.01, 0.05, 0.10 and 0.15.

1. INTRODUCTION

The superconductor $YBa_2Cu_3O_7$ oxide, with orthorhombic structure presents a transition to a tetragonal insulator phase when the oxygen content goes below 6.3. The Mössbauer studies for Cu substituted, in low concentration, by ^{57}Fe in both phases allowed the determination of preferential occupation of Cu1 site. Moreover the iron species, characteristic of tetragonal phase, may be present in the orthorhombic $YBa_2Cu_3O_7$ which has not the proper oxygen content, and this fact lead several authors to attribute its Mössbauer parameters as due to iron in Cu2 site.

The Mössbauer studies for higher iron concentration in the $YBa_2(Cu_{1-x}Fe_x)_3O_7$, x = 0.10 and 0.15 do not indicate the presence of new absorption lines which could be attributed to the occupation of Cu2 sites. This result is difficult to understand and to obtain additional information about this point, carefull studies were made on the correspondent oxygen deficient phases x = 0.005, 0.01, 0.05, 0.10 and 0.15.

2. EXPERIMENTAL

The samples were prepared by reducing the oxygen content of $YBa_2(Cu_{1-x}Fe_x)_3O_7$ heating in Ar atmosphere at 600C for 3h. The Mössbauer spectra at room temperature and 4.2K were obtained with absorber and source at equal temperature. The measurements at higher temperatures were performed "in situ" using an Elscint oven for Mössbauer Spectroscopy with a ceramic sample holder; in this case the ^{57}Co source was kept at room temperature.

3. RESULT AND DISCUSSION

The Mössbauer spectra obtained at room temperature for the oxygen deficient phase $YBa_2(Cu_{1-x}Fe_x)_3O_6$ samples with different iron concentration is presented in Fig. 1. For the samples with low iron concentration there is a predominance of the species with QS ≅ 2.0mm/s (A) which is assumed to be characteristic of this phase. For higher x values another iron species with QS = 0.70mm/s (B) appears together with a magnetic splitted spectrum which is better resolved for x ≧ 0.10, indicating the formation of a magnetic phase (MP)

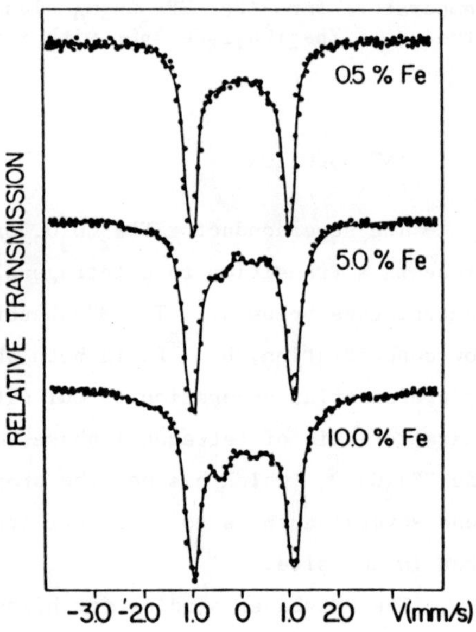

FIG. 1 - Mössbauer spectra at RT with different iron concentration.

at RT. The two quadrupole doublets (A and B) may be attributed to Fe on Cu1 site with two different oxygen coordination.

The fact that the copper ions in Cu2 site in the $YBa_2Cu_3O_6$, present an antiferromagnetic order with $T_N \sim 450K$[1,2] may lead to the conclusion that the MP is due to iron in this site. However the Mössbauer spectra for samples with different x values taken with high velocity scale (Fig. 2) indicate that the effective magnetic hyperfine field, H_{eff}, at room temperature depends on iron concentration. The analysis of the Mössbauer spectra at 4.2K (Fig. 3) shows that the saturation value of H_{eff} correspondent to the MP is independent of x value. Consequently the difference observed for H_{eff} at room temperature can be due to a dependence on iron concentration of the magnetic ordering temperature (T_0) of this phase.

To confirm this result the Mössbauer spectra have been taken at fixed temperature in an interval from 295 to 458K. It was observed a gradual decreasing of the magnetic splitted spectra due to MP, which collapse to a quadrupole doublet with QS = 0.55mm/s. The analysis of the Mössbauer spectra obtained for x = 0.15, 0.10, 0.05 indicates an upper limit of $T_0 \sim 420$, 370 and 350K respectively.

The dependence of relative intensity of the magnetic phase and the T_0 value on iron concentration suggests that one should be carefull in atributting the MP to the occupation of Cu2 site.

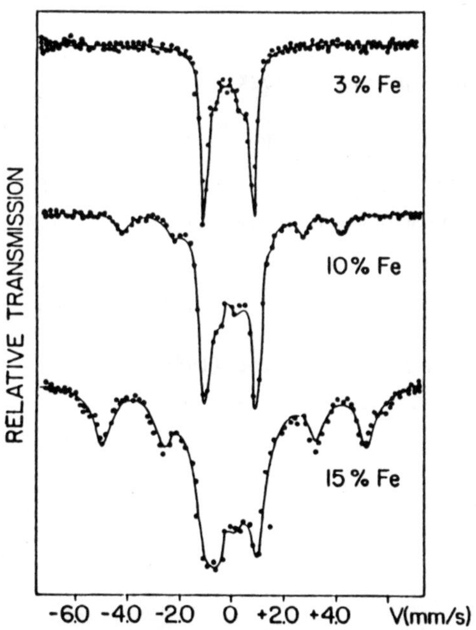

FIG. 2 - Mössbauer spectra at RT with different iron concentration.

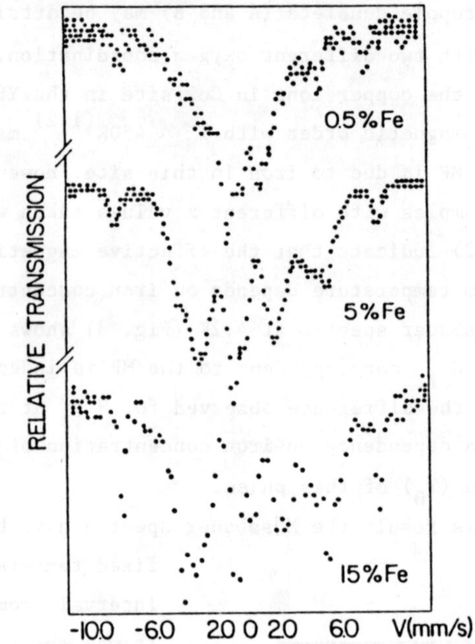

FIG. 3 - Mössbauer spectra at 4.2K.

The authors would like to acknowledge discussions with Dr. A.A. Gomes. This work was partly supported by FINEP and CNPq

1) J.M. Tranquada et al., Phys. Rev. Lett., 60, 156 (1988).

2) J. Rossat-Mignod et al., Physica C 152, 19 (1988).

INTRINSIC JOSEPHSON JUNCTIONS IN HIGH-T_c SUPERCONDUCTORS:

Roberto NICOLSKY
Instituto de Fisica, Universidade Federal do Rio de Janeiro
C. P. 68528, 21945 Rio de Janeiro, Brazil

ABSTRACT

The magnetization experiments indicate that the high-T_c superconductors have intrinsic Josephson networks between grains, and also within the grains and single crystals. We establish a criterion for categorizing the Josephson junctions as tunneling or metellic. We analyse the I-V curves of single crystals, granular ceramics e point contacts and conclude that those intrinsic junctions are metallic. We discuss the consequences of that conclusion for critical currents in the metallic oxide superconductors.

CRITERION FOR CATEGORIZING JOSEPHSON JUNCTIONS

The magnetization experiments with high-T_c superconductors (HTS) have been interpreting by the authors assuming that there are weak links connecting not only grains in the ceramics samples[1], but also connecting domains inside either grains or single crystals[2], sue to the small size of a typical Josephson loop compared to the dimensions of the grains. As those junctions are not constructed intensionally we do not know, in principle, how that network has been made during the formation process of the superconducting metallic oxides. We have previously proposed a criterion[3] to identify the nature of such junctions: tunneling or metallic. In this work we summarize it and note that a criterion is not an exact rule, so its should be applied cautiously always looking at what is general and constant in the varied experimental data.

The criterion is based on the current-voltage characteristic (CVC) for both kinks of Josephson junctions. It is easy to show that there are some peculiar features which are common to each type of junctions. The standard theory of tunneling Josephson junctions (TJJ), which can be found in current textbooks[4], solves a tunneling Hamiltonian for Cooper pairs across a barrier (Fig. 1,a) and yields us

$$J = J_c \sin \chi \ , \quad \frac{\partial \chi}{\partial t} = \frac{2eV}{h} \quad \text{and } J_c = \frac{\pi \Delta(T)}{2eR} \tanh \frac{\Delta(T)}{2kT} \qquad \text{(1a, 1b, 1c)}$$

where J is the current density, χ is the phase difference related to the voltage and $\Delta(T)$ is the temperature dependent gap. From the above expressions we note that for V=0 we have a temperature dependent DC current, provided by a constant phase difference. For finite small voltages, the current is AC with the Josephson frequency as above and its time average is essentially zero. At voltages above $V_J = 2\Delta/e$ the current has the ohmic V/R value. Thus, we conclude that the current in a TJJ in the presence of a voltage is nearly zero for $V < V_J$ and metallic (ohmic) for $V > V_J$. The other important feature is the strong hysteresis as we can see in the Fig. 2,a) for an ideal TJJ for $T << T_c$.

For metallic junctions (MJJ) we based our considerations on a recent theory[5] for SNS junctions, which also models weak links, because the MJJ have the same quantum mechanical model - a potential well, instead of the barrier corresponding to the tunnel junctions. That theory solves the time dependent Bogoliubov-de Gennes equations for a potential well (see Fig. 1,b) in the presence of a voltage and with those solutions calculates the CVC of a MJJ in the relaxation time approximation.

The solutions of the time dependent Bogoliubov-de Gennes equations for a spatially inhomogeneous pair potential (related to the gap) yields an excitation spectrum in the normal metal layer with bounded quantized states, the Andreev states[5], in the range $(-\Delta,\Delta)$ around the Fermi surface. Each state is a superposition of an electron and a hole of the same energy, which are responsible for the phase coherence between both superconducting sides. At the NS interface a peculiar scattering takes place: an electron-like quasiparticle of an excitation energy smaller than the gap Δ is reflected back as a hole with a reversed velocity but with the same momentum, and two electrons forming a Cooper pair are transfered to the condensate. Symmetrically, a hole is reflected back as an electron. This is the so-called Andreev reflection (AR) mechanism. If the scattered quasiparticle has been contributing to the current,the opposite quasiparticle reflected back also contributes additionally,due to momentum conservation. Thus, these processes increase the current over the ohmic contribution of the original quasiparticles.

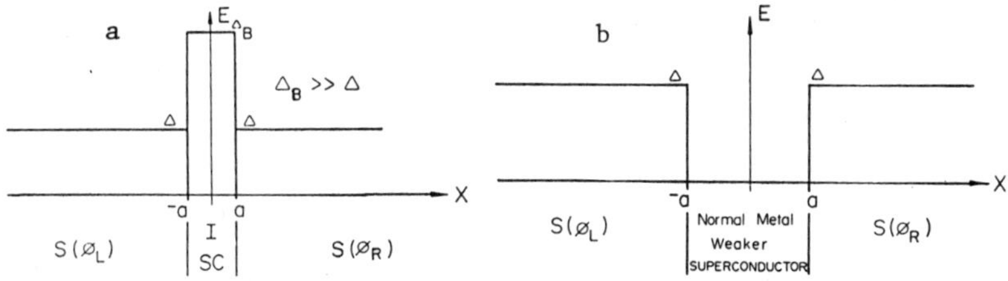

Figure 1. a) Energy model for a TJJ, which barrier can be insulator (I) or semi-conductor (SC). b) Energy model for a MJJ, which has in the middle a normal metal or a weaker superconductor.

The current density in this theory (Eq. 1 in Ref. 5) is

$$J = J_N + \sum_{n>0}^{\infty} J^{\pm}_{AR,n}(|A_n^+|^2 - |A_n^-|^2)\exp\frac{(2n-1)L^*}{2l}, \quad L^* = L_N + 2\frac{\Delta(0)}{\Delta(T)}\xi_0 \qquad (2a,\ 2b)$$

where J_N is the ohmic component proportional to the voltage, due to the Sharvin resistance of the junction. $J^{\pm}_{AR,n}$ is the summation over the Andreev states of the contributions due to the quasiparticles with n AR which momentum component in the current direction increases (+) or decreases (-) the current. L^* is the temperature dependent effective length of the potential well, L_N the length of the normal region and ξ_0 is the coherence length of the superconductor. A_n^+ and A_n^- are respectively the probability of the AR of order n for quasiparticles having their momentum component in the electric field direction in favour (+) to the current or against (-) it. Due to the shift of the Fermi surface (which accounts for the occupation in the system) in the presence of an electric field, the summation over the (+)-states in not balanced by the (-) states. Thus, the term due to AR always increases the absolute value of the current. The Fig. 2,b) shows the typical CVC of a MJJ far from T_c,

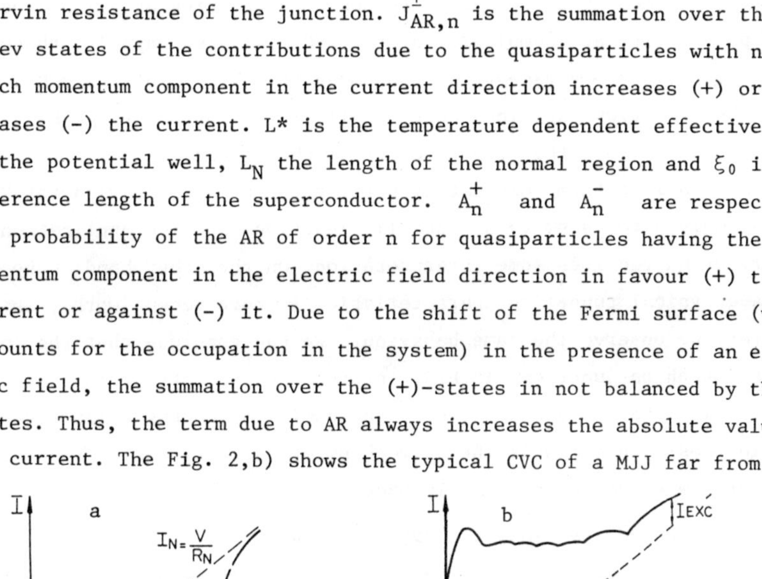

Figure 2. a) Typical CVC of an ideal TJJ with $I<I_N$, the extrapolation to zero of the linear (ohmic) part of the current;note the strong hysteresis. b) Typical CVC of an ideal MJJ with $I>I_N$, the extrapolation to zero of the linear part or a parallel to it; the additional contribution is due to AR; there is no hysteresis.

which are reversible. Note that the current often rises fast at low voltages, has much structures and in the linear part sometimes does not coincide with the ohmic current but is parallel to it (excess current).

To summarize, we define as the criterion for categorizing an unknown Josephson junction by its CVC curves the following inequalities

$$\text{TJJ: } I(V) < I_N = V/R, \quad \text{for } V < V_J = 2\Delta(T)/e \tag{3a}$$

$$\text{MJJ: } I(V) > I_N = V/R \tag{3b}$$

This applies only for an ideal junction far from T_c. Of course a real junctions could be much more complicated. The same occurs with a network which sometimes is a very complex curve. In all those complicated and unclear cases we can use additionally a very important information given by the CVC namely the hysteresis, which is associated with tunneling. However, sometimes is not possible to conclude anything at all because of the peculiarity and complexity of the CVC analysis. Happily, these complex situations are exceptions and among the CVC of intrinsic MJJ of HTS the criterion explained here usually works, as we see in Ref. 3 where we analyse more than twenty CVC from the literature.

Almost all of them immediately satisfy the above criterion for MJJ systems. Only a few seem to be metallic junctions and the application of such criterion requires some interpretation. On the other hand, none of them have typical tunneling characteristic. In more recent works, as we see later, we observe the same bahaviour. We can conclude that the intrinsic Josephson junctions in HTS, intergrains and intragrains, are of the metallic kind. As a consequence, the dominant process which takes place in those junctions is the AR mechanism. Further, a direct observation of the AR has been recently made[6].

METALLIC JOSEPHSON JUNCTIONS AND CRITICAL CURRENTS

In the polycrystalline and granular samples it is obvious that part of those MJJ are localized in intergrain boundaries. The points of contact between grains act as strong variable thickness constrictions[7], also called microbridges. In the current flow regime the current density increases enormously in each constriction and is limited by its local critical value, which depends on the particular geometry, temperature and magnetic field. In the whole volume, the sample forms a three-dimensional network of constrictions. The metallic character of those junctions indicates that the surfaces of the grains are neither insulating nor semiconducting phases. The surfaces should consist of a weaker superconductor (with an intrinsically smaller or depressed gap) or a normal metal phase. This system restrics appreciably the transport properties of the ceramic samples of HTS.

In the single crystals or grains the (110) and (001) twin planes[2] are good candidates to be the junctions. Another possibility could be other non-superconducting planes[2], since the gap varies significantly over a coherence length ξ_0, which is smaller than the lattice parameter c in this axis. This picture suggests that the superconductivity should be in the blocks of Cu-O planes and chains, which should be weakly coupled to the next blocks throuch the depressed gap planes. However, if we calculate the value of the critical current using the Bardeen-Johnson expression[8] ($n_e \simeq 10^{21}/cm^3$, $m=6m_e$ and $L^* \simeq 2\xi_0 \simeq 6nm$)

$$J_m = \frac{n_e \pi \hbar e}{2mL^*} \simeq 10^8 \ A/cm^2 \qquad (4)$$

which is much larger (two orders of magnitude) than the best reported measurements. This indicates that the intragrain intrinsic junctions should not be the most important mechanism in the limitation of the transport properties. Probably the flux dynamics plays a more important role in those junctions, as discussed in the literature[9].

As a conclusion, the CVC fo the intrinsic Josephson junctions, intergrain and intragrain, do not exhibit any evidence of intrinsic tunneling, but suggest that these junctions are of metallic nature. The intergrain junctions represent a strong limitation to the current-carrying properties but the intragrain do not. In this case, it seems that the limitation is the flux dynamic mechanism.

REFERENCES

1. S. Senoussi et al., this volume.
2. G. Deutscher and K. A. Müller, Phys.Rev.Lett.59,1745 (1987).
3. R. Nicolsky, Prog. High Temp. Sup.-v.5,p.219,W.Scien.Pub.1988.
4. M.Tinkham,Introd.to Superc.,Mc Graw-Hill, 1975.
5. R.Kümmel et al.,S.S.Com.65,1567.
6. H.van Kempen et al.,this volume.
7. R. Nicolsky et al., J. Low Temp. Phys.58,11(1985)and ref.therein.
8. J.Bardeen and J.L. Johnson,Phys. Rev. B5, 72 (1972).
9. A.P.Malozemoff et al.,this vol.

MÖSSBAUER STUDIES OF $(La_{1-x}Ba_x)_2CuO_4$ SUPERCONDUCTING OXIDES

E. Baggio Saitovitch, I. Souza Azevedo, R.B. Scorzelli,
H. Saitovitch, S.F. da Cunha, A.P.Guimarães, P.R.Silva and L.C.Sampaio

Centro Brasileiro de Pesquisas Físicas
Rua Dr. Xavier Sigaud, 150 - Urca
22290 Rio de Janeiro, RJ - Brasil

Samples with Cu substituted by small amount of iron (less than 1%) were prepared for x = 0.15, 0.25 and 0.35 mixtures. Despite the fact that there is only one Cu site in the $(La_{1-x}Ba_x)_2CuO_4$ compounds the Mössbauer spectra at room temperature indicated the presence of two quadrupole doublets whose behaviour were followed in different steps of samples preparation. They are mostly characteristic of Fe^{+3} and one of them may be due to the presence of Oxygen or Lanthanium vacancies. The existence of magnetic ordering is investigated by ^{57}Fe Mössbauer measurements at 4.2K for samples with 0.5% Fe.

1. INTRODUCTION

The $La_{1.85}Ba_{0.15}CuO_4$ superconductor oxide has lower T_c as compared with $YBa_2Cu_3O_7$ oxides, however both are antiferromagnetically ordered in the corresponding insulator phase; i.e. La_2CuO_4 and $YBa_2Cu_3O_6$. The effect of several substituents in Cu sites is different in the two systems; for example 3% Fe in the $La_{1.85}Ba_{0.15}CuO_4$ destroys completely the superconductive transition[(1)], while in the $YBa_2Cu_3O_7$ even with 10% of Cu substituted by Fe it is possible to measure, by DC magnetization, a T_c onset of $\sim 54K$[(2)]. The local studies based on the Mössbauer Spectroscopy of ^{57}Fe: $YBa_2Cu_3O_y$ oxides allowed us to establish the preferential occupation of the Cul site by Fe with different oxygen coordination. We report here ^{57}Fe Mössbauer studies in $La_{2-x}Ba_xCuO_4$ oxides (x = 0, 0.15, 0.25 and 0.35) focusing on the quality of sample preparation and the presence of a magnetic ordered state mainly for 0.5% Fe concentration.

2. EXPERIMENTAL

The samples of ^{57}Fe: $La_{2-x}Ba_xCuO_4$, with x=0, 0.15, 0.25 and 0.35, were prepared by sintering in air pellets of appropriated amounts of La_2O_3, $BaCO_3$, CuO and $^{57}Fe_2O_3$ at 900C for 20h followed by regrinding and sintering two times at the same conditions. The next step involves the regrinding and heating at 1100C for 20h and cooling down to 600C at rate of 50C/h. The samples were kept at this temperature for 50h and then cooled down to room temperature (RT) at the same rate[3].

The X-ray analysis and Mössbauer spectra at RT were taken after the first sintetization and after the final preparation.

3. RESULTS AND DISCUSSION

The $La_{2-x}Ba_xCuO_y$ oxides have a single site for Cu with a distorted octahedral oxygen coordination. The Fe ions are expected to occupy the Cu site and in the case of La_2CuO_4 it is possible to introduce 10% Fe without disturbing the orthorhombic structure. The substitution of Cu by ^{57}Fe in very low concentration as 0.5% can be used as a probe to detect by means of ^{57}Fe Mössbauer Spectroscopy how the Ba and oxygen content influence the electronic charge distribution in Cu site.

The Mössbauer spectra at RT obtained after two treatment at 900C for 20h (sample I) are shown in Fig. 1, for x = 0.15, 0.25 and 0.35. The corresponding Mössbauer parameters are listed in Table I. The doublet D1 can be associated with Fe^{+3} at Cu site, while doublet D2 responsible for the asymmetry of the spectra, can be attributed to an impurity phase always present with broad lines and low relative intensity.

After the final treatment the sample II exhibited a better resolved Mössbauer spectrum (Fig. 2) with symmetrical lines, since D2 is no more present. The X-ray diffraction studies revealed also the improvement of the tetragonal structure.

The comparison of Mössbauer parameters between La_2CuO_4 (IS = 0.30mm/s, QS = 1.60mm/s) and $La_{2-x}Ba_xCuO_4$ (Table I) indicated that

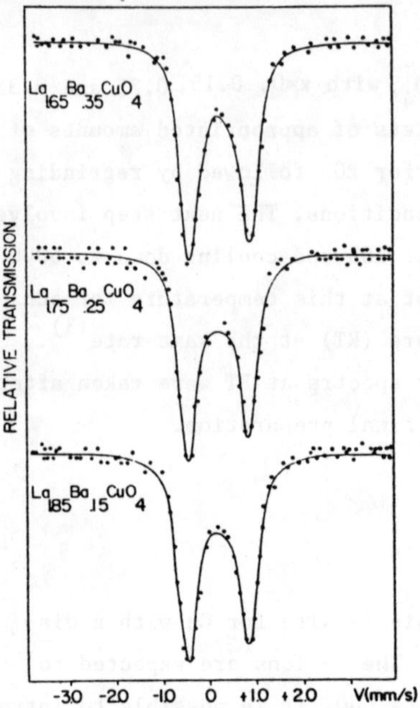

FIG. 1 - Mössbauer spectra at RT after two treatment at 900C for 20h.

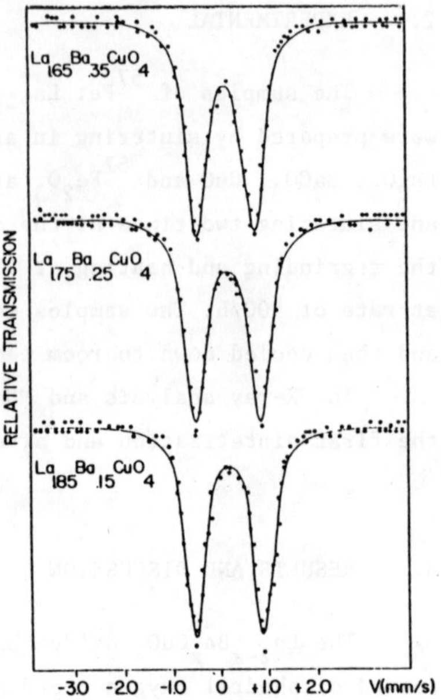

FIG. 2 - Mössbauer spectra at RT after the final treatment at 1100C for 20h.

$La_{2-x}Ba_xCuO_4$		IS1 mm/s	QS1 mm/s	IS2 mm/s	QS2 mm/s
x =0.15	I	0.28	1.32	0.15	0.56
	II	0.30	1.43	–	–
x =0.25	I	0.28	1.30	0.11	0.49
	II	0.29	1.37	–	–
x =0.35	I	0.28	1.30	0.18	0.62
	II	0.25	1.19	–	–

TABLE I: Fitted hyperfine parameters IS: isomer shift (ref. Fe-metal). QS: quadrupole splitting.

there is a clear decrease of the quadrupole interaction with the Ba content which may reflect the more symmetric electronic distribution at Cu site in the tetragonal symmetry.

The phase diagram of $La_{2-x}Ba_xCuO_4$[3] is known to go from an anti-ferromagnetic (x = 0) through a spin-glass-like region ($0 \leqq x \leqq 0.15$) to the superconducting non magnetic region ($0.15 \leqq x \leqq 0.35$). The Mössbauer measurements performed at 4.2K with samples containing 0.5% Fe may reflect the ordered state of the Cu magnetic moments. Fig. 3a presents the spectra corresponding to 0.5% Fe in La_2CuO_4 displaying a magnetic pattern with internal field of $\sim$48T which is perpendicular to the principal component of the electric field gradient (EFG) beeing parallel to the c direction at the iron site. This result is in agreement with neutron diffraction data[4] for pure La_2CuO_4 and is an indication that iron probe can provide important informations about the ordered state of Cu moments.

The broad absorption lines of the magnetically splitted Mössbauer spectra at 4.2K obtained for 0.5% Fe in $La_{1.85}Ba_{0.15}CuO_4$ (Fig. 3b) are typical of a distribution of H_{eff} at the iron site with the mean value very close to the one obtained for the ^{57}Fe: La_2CuO_4. This distribution of H_{eff} may be due to different orientation of the principal axis of EFG and the iron magnetic moments induced by the Ba atoms. Mössbauer measurements as a function of temperature may give information about the freezing mechanism of the iron magnetic moments.

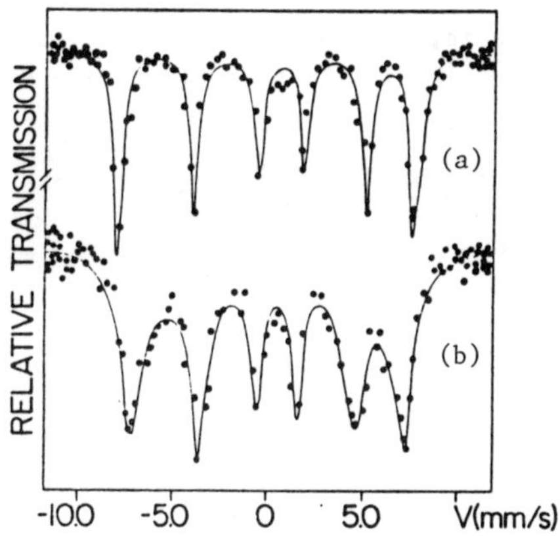

FIG. 3 - Mössbauer spectra at 4.2K:
a) La_2CuO_4 with 0.5% Fe.
b) $La_{1.85}Ba_{0.15}CuO_4$ with 0.5% Fe.

The origin of this magnetic ordering of iron moments is not clear since it may arise from long range correlation of iron moments or can be established through magnetic moments induced in the Cu ions. Anyhow, this result implies again the coexistence of magnetic and superconducting states in the same sample of superconducting oxide, similar to what was observed for x ≧ 0.03 Fe in $YBa_2Cu_3O_7$ compounds.

The authors would like to acknowledge helpfull discussions with Dr. C.A. dos Santos and M.R. Abraão for the X-ray difractogram. This work was partly supported by FINEP and CNPq.

1) J.M. Matykiewicz, C.W. Kimball, J. Giapintzakis, A.E. Dwight, M.B. Brodsky, B.D. Dunlap, M. Slaski and F.Y Fradin, Submitted to Physics Letters.

2) E. Baggio Saitovitch, I. Souza Azevedo, R.B. Scorzelli, H.Saitovitch, S.F. da Cunha, A.P. Guimarães, P.R. Silva and A.Y. Takeuchi, Phys. Rev. B 37(13), 7967 (1988).

3) M. Kato, Y. Maeno and T. Fujita, Physica C 152, 116 (1988).

4) D. Vaknin, S.K. Sinha, D.E. Moncton, D.C. Johnston, J.M. Newsman, C.R. Safinya, M.E. King Jr., Phys. Rev. Letters 58, 2802 (1987).

MOESSBAUER STUDY OF OXYGENATED AND DEOXYGENATED $YBa_2(Cu_{1-x}Fe)_3O_{7+y}$ FOR DIFFERENT X VALUES.

C.Saragovi-Badler y F.Labenski de Kanter

Departamento de Física, Comisión Nacional de Energía Atómica, Av.del Libertador 8250, 1429 Buenos Aires, Argentina

We report a Moessbauer study of $YBa_2(Cu_{1-x}Fe_x)_3O_{7+y}$ for 0.1 $\leq$ x 10% and for different y values. The oxygenated series which correspond to the higher values of y, were prepared under an oxygen flow and the deoxygenated one was obtained from the previous one. Spectra are fitted with three doublets with different relative intensities. The measured QS values, compared with those from theoretical calculations are used to assign different oxygen environment of Fe ions. The behaviour of the doublet intensities and their relationship with changes in the superconducting transition as well as the structures modifications are discussed.

In a previous paper[1], $YBa_2(Cu_{1-x}Fe_x)$ O_{7+y} samples for several concentrations have been studied by Moessbauer spectroscopy. In order to clarify the understanding of the microscopic structure of the Fe sites in these oxides the x interval has been completed. Besides, variation of oxygen content studies have been started.

The ceramic samples which will be called "oxygenated series" were prepared sinterizing $BaCO_3$, Y_2O_3 and a mixture of Cu and Fe oxides coprecipitated from an aqueous solution of the corresponding nitrates, all of them of high purity. These samples were prepared using the same procedure as in (1). Samples from batchs x=0.001 and 0.05 were fired at 530° in Ar atmosphere during 3 hours, and then slowly cooled down during 5 hours under the same atmosphere. These will be called "deoxygenated series". The decrease of the oxygen content of the mentioned samples was estimated to be about 1.3 and 1.6 respectively[2].

The variation of the characteristic temperatures of the superconducting transition for the oxygenated samples with the Fe concentration were studied in detail by electrical resistivity and magnetic susceptibility[3]. The results are presented in Fig.1. The lattice parameters a, b, studied by X-Ray powder diffraction[4] versus the Fe concentration for the same series are shown in Fig.2. Both figures have been completed with the values corresponding to the new "oxygenated" samples.

Moessbauer studies were made by using a conventional constant-acceleration spectrometer in a transmission geometry. Moessbauer absorbers (8-9 mg Fe/cm²) in powdered form were prepared mixing the samples under study with boron nitride powder in order to avoid texture effects. Spectra were least square fitted assuming symmetrical doublets of pure Lorentzian peaks, and the quality of each computer fit was checked by a X^2 test. As the ^{57}Fe is present in its natural abundance it was necessary to make a compromise between improving the signal-to-noise ratio and keeping the counting time within reasonable limits. The typical number of counts per channel was about 8×10^{6}.

The samples now measured could be fitted with the same doublets A, B and C as in (1). Summarizing the obtained results we can say that Y-Ba-Cu-O series spectra for the different x and y values are composed by the three mentioned doublet. The measured quadrupole splittings, QS, and Isomer shifts, IS, referred to that of Fe metal are as follows:

	A	B	C
QS(mm/s)	1.91(6)	1.10(6)	0.55(6)
IS(mm/s)	-0.07(2)	-0.09(6)	0.03(3)

The parameters have similar values for the samples with different Fe concentrations and different y values. Our spectra show a mixture of the three doublets with different intensities. In Fig.3, spectra of the oxygenated and deoxygenated samples for x=0.05 are shown.

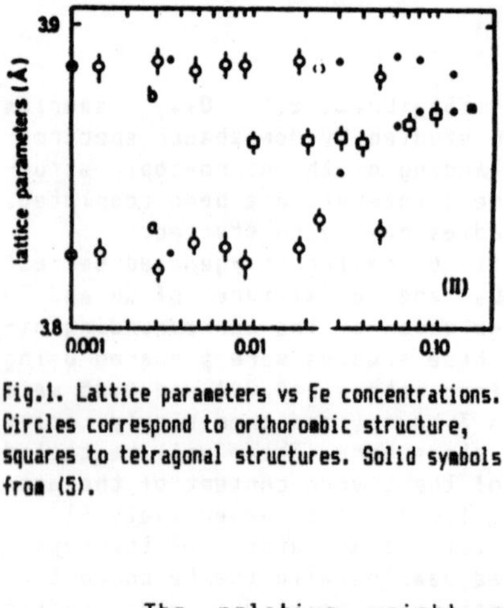

Fig.1. Lattice parameters vs Fe concentrations. Circles correspond to orthorombic structure, squares to tetragonal structures. Solid symbols from (5).

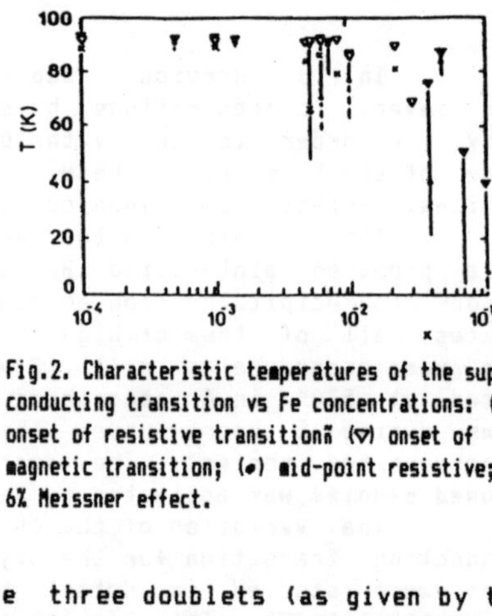

Fig.2. Characteristic temperatures of the superconducting transition vs Fe concentrations: (▼) onset of resistive transition; (▽) onset of magnetic transition; (•) mid-point resistive; (x) 6% Meissner effect.

The relative weight of the three doublets (as given by the area under each curve in the fitted spectra) is shown in Fig.4 where we have completed values of the oxygenated series. The x=0.05 "deoxygenated sample" values are shown in the right extreme of the figure. In the "deoxygenated series" samples doublet C is the predominant one for the x=0.001 sample but in the x=0.05 sample doublets C and B decrease to about 40%. Doublet A is now predominant, its intensity has been increased by a factor 3.

The IS values found are smaller that the usual ones for ionic Fe^{3+} in high spin configuration. We interpreted these values as indicative of a large degree of covalency and probably a metallic character of the Fe bonds.

The measured QS values have been compared with calculation of the electric field gradient for different oxygen configurations. Simple point charge calculations reported in (6) and a cluster model for the Fe impurity and the surrounding oxygen ions reported in (7) have been considered in order to help in the assignement of the observed doublets to different microscopic structures created by Fe substitution.

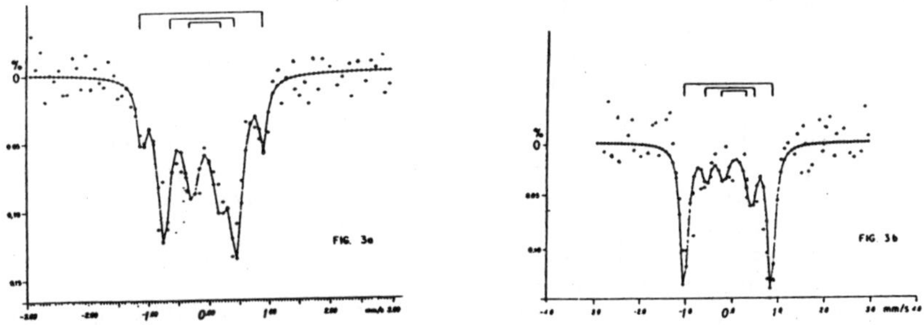

Fig.3. Moessbauer spectra of 5% Fe in $YBa_2(Cu_{1-x}Fe_x)_3O_{7+y}$, a) "oxygenated"; b) "deoxygenated".

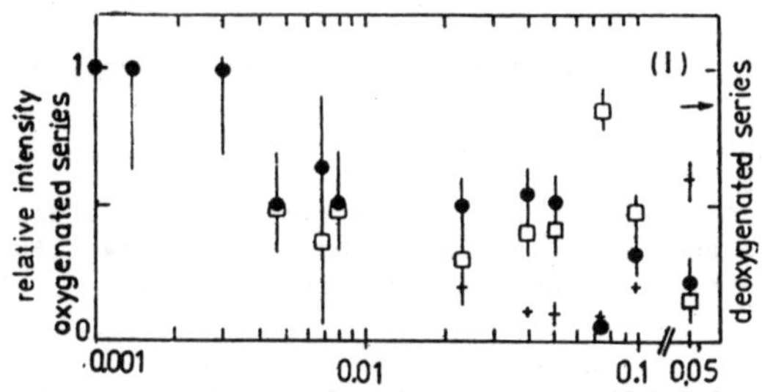

Fig.4. Relative intensity of Moessbauer doublets vs Fe concentrations: (+) doublet A, (O) doublet B, (•) doublet C

At low Fe concentrations, the lattice parameters and the superconducting transition temperature of the "oxygenated" samples are similar to those of the undoped sample. In the "deoxygenated sample" the structure has become tetragonal (a = b = 3.86(1) A) and the superconducting transition has been depressed. The Moessbauer spectra of both samples are very similar. They present only a prominent C doublet. Doublet C can be assigned to Fe in Cu(1) sites with 3 or 4 oxygens in the ab plane or eventually two in perpendicular axes. We cannot completely discared a Fe in Cu(2) sites signal. At high Fe concentration (x=0.05) the "oxygenated" and "deoxygenated" samples present different lattice parameters corresponding to a tetragonal structure (a = b = 3.86(1)A) in the latter sample. The onset of the superconducting transition in the former sample departs very little from values of lower Fe concentration and presents a small broadening of the transition. On the contrary, the superconductivity of the "deoxygenated" sample has been depressed. Moessbauer spectra are now different. Doublet A contribution has changed from 15% of the total intensity in the "oxygenated sample" to be the predominant one with 61% of the total intensity, in the "deoxygenated" one. This doublet, which is related to the high temperature tetragonal phase (6), can be assigned to Fe in Cu(1) sites having zero or one oxygen ion.

It is seen that the structures and the superconductive properties have been modified with the y variation: the oxygen depopulation has depressed the superconducting transitions.

It is possible to suggest that for low Fe concentration, Fe ions occupy Cu(1) sites with a higher degree of neighbouring oxygen sites occupancy compared with the undoped sample and this fact appears to be independent of the sample oxygen content.

When the oxygen content is appropiate for the existence of superconductive transitions (e.g. "oxygenated series"), the breakdown of the long range order of the oxygen vacancies is due to the progressive occupancy of the mentioned vacancies (1). When the oxygen depopulation is high (e.g. "deoxygenated series") the superconductivity transition is totally depressed.

When the Fe concentration is increased (x=0.05) the oxygen depopulation and the disappearance of the superconducting transition is perceived by the Fe ions which detect a predominant environment of a Cu(1) site surrounded by one or no oxygen, characteristic of the non superconducting tetragonal phase.

Further studies varying the oxygen content, of the samples are in progress.

ACKNOWLEDGEMENT

The authors are indebted to the members of the Magnetic Resonances Laboratory, Centro Atómico Bariloche, Argentina, for providing the samples and their published and unpublished results.

REFERENCES

1) C.Saragovi-Badler, F.Labenski de Kanter, M.T.Causa, S.M.Dutrus, C.Fainstein, L.B.Steren, M.Tovar, R.Zysler, Solid State Comm. 66, 381 (1988).
2) P.Satti, D.Pasquevich, A.Caneiro, J.Abriata, C.Fainstein, presented at Latin American Conference on High Temperature Superconductivity, Rio de Janeiro (1988).
3) M.T.Causa, S.M.Dutrus, C.Fainstein, G.Nieva, H.R.Salva, R.Sanchez, L.B.Steren, M.Tovar, R.Zysler, Progress in High Temperature Superconductors 1, 341 (1987).
4) M.T.Causa, S.M.Dutrus, C.Fainstein, H.R.Salva, L.B.Steren, M.Tovar, R.Zysler. 32nd Conference on Magn.and Mag.Mat.,Chicago (1987), and J.Applied Physics (in press).
5) T.J.Kistenmacher, W.A.Bryden, J.S.Morgan, K.Moorjani, Y.W.Du, Z.Q. Qui, H.Tang, J.C.Walker, Phys.Rev.B.36, 8877 (1987)
6) E.R.Bauminger, M.Kowitt, I.Felner, I.Nowik, Solid State Comm. 65, 123 (1988).
7) A.Saul, A.M.Llois, A.Levy Yeyati, M.Weissmann, Solid State Comm. 66, 471 (1988).

ANGULAR CORRELATION STUDY OF HYPERFINE INTERACTIONS IN $YBa_2Cu_3O_y$

Saitovitch H., Silva P.R.J., Saitovitch E.B., Azevedo I.S., Scorzelli R.B., Cunha S.F., Guimarães A.P., Sampaio, L.C.
Centro Brasileiro de Pesquisas Físicas
Rua Dr. Xavier Sigaud 150,
22290 Rio de Janeiro, RJ, Brazil

and

Martinez Jorge A.
Dept. Fisica UNLP - C-C67,
1900 La Plata, Argentina

1. INTRODUCTION

Since the discovery of high T_c superconductivity in Y-Ba-Cu-O much effort has been made to characterize the structure and composition of both phases, normal and superconducting[(1)]; they show, for instance, the existence of two copper sites: Cu1 has a square planar coordination due to the presence of two oxygen vacancies in the $\hat{a}$ axis, whereas Cu2 has a tetragonal pyramidal coordination due to an oxygen vacancy in the $\hat{c}$ axis. Additional information, with extension to dynamical effects, as those obtained from hyperfine interactions (HFI), revealed by local spectroscopies such as Perturbed Angular Correlations (PAC), may be of great interest since they reflect the microscopic charge distribution at the probe nucleus. We have investigated by (PAC), in it's time differential mode, above and below T_c, in oxygenated as well as argonized samples, the electric field gradients (EFG) at the site of dilute Cd^{111} impurities ($\sim 10^{-5}$ at.%) in Y-Ba-Cu-O.

2. EXPERIMENTAL DETAILS AND RESULTS

The measurements were carried out with the (172-247) keV cascade of Cd^{111} which is populated by the EC decay of the 2.8 days In^{111} isotope. For the measurements we used a four detector apparatus, each detector being 90° apart from the other. The normalized data were least-square fitted with the function

$$G_{kk}(t) = \sum_{n=0}^{3} \sigma_{Kn} \cos(\omega_n t)^{(2)} \quad , \quad \text{with } K = 2 \qquad (1)$$

appropriate for a static quadrupole interaction in a polycristaline source with a nuclear spin I=5/2[2].The finite time resolution was about 4.0nsec.Polycristaline samples(named Sample I and Sample II) of Y-Ba-Cu-O were prepared[3] from a mixture of pure Y_2O_3, $BaCO_3$ and CuO; the activity was introduced in the sample during sinterization.The argonization was done at 500C, under argon flux, during 2hs. From the temperature dependence of the magnetic susceptibility, $\chi_g = M_g/H$, the values of T_c (here defined as the temperature at which the susceptibility crosses the $\chi_g = 0$ line) were derived[4]. (PAC) measured with these samples were characterized by non-axially symmetric pure quadrupole interactions related to two sites in Sample II and to three sites in Sample I. X-ray analysis yielded ortorhombic structure to the oxygenated samples and tetragonal structure to the argonized ones. In all samples was detected the presence of an ∿ 5% admixture of In_2O_3, as a precipitated phase, which d'ont interfieres in the measurements as far it don't includes any of the Y-Ba-Cu atoms in it's structure. Fig. 1 show (PAC) spectra obtained with such samples in various temperatures. The values of the quadrupole frequencies (QF), of the (EFG), of the assymetry coefficients η, as well as of some of the fit parameters, including their definitions formulae, are displayed in Table I.

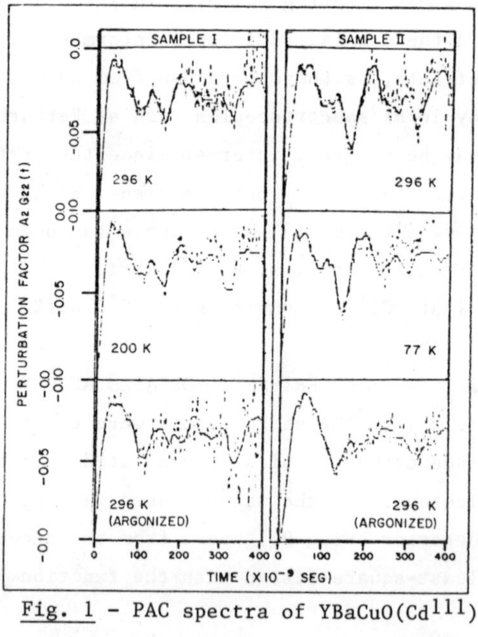

Fig. 1 - PAC spectra of YBaCuO(Cd^{111})

3. DISCUSSION

The most important features of the (HFI) measured in samples I and II are: (i) quadrupole frequencies (QF) of Sample I are quite similar to those

Table I. PAC measured parameters for $YBa_2Cu_3O_y$ doped with Cd^{111}; s^i (site fractions in %); ω^i (quadrupole frequencies in MRad); η^i (assymetry coefficients = $V_{xx}-V_{yy}/V_{zz}$); δ^i (quadrupole frequencies distributions in %); $V_{zz}\times 10^{17}$ (electric field) gradients = $20/3 \frac{\omega^i \hbar}{eQ}$ V/cm^2; Q = Cd^{111} nuclear quadrupole moment; e = electron charge).

T(K)	s^1	ω^1	η^1	δ^1	V^1_{zz}	s^2	ω^2	η^2	δ^2	V^2_{zz}	s^3	ω^3	η^3	δ^3	V^3_{zz}	In_2O_3 s	ω	η	δ	V_{zz}
Sample I																				
296	31	37.2	.314	1	1.966	22	42.0	.49	5	2.220	42	54.6	.36	11	2.886	5	132.0	.35	0	6.978
200	27.6	37.8	.23	0	1.998	16	40.8	.49	4	2.157	41	52.2	.39	8	2.759	5	132.6	.38	0	7.010
296 (argonized)	32	34.2	.36	2	1.808	18	46.8	.38	0	2.474	46	55.8	.0	11	2.950	4	127.8	.58	0	6.756
Sample II																				
296	76	36.0	.34	4	1.903	15	40.2	.58	1	2.125						9	132.6	.37	1	7.010
77	33	37.0	.0	4	1.956	57	39.0	.47	7	2.062						10	130.2	.45	2	6.883
296 (argonized)	94	27.6	.61	13	1.459											6.1	117.0	.30	1	6.185

of Sample II, beside a third (QF) which appears in Sample I, inexistent in Sample II; (ii) presence of In_2O_3 phase in all samples; (iii) small variation of (QF) with temperature in both samples; (iv) $\eta = 0$ for the 77K Sample II measurement; (v) remarkable difference between oxigenated and argonized samples spectra.

Concerning Sample II, the two measured (QF), beside the In_2O_3 assigned frequency, are quite close each other and with the lower one, at 296K, at a site rate about five times higher than the second one: this site rate discrepancy seems to avoid the possibility of assigning each of the (QF) to the two different copper sites, which are known to appear at a site rate 2:1 in the Y-Ba-Cu-O structure. By other side, (EFG) calculations performed using a point-charge model(*) displayed a much higher (QF) intensity ratio than the ω^1/ω^2 = .85 we obtained. At this moment we could like to advance that we are dealing with one (QF) related to only one probe-site, probably with slight site distortions and/or oxygen content apart from δ = 7. It remains to clear out the nature of this site. In this sense, the argonized sample displays one site with only one (QF), and, as the Cu oxygen environment content is highly changeable upon argon substitution, we would be tempted to say that the site we are dealing with should be the Cul site. A remarkable figure concerned with the 77K result is, beside the equalization of the two (QF), even with different sites assignements, that one of the sites η drops to zero ($V_{xx}=V_{yy}$), configurating an axial simmetry, and this, even cautiously, may be tributed to a higher

(*) S. Vries - Private communication.

occupancy of the vacancies around Cul, which are known to exist, or by a redistribution of density changes below T_c or/and by a higher moving of the O_2 themselves, with a consequent simmetry improving (tetrahedral → octahedral). As a counterpoint, we will not extend about Sample I, as far as we may notice that the corresponding (QF) results are quite similar to those of Sample II, beside the third (QF) that, we believe, may be due to a crystaline local misformation. The above results seems to confirm that the investigation of (HFI) in the Y-Ba-Cu-O compounds can supplement structural, as well as dynamical, results obtained by other techniques.

One of the authors (H.Saitovitch) would like to fully acknowledge Dr. S. Vries, from IFUFRGS, for to let him know (EFG) calculated results on Y-Ba-Cu-O prior to publication, as well as for fruitfull discussions. We also acknowledge the Cyclotron/CNEN Staff for the irradiations. This work was partly supported by Financiadora de Estudos e Projetos (FINEP).

4. REFERENCES

(1) The Crystal Structure of the Ceramic Superconductor of the Y-Ba-Cu-O Compounds - S.F. Schu et al.; Z. Phys. B 69 (1987) 29.

(2) Perturbed Angular Correlation - E. Karlsson et al. ed.; North-Holland, Amsterdam.

(3) Studies on the Thermal Behaviour of $Ba_2YCu_3O_{7-x}$ by X-Ray Powder Diffraction Method - K. Yukino et al.; Jpn. Journal of Applied Physics 26 nº 5 (1987) L869.

(4) Fe^{57} Mössbauer Study of the Superconductor $YBa_2(Fe_xCu_{1-x})_3O_y$ - H. Saitovitch et al.; Phys. Rev. B 37, nº 13 (1988) 7967.

MAGNETIC RESONANCE AND RESISTIVITY IN $La_{1.8}Sr_{0.2}Cu_{1-x}M_xO_{4-\delta}$ WITH M = Fe, Mn AND Co.

M. A. Continentino
Universidade Federal Fluminense, Instituto de Fisica,
Niteroi, RJ, Brasil
B. Elschner, G. Kruschel
Technische Hochschule Darmstadt, Institut für Festkörperphysik
Darmstadt, West Germany
and S. Kemmler-Sack
Institut für Anorganische Chemie der Universität Tübingen,
Tübingen, West Germany

ABSTRACT. We study the Electron spin resonance (ESR) and the resistivity of the high T_c superconducting ceramics $La_{1.8}Sr_{0.2}Cu_{1-x}M_xO_{4-\delta}$ doped with the transition metals M = Fe, Mn and Co. for different concentrations of dopants. The linewidth, g-values and the magnetic susceptibility are obtained in a wide temperature. The g-values provide information on the the states and on the different neighbourhoods of the transition metals dopants. The resistance of the samples are measured and their behavior correlated to that of the EPR parameters.

In this paper we report measurements of the resistivity $\rho(T)$ the Electron spin resonance (ESR) and magnetic susceptibility $\chi(T)$ of $La_{1.8}Sr_{0.2}Cu_{1-x}M_xO_{4-\delta}$ (M = Fe, Mn and Co) for different concentrations x. The ceramics were prepared starting with La_2O_3, SrC_3, CuO and M-oxid. The powders were heated at 1220 K for ≈ 6 hours in air, grinded several times and heated repeatedly until single phase ceramics were obtained as prooved by X-ray diffraction. We pressed pellets from the powder (≈ 70% of the X-ray density) and heated them up to 1100 K for 24 hours in air allowing them to cool down to room temperature (≈ 300 K/hour). From the pellets we cut small pieces ($8x5x1.5\ mm^3$) for measuring the DC-resistivity by a conventional four probe technique. From another part of the same pellet we grinded again a powder (particle size ≈ 1μm) for the ESR measurements. The ESR measurements were performed with a VARIAN-E- line spectrometer using an Helium-flow- system of Oxford Instr. for temperature control.

The powder was suspended in paraffin inside a quartz tube in the microwave cavity. From the intensity of our ESR signals we inferred the magnetic susceptibility χ . This procedure does not give the absolute values but accounts for the temperature dependence of the susceptibility of the entities which are responsible for the ESR signal. For samples with $x \geq 0.035$ the superconductivity was suppressed. This is consistent with the observation of Malykiewicz et al[1]. For $x > 0.035$ the resistivity raises at low temperatures. It has a minimum around $T_m \approx 80$ K for M = Co ($x \approx 0.035$) and shifts to ≈ 180 K for M = Fe ($x = 0.10$). At higher temperatures the resistivity shows metallic behaviour, i.e. it rises with a positive slope dR/dT, which depends on the concentration. The low temperature behaviour of the resistivity, below the minimum, can be well described by Mott's variable range hopping law (VRH). In the undoped samples we observe small ESR signals with $g \cong 2$ at low temperatures. The origin of this signal is unknown; it may be that spurious phases like $SrCuO_2$ are responsible for them[2]. The doped samples (M = Fe or Mn) with $x = 0.05$ or 0.10 show clear ESR absorption whereas in the samples with M = Co we could not detect any resonance absorption. The ESR spectra for M = Fe or Mn are very similar. They show always three ESR-absorption lines:

Line A ($g \approx 2$). This line is always prominent. It becomes broad at low temperatures and shows a Korringa like behaviour above its minimum value. Qualitatively this line shows a similar temperature dependence to the resistivity $\rho(T)$.

Line B ($g \approx 4.26$). This line is much less intense than line A and we can detect it only at temperatures where line A becomes narrowed. At higher and lower temperatures it is masked by line A.

Line C ($g \approx 6$). The intensity of this line is comparable to that of line B and we can detect it in the same temperature range. Due to the overlap with the strong and broad line A we could not follow the temperature dependence of line B and C.

It is worth mentioning that for M = Mn ($x = 0.05$ or 0.1) we could not observe any hyperfine structure even for line A close to its minimum ($\Delta H(T_{min}) \approx 280$ Gauss). As far as we could analyze our lines, they have always Lorentzian-shape sometimes with a dispersion part as one observes in metallic hosts for penetration depth smaller than the particle diameter. Due to the experimental technique (lock-in) we get $d\chi/dH$ as ESR signal. From these curves we infer a measure for the intensity of the absorption by $\Delta H_{pp}^2 \cdot (A + B)$, where A and B are the extreme values of $d\chi/dH$ and ΔH_{pp} the peak-to-peak linewidth. This quantity is proportional to the magnetic susceptibility χ.

DISCUSSION

The explanation of the ESR-lines B and C seems straightforward[3]. The Mn-ions and the Fe-ions are in the (3d) state. Let us consider the following Spin-Hamiltonian[4]:

$$H = g_0 \mu_B H.S_{z'} + D\,S_z^2 + E\,(\,S_x^2 - S_y^2\,) \tag{1}$$

where D and E are constants, z' is the axis of the static magnetic field H and g_0 is the g-value for the case D = E = 0. The total spin for Mn^{++} and Fe^{+++} is 5/2 as well. If D ≫ E and D ≫ $g_0 \mu_B H\, S_z$, the oscillator frequency will not be high enough to induce transitions between different values of m. For H parallel to the z axis we get only the transition $\Delta m = \pm 1/2 \Leftrightarrow \mp 1/2$, with g-value g = 2. If the magnetic field is perpendicular to the z-axis than the ± 1/2 states split in the external field with g ≈ 6. In a powder sample one gets all orientations i.e. the g-values are between g = 2 and g = 6. But there are more possibilities of H being perpendicular to the z-axis, i.e. the resonance line peaks at g = 6. So we have to assume that line C originates from a center [Fe^{+++} – □] or [Mn^{++} – □] where the Oxigen vacancy is on the z axis of the octahedron. This configuration gives one of the largest D-value e.g. for [Fe^{+++} – □] in $SrTiO_3$[5]. For the case E ≫ D the eigenfunctions of the Hamiltonian E ($S_x^2 - S_y^2$) give an isotropic g-value g = 30/7 = 4.28 for the unshifted energy levels[3]. The isotropy of this g-value allows to observe the resonance line in powdered samples. The centers which are responsible for this line B are $(3d)^5$ ions in a crystal electric field with a strong orthorombic perturbation i.e. [Fe^{+++} – □] with the Oxigen vacancy on the x or y axis.

We attribute line A to $(3d)^5$ ions in Oxygen octahedrons which can be distorted but have no Oxygen vacancies in the next neighbourhood. The line A shows a Lorentzian shape for all temperatures 20 < T < 300 K. It becomes rather narrow ($\Delta H(T_{min})$ = 300 G for 10% Fe or $\Delta H(T_{min})$ = 290 for 5% Mn) and shows a conventional linewidth variation, i.e. the linewidth increases linearly with temperature as in a metallic host due to the Korringa interaction[6] for $T > T_{min}$. This is consistent with the Barnes-Plefka[7] mechanism of narrowing of a single ion spectrum. If we compare the Korringa slopes for two Fe^{+++} concentrations (see Table 1), than it follows that the sample with the higher concentration has the lower slope. This may be an indication that we are in the bottleneck regime. This point of view is supported by the fact that in our measurements we could not see any hyperfine splitting of the g=2 Mn line even for temperatures where the linewidth assumed its minimum value of ≈ 290 G. We have fitted the linewidths of the g=2 resonance for Fe and Mn according to the following expression:

$$\Delta H = A \exp[(T_0/T)^{1/4}] + B\ T \exp[-\Delta/T] \quad (2)$$

The parameters A, B and Δ for two Fe concentrations are shown in Table 1 together with the temperature range used in the fit. The resistances of these samples below the minimum could be fitted to an expression like:

$$\rho = \rho_0 (T/T_0)^{1/2} \exp[(T_0/T)^{1/4}] \quad (3)$$

The parameters used for the fit are given in Table 1.

Table 1. Parameters used to fit the resistance ρ, the linewidth ΔH and the susceptibility χ (see the expressions in the text).

	Fe 5%	Fe 10%
ρ	$T \in [40,150\ K]$ ρ_0 = 1.22 mOhm T_0 = 7615 K	$T \in [25,150\ K]$ ρ_0 = 18.8 mOhm T_0 = 31362 K
ΔH	$T \in [43,310\ K]$ A = 1.42 Oe T_0 = 63760 K Δ = 191 K B = 11 Oe/K	$T \in [21,300\ K]$ A = 7.2 Oe T_0 = 15000 K Δ = 163 K B = 5.15 Oe/K
χ	θ = 39 K	θ = 4 K

In the bottleneck regime[6] one has $\Delta H \propto 1/x$, and B(5%)/B(10%) = 2 as obtained experimentally (see Table 1). The same relation is found for Mn. Also it can be shown that in the same regime $\Delta H \propto \rho^{\uparrow\downarrow}$, where $\rho^{\uparrow\downarrow}$ is that part of the resistivity associated with spin-flip scattering. So we expect to find a relation between the resistivity and the linewidth as in fact is observed at least for the Fe 10% sample. The susceptibilities can be fitted to a form $\chi = P + (C/T+\theta)$ where the fitting parameters are shown in Table 1. The Fe(5%) sample has a cusp on the susceptibility close to 38 K. which is most probably signaling the onset of antiferromagnetic order[1]. The Fe 10% sample is also ordered at Helium temperature[8]. The existence of magnetic ordering gives rise to an additional contribution to the low temperature linewidths due to critical fluctuations. This may be the reason for the discrepancy on the values of T_0 obtained from resistivity and ESR in particular for the Fe (5%) sample.

Acknowledgements: We would like to thank Prof.E. Anda for useful discussions. This work was partially supported by the CNPq-KFA Julich agreement.

REFERENCES

1.Matykiewiecz J.M., Kimball C.W., Giapintzakis J., Dwight A.E., Brodsky M.B., Dunlap B.D., Slaski M. and Fradin F.Y., Phys. Letters A 124, 453 (1987); Nishihara Y., Tokumoto M., Murata K. and Unoki H., Jap.Journal of Appl. Phys., 26, L1416 (1987).

2.Vier D.C., Oseroff S.B., Salling C.T., Smyth J.F., Schultz S., Dalichaouch Y., Lee B.W., Maple M.B., Fisk Z. and Thompson J.D., Phys.Rev.B36, 8888 (1987).

3.Castner Th., Newell G.E., Holton W.G. and Slichter C.P., Journal of Chemical Physics 32, 668 (1960).

4.Abragam A.and Bleaney B., Paramagnetic Resonance in Transition Metal Ions, Clarendon Press, Oxford 1970.

5.v.Waldkirsch Th., Muller K.A. and Berlinger W., Phys.Rev.B5, 4324 (1972); ibid B7, 1052 (1973).

6.Barnes S.E., Adv. in Physics 30, 801 (1981).

7.Barnes S.E., Phys.Rev. B9, 4789 (1974);Plefka T. Phys.stat.sol.(b) 55, 129 (1973).

8.Saitovich Elisa B., private communication.

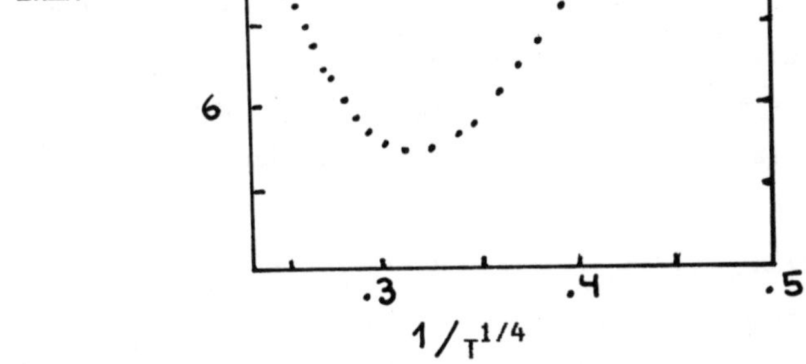

Figure 1. Plot of the linewidth of the Fe 10% sample showing explicitly the exp[$T^{-1/4}$] behavior at low temperatures.

RAMAN MEASUREMENTS IN $YBa_2Cu_3O_{7-\delta}$: A CRITICAL REVIEW

J.A. Sanjurjo
Instituto de Física "Gleb Wataghin", Universidade Estadual de Campinas,
13081 - Campinas, S.P., Brasil

ABSTRACT. A critical comparison of several papers published on Raman measurements of $YB_2Cu_3O_{7-\delta}$ type-compounds is presented. This work was motivated by the contradictory results reported by different authors. My analysis yield the following conclusions: a) Among the different features that appear in the Raman spectra only those at about 150, 335, 435 and 500 cm^{-1} are "true" Raman modes of the superconducting $YBa_2Cu_3O_{7-\delta}$ material; b) All these modes belong to the A_g symmetry representation; c) Other features that appear only in ceramic samples can be identified as Raman modes of "impurity" phases like, among others, Y_2BaCuO_5, Cu_2O and $BaCuO_2$. The variation in the Raman frequencies as of oxygen content is also analyzed.

Since the discovering of $YBa_2Cu_3O_{7-\delta}$ a rapid increasing amount of work has been done in Raman scattering in this high T_c superconductor[1-12]. However, the reported Raman spectra show marked differences and the assignments of the vibrational modes in the various studies disagree. The purpose of the present work is to make a critical comparison between the results of different authors.

$YBa_2Cu_3O_{7-\delta}$ is orthorhombic (see Fig. 1) D^1_{2h}(Pmmm) with one formula per primitive unit cell. The number of optical normal modes can be classified as follows:

$$\Gamma_{op} = 5A_g(z)+5A_{2g}(y)+5B_{3g}(x)+7B_{1u}(z)+7B_{2u}(x)+7B_{3u}(y) \qquad (1)$$

In the Raman active modes (gerade type) only the Cu(2), Ba and O(1), O(2) and O(3) atoms participate. The Cu(1) atoms in the oxygen deficient plane as well as the Yttrium atom being at the centre of inversion, do not contribute to the Raman active modes.

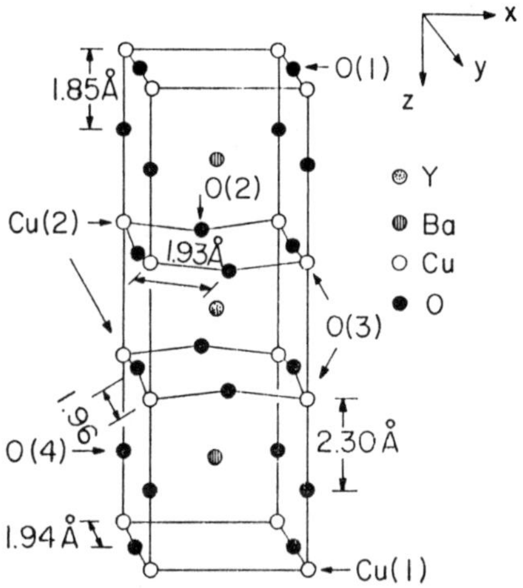

Figure 1. $YBa_2Cu_3O_{7-\delta}$ unit cell structure (Ref. 6).

In Table I we show a list of the Raman features obtained by different authors. It can be seen that the results differ not only in the number of peaks in the spectra but also in the position of corresponding lines. It is suggestive that the single crystal results show less structure than the ceramic ones and several peaks of the last ones coincide within ± 2 cm^{-1} with peaks of possible impurity phases listed in the last four columns in Table I. Some of these phases $Y_2Cu_2O_5$, $BaCuO_2$ and the most likely the "green" Y_2BaCuO_5 are final products of the phase diagram of the system[2]. It is important to point out that minor contaminations of these impurity phases (less than 1%, then not detected by X-ray diffraction) can have strong Raman intensities due to a large Raman cross section and their transparency to the 5145 Å laser excitation used in most of the works. Polarized Raman measurements in single crystals show that only the A_g symmetry modes are seen in the spectra[11,12]. Lattice dynamics calculations show that the measured A_g modes at 500, 440, 330 and 150 cm^{-1} correspond to: O(1) axial symmetric stretching; O(2), O(3) in-phase bond bending; O(2), O(3) out-of-phase bond bending and a symmetric stretching of Ba planes respectively[11,15]. The phonon at ∿ 340 cm^{-1} shows an anomalous temperature dependence at the superconducting phase transition[1,16]. This phenomenon, a phonon softening with decreasing T for $T < T_c$, is also observed in the ir-active Davidov counterpart (∿ 313 cm^{-1}) and in another ir-active phonon at slightly lower frequency ($\sim$ 278 cm^{-1})[16].

As a function of oxygen content $YBa_2Cu_3O_{7-\delta}$, with δ between 0 and 1, the Raman peaks associated with Cu-O stretching vibration (502 cm^{-1}, δ=0) and the Ba-O vibrations (154 cm^{-1}, δ=0) decrease in frequency when oxygen is removed, while the bending-stretching modes of the Cu-O frames (334 and 438 cm^{-1}, δ=0) harden under the same conditions[8,17]. The behavior of the first two modes is consistent with the decrease of the length of the unit cell in z-direction (c) when oxygen is built into the structure[8]. On the other hand, the softening of the 334 and 438 cm^{-1} phonons as oxygen is added, can be due to an increase in electronic shielding (through electron-phonon interaction) as the superconducting composition ($\delta \simeq 0.5$) is apprached[8].

As a conclusion of this work we can say: a) *impurity phases*, not detected in the X-ray spectra, can be observed in Raman measurements, particularly in ceramic samples. Then, the Raman technic is a very useful tool to characterize the quality of the samples; b) Only the modes at *150, 330, 440 and 500 cm^{-1}, all of A_g symmetry*, certainly belong to the superconducting phase; c) Only a *small anomaly* is detected in the 340 cm^{-1} mode at T_c; d) The behavior of the modes as a function of oxygen content can be explained in terms of the *variation of the cell parameters* and the *change of the electronic shielding* across the metal to semiconductor transition at about δ = 0.5.

REFERENCES

1. Mcfarlane, R.M., Rosen, H. and Seki, H., Solid State Commun. 63, 831 (1987).
2. Rosen, H., Engler, E.M., Strand, T.C., Lee, V.Y. and Bethune, D., Phys. Rev. B36, 726 (1987).
3. Iqbal, Z., Steinhauser, S.W., Bose, A., Cipollini, N. and Erhardt, H., Phys. Rev. B36, 2283 (1987).
4. Sanjurjo, J.A., Lopez-Cruz, E., Katiyar, R.S., Torriani, I., Rettori, C., Davidov, D. and Felner, I., Solid State Commun. 64, 505 (1987).
5. Batlogg, B., Cava, R.J., Jayaraman, A., van Dover, R.B.,Kourouklis, G.A., Sunshine, Murphy, D.W., Rupp, L.W., Chen, H.S., White, A., Short, K.T., Mujsce, A.M. and Rietman, E.A., Phys. Rev. Lett. 58, 2333 (1987).
6. Liu, R., Merlin, R., Cardona, M., Mattausch, H.J., Bauhofer, W., Simon, A., García-Alvarado, F., Moran, E., Vallet, M., Gonzalez-Calbet, J.M. and Alario, M.A., Solid State Commun. 63, 839 (1987).
7. Kirby, P.B., Harrison, M.R., Freeman, W.G., Samuel, I. and Haines, M.J., Phys. Rev. B36, 8315 (1987).

8. Thomsen, C., Liu, R., Bauer, M., Wittlin, A., Genzel, L., Cardona, M., Schönberr, E., Bauhofer, W. and König, W., Solid State Commun. 65, 55 (1987).
9. Collman, J.P., McDevit, J.T., Yee, G.T. and Holcomb, M.J., Phys. Rev. B37, 3660 (1988).
10. Hemley, R.J. and Mao, H.K., Phys. Rev. Lett. 58, 2340 (1987).
11. Krol, D.M., Stavola, M., Weber, W., Schneemeyer, L.J., Waszcsak Zahurak, S.M. and Kosinski, S.G., Phys. Rev. B36, 8325 (1987).
12. Liu, R., Thomsen, C., Kress, W., Cardona, M., Gegenheimer, B., deWette, F.W., Prade, J., Kulkarni, A.D. and Schröder, U., Phys. Rev. B, in press.
13. Udagawa, M., Ogita, N., Fukumoto, A., Utsunomiya, Y. and Ohbayashi, K., Jap. J. of Appl. Phys. 26, L858 (1987).
14. Compaan A. and Cummins, H.J., Phys. Rev. B6, 4753 (1972).
15. Bates, F.E. and Eldridge, J.E., Solid State Commun. 64, 1435 (1987).
16. Wittlin, A., Liu, R., Cardona, M., Genzel, L., König, W., Bauhofer, W., Mattausch, H.J., Simon, A. and García-Alvarado, F., Solid State Commun. 64, 477 (1987).
17. Kourouklis, G.A., Jayaraman, A., Batlogg, B., Cava, R.J., Stavola, M., Krol, D.M., Rietman, E.A. and Schneemeyer, L.F., Phys. Rev. B36, 8320 (1987).

TABLE I: Raman frequencies in for $YBa_2Cu_3O_{7-\delta}$. The numbers in the top The numbers in the top of each column labels the corresponding reference in the text. The frequencies are in wavenumbers.

CERAMICS						CRYSTALS								
1	2	3	4	5	7	8	9	10	11	12	13*	14†	2★	2●
153						154	154	142	140	150	181	151		
217							220		230		224	220	219	243
291											265			265
309											317	308		328
335	337		342		337	334	336	338	335	330	389		344	357
441	438	430	428	433	450	438	438		435	440	439		428	408
493								483			478			
506	502	497	510	501	502	502	500		500	500		515		518
601		583	589	592			580	595			604		583	592
640		632	645	644			632					665	637	

* : Y_2BaCuO_5 ; † : Cu_2O; ★ : $BaCuO_2$; ● : $Y_2Cu_2O_5$

INFRARED AND RAMAN SPECTRA OF SUPERCONDUCTING $Bi_4(Ca,Sr)_6Cu_4O_{16+x}$

Néstor E. Massa, Jorge Güida, Oscar E. Piro and Pedro J. Aymonino, Departamentos de Física y de Química,Universidad Nacional de La Plata, C.C. 67, (1900) La Plata, Argentina and Eduardo E. Castellano, Heitor C. Basso, José N. H. Gallo, Airton A. Martin and José P. Andreeta, Instituto de Física e Química de Sao Carlos, Universidade de Sao Paulo, C.P. 369, 13560 Sao Carlos, Sao Paulo, Brazil.

We present temperature dependent infrared and Raman spectra of $Bi_4(Ca,Sr)_6Cu_4O_{16+x}$.We find phonon features similar to the ones in other high T_c superconductors. In contrast with infrared results by Popovic et al (Sol. State Commun., to be published) in $Bi_2(Ca_xSr_{1-x})CuO_{6+x}$, we do not observe the mode at 490 cm-1 that softens. We also discuss the electronic response of our system to the 5145 A laser line.

$La_{2-x}Ba_xCuO_4$, $MBa_2Cu_3O_7$(M=Y,Sm,Eu,Gd,Ho) and $Bi_4(Ca,Sr)_6Cu_4O_{16+x}$ belong to the new class of high T_c superconductors first studied by Bernodz and Müller(1). These compounds present the onset of critical temperature (T_c) at about 40K, 90K, and 100K respectively, and have the common feature of the presence of CuO_2 layers to which temperature superconductivity seems to be associated.

Here, we present temperature dependent infrared and Raman scattering measurements of pellets of $Bi_4(Ca,Sr)_6Cu_4O_{16+x}$. Resistivity measurements show the superconducting onset at about 100K, a sharp drop and zero resistance at T_c=82K. The infrared reflectivity spectra are the average of four scans in a near normal incidence configuration run in a Perkin-Elmer 580B spectrophotometer. Raman scattering was obtained in an off-axis backscattering geometry in a Spex Ramalog spectrometer with conventional photon counting techniques using the 5145 A argon-ion laser line.All measurements were carried out with the sample mounted on the cold finger of a Displex cryostat.

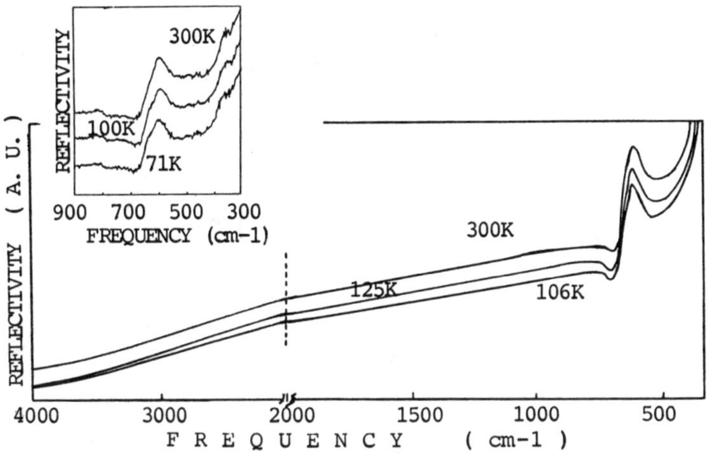

Figure 1. Infrared temperature dependent spectra of nominal $Bi_4(Ca,Sr)_6Cu_4O_{16+x}$. The detailed spectra between 900 cm-1 and 300 cm-1 are shown in the inset.

The infrared reflectivity spectra, shown in figure 1, display overall profiles similar to spectra of other high T_c compounds. In the region from 4000 cm-1 to 200 cm-1, they show a low reflectivity broad flat band that appears as part of a plasma like structure with "edge" at much higher wavenumbers. There is a prominent band at 600 cm-1. This has been assigned by Ohbayashi et al (2), and by Cardona et al (3) to a CuO streching mode. At lower frequencies, after a reflectivity dip, a band with phononic substructure is seen. Popovic et al (4) have measured the infrared spectrum of $Bi_2(Sr_{1-x}Ca_x)_2CuO_{6+x}$ finding bands at 395 cm-1, 490 cm-1 and 590 cm-1. Although these bands have clearly the same origin than ours, in our samples we fail to detect the band at 490cm-1. A factor group analysis of phonon modes in the bismuth superconductor and its correlation with the vibrational behavior of other structurally related cuprate oxide superconductors are given elsewhere along with a tentative assigment of infrared active modes of $Bi_2Sr_2CaCu_2O_{8+x}$ in the range of 200-600 cm-1(5).It should also be pointed out that such bands may indicate a macrospic field not entirely vibrational but rather vibronic, i.e.,excitonic.

Our Raman measurements span the spectral range up to 7000 cm-1. Figure 2(a) shows typical phonon lines weakly detected over the background tail of other collective excitations centered at about 2500-3000 cm-1 (see figure 2(b), where spectra from different spots of the same pellet are shown). This strong band has an integrated area that remains practically constant

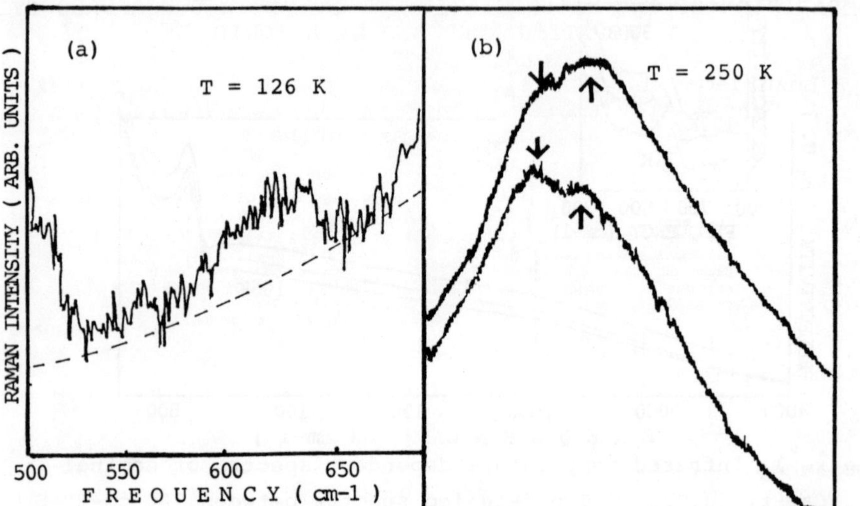

Figure 2. (a)Phonon lines of nominal $Bi_4(Ca,Sr)_6Cu_4O_{16+x}$
(b) Electronic response to the 5145 Å argon-ion laser line.

with temperature. In single crystals of La_2CuO_4 and $Ba_2YCu_3O_{6+x}$ weaker similar features have been interpreted by Lyons et al (6) as due to scattering by magnons pairs. In our case, being handicapped by the use of pelleted samples and keeping in mind that an exciton related explanation has not been ruled out, no assigment was attempted. We find that the main feature of this broad band is a double peak similar to that found in $Ba_2YCu_3O_{6+x}$ single crystals for the B_{1g} polarization that shows zone boundary contributions.

Support by CONICET, CICPBA (Argentina), and FAPESP, CNPq (Brazil) is gratefully acknowledged.

1. J. G. Berdnorz and K. A. Müller. Z. Phys. B64, 189 (1986).
2. K. Ohbayashi, N. Ogita, M. Udagawa, Y. Aoki, Y. Maeno and T. Fujita. J. of Appl. Phys. (Japan) 20, L420 (1987).
3. M. Cardona, L. Genzel, R. Liu, A. Wittlin, Hj. Mattausch, F. García-Alvarado and E. García-González. Sol. State Commun. 64, 727 (1987).
4. Z. V. Popovic, C. Thomsen, M. Cardona, R. Liu, G. Stanosic, R. Kremer and W. König and references therein. Sol. State Commun., accepted for publication.
5. O. E. Piro. J. Güida, N. E. Massa, P. J. Aymonino, E. E. Castellano, H. C. Basso, J. N. H. Gallo and A. A. Martin. to be published.
6. K. B. Lyons, P. A. Fleury, L. F. Schneemeyer and J. V. Waszczak. Phys. Rev. Lett. 60, 732 (1988).

COMPARISION OF THE MAGNETIZATION OF La Sr Cu O MEASURED BY A VIBRATING REED AND BY A SQUID

J. Luzuriaga, L. Civale, F. de la Cruz and E. Rodriguez
Centro Atómico Bariloche-Instittuto Balseiro
8400 Bariloche Argentina

ABSTRACT

The magnetization state of La Sr Cu O is studied by two different techniques. Using a vibrating reed made of the material moving in a magnetic field parallel to the long face of the reed, a change of regime is seen in the attenuation, while no extra dissipation due to the field is seen at low fields. The results are correlated with the behaviour seen in a SQUID magnetometer and a granular picture in which the order parameter is zero in some regions of the sample appears to be the model that can best describe the results.

Measurements of the magnetization in high Tc ceramic compounds, show a behaviour that is markedly different from that seen in bulk superconductors, and a granular picture seems to describe the results[1,2,3]. In this paper we hope to clarify the nature of this inhomogeneous superconducting state by measuring the response of a reed of the material that is clamped at one end and made to perform flexural vibrations in a magnetic field parallel to the long face of the sample. In such a setup (Fig.1) the magnetic field produces an increase in the resonance frequency when the sample is superconducting and a change of regime is seen at a field corresponding to Hc1 of the islands[4,5]. In measurements performed with a SQUID, on the other hand, field penetration is seen in 50% of the sample at fields two orders of magnitude lower. Two pictures can be invoked to account for the behaviour of this trapped flux. In one, the flux can be regarded as forming a vortex lattice (however distorted it may be) in a material that is a weak superconductor forming the interisland material. In the other, the interislands material can be regarded as

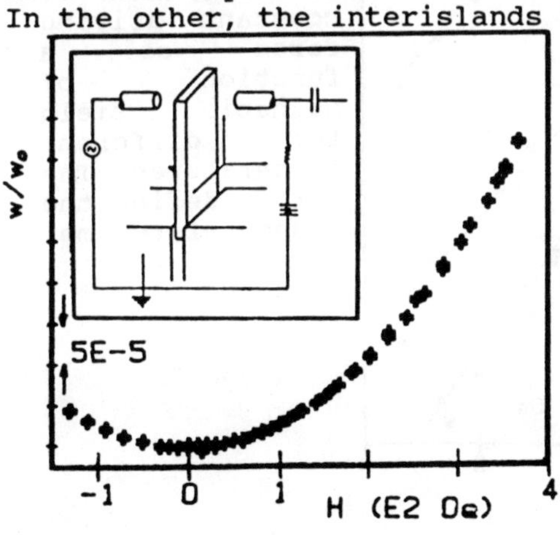

Fig.1. a) Experimental setup: The material in the form of a thin slab is made to oscillate at its resonance frequency. In a parallel field, bending of the magnetic lines of force when the sample is superconducting produces an extra restoring force that increases the resonance frequency.

b) Change in frequency as a function of field.

essentially normal, with zero order parameter, but in some regions superconductivity is induced by the proximity effect and a multiply connected structure results. The flux is therefore not in superconducting regions driven normal by the field, but in normal regions that can exist in a very small scale due to the extremely short coherence length of the material, surrounded by superconducting circuits that can sustain rather weak superconducting currents.

The measurements were performed in a $La_{1.8}\ Sr_{0.2}\ Cu\ O_4$ ceramics sample that was prepared from the appropiate amounts of the oxides of the material reacted in air. It was later fully oxidized by annealing for 12 hours in a pure oxygen atmosphere at 600 C. Sample dimensions where .915 x .405 x .024 cm and its mass density was 6.08 g cm^{-3}, which indicates a low porosity, as was chequed also by inspection in an optical microscope.

The change in resonance frequency as a function of magnetic field is shown in Fig.1, and an almost purely parabolic behaviour can be seen, as could be expected for either purely diamagnetic response (Meissner state) or for infinite pinning of the flux line lattice[4,5].

The amplitude of vibration of the reed at resonance was also measured and for constant drive amplitude this is inversely proportional to the attenuation Q^{-1}. This is almost constant and equal to the zero field value until at fields of a few hundred Oe a decrease in the amplitude is seen indicating an increase in the dissipation (Fig.2).

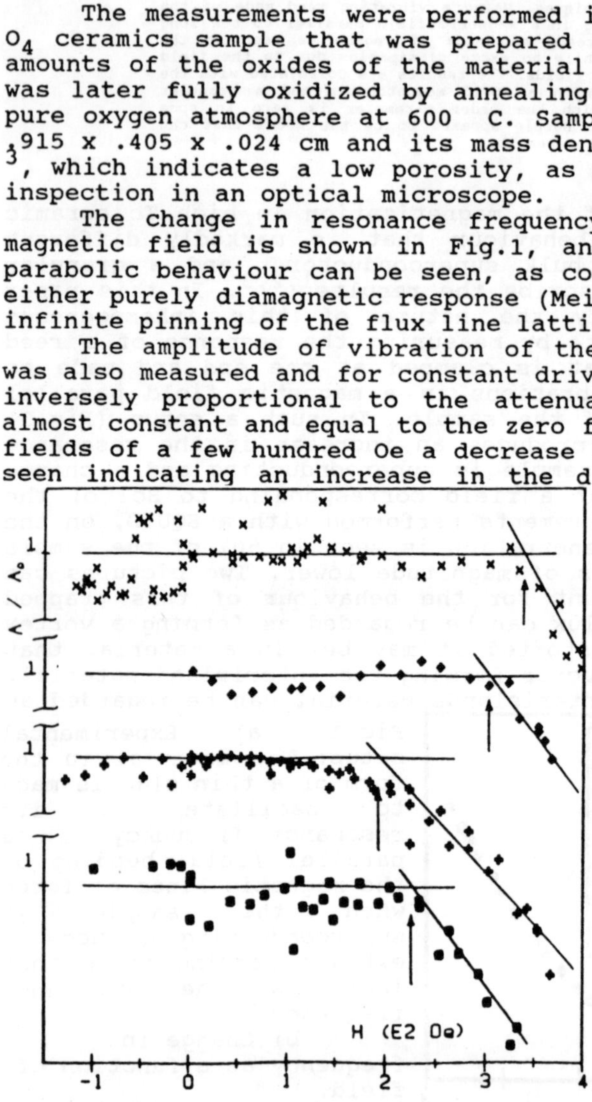

Fig.2. Amplitude of vibration at constant driving force as a function of magnetic field for different temperatures. The arrows define the change of regime.

The point at which this change of regime occurs shifts with temperature as shown in Fig. 3.

Measurements performed with the SQUID (Fig.4) show a complex behaviour of the magnetization. For very low fields (H<1 Oe), the expulsion of magnetic flux is almost complete, although irreversibilities are seen and careful measurements of the scaling of the expulsion with applied field indicate that this is not a true Meissner state. This regime can be best described with a critical current model[6]. For field between 1 and 10 Oe, flux starts to penetrate and this region can also be explained using the same critical current model. By increasing the field further one reaches a region where scaling is seen with 50% flux expulsion. This is interpreted as coming from the Meissner expulsion of the islands. Finally, Hc1 of the islands is observed when there is a breakdown of the scaling behaviour[6].

The fact that the attenuation remains constant and equal to the zero field value in the vibrating reed measurements, seems to indicate that the flux trapped in the sample does not provide an extra mechanism for dissipation. This is in contrast to what is observed for pinning of vortices where an infinite pinning is necessary to insure no dissipation. However, such strong pinning of vortices should prevent flux from penetrating the sample, and should produce almost total screening of the flux in the SQUID instead of the 50% screening observed above 10 Oe.

Because of the way the experiments of flux screening are performed, with the sample cooled in zero field and magnetic field being applied at low temperature, the change in field is partially compensated by the Meissner current in the islands and by the persistent critical current induced by Lenz's law in the superconducting links. The change in flux is strong enough so that the critical current of the links is exceeded and flux

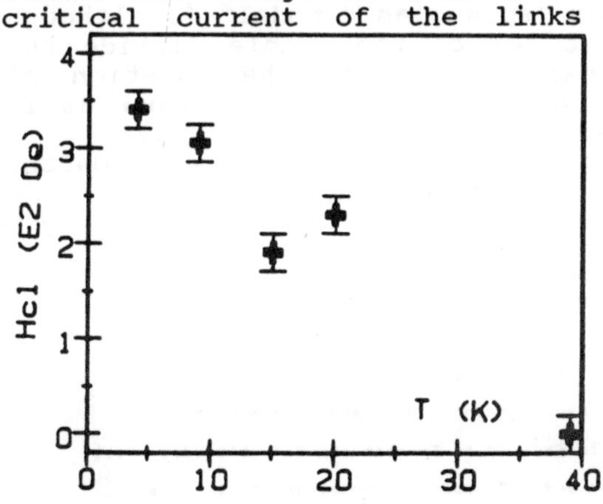

Fig.3. Magnetic field at which the amplitude starts to decrease as a function of temperature.

penetrates, although it is still excluded from the islands by the Meissner effect. In this picture, vibrating the sample will not induce currents strong enough to surpass the critical current of the links, since the amplitude of the vibration at the end of the sample is less than a micron, and therefore no dissipation will be observed. The flux trapped inside a normal hole surrounded by a superconducting region has to move rigidly with this region, until Hc1 is exceeded, in which case vortices can penetrate into the superconducting region i.e. inside the islands. The pinning force causes them to follow the motion of the reed although in a less rigid way than in the previous case,. We believe this is the reason of the change in attenuation since a less rigid coupling of the flux and the vibration of the reed produces more dissipation.

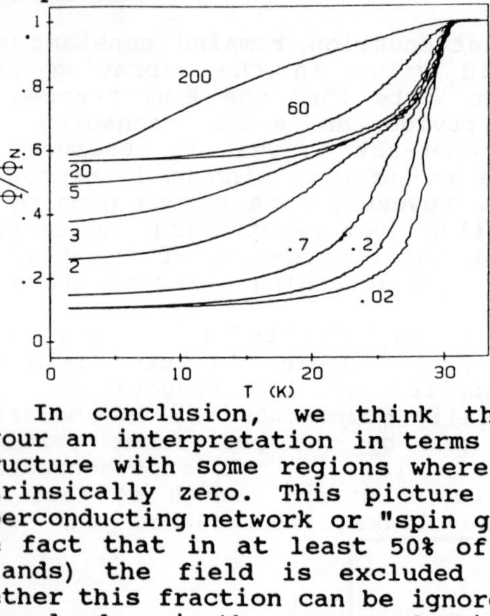

Fig.4. Behaviour of the relative flux expulsion in measurements performed with a SQUID.

In conclusion, we think the present measurements favour an interpretation in terms of a multiply connected structure with some regions where the order parameter is intrinsically zero. This picture is similar to a random superconducting network or "spin glass" model[2,7], although the fact that in at least 50% of the sample (inside the islands) the field is excluded raises the question of whether this fraction can be ignored in calculations as is commonly done in the superconducting network problem.

Support of this work by TWAS grant 87-63 and CONICET of Argentina is gratefully acknowledged.

References

1) D. A. Esparza, C.A. D'Ovidio, J. Guimpel, E.O. Osquiguil, L. Civale F.de la Cruz, Solid State Commun. 63,137(1987)
2) K.A. Muller, M. Takashige, J.G. Bednorz, Phys.Rev.Letters 58,1143(1987)
3) P.Peyral, J. Rosenblatt, A. Rabotou and C. Lebeau, Interlaken Conference Abstract C293
4) E.H. Brandt, P. Esquinazi, H. Neckel and G. Weiss Phys.Rev.Lett. 56,89 (1986)
5) C. Duran, P. Esquinazi, J. Luzuriaga and E.H. Brandt, Phys.Lett.A123, 485(1987)
6) L. Civale, H. Safar, F. de la Cruz, D.A. Esparza and C.A. D'Ovidio, Soli State Commun. 65,129(1988).
7) I. Morgenstern, K. A. Muller and J.G. Bednorz, to appear in Z. Phys.B.

MAGNETIC CHARACTERISTIC OF Y-Ba-Cu-O COMPOUNDS

V. Sagredo and M. Chourio

Physics Department, Faculty of Sciences
Universidad de Los Andes, Mérida, Venezuela

INTRODUCTION

Recently, several research groups have got the single phase oxide $YBa_2Cu_3O_{7-y}$ with T_c above 90 K, and many researches have been made on characterizing such materials and on understanding their associated high T_c superconductivity.[1,2,3]
The high T_c superconducting phase in this system has been identified to the orthorhombic, oxygen deficient perovskite type structure with ordering of the ytrium and barium atoms.[4,5]
2:7. Wang et al.[6] have studied how the superconducting properties change with carrier concentration when the oxygen content in $YBa_2Cu_3O_{7-y}$ is altered. They found that the Hall coefficient, the volumen fraction showing the Meissner effect and the superconducting transition temperature are insensitive to y over the range $0.1<y<0.5$.
$YBa_2Cu_3O_{7-y}$ is formed with a stoichiometric composition in the ternary Y_2O_3-BaO-CuO. The existence of solidisolutions for small deviations from the 1:2:3 stoichiometry would make it possible to observe a dependence of superconducting and normal state properties upon the composition.[7]
In the Communication we present the results of magnetic susceptibility measurements for four samples of $YBa_2Cu_3O_{7-y}$ prepared in different conditions.

EXPERIMENTAL DETAILS

Samples of composition $YBa_2Cu_3O_{7-y}$ were prepared by using high purity powders Y_2O_3, CuO, Ba_2O. Samples A and B were obtained heating the mixture in air at 900 C for 10 hours and then pressed into disk shape. Such disks were sintered at 900 C in an oxygen rich atmosphere for 6 hours.[8] Samples were also annealed in air at 900 C for 6 hours and then cooled without control. Samples C and D were prepared following similar procedures without further treatment.

Measurements of the static magnetic susceptibility were carried out using a sensitive Faraday system with a Cahn RG electrobalance in conjuction with the Lewis gradient coils. The temperature was stabilized and measured by a Lake Shore system, by using as temperature sensor the diode DT-500[9]

RESULTS AND DISCUSION

The zero field cooled magnetic susceptibility of four samples of $YBa_2Cu_3O_{7-y}$ have been measured in a magnetic field of 200 and 1000 Oe in the $70 \leq T \leq 300$ K temperature range and the obtained results for samples A and B are shown in Fig. 1 and Fig. 2.

The d.c susceptibility measurements showed a weak paramagnetism above the transition temperature and starting at about 92 K there was a net diamagnetism that increased with decreasing temperature. The susceptibility is found to decrease linearly with temperature below the transition with a small amount of rounding within 1-2 degrees of T_c. This linear behavior indicates strong magnetic field penetration.

The susceptibility for $YBa_2Cu_3O_{7-y}$, samples A (annealed sample) between 70 and 300K as a function of magnetic field is shown in Fig. 1. As can be seen the slope of the susceptibility near the superconducting transition is very sharp at H = 200 Oe. Strong field dependence of the susceptibility is observed, as the magnetic field increases, the fully developed diamagnetic response is reduced.

Above the transition temperature, the susceptibility becomes positive in every sample and independent of magnetic field. Annealing in air reduces the temperature dependence of χ in the normal state. This effect could be associated with the type and homogeneity of the oxygen vacancies in these compounds.

In order to interpret the paramagnetic susceptibility above T_c for the studied samples several contribution to the susceptibility must be considered. We may write:

$$\chi_e = \chi_c + \chi_p + \chi_{vv} + \chi_1 + c/T-\theta$$

where χ_c is the ion core diamagnetim, χ_p Pauli paramagnetism χ_{vv} Van Vleck paramagnetism, χ_1 Landau diamagnetism due to the conduction electrons and the last term is the Curie-Weiss paramagnetism. Only the Curie-Weiss contribution should be temperature dependent unless there is an anomalous temperature dependence for the electron density of states.

The temperature dependence of inverse susceptibility is shown in Fig.3. Curie-Weiss behavior with a negative Curie temperature is observed,suggesting the existence of antiferromagnetic interaction. If the temperature dependence is attributed solely to the copper one abtains an effective moment $P_{ef} = 0.25\ \mu_B$/Cu ion.

Calculating the core diamagnetic contribution, of -14×10^{-16} cm^3/g-at[10] and assuming that the sum of $\chi_{vv} + \chi_1$ is neglected we have the possibility to calculate χ_p.

We can obtain an estimate for the Fermi-level density of states from the constant susceptibility term once we subtract the Curie term and core diamagnetic correction. The remaining term was assigned to the Pauli susceptibility, with $\chi_p = \mu_B^2 N(E_F)$[11]. This yield 5.1 states/ev-Cu which is to be compared with ~ 3 states/ev-Cu obtained from resistivity and critical field measurements[12].

In summary, we have presented current measurements of the magnetic susceptibility of a $YBa_2Cu_3O_{7-y}$ compounds. The data reveal a material which can be understood within the classical mechanism of superconductivity.

ACKNOWLEDGEMENT

The authors would like to thank Prof. A. Mariño, Universidad Nacional de Colombia and Prof. L. Martínez whose kindly supplied the samples. We should also like to thank CDCH of Universidad de Los Andes for the provision of Grants C-204.

REFERENCES

1. C. Allgeier, J. S. Shilling and E. Amberger., Phys. Rev. (B) 35, #16, 8791 (1987).
2. L. R. Testardi, W. G. Moulton, H. Mathias, H. K. Ng., C. M. Rey and R. J. Kennedy. (Preprint)
3. Y. Yeshurum, I. Flener, and H. Sompolisky., Phys. Rev. (B), 36, #1, 840 (1987).
4. T. Siegrist, S. Sunshine, D. W. Murphy, R. J. Cava, and S. M. Zahurak (preprint).
5. R. Beyers, G. Lim, E. M. Engler, R. J. Savoy, T. M. Shaw, T. R. Dinger, W. J. Gallagher and R. L. Sandstrom (preprint).
6. Z. Z. Wang, J. Clayhold, N. P. Ong., J. M. Tarascon, L. H. Greene, W. R. McKinnon and G. W. Hull. Phys. Rev. (B) 36, # 13, 7222 (1987).
7. G. R. Wagner, A. J. Panson and A. I. Braginski, Phys. Rev. (B) 36, #13, 7124 (1987).
8. E. Posada, G. Fraco, G. Holguin, A. Jimenez, A. Mariño, V. Niño and D. Rodríguez.
9. V. Sagredo, A. Aristimuño and A. López Rivera. Il Nuovo Cimento 2D, #6, 1900 (1983).
10. A. Bezinge, J. L. Jorda, A. Junod and J. Muller. Solid State Comm. 64, #1, 79 (1987).
11. A. Junod, A. Bezinge and J. Muller. Physica C, 52, #1, March (1988).
12. R. J. Cava, B. Batlogg, R. B. van Dover, D. W. Murphy, S. Sunshine, T. Siegrist, J. P. Remeika, E. A. Rietman, S. Zahurak and G. P. Espinosa. (Preprint)

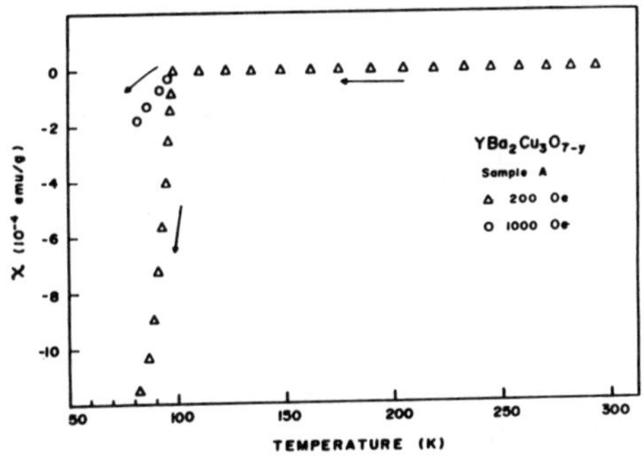

Fig.1 Static magnetic susceptibility of $YBa_2Cu_3O_{7-y}$ as a function of temperature in magnetic fields of 200 and 1000 Oe.

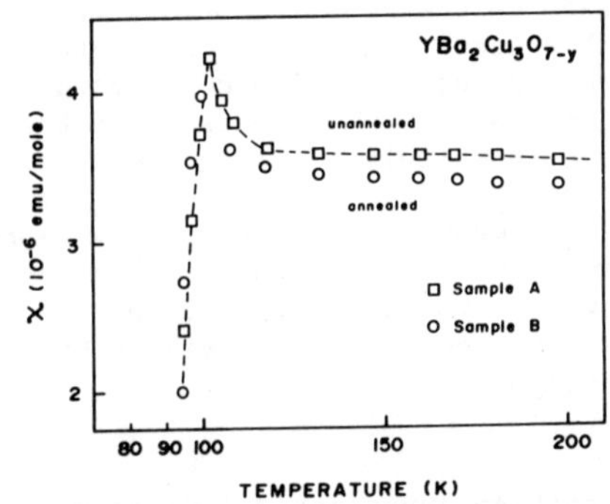

Fig.2 Normal-state susceptibility vs temperature for samples A and B.

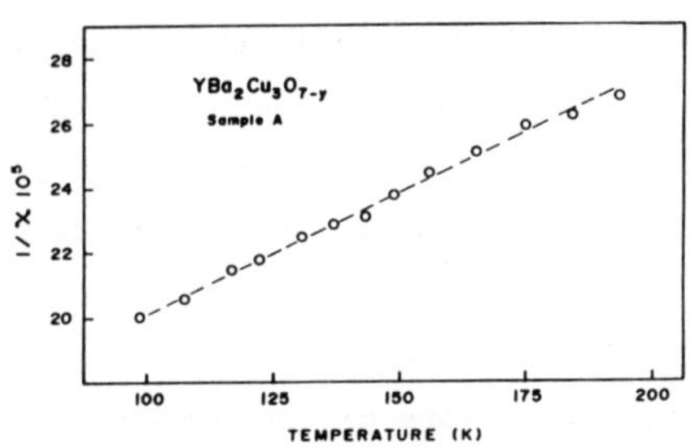

Fig.3 Variation of inverse susceptibility with temperature above Tc.

STRUCTURAL TRANSITION AND OXYGEN ORDERING IN $YBa_2Cu_3O_{6+x}$

J.M. SANCHEZ
HENRY KRUMB SCHOOL OF MINES
COLUMBIA UNIVERSITY
NEW YORK, NY 10027, U.S.A.

and

F. MEJIA-LIRA and J.L. MORAN-LOPEZ
INSTITUTO DE FISICA
UNIVERSIDAD AUTONOMA DE SAN LUIS POTOSI
7800 SAN LUIS POTOSI, SLP, MEXICO

ABSTRACT

The structural transition observed in $YBa_2Cu_3O_{6+x}$ is modeled by means of the ordering of oxygen atoms and vacancies on the basal plane of the tetragonal and/or the orthorhombic structures. The corresponding phase diagram and the temperature dependence of the oxygen order parameter is calculated using the cluster variation method in a four point cluster approximation. Most of the experimental observations are reproduced well by a simple model including only nearest neighbor repulsion between oxygens. In particular, we emphasize the importance of a percolation limit in the order-disorder transition.

Recent neutron diffraction experiments [1,2] have shown that the tetragonal to orthorhombic transition seen in the high temperature superconductor $YBa_2Cu_3O_{6+x}$ is driven by the ordering of oxygen atoms in the basal plane of the structure. The neutron diffraction experiments have established that: 1) the transition is second order, 2) it appears to occur always at an oxygen stoichiometry of approximately 6.5 (x = 0.5), and 3) the transition temperature changes appreciably with the oxygen partial pressure. We present here a thermodynamic model[3] for the oxygen order-disorder transition that explains all the experimental observations mentioned above. The model includes only nearest-neigbhors oxygen-oxygen repulsion and the free energy is calculated in the cluster variation method (CVM) using two types of rectangular clusters as the basic units. For the present case of nearest neighbor interactions, the results are the same as in the square approximation.

The ordered ground state in the present model correspond to a structure where oxygen atoms preferentially occupy (1/2,0,0) sites and where the (0,1/2,0) sites are vacant. The possibility of lower symmetry oxygen arrangements for low oxygen stoichiometries have also been proposed in the literature [4,5]. The stability of these structures, however, require that additional and mostly unknown interaction parameters among second and/or higher neighbors be introduced. Whereas such parameters might be needed for a detailed description of the system, we show here that an essentially correct phenomenology can already be obtained with only nearest-neighbor interactions.

The calculated phase diagram is shown in Fig. 1 for a value of an effective nearest neighbor oxygen-oxygen interaction energy of $V/k = 10^3$ K. The order-disorder transition line is second order in agreement with the experimental observation 1) mentioned above. In Fig. 1, the high temperature disordered phase corresponds to the tetragonal phase and the low temperature ordered phase to the superconducting orthorhombic phase. We see from the figure that there is a percolation limit for the ordering of oxygen corresponding approximately to an oxygen stoichiometry of 6.62. This value is close to the experimetally observed oxygen stoichiometry at which the structural transition in $YBa_2Cu_3O_{6+x}$ takes place. Since the order-disorder transition line below 1000 K is essentially vertical, i.e. independent of composition, we see that the model is also in semiquantitative agreement with experimental observation 2) above. In Fig. 1 we also show the average concentration of oxygen in the basal plane for two different oxygen partial pressures. These lines are calculated assuming a temperature independent activity of oxygen in the solid phase. As seen from the figure, variations in oxygen partial pressures result in signicant changes in the transition temperature, in agreement with experimental observation 3).

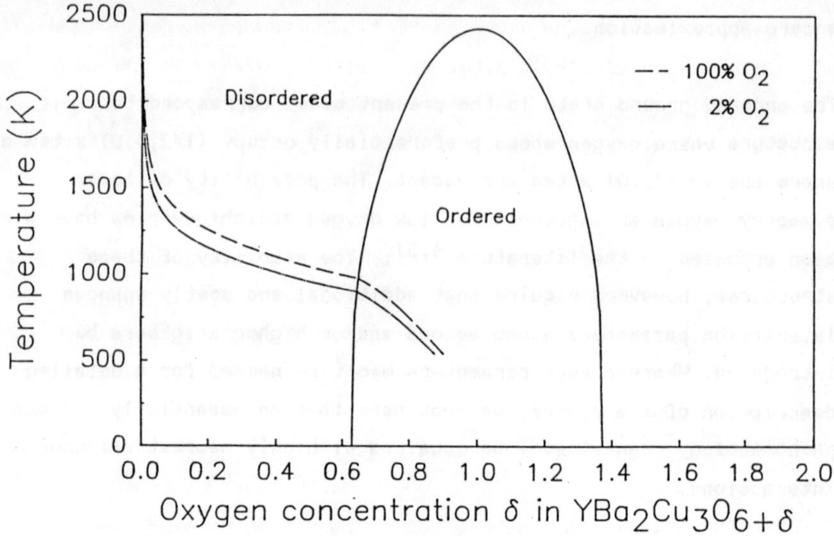

Figure 1
Oxygen order-disorder phase diagram (full lines). The average concentration of oxygen at 1 atmosphere are shown for 100% (dashed lines) and 2% oxygen-98% argon (dash-dotted lines

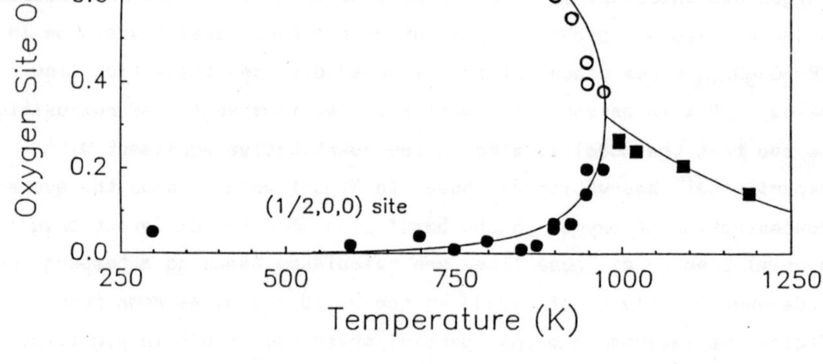

Figure 2
Comparison between the calculated and the experimentaloxygen site occupancies as a function of temperaturefor 100% oxygen

A comparison between the experimentally determined oxygen occupancies in (1/2,0,0) and (0,1/2,0) sites and those predicted by the model are shown in Fig. 2 for the case of one atmosphere of oxygen. As seen, the model can reproduce accurately the results ofthe neutron diffraction experiments.

In conclusion, a simple thermodynamic model involving only nearest neighbor oxygen-oxygen repulsion explains accurately the most important aspects of the oxygen ordering and the related tetragonal to orthorhombic structural transition observed in the $YBa_2Cu_3O_{6+x}$ high temperature superconductor.

This work was funded by the National Science Foundation under Grants No. DMR-8510594 and INT-8409776 and in Mexico by Consejo Nacional de Ciencia y Tecnologia (Grant PCCBBEU-022007) and Direccion General de Investigacion Cientifica y Superacion Tecnica de la Secretaria de Educacion Publica (Grant 87-01-0300). One of the authors (JMS) whishes to acknowledge the hospitality and support of FAPESP (Brazil), IFUSP (Brazil) and the Laboratorio Nacional de Luz Síncrotron (Brazil).

REFERENCES

1. Jorgensen, J.D., Veal, V.W., Kwok, W.K., Crabtree, G.W., Umezawa, A., Nowicki, L.J., and Paulikas, A.P., Phys. Rev. B36, 5371 (1987).
2. Jorgensen, J.D., Beno, M.A., Hinks, D.G., Soderlhom, L., Volin, K.J., Hitterman, R.L., Grace, J.D., Schuller, I.K., Segre, C.U., Zhang, K., and Kleefish, M.S., Phys. Rev. B36, 3608 (1987).
3. Sanchez, J.M., Mejia-Lira, F., and Moràn-Lopez, Phys. Rev. B, March 1, (1988).
4. de Fontaine, D., Willie, L., and Moss S.C., Phys. Rev. B36, 5709 (1987).
5. Van Tandeloo, G. and Amelinckx, S., Journal of Electron Microscopy Technique 8, 285 (1988).

ORDERING OF O ATOMS IN $YBa_2Cu_3O_{7+x}$

A.A. Aligia, A.G. Rojo and B.R. Alascio
Centro Atómico Bariloche, 8400 Bariloche, Argentina

Oxygen vacancies in $YBa_2Cu_3O_{7+x}(-1\leqslant x\leqslant 1)$ fall in one of the two types of Cu-O planes perpendicular to the c axis of the structure. Assuming that the relevant interaction of the oxygen ions is a screened Coulomb repulsion between those contained in the same plane, we determine the minimum energy order for each composition. For small $|x|$, $x<0$ we find long range order of the vacancies. For $|x|=1/2$ order persists although it may not be completely apparent in X ray or neutron diffraction patterns.

The oxygen(O) content in the $YBa_2Cu_3O_{7+x}$ type compounds depends sensitively on temperature(T) and O partial pressure. x is determined by the quantity of O in the Cu-O planes perpendicular to the c axis which lie between two Ba planes[1]. The O atoms can occupy any site lying between two Cu atoms of this plane. If p is the probability that any of these sites is occupied, one has: $x = 2p-1$.
As a consequence of the low concentration of current carriers,the screening length of the interatomic Coulomb energies (ICE)is expected to be a few Å or larger. Since the magnitude of the unscreened ICE reaches values of order of 1Ry for the shortest distances, the ICE are expected to play a dominant role in determining the structure. Here we study the structure of minimum energy for each composition taking into account only ICE and assuming a constant charge on each ion. Under these assumptions we need to consider only the repulsion between O ions. We take for this the screened form:

$$V_{ij} = (A/R_{ij}) \exp(-R_{ij}/\lambda)$$

where R_{ij} is the distance between the O ions, λ the screening length and A a constant of order $\sim 4e^2$. Neglecting interactions between Cu-O planes, the problem is isomorphic to an Ising model with long range exchange interactions, all antiferromagnetic, in the square lattice spanned by all sites which can be occupied by O ions. A spin up (down) represents an occupied (vacant) site. We call $a_0 = a/\sqrt{2}$ the lattice parameter of the O lattice.

We have searched for the ground state for fixed x, using Monte Carlo methods with Kawasaki spin exchange[2] between nearest neighbours and next nearest neighbours. In general we took a random configuration with a given composition, we let it thermalize at $T\sim 0.2\ A/a_0$ and we decreased T slowly to zero. Results for typical T=0

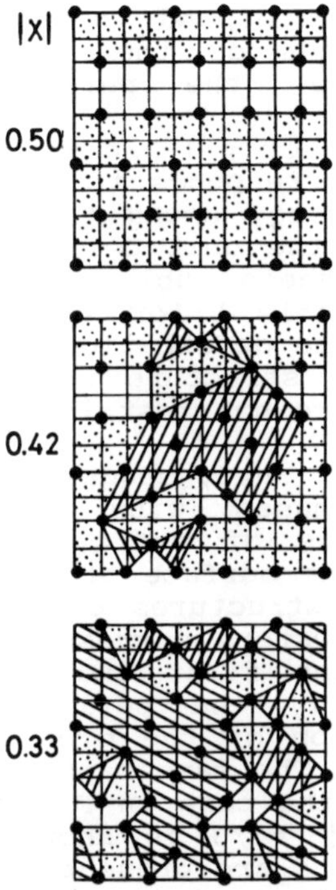

FIG.1-Typical T=0 configurations obtained from the Monte Carlo calculations for $\lambda=a_0$ and various compositions shown at the left of each figure. The dots denote the positions of the 0 atoms(vacancies) if $x<0$ (>0).

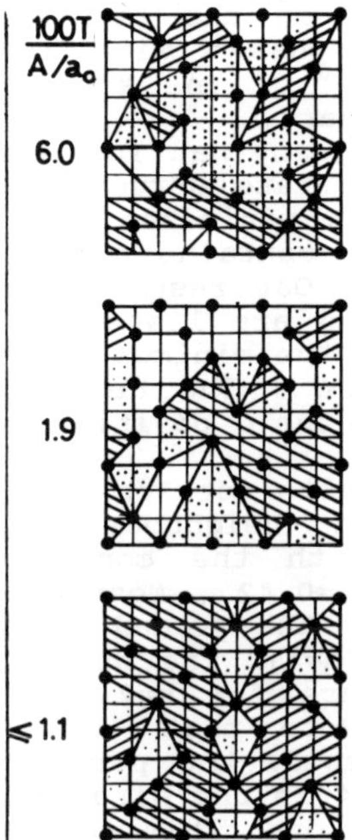

FIG.2-Typical configurations obtained from the numerical calculations for $|x|=1/3$, $\lambda=a_0$ and different temperatures shown at the left of each figure. The dots denote the positions of the 0 atoms (vacancies) if $x<0(>0)$.

TABLE I

n	$\vec{V}_1$	$\vec{V}_2$
2	(1,1)	(-1,1)
3	(-1,1)	(1,2)
4	(1,2)	(2,0)
5	(1,2)	(2,-1)
6	(-2,2)	(2,1)
7	(-3,1)	(1,2)
8	(-2,2)	(1,3)
9	(0,3)	(3,1)
10	(-3,1)	(1,3)

TABLE I-Basis vectors describing the structure of minumum energy among those correspondng to simple Bravais sublattices. The structures for different compositions are described as follows: a) for $x=\pm(2/n-1)$ the lattice describe the position of the 0 atoms (vacancies). (1,0) and (0,1) are the unit vectors of the 0 lattice. b) for $x=\pm 1/n$ ($n>3$) the lattice describe the position of the additional 0 atoms (vacancies) added to the structure of x=0. Here (1,0) and (0,1) are the unit vectors of the Cu lattice.

configurations are shown in Fig.1 and the bottom of Fig.2. In general, depending on the initial configuration, different metastable states near the ground state are reached at T=0. This is because the difference between the energies of any two of these states is very small compared with the energy barrier along a path between them compatible with the Kawasaki dynamics in the space of configurations. Nevertheless, for $|x|$ = 0,1/3 and 1/2 (corresponding to 1/p or 1/(1-p) being an integer), at least 60% of the cluster is occupied by a structure corresponding to a simple Bravais sublattice. The result for x=0 , not shown in the figures, correspond to the observed orthorrombic phase[1] with islands of the same structure rotated 90°. In any case, the perfect sublattice has lower energy.

These results suggest that for p=1/n or p=1-1/n with n integer, the ground state structure corresponds to the simple sublattice which minimizes the energy among those in which the unit cell contains n times the unit cell of the square O lattice[3]. Our results for these sublattices are independent of λ and are listed in table 1 for n⩾10. They seem to represent the minimum possible deformation of the hexagonal structure . The hexagonal structure would minimize the energy if the O positions were not restricted to a lattice.

Our numerical results for 1/n>p>1/n+1, n⩾2, suggest a segregation into the neighbouring phases with 1/p integer. This agrees with the conclusions of ref.4. As shown in Fig.1 for |x|=0.42 , most of the surface of the cluster is occupied by the ground state structures of |x| =1/3 (n=3) and |x|=1/2 (n=4). The other clusters correspond to the ground structure for n=2 and the first excited structure for n=4. Note that in Fig.1, the periodic boundary conditions impede the percolation of the ground state structure for n=4 in one direction. The same happens for n=3 in both directions. This is not the case in Fig.2 where the temperature evolution for |x| =1/3 is shown. Although for T=0.06 A/a_0 large fluctuations are present, a tendency of the system to form islands with concentrations 1/n (1-1/n) seems apparent. Clusters that resemble the ground state for n=2 to 6 and excited structures for n=2 and 4 can be seen. For T<0.011 A/a_0 , except for small fluctuations, the configuration is frozen.

For x ~0, we can take advantage of the fact that the ground state for x=0 (p=1/2) consist of two equal interpenetrating square sublattices: one empty and the other filled with O atoms. Then, assuming that the occupied sublattice remains frozen as new atoms are added, the problem of finding the ground state for p=1/2+1/2n can be mapped into the one corresponding to p=1/n. An analogous argument can be made for p=1/2-1/2n interchanging O atoms and vacancies. The assumption made is valid for large enough n and breaks down for n~3.

The ground state structures which we have obtained for x=0 and x=-0.5 agree with the structural analysis[1]. In ref.1 the structure for x=-0.5 was interpreted as a totally disordered one, but since distances up to 3.95 Å were investigated in ref.1 and disorder along the c axis can exist, both structures cannot be discerned in the experiment[3]. Another difference in interpretation is that we do not view the orthorrombic-tetragonal transition as an order-disorder one, but as a transition between different, more complex ordered phases.

Our predicted structure for x=-1/3 should present an orthorrombic distortion, but in disagreement with experiment[1,5], the occupation probability of both unequivalent sites of the unit cell of $YBa_2Cu_3O_7$ is the same. The distortions of the lattice neglected in our calculations would make these probabilities different. The same result would be caused by a chemical energy that destabilizes the three-fold coordinated Cu atoms[3]. Both effects have no relevance for x=0. For $|x|\geqslant 0.5$, the last effect could only change the ground state if the destabilization energy were of the order of the repulsion between oxygen atoms at a distance a (next nearest neighbours).

Our results for x ~0 seem to agree with the order of vacancies observed by electron diffraction[6]. For x=-0.125 the ground state is described by the vectors given in table 1 in the entry n=8. The first excited structure with unit vectors (-2,2) and (2,2) is the structure proposed in ref.6. At the quenching temperature of the samples[6], clusters of both structures should be present, and both of them are contained in a unit cell of parameters $4\sqrt{2}$ a x $2\sqrt{2}$a.

An interesting fact for x~0 is that if $\lambda < a_0/2$, it is more convenient to arrange the additional O atoms or vacancies in twin boundaries separating two regions with x=0 that differ in the orientation of the Cu-O chains. The most stable twin boundaries are parallel to the chains lying at one side . This situation should change if elastic energies are included.

We thank Drs. E. Gagliano, O. Mouritsen an N. Parga for helpful discussions concerning the numerical methods.

References

1- J.D. Jorgensen, M.A. Beno, D.G. Hinks, L. Soderholm, K.J. Volin, R.L. Hitterman, J.D. Grace, I.K. Schuller., C.U. Segre, K. Zhang, and M.S. Kleefisch, Phys.Rev.B 36,3608 (1987).

2- O.G. Mouritsen, "Computer Studies of Phase Transitions and Critical Phenomena" (Springer Verlag, New York, 1984); K. Kawasaki in "Phase Transitions and Critical Phenomena", C. Domb and M.S. Green, eds, Vol.II, p.443 (Academic Press, London, 1972).

3- A.A. Aligia, A.G. Rojo, and B.R. Alascio, Phys.Rev.B, to be published.

4- P. Satti, D. Pasquevich, A. Caneiro, J.P. Abriata and C. Fainstein, this conference.

5- C. Chaillout, M.A. Alario Franco, J.J. Capponi, J. Chenavas, P.Strobel, and M. Marezio, Solid State Commun. 65,283 (1988).

6- M.A. Alario Franco, C. Chaillout, J.J. Capponi, and J. Chenavas, Mat.Res. Bulletin, to be published.

ELECTRONIC STRUCTURE OF PEROVSKITE TYPE COMPOUNDS

J.P. Julien, D. Mayou, F. Cyrot-Lackmann

Laboratoire d'Etudes des Propriétés Electroniques des Solides
associated with Université Joseph Fourier
Centre National de la Recherche Scientifique
B.P. 166, 38042 Grenoble Cedex, France

ABSTRACT - Electronic structure of perovskite type compounds, such as $SrTiO_3$ and La_2CuO_4 is calculated within a tight binding scheme and a recursion method. Hybridization effects between d Ti or Cu orbitals and p oxygen orbitals are included. The alternating structure of the lattices which do not couple directly Ti or Cu allows a transformation of the real lattices into effective lattices where the oxygen atoms are removed. The DOS of the original lattice can be deduced from that of the effective one. This permits a detailed study of the singularities of the DOS and of the different contribution of each of the orbitals on the different atomic sites. For La_2CuO_4, the splitting between the $3z^2$-r^2 and x^2-y^2 bands associated with the tetragonal distortion, is mainly a consequence of the covalency between copper and oxygen.

Since the discovery of high T_c superconductivity by Bednorz and Muller in Ba-La-Cu-O systems[(1)], a lot of attention has been focused on mechanisms for the origin of this kind of superconductivity. In particular, several electronic band structure studies[(2,3)] have been performed to gain insight into the structure of La_2CuO_4 and its consequence on possible mechanisms. The results of these studies show that the σ band between Cu $d(x^2$-$y^2)$ and O(2p) states in tetragonal La_2CuO_4 perovskite K_2NiF_4 type structures exhibit a two-dimensional character at the Fermi level. Many conjectures have been made on the importance of this unique electronic structure feature for the superconductivity. However, at this point, there does not exist a detailed analysis of the interplay between the Cu(3d) and O(2p) hybridization and covalency effects, and their role in lifting the degeneracy of d-subbands in various perovskite structures.

In this paper, we propose a study of the electronic structure of different perovskite type compounds, such as $SrTiO_3$ and La_2CuO_4 and show the importance of the (Ti,Cu)-O octahedron distortion on the splitting of the hybridized 3d-O(2p) subbands.

For La_2CuO_4, the usual tetragonal phase is compared to an ideal K_2NiF_4 structure with no axial distortion. We also point out the importance of hybridization between pσ oxygen orbitals and both $x^2 - y^2$ and $3z^2 - r^2$ d orbitals of copper.

We have been able to make a systematic study of the location and the shape of the five d subbands hybridized with O(2p) due to the alternating geometry of the perovskite type lattice. Indeed, the lattice is such that, in a first nearest neighbour approximation, one can neglect the Cu-Cu, Ti-Ti and O-O hopping integrals because the nearest atoms of copper or titanium (oxygen) are oxygen (copper or titanium). The tight-binding Hamiltonian thus reduces to an effective Hamiltonian $\tilde{H}$ on a reduced lattice. This effective Hamiltonian is such that transition metal states are not coupled to oxygen states.

This effective Hamiltonian determines effective density of states (DOS) $\tilde{N}$ which have a universal character as they do not depend on the electronic parameters (hopping integrals and the atomic levels of the transition metal E_d and the oxygen E_p) but only on the structure of the lattice. Thus, for $SrTiO_3$, the effective DOS are calculated on a simple cubic lattice. The DOS N(E) of the original lattice can be easily deduced from these effective DOS $\tilde{N}$ through a simple transformation[4]. For example, one can write for the local DOS on the d orbital of the Cu(Ti) atomic sites :

$$N^{\alpha}_{Cu}(E) = |E - E_p| \, \tilde{N}^{\alpha}(\tilde{E}) \qquad (1)$$

where α denotes one of the five d orbitals and $\tilde{E}$ an effective energy :

$$\tilde{E} = (E - E_d)(E - E_p) \qquad (2)$$

One can show that $\tilde{E}$ is always positive and thus there is a gap between E_p and E_d.

One can also relate the local DOS on the oxygen sites to the local DOS on the copper sites :

$$N^{p}_{O}(E) = N^{d}_{Cu}(E) \left(\frac{E - E_d}{E - E_p}\right) \qquad (3)$$

This relation holds for the partial pσ or pπ bands of oxygen corresponding to the e_g and t_{2g} d copper bands.

Let us first describe the electronic structure of the $SrTiO_3$ lattice. It is well-known that the DOS of crystals having the cubic perovskite structure is unusual in that it displays van Hove singularities characteristic of two dimensional energy bands due to the hybridization between the d orbitals of the transition metal and the p orbitals of the oxygen[5,6]. Using the properties of the alternating lattice and the tight binding fit of the APW band structure calculation of Mattheiss[5] given in Table I, we derive all the local DOS on the different orbitals through two universal effective DOS on the t_{2g} and e_g orbitals of Ti. These DOS have the van Hove singularities of a square lattice or of a simple cubic lattice (Fig. 1). Figure 2 represents the total DOS of $SrTiO_3$ which compares well to the APW result[5], if one adds at E_p a broadened δ peak with 4 states per Ti atom corresponding to the oxygen-oxygen interactions.

Table I

Tight binding parameters (in eV)

Tetragonal La_2CuO_4		Pseudo La_2CuO_4 with no axial distortion	$SrTiO_3$
E_p	4.5	4.5	- 0.9
E_{t2g}	4.9	4.9	3
E_{eg}	6.25	6.25	3
pd σ	in plane - 1.4 ⊥ plane - 0.7	- 1.4	- 2.4
pdπ	in plane 0.8 ⊥ plane 0.4	0.8	1.1

Let us now describe the electronic structure of the tetragonal La_2CuO_4. We will compare it to an ideal K_2NiF_4 structure with no axial distortion to show the importance of the octahedron distortion on the separation of the $3z^2 - r^2$ and $x^2 - y^2$ d orbitals bands. The tight binding parameters used are those of a fit to LAPW band structure calculation of Papaconstantopoulos et al.[3] given in Table I. As for $SrTiO_3$, we use the property of the alternating lattice and calculate first the effective DOS of the reduced lattice for both structures and then calculate the local t_{2g} and e_g DOS (Figure 3). As we can see, the t_{2g} band is completely filled and the Fermi level is within the e_g band just below the

2 dimensional singularity peak. Figures 4, 5 and 6 show respectively the local $x^2 - y^2$ and $3z^2 - r^2$ subbands on the Cu sites, and the corresponding pσ band on the O sites for both structures. As expected, the distortion of the octahedron lifts the degeneracy between the 2 subbands. Indeed the $3z^2 - r^2$ states are less coupled to the pσ orbitals than the $x^2 - y^2$ states. Their repulsion with the oxygen level is thus less important, and as a consequence the $3z^2 - r^2$ band is lower in energy than the $x^2 - y^2$ band.

We note also that the non-bonding pσ states which are invoked in some theories (see Varma in the same volume) do not appear in our calculation. This is due to the coupling of pσ with $3z^2 - r^2$ which is still important.

As a conclusion, we recall the importance of the alternating structure of the perovskite compounds. It allows a simple description of their electronic structure, in terms of universal DOS which depend on the geometry but not on the tight binding parameters. We stress also the coupling between the p_σ and $3z^2 - r^2$ states which destroys the non-bonding p_σ states invoked in some models. A more detailed description of the covalency between copper and oxygen in $La_{2-x}A_xCuO_4$ (A = Sr, Ba) will be given elsewhere.

REFERENCES

(1) Bednorz, J.G., Müller, K.H., Z. Phys. B64, 189 (1986).

(2) Mattheiss, L.F. : Phys. Rev. Lett. 58, 1028 (1987) ;
Yu, J.J., Freeman, A.J., Xu, J.H., Phys. Rev. Lett. 58, 1035 (1987) ;
Takegahara, K;, Harima, H., Yanase, A., Jap. J. Appl. Phys. 26, L352 (1987) ;
Hatsugai, Y., Fujiwara, T., Solid State Comm. 65, 1271 (1988).

(3) Pickett, W.E., Krakauer, H., Papaconstantopoulos, D.A., Boyer, L.L., Phys. Rev. B35, 7252 (1987) ; Papaconstantopoulos, D.A., Deweert, M.J., Pickett, W.E., Mat. Res. Soc. Meeting Boston (1987).

(4) Julien, J.P., Mayou, D., Cyrot-Lackmann, F., to be published.

(5) Mattheiss, L.F., Phys. Rev. B6, 4718 (1972).

(6) Ellialtioglu, S. Wolfram, T., Phys. Rev. B15, 5909 (1977).

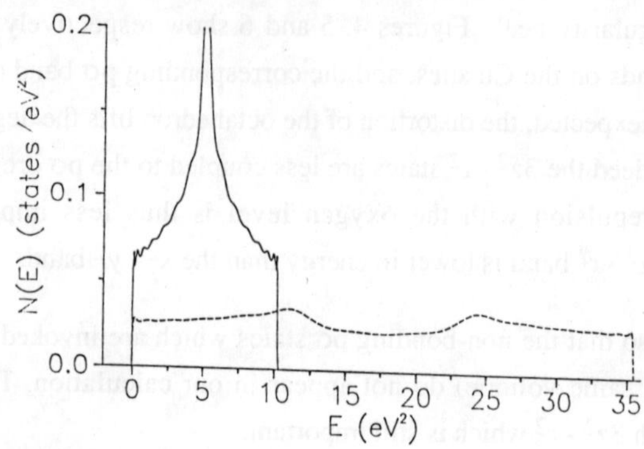

Fig. 1 : Effective DOS for $SrTiO_3$. Full line : t_{2g} orbitals ; dashed line : e_g orbitals.

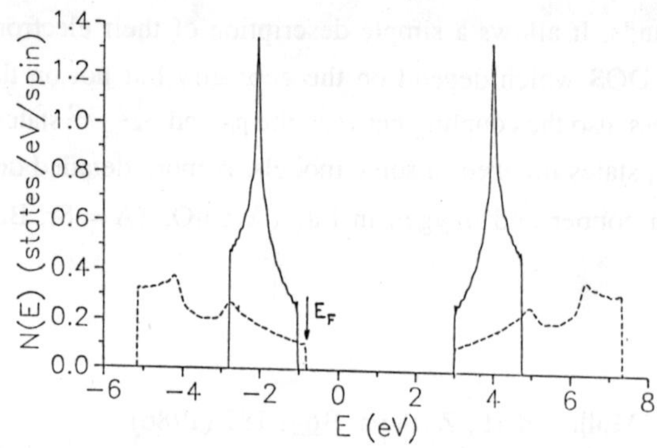

Fig. 2 : Total DOS for $SrTiO_3$. Full line : t_{2g} orbitals ; dashed line : e_g orbitals. The arrow indicates the Fermi level.

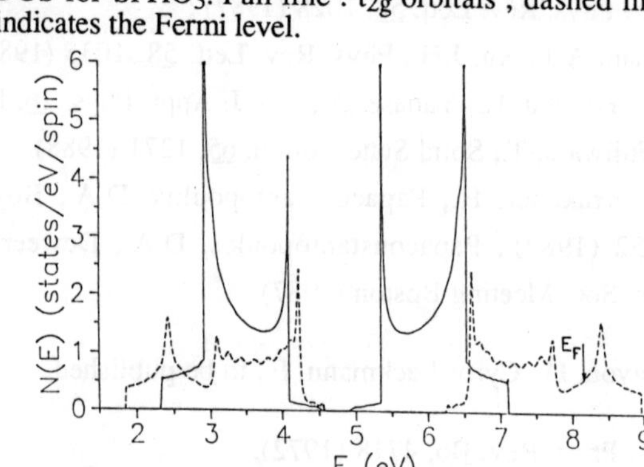

Fig. 3 : Total DOS for tetragonal La_2CuO_4. Full line : t_{2g} orbitals ; dashed line : e_g orbitals.

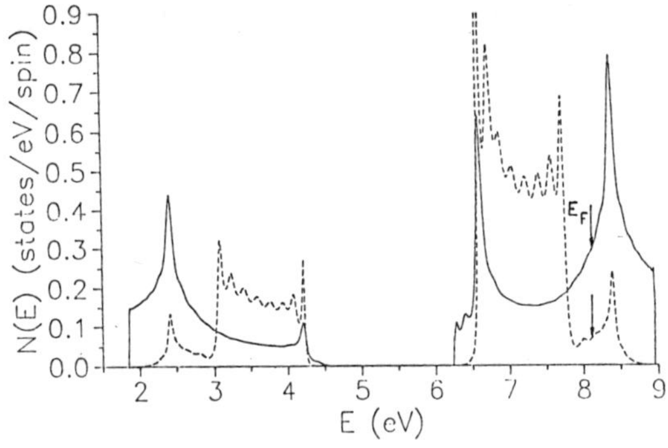

Fig. 4 : DOS for tetragonal La_2CuO_4 on a Cu-site. Full line : $x^2 - y^2$ orbital ; dashed line : $3z^2 - r^2$ orbital.

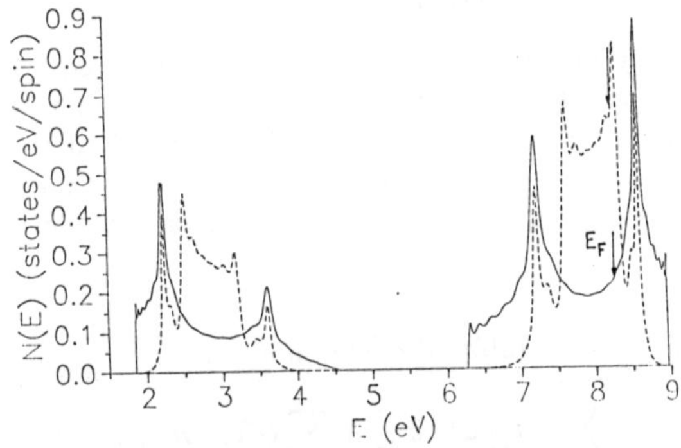

Fig. 5 : DOS for La_2CuO_4 without John-Teller distorsion on a Cu-site. Full line : $x^2 - y^2$ orbital ; dashed line : $3z^2 - r^2$ orbital.

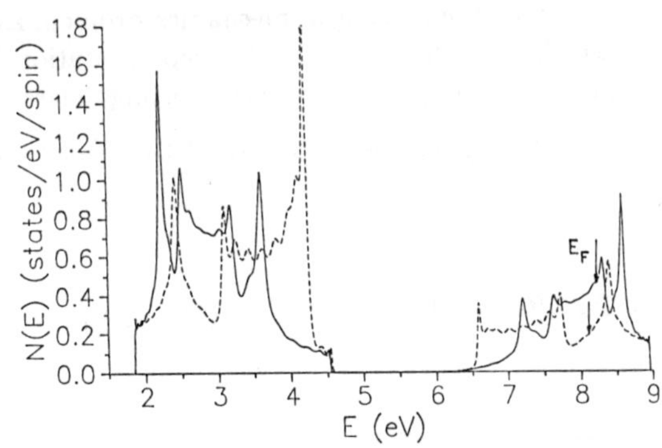

Fig. 6 : DOS for La_2CuO_4, pσ oxygen orbital. Full line : structure without John-Teller distortion ; dashed line tetragonal La_2CuO_4.

QUANTUM FLUCTUATIONS IN THE HUBBARD CHAIN[1]

C.M. Chaves

Department of Physics, PUC/RJ, Rio de Janeiro, Brazil

ABSTRACT

We show that the one-dimensional Hubbard model is equivalent to a two-dimensional *(Minkovskian)* non-linear σ–model. The result holds when the fluctuactions in the magnitude of the field associated to the spin are frozen. The same procedure applied to the two-dimensional *classical* Heisenberg chain gives the *Euclidean* version of the non-linear σ–model.

1. We start with the simple Hubbard model in $d = 1$

$$H = \sum_{\substack{<ij>\\ \sigma}} t_{ij} \hat{a}^{+}_{i\sigma} \hat{a}_{j\sigma} + \sum_{i} U \hat{n}_{i\uparrow} \hat{n}_{i\downarrow} . \qquad (1)$$

A convenient way to treat the statistical mechanics problem associated with the model is to use the so called holomorphic representation in which the operators $\hat{a}_{i\sigma}$ and $\hat{a}^{+}_{i\sigma}$ become Grassman anticomuting "classical" variables, $a_{i\sigma}(\tau)$, $a^{+}_{i\sigma}(\tau)$ depending on time. The partition function in this representation is [1]

$$Z = \int \mathcal{D}a^{+}\, \mathcal{D}a \exp \int_{0}^{\hbar\beta} d\tau \left\{ a^{+}(\tau)\, \dot{a}(\tau) - \frac{H}{\hbar} \right\} \qquad (2)$$

[1] Part of this work was done in the Max-Planck Institut, Stuttgart, in collaboration with Dr. Roland Zeyher

where β is the inverse of temperature and, for simplicity, all indices in a and a^+ were omited.

Before introducing (1) into (2) we make use of the identity [2]

$$\hat{n}_{i\uparrow}\hat{n}_{i\downarrow} = \frac{\hat{n}_i^2}{4} - \frac{(\vec{S}_i \cdot \vec{e}_i)^2}{\hbar^2} \tag{3}$$

where $\vec{e}_i$ is an arbitrary unit vector and $\vec{S}_i = \frac{\hbar}{2}\hat{\psi}_i^+ \vec{\sigma}\hat{\psi}_i$ is the spin operator. The $\hat{\psi}_i$ are the spinors

$$\psi_i = \begin{pmatrix} \hat{a}_{i\uparrow} \\ \hat{a}_{i\downarrow} \end{pmatrix}$$

and $\vec{\sigma}$ the Pauli matrices and $\hat{n}_i = \hat{n}_{i\uparrow} + \hat{n}_{i\downarrow}$.

Because (3) holds for any direction we integrate over the $\vec{e}_i$ in Z:

$$D\vec{e} = \prod_{i,\tau} \frac{d\vec{e}_i(\tau)}{4\pi}\,\delta(\vec{e}_i^2 - 1)\,.$$

The quadratic terms in the identity (3) are further transformed through the Hubbard-Stratonovich (HS) relation

$$\exp\int d\tau\,\frac{U}{\hbar}\,\frac{(\vec{S}_i\cdot\vec{e}_i)^2}{\hbar^2} = \int_{-\infty}^{\infty} dv_i\,\exp\int d\tau\frac{U}{\hbar}\left(-\frac{v_i^2(\tau)}{4} + v_i(\tau)\,\frac{(\vec{e}_i\cdot\vec{S}_i)}{\hbar}\right) \tag{4}$$

thereby introducing an auxiliary classical boson field $v_i(\tau)$. (An unimportant constant in (4) and in some expressions below has been omited). We now define $\vec{v}_i(\tau) = v_i(\tau)\,\vec{e}_i(\tau)$ but notice the integration over only the magnitude in (4). It can be shown that the physical meaning of the $\vec{v}_i$ is the following: its correlation functions are proportional to the correlation functions of the original spins $\vec{S}_i$. In particular the magnetization $< \vec{S}_i >$ is proportional to $< \vec{v}_i >$ and that is why we refer to it as the spin field.

An analogous (HS) transformation for the charge term can be made and another field introduced. Giving that the charge excitations play a subsidiary role in the Hubbard model when $U > 0$ – the case we are interested here – we treat this term in a mean field aproximation putting $n_i \equiv < n_i > = n = 1$ (for half filled band). As a result this term is cancelled by one involving the chemical potential. We are then left with

$$Z = \int \mathcal{D}v\,\mathcal{D}\vec{e} \prod_{i,\tau} \delta(\vec{e}_i^{\,2} - 1)\, e^S \tag{5a}$$

$$S = -\frac{U a_\tau}{4\hbar} \sum_{k\omega} \vec{v}_k(\Omega) \cdot \vec{v}_{-k}(-\Omega) + \operatorname*{tr}_{k\omega\sigma} \ell n \frac{a_\tau}{\hbar} G^{-1} \tag{5b}$$

$$\begin{aligned}(G^{-1})_{k_1\omega_1\sigma_1, k_2\omega_2\sigma_2} &= \left[(G_0^{-1})_{k_1\omega_1, k_2\omega_2} + \frac{U\sigma_1}{2\sqrt{N\bar{N}}} v^z_{k_1 - k_2}(\omega_1 - \omega_2) \right] \delta_{\sigma_1,\sigma_2} \\ &+ \frac{U}{2\sqrt{N\bar{N}}} \left(v^x_{k_1 - k_2}(\omega_1 - \omega_2) - i\sigma_1 v^y_{k_1 - k_2}(\omega_1 - \omega_2) \right) \delta_{\sigma_1,\bar{\sigma}_2}\end{aligned} \tag{5c}$$

where $(G_0^{-1})_{k_1\omega_1, k_2\omega_2} = \dfrac{\delta_{k_1 k_2} \delta_{\omega_1\omega_2}}{i\hbar\,\omega_1 - \varepsilon_{k_1}}$ is the free propagator. In (5) we used Fourier representation and discretized time by introducing a time lattice with $\bar{N} a_\tau = \hbar\beta$. N is number of sites in the space lattice.

2. Before proceeding let us recall how the (it Euclidean) non-linear σ–model is obtained for the two-dimensional *classical* (say, antiferromagnetic) Heisenberg model

$$H = J \sum_{<i,j>} \vec{S}_i \cdot \vec{S}_j \tag{6}$$

There is an intrinsic difficulty in describing low temperature properties of the $O(3)$ model within the convertional $\varepsilon = 4 - d$ expansion in the ϕ^4 theory because of the infrared divergences introduced by the massless Goldstone bosons. We clearly need an interpolation between the non-trivial fixed point near 4-dimensions and the zero temperature fixed point.

An original approach to this interpolation has been suggested by Polyakov and further extended by Brezin and Zinn Justin [3]. Giving that the lower critical dimension for systems with global continuous symmetry is 2 we expect that $T_C = T_c(d-2)$, vanishing at $d = 2$. This would give $T_C(\varepsilon)$ with $\varepsilon = d - 2$; around two dimension the system would be described by an ε-expansion. Many indications suggested that the major agent destroying order in two dimensions is the fluctuating phase rather than the amplitude. We then impose $\vec{S}_i^{\,2} = 1$ and write

$$\vec{S}_i \cdot \vec{S}_j = \frac{1}{2}(\vec{S}_i + \vec{S}_j)^2 - 1$$

the $\vec{S}_i + \vec{S}_j$ becomes, in the continuum limit, the derivatives $\partial S/\partial x$ and $\partial S/\partial y$, after changing the sign of spin in one sublattice ($\vec{S}_n \to (-1)^n \vec{S}_n$). The partition function is then

$$Z = \int D\vec{S} \prod_{\vec{r}} \delta(S^2(\vec{r}) - 1)\, e^{A \int d^2 r (\nabla \vec{S})^2} \tag{7}$$

where $A =$ constant and

$$(\nabla \vec{S})^2 = \left(\frac{\partial \vec{S}}{\partial x}\right)^2 + \left(\frac{\partial \vec{S}}{\partial y}\right)^2$$

From this connection critical exponents and correlation functions can be found (see ref. [3])

3. We tentatively make the same hypothesis here, that is, we argue that the physics of the $d = 1$ Hubbard model is captured by freezing the fluctuations in the magnitude of $\vec{v}$ and keeping only the transversal fluctuations in the vector $\vec{e}_i$. We choose $v = |\vec{v}|$ to be the one that minimizes the action S

$$\left.\frac{\delta S}{\delta v_i(\tau)}\right|_{\bar{v}} = 0 \tag{8}$$

The trace log in (5) is expanded and the $\vec{e}_i$ allowed to perform small fluctuations with wave vector $q \simeq Q = \pi/a$ and $\omega \simeq 0$. The details of the calculation will not be displayed here. It suffices to say that the expansion of the trace log generates terms like $\vec{e}_{q_i+Q} \cdot \vec{e}_{q_j+Q}$ or $(\vec{e}_{q_i i+Q} \wedge \vec{e}_{q_j+Q}) \cdot (\vec{e}_{q_i+Q} \wedge \vec{e}_{q_j+Q})$ as it is obvious from the fact that the action is an scalar. The coeficients of these terms depend on the susceptibility, which is expanded around $q = Q$ and $\omega = 0$ up to second order terms in frequency and in wave-vector. When we Fourier transform back to real space every term of the trace log are seen to be, in the continuum limit, proportional to

$$(\nabla \vec{e}) = \left(\frac{\partial \vec{e}}{\partial x}\right)^2 - \frac{1}{v_F^2}\left(\frac{\partial \vec{e}}{\partial \tau}\right)^2$$

where v_F = Fermi velocity = $\frac{2ta}{\hbar}$ (a is the lattice constant and t the nearest neighbor hopping). We are then left with

$$Z = \int \mathcal{D}\vec{e} \prod_{x,\tau} \delta(\vec{e}^2 - 1)\, e^S \tag{9a}$$

$$S = A \int d\vec{r}\, (\nabla \vec{e})^2 \tag{9b}$$

where A is a constant and $v_F\tau$ is the second dimension. (In the $T = 0^o K$ limit the crystal becomes infinite in the second dimension).

The consequences of this connection will be fully explored in the future but it is clear that the quantum fluctuactions are so important that they do *more* than just increase by one the effective dimensonality of the system.

References

[1] L. Faddeev and A. Slavnov, Gauge Fields. Introduction to Quantum Theory, Benjamin, 1980

[2] R. Prange and V. Koreman, *Phys. Rev.* **B19**, 4691 (1978)

[3] E. Brézin and J. Zinn-Justin, *Phys. Rev.* **b14**, 3110 (1976). See also D.Amit, Field Theory, The Renormalization Group, and Critical Phenomena, 2nd edition, part II, chap.6, and references therein.

EXCITONIC MECHANISM FOR ELECTRON PAIRING: A MANY BODY CALCULATION

Horacio Castillo+, Carlos A. Balseiro*, Blas Alascio*, Horacio Ceva+

+ Departamento de Fisica, Comision Nacional de Energia Atomica, Avda. del Libertador 8250, 1429 Buenos Aires, Argentina

* Centro Atomico Bariloche e Instituto Balseiro, Comision Nacional de Energia Atomica, 8400 Bariloche, R.N., Argentina

ABSTRACT

We apply many-body techniques to the study of a spinless interacting electron model of Cu-O planes in high T_c superconductors (E. Gagliano, A. Rojo, C. Balseiro, B. Alascio, Solid State Commun. 64, 901(1987)). We calculate polarization propagators and effective interactions. In some cases we find attractive interactions that could generate a mechanism for pairing.

INTRODUCTION

Since the discovery of high T_c superconductivity most of the theoretical work in the field refers to new mechanisms. It has been shown that conventional electron-phonon interactions can not explain the high values of T_c observed (Ref. 1). This supports the idea that these materials are either new -non BCS-superconductors or that they have a BCS-like ground state in which the interactions are mediated by collective excitations other than phonons (Ref. 2).

The problem of the superconductivity in the new compounds is complicated by the existence of high correlations. Both the existence of magnetic phases in related compounds and the spectroscopic data indicate strong intra-atomic interactions. In consequence conventional band structure calculations for these systems are not reliable and can not be used as a starting point for a model calculation.

Numerical simulations including large intra-atomic and moderate interatomic interactions indicate that charge fluctuations can give rise to attractive interactions between electrons in the Cu-O planes of these materials (Ref. 3).

In this communication we present a perturbative calculation to evaluate the effective interactions mediated by charge fluctuations. We show that an interatomic repulsion can give rise in these systems to an effective atractive interaction.

MODEL AND CALCULATIONS

Our starting point is the model Hamiltonian which describes the dynamics of electrons in the Cu-O planes. The unit cells of these planes contain one Cu and two oxigen atoms. We consider only one 3d orbital per Cu atom, and one 2p orbital per O atom. It has been argued (Ref. 4) that the limit of large intra-atomic interactions can be described with the following spinless Hamiltonian

$$H = \sum_{i,\alpha} \Delta_\alpha a^+_{\alpha i} a_{\alpha i} + t \sum_{\langle i,j \rangle} (a^+_{di} a_{pj} + a^+_{pj} a_{di}) + G \sum_{\langle i,j \rangle} n_{di} n_{pj} \quad \text{Eq.1}$$

where $\alpha = p,d$, $\Delta_p = -\Delta_d = \Delta$, and $a^+_{\alpha i}$ creates an α-type hole in the i-th site and $n = a^+ a$.

The vacuum of holes is the state Cu^+, O^{2-}. We are interested in the case of one hole per unit cell, which corresponds to undoped La_2CuO_4. A Hartree-Fock study of the model indicates that this situation corresponds to a semiconducting state. Doping with Sr introduces extra holes.

We evaluate the effective interaction between two added holes as

$$V^{\alpha\alpha}_{eff}(\vec{q},w) = V^{\alpha\alpha'}_{o}(\vec{q}) + \sum_{\beta\beta'} V^{\alpha\beta}_{o}(\vec{q})\, \Pi^{\beta\beta'}(\vec{q},w)\, V^{\beta'\alpha'}_{o}(\vec{q}) \quad \text{Eq.2}$$

where $V_o(\vec{q})$ is the Fourier transform of the bare interaction and $\pi(\vec{q},w)$ is the polarization propagator. We calculated this propagator in the RPA approximation (Ref. 5), including the diagrams shown in Fig. 1. The direct and exchange diagrams give the electron-hole and excitonic contributions, respectively. One particle propagators are obtained in the Hartree-Fock approximation, which produces renormalized parameters $\tilde{t}$, $\tilde{\Delta}$.

FIGURE 1

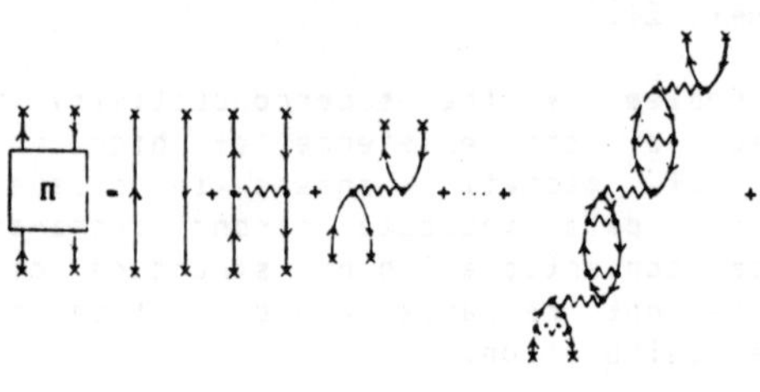

Results for the effective interaction and the imaginary part of the polarization propagator are summarized in Figs. 2 and 3. All results were obtained with $\tilde{t}$=1.6 eV and $\tilde{\Delta}$ =1.75 eV (Ref. 6).

The effective interaction for $w \approx 0$ is shown in Fig. 2. V_{Cu-O} is always positive. Note that V_{Cu-O} includes the bare interaction, which is a short range repulsion. V_{Cu-Cu} and V_{O-O} are attractive for all considered values of q in the first Brillouin zone.

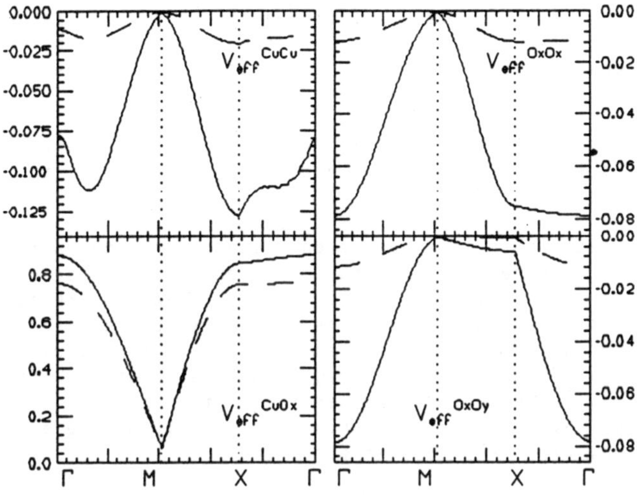

FIGURE 2 Dashed line: $5 \times V_{eff}(\vec{q}, w \approx 0)$ in eV, G=0.15 eV
Full line: $V_{eff}(\vec{q}, w \approx 0)$ in eV, G=0.8 eV

The absolute value of all effective interactions increase as G increases. As a function of w, $V_{eff}^{\alpha\alpha'}(\vec{q}, w)$ are nearly constant for small w, and present a singularity near the band edge.

It is instructive to analize the behaviour of the polarization $\Pi(\vec{q}, w)$. Its imaginary part is a product of some matrix elements times the density of states with equal number of particles as the ground state. In Fig. 3 it can be seen that, as G increases, Im $\Pi(\vec{q}, w)$ for $\vec{q} = (\pi/2, \pi/2)$ develops an excitonic peak (Ref. 7).

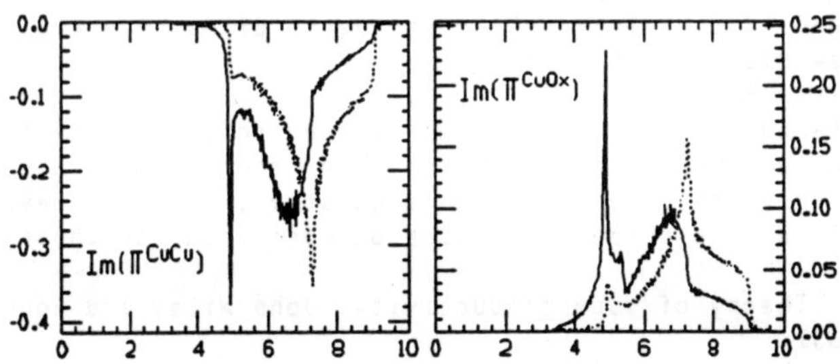

FIGURE 3

Im Π, arbitrary units.
Dots: G=0.15 eV
Full line: G=0.8 eV

CONCLUSIONS

In conclusion, we have shown that a repulsive interatomic potential can produce an effective atractive interaction.

In the present calculation we have considered two types of polarization excitations: electron-hole pairs and excitons. An analysis of the results shows that excitons play an important role in the effective interactions. In fact, in a similar calculation in which exchange diagrams are neglected, significantly different results are obtained.

For the set of parameters that we have considered, and $w \approx 0$, all $V_{eff}(\vec{q},w)$ are negative, except V_{Cu-O}, which includes the bare repulsive interaction.

All these results have been found for the semiconducting regimen. A more interesting case for high T_c superconductivity is the one in which a small number of holes are added to the system. For only two added holes, their effective interaction will essentially be the one presented here. These two holes will form a bound state (Cooper pair); neglecting retardation effects of the interaction, the binding energy of the pair can be easily calculated (Ref. 8). These results, as well as the effect of the retardation, will be published elsewhere.

REFERENCES

1 W. Weber, Phys. Rev. Lett. 58, 1371 (1987); L.F. Mattheiss Phys. Rev. Lett. 58, 1028 (1987)

2 P.W. Anderson, G. Baskaran, Z. Zou, T. Hsu, Phys. Rev. Lett. 58, 2790 (1987); V.J. Emery, Phys. Rev. Lett 58, 2794 (1987); C.M. Varma and S. Schmitt-Rink, Solid State Commun. 62, 681 (1987), among others.

3 J.E. Hirsch, S. Tang, E. Loh Jr., D.J. Scalapino. Preprint 1987. C.A. Balseiro, A.G. Rojo, E.R. Gagliano, B. Alascio, Int. Conf. on High T_c Superconductors and Materials and Mechanisms of Superconductivity, Interlaken (1988)

4 E.R. Gagliano, A.G. Rojo, C. A. Balseiro y B. Alascio, Solid State Commun. 64, 901 (1987).

5 A.L. Fetter and J.D. Walecka, Quantum theory of many-particle systems. Mc Graw-Hill Inc, N.Y. (1971)

6 See L.F. Mattheis, Ref.1

7 In fact, for larger values of G it can be seen that the density of states has a strong peak inside the energy gap. This, however, can only be considered as suggestive, because these big values of G are probably outside our perturbation theory approach.

8 G. Rickayzen, Theory of superconductivity, John Wiley and Sons Inc., N.Y. (1965)

ANISOTROPIC SUPERCONDUCTIVITY IN HIGH T_c CERAMICS

M.E. Foglio

I.F.G.W. - UNICAMP, 13081 Campinas, SP, Brasil

and

C.A. Balseiro

Centro Atomico Bariloche, 8400 Bariloche, Argentina

ABSTRACT

A pairing mechanism in which an attractive intersite interaction arises from the highly correlated electronic structure of the superconducting ceramics has been proposed recently by Balseiro ed.al (Inf. Tec. CNEA-CAB 88/003). In a previous work (Balseiro and Foglio, Inf.Tec. CNEA-CAB 88/009) the superconducting state resulting from such interactions in the Cu-O planes of SrCuO and YBaO was analized. Here the latter work is extended to include inter-plane interactions. The critical temperature is obtained as a function of electron concentration and attractive coupling from a pair of coupled equations connecting interplane and intraplane order parameters.

1. INTRODUCTION AND REVIEW OF THE MODEL

It is now generally accepted that the superconducting excitations in the high T_c ceramics that have been recently discovered are mainly confined to the CuO planes. The antiferromagnetic phases observed in these compounds lend support to the idea that intra-atomic correlations are large, at least in the Cu ions. Varma ed al[1)] have proposed that the interatomic repulsion between holes in neighbouring Cu and O ions plays an important role in the superconductivity of these compounds. It has been recently shown[2,3)] that the interatomic repulsion together with strong intra-atomic correlations and a quasi degeneracy of the d(Cu) and p(O) orbitals leads naturally to a pairing mechanism in the CuO planes.

In a recent work[4] the present authors have adopted this point of view to describe the dynamics of the electrons in the CuO planes, including both intra and interatomic correlations. In the limit of infinite intra-atomic repulsion, and using standart decouplings, they obtained T_c and the superconducting gap as a function of the model parameters. Although the coupling between nearest neighbour planes was considered in that formalism, no explicit calculation of their effect was presented.

In what follows all the out-of-plane (in-plane) parameters are denoted with primed (unprimed) variables. In the effective Hamiltonian there are attractive intersite interactions with strength constants G and G'. Two superconducting order parameters Δ_σ and Δ'_σ were employed when the Green's function equations of motion were decoupled by standardt techniques. It was shown that at the critical temperature the following system of equations has non-trivial solutions:

$$\begin{aligned} (G.I - Z)\Delta_\sigma^* + G'.I''(\Delta'_\sigma)^* &= 0 \\ G.I''\,\Delta_\sigma^* + (G'.I' - Z')(\Delta'_\sigma)^* &= 0 \end{aligned} \qquad (1)$$

where z and z' are the coordination numbers.

The I, I', and I", are integrals over the Brillouin zone that involve products of in-plane, out-of-plane and cross products respectively of typical structure factors s(k) and s'(k). One should also satisfy the equation that gives the number of electrons N_σ as a fucntion of the effective chemical potential $\tilde{\mu}$:

$$N_\sigma = (1 - N_\sigma)\Sigma_k\, f(\varepsilon_{k\sigma})/N \qquad (2)$$

where f(x) = 1/(1+exp x) is the Fermi function, N the number of sites and $\varepsilon_{k\sigma} = (1-N_\sigma).(t.s(k) + t'.s'(k)) - \tilde{\mu}$, where t and t' are the hopping parameters.

2. RESULTS AND DISCUSSION

Fixing T_c, t, $R_t = t'/t$ and $R_g = G'/G$, one finds the $\tilde{\mu}$ and G that satisfy Eqs. (1.2). Two values of G are obtained for each T_c, and one should select the smaller one; one can then plot T_c . vs. G.

To calculate the I, I' and I" a rectangular density of k vectors was assumed in the k plane associated to the CuO plane: this approximation is adequate for a nearly filled band (i.e. near stoichiometry).

In Fig. 1 kT_c/t . vs. N_σ was plotted for several values of G/t (0.75, 1.0, 1.25, 1.5) and for a weak interplane coupling ($R_g = R_t$ = 0.1).

The curves in Fig. 1 are essentially the same obtained[4)] when all interplane interactions are neglected: only for $N_\sigma > 0.49$ there are appreciable differences between the curves in Fig. 1 and those in Ref.[4)], and they are more noticeable when T_c decreases. At T_c, the two Eq. (1) have non-trivial solutions, and although both Δ_σ and Δ'_σ tend to zero at T_c, one can interpret their ratio as the limit when $T \to T_c$. In Fig. 2 $\Delta'_\sigma/\Delta_\sigma$ is plotted as a function of N_σ for $kT_c/t = 0.001$, 0.05, 0.1. All these curves present a common feature: the ratio $\Delta'_\sigma/\Delta_\sigma$ changes sign at a value of N_σ close to 0.5. This point moves from $N_\sigma = 0.47$ (for $kT_c/t = 0.1$) to $N_\sigma = 0.49$ (for $kT_c/t = 0.001$) and $\Delta'_\sigma/\Delta_\sigma$ reaches a maximum fairly close to $N_\sigma = 0.5$.

Although both Δ_σ and Δ'_σ are zero at T_c, the change of sign of their ratio indicates that at least close to T_c there is a change in the symmetry of the order parameters as a function of N_σ: from d-like at lower N_σ to s-like at higher N_σ.

Acknowledgements. The authors would like to acknowledge the finantial help of FAPESP Brasil that made possible the cooperation that resulted in the present work. One of us (M.E.F.) would also like to acknowledge the finantial help of CNPq Brasil, and the hospitality of Instituto de Fisica Balseiro, CNEA-CAB.

[1] C.M. Varma, S. Schmitt-Rink and E. Abrahams, Solid State Comm. 62, 681 (1987).
[2] C.A. Balseiro, A.G. Rojo, E.R. Gagliano and B.R. Alascio, Informe Tecnico CNEA-CAB 88/003. See also the same authors at the Proc. Interlaken Conference (1988).
[3] J.E. Hirsh, S. Tang, E. Loh Jr., and D.J. Scalapino, Phys. Rev. Letters, issue of april 18, 1988.
[4] C.A. Balseiro and M.E. Foglio, Informe Tecnico CNEA-CAB 88/009.

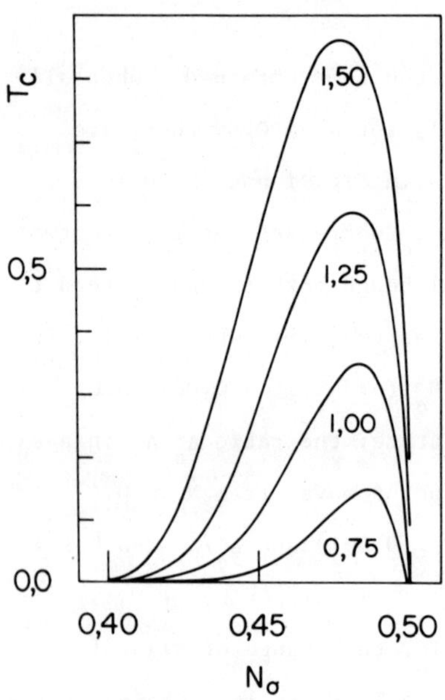

Figure 1

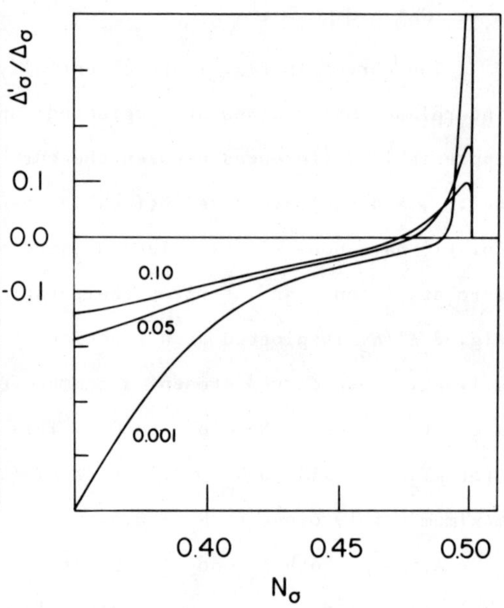

Figure 2

NEW ANALYTIC APPROACH TO MOTT INSULATORS

Miguel Lagos and Miguel Kiwi
Facultad de Física
Universidad Católica de Chile
Casilla 6177, Santiago 22, Chile

The lack of a full understanding of the physics of Mott insulators acquired dramatic relevance after the discovery that the parent compound La_2CuO_4 of a class of ceramic superconductors belongs to the family.[1,2] A close relation between doped Mott insulators and high temperature superconductivity is apparent from experiment.

The Hubbard Hamiltonian with a half-filled band,[3] which is believed to describe quite accurately the Mott antiferromagnetic insulating phase of a number of metal oxides,[4,5] is studied. We obtain analytic solutions for the spectrum and stationary states which are valid in the limit of large antiferromagnetic correlations. We write the Hubbard Hamiltonian as

$$H = t \sum_{\vec{l}\vec{\delta}s} c^{\dagger}_{\vec{l}+\vec{\delta},s} c_{\vec{l},s} + U \sum_{\vec{l}} n_{\vec{l},\uparrow} n_{\vec{l},\downarrow} \tag{1}$$

where $\vec{l}$ and $\vec{l}+\vec{\delta}$ are neighboring lattice sites, $s = \uparrow, \downarrow$ are spin indices, $c^{\dagger}_{\vec{l},s}$ creates a Wannier state, $n_{\vec{l},s} = c^{\dagger}_{\vec{l},s} c_{\vec{l},s}$, and we assume large U/t. It is well known that the on-site repulsion term, proportional to U, induces antiferromagnetic order, eventually long-ranged. We restrict ourselves to lattice geometries which do not frustrate antiferromagnetic correlations. Thus we can distinguish two crystallographycally equivalent sublattices, denoted a and b. Any neighbor of a lattice site $\vec{l}_a$ in a is in the sublattice b, and conversely, so that

$$\vec{l}_a + \vec{\delta} = \vec{l}_b, \quad \vec{l}_a + \vec{\delta} + \vec{\delta}' = \vec{l}'_a. \tag{2}$$

We start assuming antiferromagnetic order in the sense

$$\langle n_{\vec{l}+\vec{\delta},s}(1-n_{\vec{l},s})\rangle \approx \begin{cases} 1, & \text{if } \vec{l} \text{ is in the sublattice with spins } -s \\ 0, & \text{otherwise} \ . \end{cases} \tag{3}$$

Notice that the assumption of long-range antiferromagnetic order is implicit in the statement of the right hand side of Eq. (3), which assigns a definite spin to each sublattice. We define sublattice a as the one with $s=\uparrow$.

Using the identity

$$\left[\sum_{\vec{l}} n_{\vec{l},\uparrow}n_{\vec{l},\downarrow}\ ,\ c^{\dagger}_{\vec{l}+\vec{\delta},s}c_{\vec{l},s}\ n_{\vec{l}+\vec{\delta},-s}(1-n_{\vec{l},-s})\right] = c^{\dagger}_{\vec{l}+\vec{\delta},s}c_{\vec{l},s}\ n_{\vec{l}+\vec{\delta},-s}(1-n_{\vec{l},-s}) \tag{4}$$

and taking into consideration relations (3), one can easily show that the operators

$$\phi^{\dagger}_{\vec{\delta}\uparrow} = \sqrt{\frac{2}{N}}\sum_{\vec{l}_a} c^{\dagger}_{\vec{l}_a+\vec{\delta},\uparrow}c_{\vec{l}_a,\uparrow} + \frac{t}{U}\sqrt{\frac{N}{2}} \tag{5}$$

$$\phi^{\dagger}_{\vec{\delta}\downarrow} = \sqrt{\frac{2}{N}}\sum_{\vec{l}_a} c^{\dagger}_{\vec{l}_a,\downarrow}c_{\vec{l}_a+\vec{\delta},\downarrow} + \frac{t}{U}\sqrt{\frac{N}{2}}\ , \tag{6}$$

satisfy

$$\left[\sum_{\vec{l}} n_{\vec{l},\uparrow}n_{\vec{l},\downarrow}\ ,\ \phi^{\dagger}_{\vec{\delta}s}\right] = \phi^{\dagger}_{\vec{\delta}s} - \frac{t}{U}\sqrt{\frac{N}{2}}\ ,\quad s=\uparrow,\downarrow\ . \tag{7}$$

The equation

$$t\sum_{\vec{l}\vec{\delta}s} n^{\dagger}_{\vec{l}+\vec{\delta},s}c_{\vec{l},s} = t\sqrt{\frac{N}{2}}\sum_{\vec{\delta}s}\left(\phi^{\dagger}_{\vec{\delta}s}+\phi_{\vec{\delta}s}\right) - 2zN\frac{t^2}{U} \tag{8}$$

follows directly from the definitions (5) and (6), and one also has the exact relation

$$\left[\phi_{\vec{\delta}\uparrow}\,,\,\phi^{\dagger}_{\vec{\delta}'\uparrow}\right] = \frac{2}{N}\sum_{\vec{l}_a}\left(c^{\dagger}_{\vec{l}_a+\vec{\delta}-\vec{\delta}',\uparrow}c_{\vec{l}_a,\uparrow} - c^{\dagger}_{\vec{l}_a+\vec{\delta},\uparrow}c_{\vec{l}_a+\delta',\uparrow}\right). \tag{9}$$

We approximate the commutation relation (9) by the much simpler one

$$\left[\phi_{\vec{\delta}s}\,,\,\phi^{\dagger}_{\vec{\delta}'s'}\right] = \delta_{s,s'}\delta_{\vec{\delta},\vec{\delta}'}\,,\quad s=\uparrow,\downarrow\,. \tag{10}$$

Dropping the right hand side of Eq. (9) when $\vec{\delta} \neq \vec{\delta}'$ is the main approximation of the present theory. It is not difficult to realize that Eq. (10) becomes trivially exact in the limit of perfect antiferromagnetic order. The real issue, however, is how strongly Eq. (9) converges to Eq. (10), and in which region of the parameter space of the model. We skip this discussion for the time being and justify our approximation by the accuracy of the results.

Combining Eqs. (7), (8) and (10) one can show that, in the limit of high correlations

$$H = U\sum_{\vec{\delta}s}\phi^{\dagger}_{\vec{\delta}s}\phi_{\vec{\delta}s} + E_0\,, \tag{11}$$

where E_0 is the ground state energy. The fact that the Hubbard Hamiltonian can be cast into the form above, i.e. that it can be approximately described as a free boson gas, is important in the understanding of the nature and characteristics of the relevant elementary excitations of the system in the highly correlated regime.

It can also be shown that

$$|g\rangle = exp\left[-\frac{t}{U}\sqrt{\frac{N}{2}}\sum_{\vec{\delta}s}\left(\phi^{\dagger}_{\vec{\delta}s} - \phi_{\vec{\delta}s}\right)\right]\prod_{\vec{l}_a}\prod_{\vec{l}_b}c^{\dagger}_{\vec{l}_a\uparrow}c^{\dagger}_{\vec{l}_b\downarrow}|0\rangle \tag{12}$$

satisfies

$$\phi_{\vec{\delta}s}|g\rangle = 0\,,\quad s=\uparrow,\downarrow\,. \tag{13}$$

Hence $|g\rangle$ is the ground state of H.

Closed-form expressions for the ground state energies for the various lattices satisfying the requirement (2) are obtained replacing the ket (12) and the exact expression (1) of H in

$$E_0 = \langle g|H|g\rangle \; . \tag{14}$$

The expectation value in the right hand side of Eq. (14) can be evaluated exactly. This way the results for E_0, obtained using the ket (12), turn out to be in satisfactory agreement with the exact 1-D results[6]. However, they show excellent agreement with recent variational[7] and quantum Montecarlo calculations for the square and simple cubic lattices.[8]

In conclusion, we have introduced a new analytic approach to treat the half-filled Hubbard model which leads to a complete, though approximate, solution. More importantly, we have identified the excitations which play the key role in the highly correlated regime. The theory is quite simple and transparent, and the resulting eigenvectors allow to calculate matrix elements analytically. An insight into the structure of the ground state and some of its properties are obtained.

This work was supported by DIUC and Conicyt-Fondecyt (CHILE).

References

[1] L. F. Mattheiss, *Phys. Rev. Lett.* **58** , 1028 (1987); J. Yu, A. F. Freeman, and J. H. Xu, *Phys. Rev. Lett.* **58** 1035 (1987)
[2] D. Vaknin, S. K. Sinha, D. E. Moncton, D. C. Johnson, J. M. Newsam, C. R. Safinya, and H. E. King, Jr., *Phys. Rev. Lett.* **58** , 2802 (1987)
[3] J. Hubbard, *Proc. Roy. Soc.* **A276** , 238 (1963); **A277** , 238 (1964); **A281** , 401 (1964)
[4] N. F. Mott, *Proc. Phys. Soc.(London)* **A62** , 416 (1949)
[5] D. Adler, *Solid State Phys.* **21** , 1 (1968)
[6] E. H. Lieb and F. Y. Wu, *Phys. Rev. Lett.* **20** , 1445 (1968)
[7] H. Yokoyama and H. Shiba, *J. Phys. Soc. Japan* **56** , 3582 (1987)
[8] J. E. Hirsch, *Phys. Rev. Lett.* **51** , 1900 (1983)

PREPARATION OF HIGH T_c SUPERCONDUCTORS: Y/Al-Ba-Cu-O SYSTEMS

A.V. Bellido, L.T. Auler and B.F. da Silva, Instituto de Engenharia Nuclear, Comissão Nacional de Energia Nuclear, C.P. 2186, CEP 20001, Rio de Janeiro, Brasil.

ABSTRACT

Several ceramics of the general formula $M^{III}Ba_2Cu_3O_{7-y}$ were prepared, under the same thermal treatment, from mixtures of constant Ba and Cu compounds and different Y compounds. In some experiments, Y was partial or totally substituted for Al. All the ceramics with the only exception of that starting from Al_2O_3 showed Meissner Effect with a T_c higher than the liquid nitrogen temperature.

PACS numbers: 74.10.+v, 74.70-b

1. INTRODUCTION

Bednorz and Müller[1)] reported in September 1986 that compounds of the La-Ba-Cu-O system exhibited properties of high T_c superconductors. An almost scientific competition started straight away between many laboratories of all over the world trying to get, each one, superconductors with higher T_c. Few months later, February 1987, Wu and collaborators[2)] got superconductivity in mixed oxides of the Y-Ba-Cu-O system with T_c of 93 K, i.e. higher temperature than that of the liquid nitrogen. Bhargava et al.[3)], in August of the same year, announced the preparation of superconductors with T_c up to 153 K through a cyclic thermal treatment of Y-Ba-Cu-O and Y-Ba-Cu-O:F samples. At the same time, all the available techniques were used to characterize the molecular structure and to measure the physical chemical properties of the superconducting ceramics. X-ray diffraction, transmission electron microscopy and neutron diffraction studies revealed that in the so-called

1-2-3 superconductors (Y = 1, Ba = 2, Cu = 3) the crystal structure was a derivative of the perovskite structure with a orthorhombic unit cell and with oxygen deficiency[4-8].

In view of the wide and important applications of the superconductors it was decided, in our laboratories, to start research on the preparation of those compounds. The first studies were made with the 1-2-3 ceramics and the results of using different yttrium precursors and of the partial and total substitution of Y for Al are presented in this paper.

2. EXPERIMENTAL

2.1. Sample Preparation

Five samples of ceramics corresponding to the general formula $M^{III}Ba_2Cu_3O_{7-y}$ (M^{III} = Y and/or Al) were prepared starting from mixtures of constant Ba and Cu compounds and different Y compounds. Y was also partial or totally substituted by Al.

The trivalente element compounds used were:

Sample nº 1 : $Y(NO_3)_3.6H_2O$

Sample nº 2 : Y_2O_3

Sample nº 3 : $1/3Al_2O_3 + 2/3Y_2O_3$

Sample nº 4 : $Y_2(C_2O_4)_3.9H_2O$

Sample nº 5 : Al_2O_3

$BaCO_3$ and CuO were the constant barium and copper compounds, respectively.

A.R. reagents of the mentioned compounds were weighed, in the appropriate proportions, carefully ground and mixed in an agate morter with a pestle of the same material. In the case of the sample nº 4 (oxalates sample) Y, Ba and Cu nitrates were weighed, dissolved in water and precipitated by oxallic acid. The blue precipitate was decanted, filtered and dried.

2.2. Thermal Treatment

The thermal treatment was just the same for all the samples and consisted of the following steps:

. Heating the samples, contained in porcelain crucibles, at

950° C for 6 hours in air, in a muffle furnace;

- Regrinding the solid button formed to reduce it to very fine particles and mixing them in an agata morter;
- Heating again at 975° C for 5 hours in the same crucibles and furnace;
- Regrinding, mixing, once more, and pressing the fine powder at 200 kg.cm^{-2};
- Sinterization of the pellets obtained at 975° C for a period of 4 hours in the same crucibles and furnace;
- Annealing of the pellets, placed in an alumina boat, in oxygen flow at 450° C for 4 hours in a tubular furnace.

The furnaces were always switched off at 5 pm. and the samples remained inside up to next day in order to allow a slow cooling.

3. RESULTS AND DISCUSSION

The five samples were tested for Meissner Effect before and after the annealing step. In both cases, the 1-4 samples showed superconducting properties at liquid nitrogen temperature, but not the sample nº 5 in which the Y content was totally substituted by Al.

The superconductivity observed in the sample of the Al-Y-Ba-Cu-O system means that the aluminum ions in 1/3 proportion do not alter in a great extent the crystal lattice, although more research is needed to learn about the location and influence of the Al ions.

It was found out that the superconducting properties improve with the time of heating being, then, advisable to do it for periods of about 6 hours, regrinding and mixing carefully the powder - in between the periods - to ensure an intimate contact of the components to facilitate the chemical reactions in solid state.

Heating in air can liberate the annealing step providing the experimental conditions allow enough amount of oxygen to generate the orthorhombic superconducting phase.

4. REFERENCES

1. Bednorz J.G. and Müller K.B., Z. Phys. B. 64, 189 (1986).

2. Wu M.K., Ashburg J.R., Torng C.J., Hor P.H., Meng R.L., Gao L., Huang Z.J., Wang Y.Q. and Chu C.W., Phys. Rev. Lett. 58, 908 (1987).

3. Bhargava R.N., Herko S.P. and Osborne W.N., Phys. Rev. Lett. 59, 1468 (1987).

4. Grant P.M., Beyers R.B., Engler E.M., Lim G., Parkin S.S.P., Ramirez M.L., Lee V.Y., Nassal A., Vasquez J.E. and Savoy R.J., Phys. Rev. B 35, 7242 (1987).

5. Cava R.J., Batlogg B., van Dover R.B., Murphy D.W., Sunshine S., Siegrist T., Remeika J.P., Rietman E.A., Zahurak S. and Espinosa G.P., Phys. Rev. Lett. 58, 1676 (1987).

6. Syono Y., Kikuchi M., Oh-Ishi K., Hiraga K., Arai H., Matsui Y., Kobayashi N., Sasaoka T. and Muto Y., Jpn. J. Appl. Phys., 26, L498 (1987).

7. Van Tendeloo G., Zandbergen H.W., Okabe T. and Amelinckx S., Solid State Commun. 63, 969 (1987).

8. Beyers R., Lim G., Engler E.M., Lee V.Y., Ramirez M.L., Savoy R.J., Jacowitz R.D., Shaw T.M., La Placa S., Boehme R., Tsuei C. C., Park Sung I., Shafer M.W. and Gallagher W.J., Appl. Phys. Lett. 51, 614 (1987).

PHASE DIAGRAMS OF Y-Ba-Cu-O AND Bi-Sr/Ca-Cu-O COMPOUNDS

J.Betancourt*, M.Esguerra**, G.Franco*

*Departamento de Física, Universidad Nacional de Colombia, Bogotá
**Centro Internacional de Física, A.A.49490, Bogotá, Colombia

ABSTRACT

In the phase diagram of the Y-Ba-Cu-O system (900-1000°C) we have explored regions for compatibilities among different compounds and for the existence of superconducting phases other than the 1-2-3. In the phase diagram for the Bi-Sr/Ca-Cu-O system (800-900°C) zones are shown, where it is possible to obtain superconducting phases.

The discovery of superconducting oxides containing no rare earth elements has broadened the high-Tc material base and has shown that a system can have more than one superconductive phase. For that reason we continued work on the Y-Ba-Cu-O system exploring the phase diagram for new superconducting compounds in addition to the investigation of the Bi-Sr/Ca-Cu-O phase diagram.

Fig.1 shows the phase diagram for Y_2O_3-BaO-CuO in weight percentage (900°C-1000°C) with known as well as hypothetical compounds. It is interesting to note that $BaCuO_2$(BC), $YBa_2Cu_3O_x$(123), $YBaCu_2O_x$ (112)[1], $Y_2BaCu_3O_x$(213)[2] and $Y_2Cu_2O_6$(Y_2C_2, hypothetical phase with mixed valence) lie on a straight line and this is suggestive for the search of compatibilities among these compounds. We prepared samples with nominal compositions $YBa_4Cu_{12}O_x$(A) and $YBa_6Cu_{15}O_x$(B) which lie in the triangle BC-123-CuO in order to investigate the phase formation tendency in that region. For the same reason a composition $YBa_3Cu_{8.5}O_x$(C) lying inside the triangle 123-112-CuO was chosen.

The samples were prepared in two different ways: (i)sinterization at 900°-930°C followed by annealing at the same temperature in oxygen flow and slow cooling to room temperature in the usual way[3]

2-1-1-2 and 2-1-1-4 in order to investigate their thermal behaviour: all the samples melted at 900°C, melted partially between 800° and 840°C and below 800°C no melting was observed. A line corresponding to 50 wt-% Bi was chosen to prepare samples (B-G) at 820°C, in order to study the compound-forming tendency in this system. In samples with nominal compositions E,D, and F small crystals formed and their composition from EDAX-analysis are denoted by E_1, D_1 and F_1 in the phase diagram. None of this samples showed superconducting behaviour in susceptibility measurements. It is possible that a non-superconducting phase with composition $Bi_3SrCaCuO_x$ exists, corresponding to E_1, D_1 and L_1(see below), but further work is needed to clarify this point.

A sample with composition A lying in the center of a triangle formed by the hypothetical compounds $Bi(Sr,Ca)_2Cu_3O_x$, $Bi_2(Sr,Ca)Cu_3O_x$ and $Bi_2(Sr,Ca)CuO_5$ was taken to see the possible formation of these in analogy to the Y-Ba-Cu-O system. Susceptibility measurements showed a superconducting onset at 110K for a sample prepared at 840°C. In the rather homogeneous sample black crystals with plane faces and about 1μm in size and needle-like crystals could be seen optically.

Samples with nominal composition 1-1-1-2 were prepared at different temperatures to study the effect of preparation temperature on transport properties. This composition of a possible compound was chosen in order to avoid Bi-rich phases, which seem to be not superconductive as shown by the D1,E1 and F1 compounds. After our work on the 1-1-1-2 sample was completed, we learned from several studies by other groups[5], which used this as the starting composition. We found a large influence of the preparation temperature: a change of 70°C from 910°C downwards varied the sample resistance five orders of magnitude. The analysis of two samples, one prepared at 890°C (L) and one at 850°C (K) showed the existence of quite different phases and electrical behaviour. Sample L showed no anomaly in susceptibility measurements and the analysis of black crystals yield a composition L_1 similar to E_1 and D_1. In contrast sample K has a clear change in its susceptibility behaviour at 118K and 106K. A careful analysis showed the existence of five different phases (K_1-K_5), which surprisingly lie near the binary axes Bi_2O_3-CuO and Bi_2O_3-SrO/CaO instead of near to central regions which are known to yield superconductivity. K1 corresponds to the composition of a black crystal embedded in a matrix K_2; other crystals, black but with a different shape, have composition K_3 and the surrounding matrix K_4. Composition K_5 corresponds to a porous region in the sample, where we found the missing amounts of alkaline earths. Further work is needed to find the superconducting phase. Although we are well aware of errors in compositions found by microanalysis, we cannot exclude up to now that a quasi-binary compound could be responsible for the observed behaviour.

and (ii) melting at 1000°C followed by a quick heating to 1050°-1100°C for a few minutes returning to 1000°C, annealing at this temperature between one and twelve hours and cooling either by natural cooling to room temperature or by quenching in liquid air.

Samples were analized by an electron probe microanalyzer (EDAX-9100, Philips) in a Philips SEM-505 scanning electron microscope and by optical microscopy. The superconducting properties were examined through ac-susceptibility (2kHz) and dc conductivity measurements by the standard four probe method.

Both resistivity and susceptibility measurements on all sinterized samples of composition A and B showed superconducivity around 95K. Optical microscopy revealed that samples consist of small black crystals of less than 1μm in size. Microanalysis of samples A showed regions of composition A_1 in the neighbourhood of the 123 phase. Other regions had composition A_2 which lies close to the binary axis BaO-CuO and could be a ternary eutectic composition[4]. In samples B regions with composition B_1 were found, which lies on the line connecting BC and 123, with a majority composition BC and minority composition 123. In the optical microscope no "green phase" (211) could be observed. This would indicate that the triangle BC-123-CuO is a region where these phases are compatible up to 900°C.

These subsolidus compatibilities are broken by melting of both A and B samples. Optical microscopy showed the presence of green crystals (211 pase) as well as black crystals with plain faces, 10 to 100μm in size (123 phase). This was confirmed by microanalysis. Both resistivity and susceptibility on an as-prepared, melted sample A show a superconducting transitions with an onset at 87K, which could be observed again one month later. We are now studying this preparation technique, since it has many advantages: short preparation time, no need of oxygen annealing and stability against degradation of superconducting properties.

Sinterized samples with nominal composition C displayed a step in the resistance measurement at 84K after an superconducting transition onset at 92K. This could be attributed to the 123 phase and to another phase, inestable at low temperatures, since the step at 84K almost disappeared after four days. Then we tried to increase the proportion of this possible second phase by quenching, but the resistance showed a semiconducting behaviour. However, susceptibility measurements displayed two transitions again, thus indicating that this assumed second phase is not that unstable. By optical microscopy we could only establish the presence of black and green crystals in the same proportion.

The phase diagram (fig.2) for Bi_2O_3, SrO/CaO (molar ratio 1:1) and CuO (800°C-900°C) shows the location of the nominal composition of the prepared samples and their real composition through EDAX-analysis. We prepared samples with nominal composition 4-1-1-3,

We are indebted to Prof.P.Orozco and to Ms.A.M.Castellanos from the Physics Department, and to Dr.H.Duque and Ms.B.L. Pesca from Ingeominas for their help with the electron microscope and with the EDAX analysis. We express our gratitudes to Colciencias for financial support.

REFERENCES

1. Frase, K.G. and Clerke, D.R., "Phase compatibilities in the system Y_2O_3-BaO-CuO", Adv.Cer.Mat.2(3B), 300-302 (July 1987)
2. Takayama,E., Ushida,Y., Matsui,Y., Kato,K., "Identification of the High Tc Superconductor in the system Y-Ba-Cu-O", Jpn.J.Appl.Phys.Lett.26(4), L476-L478 (April 1987)
3. Posada,E., Franco,G., Holguín,G., Jiménez,A., Mariño,A., Niño,V. and Rodríguez,D., "Composition and Annealing Effects on the Superconducting Transition of Y-Ba-Cu-O compounds", Rev.Col.Fis., to appear
4. Takekawa,S. and Iyi,N., "Single Crystal Preparations of $Ba_2YCu_3O_x$ from Nonstoichiometric Melts", Jpn.J.Appl.Phys.Lett.26 (5), L851-L853 (May 1987)
5. Maeda,H., Tanaka,Y., Fukutomi,M. and Asano,T., Jpn.J.Appl.Phys. Lett.27(2) (1988); Liu,J.Z., Crabtree,G.W., Rehn,L.E., Geiser, Urs, Young, D.A., Kwok,W.K., Baldo,P.M., Willians, J.M., and Lam,D.J., "Crystal Growth and Superconductivity in the Bi-Ca-Sr-Cu-O System", Phys.Lett., to appear

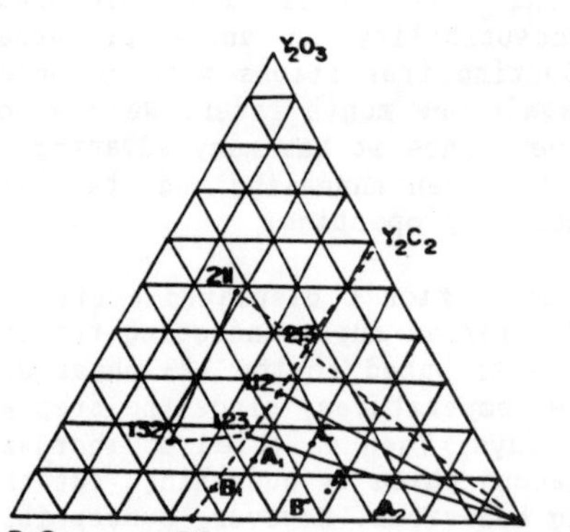

Fig.1 Phase diagram(wt.-%) for the Y_2O_3-BaO-CuO system (900-1000°C)

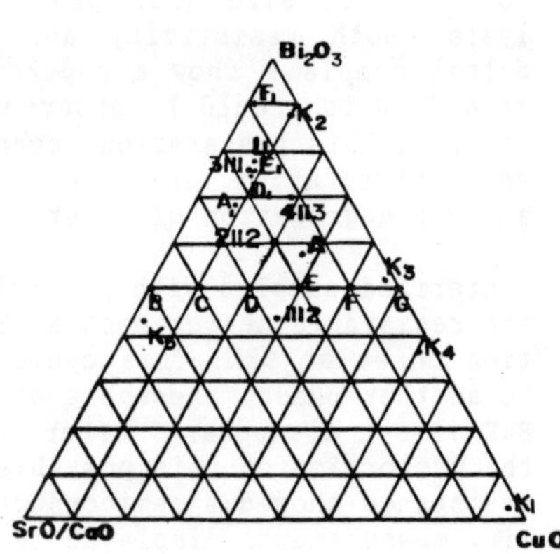

Fig.2 Phase diagram(wt.-%) for the Bi_2O_3-SrO/CaO-CuO system (800-900°C)

SUPERCONDUCTIVITY AND OXYGEN KINETICS IN THE HIGH T_c OXIDE $YBa_2Cu_3O_x$, WHERE $6.30 < x < 7.00$

O. F. de Lima, R. F. Jardim, S. Gama and I. Torriani
Instituto de Fisica, UNICAMP, 13081-Campinas, SP, Brazil.

ABSTRACT

Disk-shaped samples of $YBa_2Cu_3O_7$ were initially produced through the solid diffusion method. Following, these samples were cut into several small bars, which were then used for oxygen extraction under vaccum, in temperatures between 330 °C and 550 °C. X-ray analysis shows a gradual transformation from the very pure superconducting orthorhombic phase (x=7.0) to the very pure semiconducting tetragonal phase (x=6.3). The superconducting critical temperature (T_c), defined at the mid-point of the resistive transition, shows a well defined plateau around 70 K for $6.6<x<6.8$, though a very sharp transition ($\Delta T \sim 1$ K) was observed for the parent sample (x=7.0) at T_c=93K. The kinetics for oxygen intercalation and desorption was studied by monitoring the mass variation of the sample.

INTRODUCTION

It is very well established[1,2,3] that oxygen deficiency in the compound $YBa_2Cu_3O_{7-y}$ affect its superconductivity and structure. Despite the lack of a conclusive theory for high-T_c superconductivity, certainly one has to pay attention to: 1) The state of order-disorder for the oxygen vacancies, specially the ones belonging to the Cu-O chains[2,4]; 2) The Cu-O bond lengths[5]; 3) The electronic density of states at the Cu-O planes and chains[6]. In fact, these three properties are intimately connected, producing a consistent picture where oxygen plays a central role in the still unknown mechanism for high-Tc superconductivity.

Different methods, like annealing the samples under Ar flow[7], or using a Zr getter[3], have been tried in order to fix a desired amount of oxygen in the $YBa_2Cu_3O_{7-y}$. In this work we choose to use a method of oxygen extraction under vacuum that we found to be very appropriate.

EXPERIMENTAL PROCEDURE.

Disk-shaped samples of the parent composition $YBa_2Cu_3O_7$ were prepared by mixing Y_2O_3, $BaCO_3$ and CuO powders, which were compacted and calcined under flowing oxygen gas for 30 hr. Following, the disks were grinded, compacted again, annealed under flowing oxygen for 30 hr and cooled slowly (50°C/hr) to room temperature. This step was repeated for three times, producing samples of very pure superconducting phase with transition widths smaller than 1K; x=7.00 was then assumed for these samples. The disks were cut into thin bars (3x2x10 mm^3), which were used for oxygen extraction under a vacuum typically around 10^{-2} torr.

Fig. 1 shows three isothermal curves, that were calculated from the mass loss measured by an analytical balance at appropriate time intervals. It is remarkable that the terminal concentration of oxygen in each case (T=350°C, x=6.94; T=380°C, x=6.78; T=450°C, x= 6.37) is reached asymptotically, as the annealing time increases. This suggests that one should get very homogeneous samples through this method. Other samples not shown in Fig. 1 were: T=400°C, x=6.72; T=430 °C, x=6.57; T=500°C, x=6.30. An important detail in this method is the sample quenching to room temperature, at the end of the annealing, with the sample still inside the furnace under vaccum.

X-ray diffractograms were taken for all samples, and a gradual transformation from the pure orthorhombic phase (x=7.00) to the pure tetragonal phase (x=6.3) was observed. The resistive transition was measured through the four probe method with T_c defined at the mid-point of the transition.

RESULTS AND DISCUSSION

Fig. 1 displays also a curve corresponding to the process of oxygen intercalation for T=450°C, when the sample was submitted to 1 atm of flowing oxygen. This curve is plotted backwards in time only to

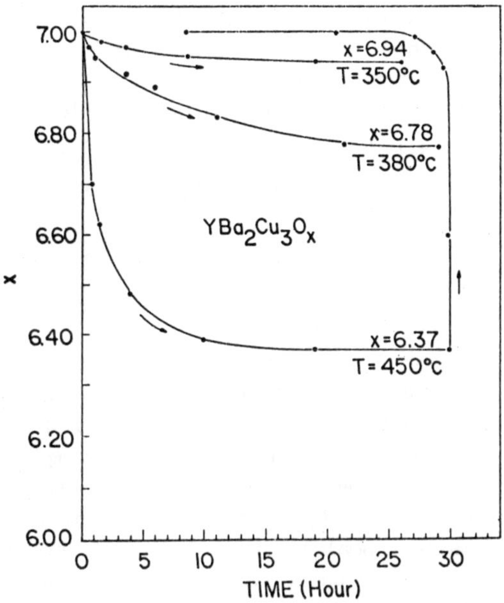

Fig.1 -
Oxygen concentration (x) as a function of time for different isothermal processes. The samples were kept under vacuum, except for the case T=450°C, where the sample was kept under 1 atm of oxygen in the final heat treatment to recovery the original oxygen content.

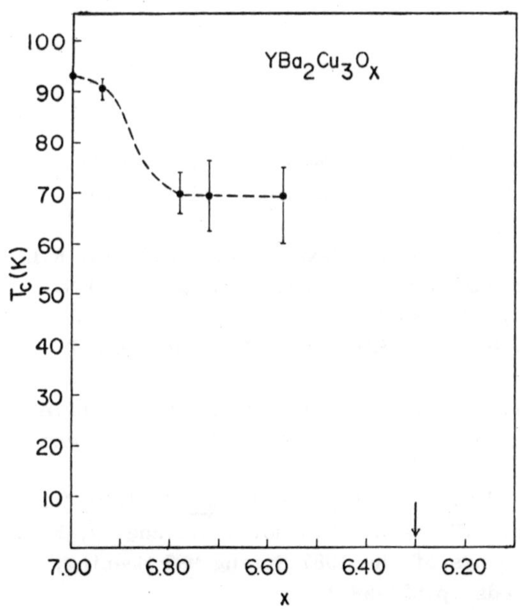

Fig.2 -
Critical temperature (T_c) against the oxygen concentration. The vertical bars represent the transition width, defined between the onset and the zero resistance points. The dashed line is only a guide to the eyes.

emphasize the recovery of the initial oxygen concentration ($x=7.00$). Moreover, we can see clearly that the kinetics for oxygen intercalation is much faster than for desorption.

Fig. 2 shows T_c vs. the oxygen concentration for five samples in the range $6.57<x<7.00$. There is a well defined plateau around $T_c=70K$, for $6.57<x<6.80$, and a very sharp transition ($\Delta T<1K$) at $T_c=93K$ for the sample $x=7.00$. The sample having $x=6.30$ (arrow in Fig. 2) was semiconducting. We have no conlusive explanation for the discrepancy between our plateau at $T_c=70K$, and the widely reported plateau at $T_c=60K$. One possible hypotesis could be that our samples had a different metastable configuration for the oxygen vacancies ordering, due to our different method of oxygen extraction. This would mean that having the same nominal concentration of oxygen for two samples, prepared through different methods, does not imply that the samples are the same. In other words, they could well have a different atomic structure (vacancies ordering), a different microstructure (homogeneity in a larger or smaller scale); therefore, different superconducting properties. More work are under way in order to check the present results and add more data points.

The authors acknowledge the partial support from FAPESP-Brazil, through a grant # 87/3102-2.

REFERENCES

1. Schuller,I.K., D.G.Hinks, M.A. Beno, D.W.Capone, L.C.Soderholm, J. P.Locquet, Y.Bruynseraede, C.U.Segre and K.Zhang, Solid State Commun. 63,385(1987).
2. Jorgensen,J.D., M.A.Beno, D.G.Hinks, L.C.Soderholm, K.J.Volin, R.L. Hitterman, J.D.Grace, I.K.Schuller, C.U.Segre, K.Zhang, and M.S. Keefisch, Phys. Rev. B36(8),3608(1987).
3. Cava,R.J., B. Batlogg, C.H.Chen, E.A.Rietman, S.M.Zahurak and D. Werder, Phys. Rev. B36,5719(1987).
4. Alario-Franco,M.A., C.Chaillont, J.J.Capponi, J.Chenavas (Preprint)
5. Miceli,P.F., J.M.Tarascon, L.H.Greene, P.Barboux, F.J.Rotella and J.D.Jorgensen (Preprint).
6. Hermann,F., R.V.Kasowski, and W.Y.Hsu, Phys. Rev. B36,6904(1987).
7. Tarascon,J.M., W.R.McKinnon, L.H.Greene, G.W.Hull, B.G.Bagley, E.M. Vogel, and Y.LePage, Proceedings of the 1987 Spring MRS Meeting, D.V.Gubser and M.Schluter, eds.,p.65(1987).

STUDY OF THE INFLUENCE OF Fe AND Zn SUBSTITUTING Cu IN $Y_1Ba_2Cu_3O_y$

D. Ríos-Jara, G. González, C. Vázquez and R. Escudero.
Instituto de Investigaciones en Materiales, UNAM.
Apdo. Postal 70-360, 04510 México, D. F.

ABSTRACT

We compare the influences of Fe and Zn substituting Cu in the $Y_1Ba_2Cu_{3-x}(Fe_x$ or $Zn_x)O_y$ compounds. Small structural changes were found in Zn substituted compounds. For Fe substitution, an orthorhombic to tetragonal phase transition was observed at $x \approx 0.07$. Loss of superconductivity was obtained for $x \geq 0.2$ and $x \geq 0.29$ in Zn and Fe compounds respectively. We discuss the effect of both types of impurities on Tc.

EXPERIMENTAL

Samples of both series of compounds were prepared in the form of pellets by grinding, mixing and compressing appropriate ammounts of Y_2O_3, $BaCO_3$ and CuO with ZnO or Fe_2O_3 powders. Calcination at 850°C, regrinding at room temperature and compactation was repeated at least twice. Sintering by solid state reaction was performed at 920°C for at least 15 hours.

Resistance was measured using the four point contact configuration and its dependence with temperature was obtained in a close cycle refrigerator down to 10K. The lattice parameters were determined with a X-ray powder diffractometer, by first using the 006+020 and 200 calibrated reflections and then adjusting those parameters to generate the measured plane spacement of other peaks of the structure with an accurancy better than ± 0.002Å.

STRUCTURAL CHARACTERISTICS

<u>$Y_1Ba_2Cu_{3-x}Zn_x$.</u> The dependence of the lattice parameters with x, for $0<x\leq 0.5$, is shown in Fig. 1. Note the small modifications on the lattice parameters up to the solubility limit of Zn in the compound, which was observed for $x<0.2$. For $x\geq 0.2$ a mixture of the orthorhombic phase and Zn-based precipitates was detected.

<u>$Y_1Ba_2Cu_{3-x}Fe_x$.</u> Fig. 2 shows the dependence of the lattice parameters with x, for $0<x\leq 0.3$. Complete solubility of Fe in the compound was observed in all this range of values of x. As reported [1,2], the transition from the orthorhombic to the tetragonal superconducting phase can be placed at $x \approx 0.07$.

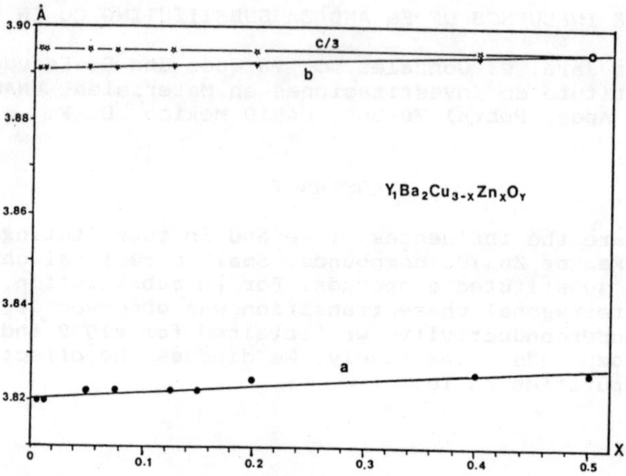

Fig. 1. Dependence of the lattice parameters with x for $Y_1Ba_2Cu_{3-x}Zn_xO_y$ compounds.

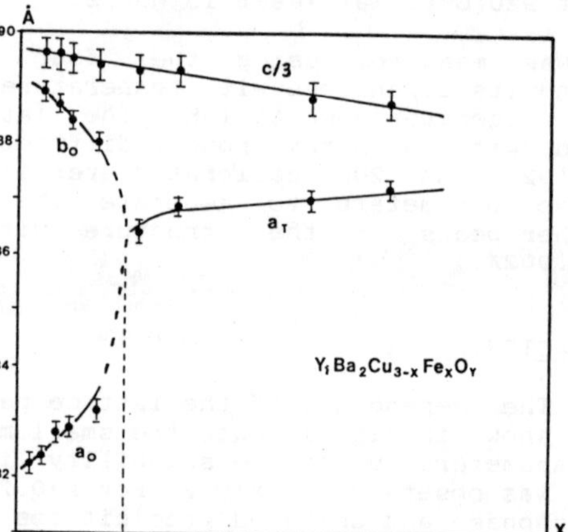

Fig. 2. Dependence of the lattice parameters with x for $Y_1Ba_2Cu_{3-x}Fe_xO_y$ compounds.

Discussion

The dependence of the critical superconducting transition temperature (Tc) with x, for both Zn and Fe compound, is shown in Fig.3. As indicated there, non superconducting transition was found for $x \geq 0.2$ and $x \geq 0.3$ for Zn an Fe compounds respectively. A stronger depletion of Tc is obtained by the addition of Zn, even if the corresponding structural modifications are smaller. On the other hand, the addition of Fe, producing strong structural changes, does not affect Tc to the same extend. Also relevant is the small effect on Tc due to the magnetic impurity.

It has been reported the preference of Fe atoms for occupying Cu(I) sites (i.e. substituting Cu in the Cu-O chains) at low Fe concentrations [2-4]. In contrast, Zn atoms are assumed to occupy indistinctly Cu(I) or Cu(II) sites. On the other hand, it has become more and more apparent that superconductivity occurs mainly in Cu(II) planes, as a two dimensional transport phenomenon in these ceramic materials [5]. Then, the rapid drop in Tc observed in Fig. 5, for Fe concentration above $x \approx 0.26$, could be associated to the begining of the occupancy of Fe atom in the Cu(II) sites. In the case of Zn atoms, occupying Cu(II) (and Cu(I)) sites for lower concentrations, a faster drop in Tc with x is observed, even if small structural modifications are introduced. It is worthwhile noting here, that these small structural modifications are consistent with the idea of a preferential substitution of Zn in Cu(II) planes only, since in these planes no additional oxygen atoms can be accommodated.

The magnetic character of Fe atoms, at concentrations $x < 0.26$, does not affect Tc strongly, given their distance from Cu(II) planes. In contrast, the drop in Tc becomes very fast as soon as Fe starts to occupy Cu(II) sites, as could be expected for a magnetic impurity in a "normal BCS" superconductor; however, additional analysis is needed to probe that this is the case. If one could force Fe to occupy Cu(II) sites at lower concentrations, then one could expect an earlier drastic drop in Tc. The same conclusions could be applied to Co impurities as reported by Maeno et al [2]. It is important to note that only minor changes in the electronic structure and valence state were found for Zn substitution in Cu(I) and Cu(II) sites by Richert et al [6]. For Fe and Co substitutions, strong effects on the electronic density of states and valence were found by these authors. Thus, a more important deplection in Tc is expected for Fe and Co than for Zn. These conclusions are only in good agreement with our data for $x \geq 0.26$, when Fe (and Co) go into Cu(II) sites. On the other hand, the mechanism for the deplection of Tc associated to Zn impurities in Cu(II) sites seems not to be related to the electronic density of states of Zn or to structural modifications, but probably to differences in electronic configuration. More work is in progress in order to elucidate this issue.

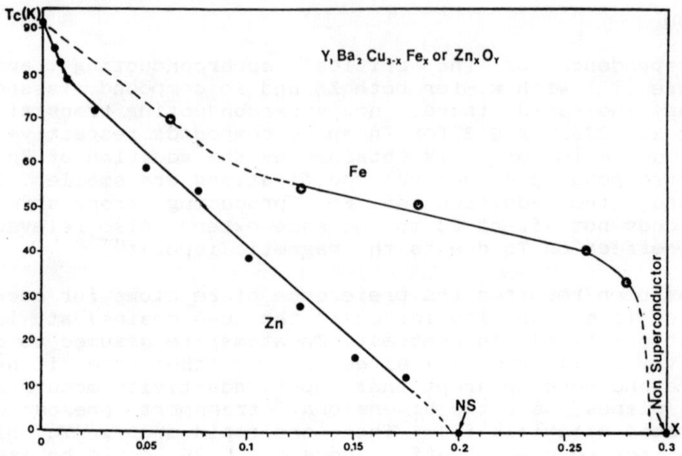

Fig. 3. Dependence of Tc with x, for Zn and Fe compounds.

REFERENCES

[1] G. Gonzalez, D. Ríos-Jara, T. Akachi, R. A. Barrio, L. Baños and R. Escudero; Proc. of the 1987-MRS Fall Meeting, Boston, Mass., pp. 899-902 (1988).

[2] Y. Maeno and T. Fujita; Proc. of the Intern. Conf. on High Temperature Superconductors and Materials and Mechanisms of Superconductivity, Interlaken, Switzerland, march 1988.

[3] R. Gomez, S. Aburto, M.L. Marquina, M. Jimenez, V. Marquina, C. Quintanar, T. Akachi, R. Escudero, R. A. Barrio and D. Rios-Jara, Phys. Rev. B, 36, No.13, 7226-7229 (1987).

[4] Y. K. Tao, J. S. Swinnea, A. Manthiram, J. S. Kim, J.B. Goodenough and H. Steinfink, J. Materials Research, 3, No.2, 248-256 (1988).

[5] Y. Tokura, J. B. Torrance, T. C. Huang and A. I. Nazzal, submmited to Physical Review Letters in april 1988.

[6] B. A. Richert and R. E. Allen, to be published in Proc. of the 34th Nat. Sym. Am. Vacuum Soc., Anaheim, C.A. (1987).

ACKNOWLEDGMENTS

Financial support from the Programa Universitario de Superconductores de Alta Temperatura de Transición of the UNAM and CONACYT are gratefully acknowledged.

INFLUENCE OF THE USE OF DIFFERENT BARIUM PRECURSORS IN THE PREPARATION OF Y-BA-CU SUPERCONDUCTORS.

C.A.D'Ovidio, D.A.Esparza, M.T.Malachevsky and E.Oliber.
Centro Atómico Bariloche and Instituto Balseiro.
8400 - S.C.de Bariloche. Rio Negro. Argentina.

We studied the influence of different barium precursors on the resistivity, density and the presence of secondary phases in the ceramic superconductor $Y Ba_2Cu_3O_{7-x}$.
The samples were made by mixing in stoichiometric proportions Y_2O_3, CuO and one of the following barium compounds: CO_3Ba, BaO_2, BaO, BaC_2O_4 or Ba $(OH)_2.8 H_2O$, and then they were processed simultaneously in a standard way.
We discuss the results obtained from resistivity and density measurements, and x-ray diffractometry.

1- INTRODUCTION:

The conventional methods of preparation of ceramic superconductor samples include several groundings and heatings, either in flowing oxygen or in air with an oxygenization treatment after sintering. Thinking in industrial applications, it would be interesting to simplify the process of preparation so as to diminish times and costs. For the making of the compound $Y Ba_2Cu_3O_{7-x}$, some authors use CO_3Ba as barium precursor (1) , other BaO_2 and BaO (2) and others Ba $(OH)_2$ (3), but up to now we haven't found publications about a systematic comparison of the use of different barium precursors. In the present work we analize comparatively the behaviour of the $Y Ba_2Cu_3O_{7-x}$ compound when made from different barium precursors.

2- PREPARATION:

Samples were made following a standard process: Thoroughly mixing stoichiometric cuantities of CuO, Y_2O_3 and one of the barium precursors (CO_3Ba, BaO_2, BaO, Ba $(OH)_2$ or BaC_2O_4) in an agate mortar with isopropylic alcohol, dried in air and press into pellets at 1000 bars. Then they were fired at 970 C (well bellow the peritectic temperature) for 16 hours in air and slowly cooled. To ensure equal conditions all samples were heated simultaneously. The pellets were then grounded in an agate mortar, pressed again and sintered in the same conditions as above. This process was repeated one more time. Samples were cut from the pellets with a diamond saw, without lubricant to avoid contamination, for resistivity measurements.

3- EXPERIMENTAL TECHNIQUES AND RESULTS:

Bulk densities of the pellets were measured by mercury immersion and were compared with the theoretical density of the stoichiometric compound (table 1).
The presintered samples and the sintered OBa (barium oxide sample) were studied by x-ray powder diffractometry with Cu $K\alpha$ radiation. We identified the same orthorhombic superconducting phase for all samples, either

alone or with secondary phases (table 2). The lattice parameters of the 1-2-3 phase were adjusted by a regresion method, giving the following values: a= 3.829 Å, b= 3.892 Å and c= 11.717 Å.
Cualitative experiments on magnetic levitation were made to the samples cooling then in liquid nitrogen and putting on top of them a small Nd-Fe-B magnet. In table 1 the samples that show this effect were labelled with an L (for levitates). In case they don't levitate they were submitted to a repulsion test by mounting the sample on a pendulum and if they were repelled by the magnet they were labelled with an R (for repels).
The four probe method was used for the resistivity measurements, that are shown in figures 1, 2 and 3. The cross section of the samples was 1 x 4 mm^2 aproximately and the current was 100 A. Temperature was measured with a platinum thermometer and a carbon one under 15 K. The sample OxaBa (barium oxalate sample) couldn't be measured because the electrical contacts failed.

4- DISCUSSION:

It is evident that the type of barium precursor used strongly affects the properties of the final ceramic. However all samples present an orthorhombic 1-2-3 phase in different proportions. The single phased ones were prepared with CO_3Ba and BaO_2 presintered or with BaO sintered. These show a transition temperature sensibly lower than a sample processed in oxygen in the standard way, though they both have the same lattice parameters. This indicates an oxygen deficiency that can also be observed in the slopes of the transition curves. Noteworthy the samples levitate in the presence of a magnetic field in spite of their depressed transition temperature, showing that their current densities are sufficiently high to produce levitation. Meanwhile, samples with "good transitions" not always show this effect. This could be due to a lack of oxygenization of the material without low current density interfases compared with grains with optimun composition that are poorly connected among them.
In the R vs T curves it can be observed that the slopes in the normal state are generally of metallic character, except the sintered PerBa (barium peroxide sample) that presents a resistivity increase when the temperature diminishes. The most metallic slope is for sintered OBa (barium oxide sample). The higher transition temperatures were get for the materials sintered OBa and presintered PerBa, that superimposes within experimental error. The worst result was observed for the sintered HiBa (barium hydroxide sample) which doesn't percolate down to 4.8 K. In this case the presence of abundant green phase ($Y_2Ba\ Cu\ O_5$) would explain this fact. The samples made with barium peroxide show a metallic behaviour for the presintered specimen and a higher transition temperature, than the sintered one that is due to an oxygen loss during the second treatment in air.
The samples densities increase with succesive sinterings but this don't necessarily improve the superconducting properties as in the case of PerBa. The best presintered densities were obtained for the samples PerBa and HiBa while the samples OBa and OxaBa cracked in this stage, possibly due to a reaction among the barium precursors and the isopropylic alco-

hol that fixed organic material that decomposes with temperature destroying the sample.

5- CONCLUSIONS:

It is possible to obtain high temperature superconductors with interesting properties using different barium precursors and with simpler processes than conventionally established.
In one case this properties has been observed even in the presintered state while further treatments degrade them.
Then materials can be made following this simple single step method using air as the firing atmosphere with useful properties that can find applications.

6- REFERENCES:

(1) M.K.Wu, J.R.Ashburn, C.J.Tong, P.H.Hor, R.L.Meng, L.Gao, Z.J.Huang Y.Q.Wang and C.W.Chu. Phys.Rev.Lett. 58 (1987) 908.
(2) K.Noto, H.Morita, K.Watanabe, T.Murakami, Y.Koyanagi, I.Yoshi, I. Sato, H.Sugawara, N.Kobayashi, H.Fujimori and Y.Muto. Physica 148B (1987) 239-242.
(3) Microstructure and properties of ceramic $Ba_2Y\ Cu_3O_7$. G.J.Fisanick et al. AT & T Bell Lab. (preprint).

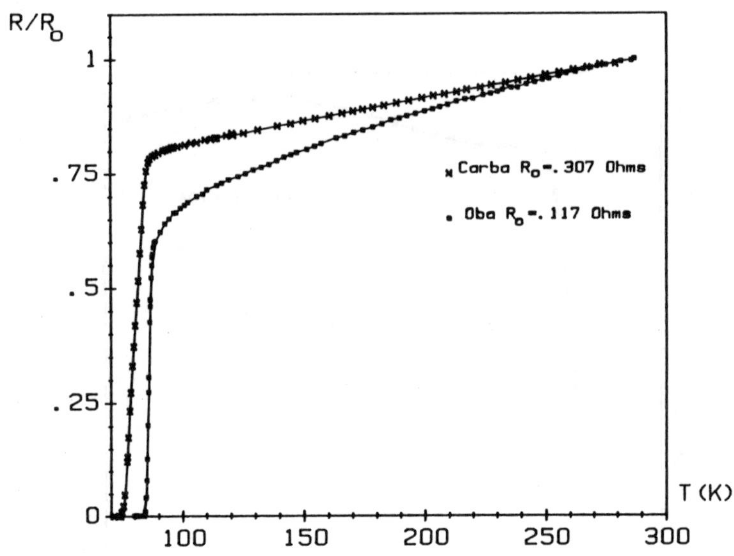

Fig.1: Superconducting transition of the CarBa (barium carbonate precursor) and OBa (barium oxide precursor) samples in the sintered state.

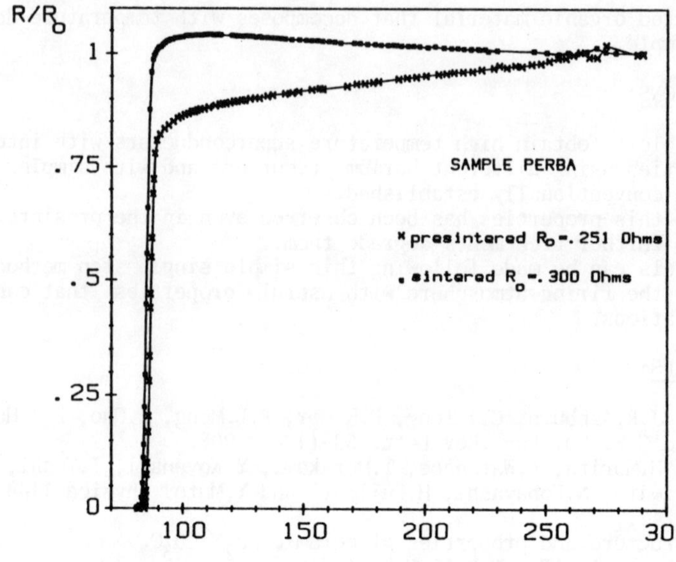

Fig.2: Superconducting transition of the PerBa (barium peroxide precursor) sample in the presintered and sintered state.

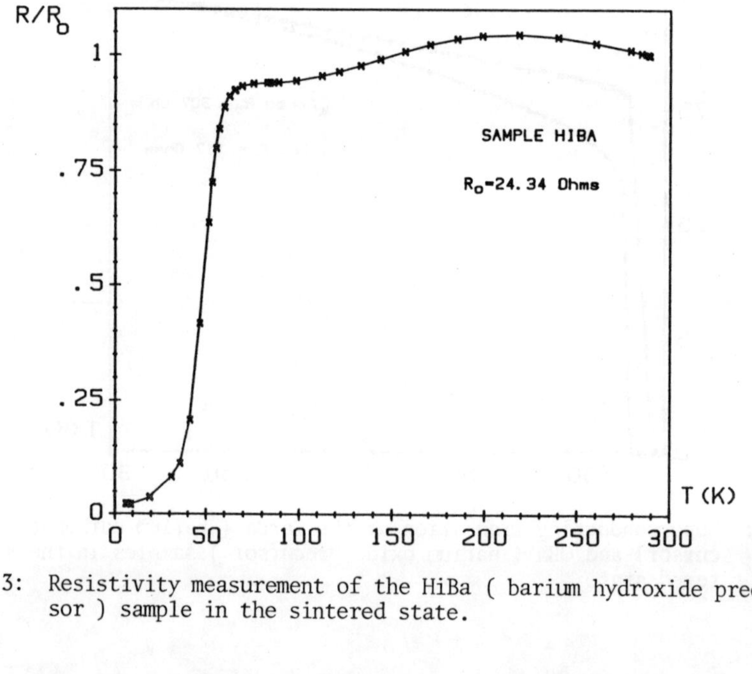

Fig.3: Resistivity measurement of the HiBa (barium hydroxide precursor) sample in the sintered state.

Table 1 - Bulk densities in gr/cm^3, comparison with the theoretical density (ρ_t= 6.6 gr/cm^3) and levitation tests (LT).

SAMPLES	PRESINTERED			SINTERED 1			SINTERED 2		
	LT	ρ	ρ/ρ_t %	LT	ρ	ρ/ρ_t %	LT	ρ	ρ/ρ_t %
CarBa	L	3.99	60.5	R	4.5	68.2	R	4.82	73
PerBa	L	4.54	68.8	L	4.88	73.9	L	5.02	76
HiBa	R	4.49	68	-	5.45	82.6	-	5.48	83
OBa	L	-	-	L	5.39	81.66		-	-
OxaBa	-	-	-	-	5.08	77		-	-

Table 2 - X-ray diffractometry results. 1-2-3 accounts for the orthorhombic superconducting phase and green phase for the $Y_2Ba\,Cu\,O_5$ compound.

SAMPLES	IDENTIFIED PHASES
CarBa	1-2-3
PerBa	1-2-3
HiBa	1-2-3 + green phase
OBa	1-2-3 + $BaCuO_2$ + non-identified peaks (presintered state) 1-2-3 (sintered state)
OxaBa	1-2-3 + green phase + CuO

THERMOGRAVIMETRIC EVIDENCE FOR A FIRST-ORDER PHASE TRANSITION IN THE ONE-ATMOSPHERE ISOBAR OF $YBa_2Cu_3O_{6+x}$

P. Satti, D. Pasquevich, A. Caneiro and J. Abriata, Metallurgy Division, and C. Fainstein, Magnetic Resonances Division; Centro Atómico Bariloche, Comisión Nacional de Energía Atómica, Bariloche, Argentina

Introduction. It has been well documented that the oxygen content decisively affects the properties of the high temperature superconductors $YBa_2Cu_3O_{6+x}$ [1,2]. Therefore, it is important to know the oxygen chemical potential for this ceramic material, μ_{O_2} , as a function of temperature, T, and oxygen content, x. Indeed, several papers have been published on the partial pressure of oxygen $Po_2(T,x)$ for $YBa_2Cu_3O_{6+x}$ [1,3-7]. Although there are discrepancies among the results reported, most of them clearly indicate that $(\partial x/\partial T)_{Po_2}$ tends to zero for x tending to 1. This can be interpreted as due to the oxygen saturation of the Cu-O chains which exist in the basal plane of the orthorhombic structure of $YBa_2Cu_3O_{6+x}$ [1]. In turn, this suggests that in the pseudo-binary system $YBa_2Cu_3O_6$-O, the oxygen rich boundary of the orthorhombic superconducting phase is situated around the x=1 composition, a conclusion which has also been arrived at in experiments performed under oxygen pressures higher than one atmosphere [8]. Therefore, for x>1 there should exist a two-phase region consisting of the superconducting orthorhombic phase in equilibrium either with the same orthorhombic structure (but with added oxygen content higher than x=1) or with another phase with a different structure. These considerations led us to make a detailed thermogravimetric study of the Po_2=1 atm isobar for $YBa_2Cu_3O_{6+x}$ paying more attention than other works to obtain the final stable equilibrium, in an attempt to detect possible structures in this isobar which might reveal the existence of underlying two-phase regions. The preliminary results of this study are described below; further work is in progress.

Experimental. The samples were prepared by conventional ceramic methods. Samples of different densities (4 to 5 g/cm^3) were produced. Powder X-ray diffraction indicated the presence of the superconducting orthorhombic structure only. The samples showed their superconducting transition to be at 89 K with a width of 1 K. The thermogravimetric measurements were performed using the equipment described elsewhere [9]. Flowing atmospheres were pure oxygen and mixtures of oxygen and air.

Results and discussion. In order to improve the kinetics leading to equilibrium, our experiments were mostly performed on a low density sample (4 g/cm^3).

The Po_2=1 atm isobar was measured under pure oxygen (isobar #1). The sample was previously cycled in temperature in order to relax unstable microstructures. The data points obtained are plotted in Fig. 1. These results are in excellent agreement with those of Jorgensen et al. [1], both being measured relative to the value of x at 20 C. For

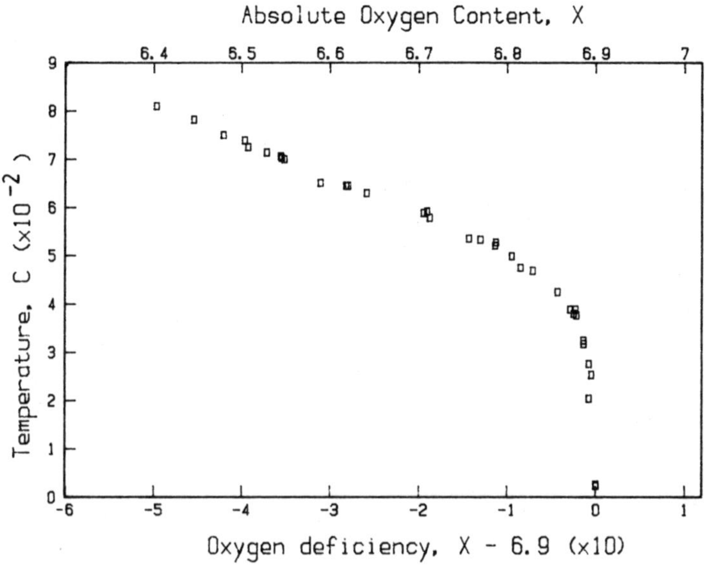

Figure 1. Isobar #1 for $YBa_2Cu_3O_x$. Po_2=1 atm.

this reason, we have adopted as reference in this work the absolute values of x given by these authors [1]. It must be pointed out that the fact that no first order transitions were observed in isobar #1 does not necessarily means that they do not exist in the stable equilibrium phase diagram (SEPD) of $YBa_2Cu_3O_6$-O. This is so because for $0<x<1$, the oxygen atoms fill the vacant sites in the basal plane of the orthorhombic structure [1] which otherwise keeps esentially rigid; thus, no doubt, severe difficulties can be anticipated for the observation of the SEPD due to the possible existence of metastable equilibriums associated with the coherent, or semicoherent, coexistence of the different phases into which eventually the wide field of nonstoichiometry ranging from x=0 to x=1 would decompose for the stable equilibrium.

Additional measurements were done using mixtures of oxygen and air. We found that, systematically, at around 520 C, and after an induction period of around 4-5 hours, the sample started to absorb important quantities of oxygen up to values of x=1.25 for annealing times of 120 hours. According to our experiments, longer annealing times would allow the incorporation of larger quantities of oxygen in $YBa_2Cu_3O_{6+x}$. It must be noticed that similar oxygen absortions have been reported in the literature [5,6,10,11]. X-ray analysis revealed the permanence of the orthorhombic structure, even in the most oxidized state. We found no indication of corrosion phenomena related to the presence of air as those described in the literature [12,13]. In fact, we think that some of the air components, in particular moisture, catalyzed the

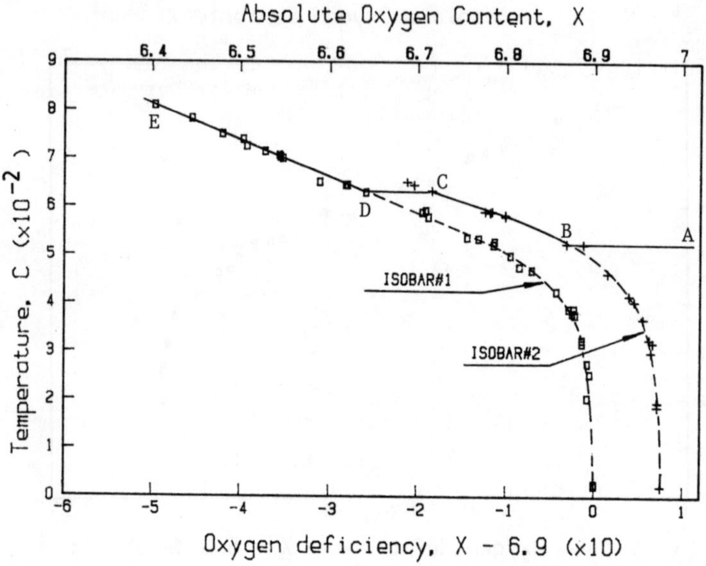

Figure 2. Isobars #1 and #2 for $YBa_2Cu_3O_x$. Po_2=1 atm. Line ABCDE corresponds to the stable equilibrium states. Data points appearing on the dashed curves correspond to metastable equilibrium states.

observed oxidation. The excess oxygen absorbed by the sample was abruptly released (2-3 minutes) when the sample was heated up to 800 C where at, under pure oxygen, the value of x which corresponds to stable equilibrium was readily recovered. This cycle of oxidation and reduction was repeated at will. We think that what was happening at 520 C was the precipitation of a variant of the orthorhombic structure with an oxygen content much higher that x=1. This would indicate the formation of a peroxide or a superoxide. Under pure oxygen, without the catalyzing action of moisture, this precipitation should also occur but, probably, rather high oxygen pressures would be necessary and a marked hysteresis would be observed.

An isobar was also measured under pure oxygen at 1 atm after oxidizing the sample to near x=1.0 (isobar #2). The results obtained are represented in Fig. 2. Isobar #2 transformed upon reduction into isobar #1 at around 620 C. This is a first order phase transition and took about 70 hours for completion. Cycling of the sample from isobar #1 to isobar #2 and back was reproducible in our experiments. For a fixed temperature, either isobar #1 or isobar #2 can correspond to the stable equilibrium. Consistent with our observations, we conclude that the isobar corresponding to stable equilibrium at Po_2=1 atm is the one represented by the line ABCDE in Fig. 2. Since X-rays always revealed the existence of esentially the same orthorhombic structure only, isobars #1 and #2 must correspond to variants of this structure depending on its oxygen content. This two variants show up separated from each other by the two phase region designated by CD in Fig. 2. The oxygen values defining this two-phase region are $x \cong 0.64$ and $x \cong 0.71$ at 620 C. Thus, we propose that the orthorhombic structure appears in the SEPD of $YBa_2Cu_3O_6$-O in the form of at least three variants, each of them stable in different composition ranges. Approximately, these ranges are as follows: $0.50 \leq x \leq 0.64$, $0.71 \leq x \leq 0.89$, and $1.25 < x < ?$. The value 0.50 corresponds to the orthorhombic-tetragonal transformation [1].

An important consequence of our results is that too long annealing times of $YBa_2Cu_3O_{6+x}$ at temperatures of about 500 C or lower, under pure oxygen or air, may result in the deterioration of the superconducting properties. This would be due to the incipient precipitation of the orthorhombic variant with oxygen content $x > 1.25$. Results reported by Kanno et al. [5] seem to support this conclusion.

References. [1] J.D. Jorgensen, M.A. Beno, D.G. Hinks, L. Soderholm, K.J. Volin, R.L. Hitterman, J.D. Grace, I.K. Schuller, C.U. Segre, K. Zhang and M.S. Kleefisch, Phys. Rev., 36B (7) 3608 (1987). [2] J.D. Jorgensen, B.W. Veal, W.K. Kwok, G.W. Crabtree, A. Umezawa, L.J. Nowicki and A.P. Paulikas, Phys. Rev., 36B (10) 5731 (1987). [3] P.K. Gallagher, Adv. Ceram. Mat., 2 (3B) 632 (1987). [4] H. Oesterreicher and M. Smith, Mat. Res. Bull., 22 1709 (1987). [5] R. Kanno, Y. Takeda, M. Hasegawa, O. Yamamoto, M. Takano, Y. Ikeda and Y. Bando, Mat. Res. Bull., 22 1525 (1987). [6] J. Hauck, K. Bickmann and F. Zucht, Z. Phys. B. Cond. Matt., 67 299 (1987). [7] K. Kishio, J. Shimoyama, T. Hasegawa, K. Kitazawa and K. Fueki, Japanese J. of Appl. Phys., 26 (7) L1228 (1987). [8] P.K. Gallagher, H.M. O'Bryan, S.A. Sunshine and D.W. Murphy, Mat. Res. Bull., 22 995 (1987). [9] A. Caneiro, P. Bavdaz, J. Fouletier and J.P. Abriata, Rev. Sci. Instrum., 53 1072 (1982). [10] Q.W. Yan, P.L. Zhang, L. Jin, Z.G. Shen, J.K. Zhao, Y. Ren, Y.N. Wei, T.D. Mao, C.X. Liu, T.S. Ning, K. Sun and Q.S. Yang, Phys. Rev., 36B (10) 5599 (1987). [11] J.P. Burger, L.K. Lesueur, M. Nicolas, J.N. Daou, L. Dmoulin and P. Vajda, J. Physique, 48 1419 (1987). [12] J.G. Thompson, B.G. Hyde, R.L. Withers, J.S. Anderson, J.D. Fitz Gerald, J. Bitmead, M.S. Paterson and A.M. Stewart, Mat. Res. Bull., 22 1715 (1987). [13] S. Davison, K. Smith, R. Kershaw, K. Dwight and A. Wold, Mat. Res. Bull., 22 1659 (1987).

EFFECTS OF HEAT TREATMENT ON THE SUPERCONDUCTING Bi-Ca-Sr-Cu-O COMPOUND

S. Moehlecke and M.C. Andrade

Instituto de Fisica, UNICAMP, Campinas 13081, SP., Brazil

ABSTRACT

Sintered samples with nominal compositions $Bi_2Ca_{0.4}Sr_2Cu_2O_x$, $Bi_2Ca_1Sr_2Cu_2O_x$ and $Bi_2Ca_1Sr_2Cu_3O_x$ were studied as a function of heat treatment. These samples are multi-phase with two superconducting onsets, one small transition around 115 - 125K and a large one at 83 - 98K. Isothermal (850°C) anneals show an improve of almost 15K in the low T_c and the appearance or disappearance with annealing time of the high T_c, depending on sample composition.

1. INTRODUCTION

Recently a high temperature superconductor without rare earths was found in the system Bi-Ca-Sr-Cu-O[1)]. Generally, the samples are multi-phase with a partial transition at 110K, a large foot ~80K and a low temperature transition near 20K[2)]. Until now, no definite agreement has been reached on the composition and structure of these phases, and also the high temperature phase could not be obtained in a pure state due to the unknow sample preparation conditions and phase diagram.

In order to get a better understanding of these problems we initiate a study of the heat treatment effects on this system, and we report here our results of isothermal anneals on a set of samples with different compositions.

2. EXPERIMENT

The samples were prepared from the starting powders Bi_2O_3, $SrCO_3$, $CaCO_3$ and CuO, mixed in the desired ratios, calcined at 800°C for 20hr, reground, pelletized and sintered in air. With this procedure three samples with nominal composition $Bi_2Ca_{0.4}Sr_2Cu_2O_x$, $Bi_2Ca_1Sr_2Cu_2O_x$ and $Bi_2Ca_1Sr_2Cu_3O_x$ were sintered for 65 hr/822°C, 4 hr/840°C and 1 hr/850°C respectively. After this, the samples were submitted to an isothermal anneal at 850°C for up to 96 hours. The resistance-temperature dependences of these samples were measured by the usual four probe DC method with silver paste electrodes or RF sputtered Pd contacts. The data were taked between room temperature and liquid nitrogen in cooling down and warming up to insure thermal equilibrium.

3. RESULTS AND DISCUSSIONS

The as-sintered samples show different resistance behaviours. The sample with nominal composition $Bi_2Ca_{0.4}Sr_2Cu_2O_x$ (2, 0.4, 2,2) shows a semiconductor behaviour (see inset A-I of Fig.1) while the sample $Bi_2Ca_1Sr_2Cu_2O_x$ (2, 1, 2, 2) shows a metallic character, with no superconducting transition until 77K. On the other hand, the as-sintered sample (1hr/850°C) with nominal composition $Bi_2Ca_1Sr_2Cu_3O_x$ (2, 1, 2, 3) is metallic and shows two transitions, one small with an onset (Tc_{on}) at 115K and another around 92K. As we do not have a full transition (R=0) until 77K in most of the samples we will compare them with their onsets.

The isothermal (850°C) anneal of sample (2, 1, 2, 3) did not show significant changes in the low Tc_{on} (92K) after 38hr or 84hr of treatment, but the higher transition could not be more observed after 38 hours. This sample (2, 1, 2, 3) melts at 870 ± 5°C.

For the sample (2, 1, 2, 2) annealed 46hr, only one Tc_{on} appears at 83K and this value improve to only 86K after 70hr/850°C.

The as-sintered sample (2, 0.4, 2, 2) annealed only 4hr at 835°C changes his behaviour from semiconductor to metallic and presents a superconducting transition with Tc_{on} ~83K (see inset A-II of Fig.1). Another piece of this same sample was also annealed for 4 hours but at 842°C and shows a Tc_{on} at 88K. It was observed that the sample with this composition melts at 858 ± 3°C.

The isothermal annealing effect on this sample is shown in Fig.1. There is an increase of Tc_{on} from 83K to 98K after 96hr, and with 74hr of anneal and after, another Tc_{on} around 125K appears (see inset B of Fig.1). These two Tc_{on}'s are really superconducting transitions as we could observe by the resistance rise when a constant magnetic field was applied to the sample at temperatures bellow the Tc_{on}'s. The occurence of superconductivity in all these samples with different compositions and the fact that the melting temperature varies with composition seems to indicate that in this Bi-system there is a homogeneity range for at least the 80K superconducting phase.

Some of these results are rather peculiar, specially the rise of almost 15K of the low Tc_{on} in sample (2, 0.4, 2, 2) or the appearance (2, 0.4, 2, 2) or disappearance (2, 1, 2, 3) of the high Tc_{on} with annealing time. Certainly structure analysis of these samples and further heat treatments, which are currently being done, will help in understand these results.

The partial transition at high temperature (Tc_{on} ~125K) observed in some of these samples, that have the same value found in the thallium system[3], may indicate that this bismuth system could also have an identical full transition of only the high Tc phase as Tℓ with an appropriate composition and heat treatment.

ACKNOWLEDGMENT

This work was partially supported by FAP-UNICAMP grant 453/87, and one of us (M.C.A.) would like to acknowledge FAPESP for the fellowship.

REFERENCES

1. Maeda H., Tanaka Y., Fukutomi M. and Asano T., Jpn. J. Appl. Phys. 27, 1209 (1988).
2. Michel C., Hervieu M., Borel M.M., Grandin A., Deslandes F., Provost J. and Raveau B., Z Phys. B - Condensed Matter. 68, 412 (1987).
3. Sheng Z.Z. and Hermann A.M., Nature 332, 55 (1988).

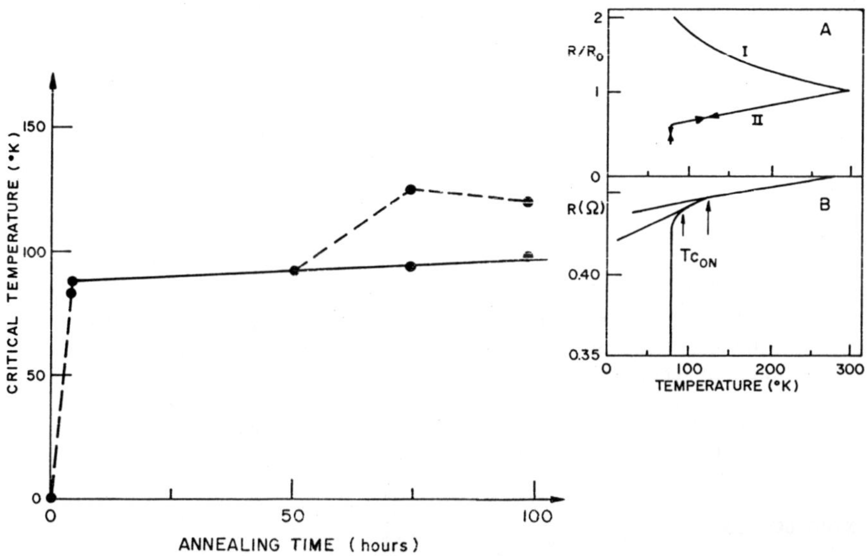

Fig. 1

Transition temperature (onset) as a function of annealing time (850°C) for sample (2, 0.4, 2, 2). Dashed line above 50 hours anneal represents the highest Tc_{on}. For the insets see text.

INDIUM SUBSTITUTION IN $YBa_2Cu_3O_7$

S. G. Fries

Instituto de Física, Universidade Federal do Rio Grande do Sul

E. R. Fraga

Instituto de Química, Universidade Federal do Rio Grande do Sul

90049 Porto Alegre, RS, Brasil

ABSTRACT

We study the $YBa_2(Cu_{1-x}In_x)_3O_7$ compositions. x=0.01, 0.03 and 0.05 by means of tne time differencial perturbed angular technique (TDPAC). All the samples are still superconductors and the electric field gradient measurements showed two sites for the In atoms suggesting that they occupy the two Cu sites of the superconductor ceramic.

INTRODUCTION

The incorporation of In on the $YBa_2Cu_3O_7$ structure allows the investigation by means of the time-differencial perturbed angular correlation (TDPAC) technique using ^{111}In atoms as a probe. This technique presents a high localized sensitivity to extra nuclear fields given information about several properties of the crystal at the probe site.

We start studying the superconductivity properties in inactive samples and after a sample characterization we evaporated radioactive ^{111}In atoms to the ceramic and perform the TDPAC measurements.

EXPERIMENTAL

Sample Preparation

The samples of $YBa_2Cu_{1-x}In_x)_3O_7$, with x = 0.01, 0.03 and 0.05, were synthesized from stoichiometric mixtures of CuO, $BaCO_3$, Y_2O_3 and In_2O_3 pressed into pellets. The pellets were sintered in air for 17 hs at 900 °C and then slowly cooled.

For TDPAC samples, minute amounts of radioactive ^{111}In were mixed with inactive compound. The radioactive In atoms incorporation was difficult and only a small quantity was taken by the sample.

Experimental Details

The EFG was measured by the TDPAC technique applied to the 173-247 keV gamma-gamma cascade of ^{111}Cd at the indium sites. The gamma-gamma coincidences were recorded at 90° and 180° angles, and from these the perturbation factor $G_{22}(t)$ was computed.

The data were fitted with the $A_{22}G_{22}(t)$ theoretical function describing polycrystalline[1)] samples of nuclei with a spin -5/2 intermediate level, subjected to an EFG, V_{zz}, with asymmetry parameter $\eta = \frac{V_{xx} - V_{yy}}{V_{zz}}$ for $0 <= \eta <= 1$ and a Lorentzian frequency distribution of width δ. The quadrupole frequency was related to V_{zz} by $\nu_Q = \frac{eQV_{zz}}{h}$ where Q is the nuclear quadrupole moment of the intermediate level.

RESULTS AND DISCUSSION

The samples were analysed by X-ray diffractometry and Figure 1 shows the diffraction pattern for the X = 0.01 sample which can be identified as being the YBCO characteristic pattern[2)]. All the samples present superconductor behavior.

Figure2 shows an experimental $A_{22}G_{22}(t)$ curve for the x = 0.01 sample. Table 1 lists the data got by a least-square fitting of the theoretical function to the experimental $A_{22}G_{22}(t)$ curve which reveal the presence of two hyperfine interactions, with amplitude ratio f1:f2 of about 2:3.

The value of the frequencies and the ratio between them found for $YBa_2(Cu_{1-x}In_x)_3O_7$, x = 0.01 are close to that got by TDPAC measurement of In_2O_3 compound[3,4)], suggesting the presence of In_2O_3 precipitates. Otherwise the high η (0.7) value associated with the lower frequency of this compound is not compatible with $YBa_2(Cu_{1-x}In_x)_3$ experimental results. Besides, the X-ray measurements did not show additional In_2O_3 on peaks even for the x = 0.05 sample.

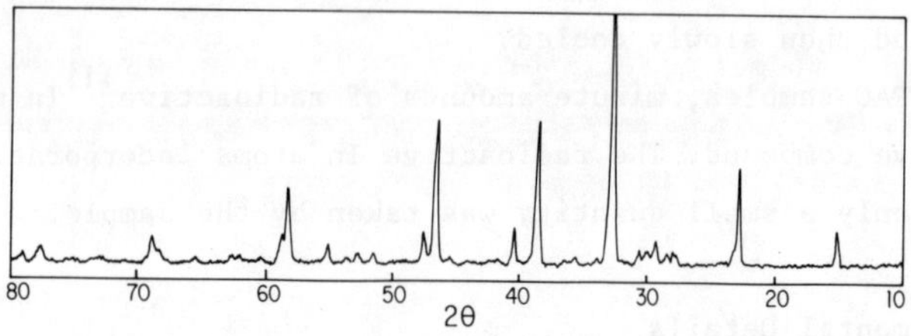

Figure 1. X-ray diffraction pattern of $YBa_2(Cu_{1-x}In_x)_3O_7$ (x = 0.01).

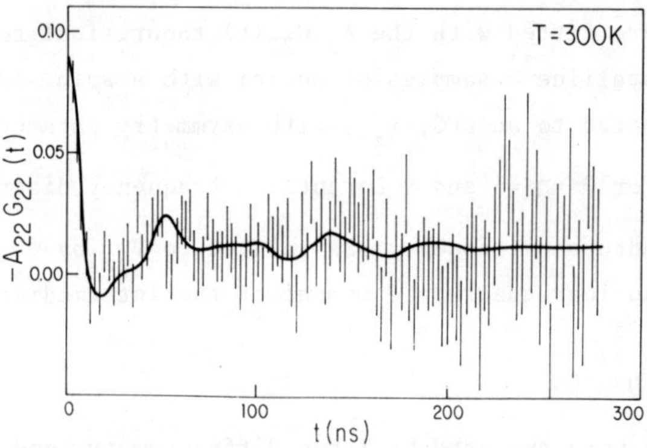

Figure 2. A room temperature TDPAC spectrum of ^{111}Cd in $YBa_2(Cu_{1-x}In_x)O_7$. The solid line shows the curve fitted to the data (x = 0.01).

	$\gamma_Q(MH_z)$	η	δ
SITE 1	107.3(7)	0.07(1)	0.15(1)
SITE 2	150.0(6)	0.29(2)	0.11(1)

Table 1. Hyperfine interaction parameters of ^{111}In in $YBa_2(Cu_{1-x}In_x)_3O_7$, x = 0.01.

In order to get more information about the asymmetry of the EFG for the $YBa_2(Cu_{1-x}In_x)_3$, we calculated the ionic contribution to V_{zz} and the η values with a Coulombian point charge model. Using the lattice parameters and atomic positions reported by Capponi et al.[5] and z=+3, +2, +2, -1.87 for Y, Ba, Cu, O atoms respectively, we found a value of 1.8 for the ratio $V_{zz}(Cu^1)/V_{zz}(Cu^2)$, $\eta(Cu^1) = 0.18$ and $\eta(Cu^2) = 0.14$. The ratio f1:f2 for the Cu sites are 1:2. Small changes in the O charges do not alter significantly the η values.

So, if we suppose that the In probes occupy Cu sites in the YBCO (because of the small η values) we must conclude that there is a preferential site for the In probe.

As shown by the data, the In atoms positions are not determined up to now. In order to elucidate this question we are performing measurements with samples containing higher In concentration. To investigate the possibility that In atoms occupy Y sites, we are preparing $Y_{1-x}In_xBa_2Cu_3O_7$ compositions.

REFERENCES

1. Steffen R.M. and Frauenfelder H., Perturbed Angular Correlations, ed. E. Karlsson et al (Amsterdam: North-Holland) 1964.
2. Cava R.J., Batlogg B., Van Dover R.B., Murphy D.W., Sun Shine S., Siegrist T., Remeika J.P., Rietman E.A., Zahurak S. and Espinosa G. P., Phys. Rev. Lett. 58, 16 (1987).
3. Bibiloni A.G., Massolo C.P., Desimoni J., Mendoza-Zélis A.L., Sanchez F.H., Pasquevich A.F., Damonte L. and López-Garcia A.R., Phys. Rev. B 32, 4 (1985).
4. Uhrmacher M. and Bolse W., Hyperfine Int. 15/16, 445 (1983).
5. Capponi J.J., Chaillout C., Hewat A.W., Lejay P., Marezio M., Guyen N., Raveau B., Soubeyroux J.L., Tholence J.L. and Tournier, Europhys. Lett. 3, 1301 (1987).

SUPERCONDUCTING FRACTION IN THE $YBa_2(Cu_{1-x}Mn_x)_3O_7$ COMPOUND

R.F.JARDIM, S.GAMA, O.F.DE LIMA AND I.TORRIANI
Instituto de Física "Gleb Wataghin", UNICAMP,
CEP 13081, C.P. 6165 - Campinas - S.P. - Brazil
And
A.PADUAN FILHO, S.ZACARELLI AND C.C.BECERRA
Instituto de Física - USP - São Paulo - Brazil

1. ABSTRACT

The superconducting transition temperatures of the $YBa_2(Cu_{1-x}Mn_x)_3O_7$, with x = 0.0, 0.025, 0.05, 0.075, 0.10, 0.15 and 0.20, had been explored by A.C. Magnetic susceptibility measurements. All samples had been caracterized by metallographic and X-ray analysis. Superconducting volume fraction of the samples measured at 10 and 30K decreases linearly with the Mn content. Metallographic analysis revealed that the amount of additional phases and porosity of the samples increase with the Mn content.

2. INTRODUCTION

In a previous paper/1/ we have described results on the influence of the substitution of Mn for Cu in the properties of the high-T_c superconductor $YBa_2Cu_3O_7$. X-ray diffraction analysis revealed the predominance of the orthorhombic structure described by Cava et al /2/ for all concentrations studied. This was accompanied by changes in the relative intensities of the diffraction peaks and the appearance of extra peaks corresponding to phases other than the $YBa_2Cu_3O_7$.

These facts motivated us to study in more detail those extra phases using other techniques, as magnetic susceptibility in the superconducting state and metallographic analysis.

3. EXPERIMENTAL METHODS

The preparation of the samples was described in ref.1.

Magnetic susceptibility was measured in an A.C. inductance bridge operating at 155 Hz; the modulation field was about 3 Oe. The temperature was measured using a calibrated carbon resistor and the critical temperature was defined at the onset of the diamagnetic transition.

The metallographic preparation of the samples followed this sequence: griding in SiC papers up to 600 grit, polishing with diamond paste of 15, 7, 3 and 1 μm sizes. The samples were analyzed by optical microscopy as polished, without chemical etching.

4. RESULTS AND DISCUSSION

Fig. 1 shows the magnetic susceptibility for all samples. It is observed that the transition broadens and there is a small decrease of the onset temperature as a function of the Mn concentration. Another important feature is that the diamagnetic signal decreases with the Mn content. To show this clearly we have plotted the relative superconducting fraction versus the Mn concentration at 10 and 30 K in Fig.2. We have defined the parent sample, $YBa_2Cu_3O_7$, as a standard 100% value of the superconducting fraction. The relative superconducting fraction decreases linearly with the Mn concentration in a rate of ∿ 4.3%/at% Mn and it would show zero fraction for a Mn concentration around 24at%Mn.

Metallographic analysis gives direct support for the results shown in Figs. 1 and 2.

Fig. 3 shows a typical microstructure of the as polished samples with x = 0.00 and 0.20. It is observed even for our $YBa_2Cu_3O_7$ sample the presence of small amounts of intragranular additional phases. This observation

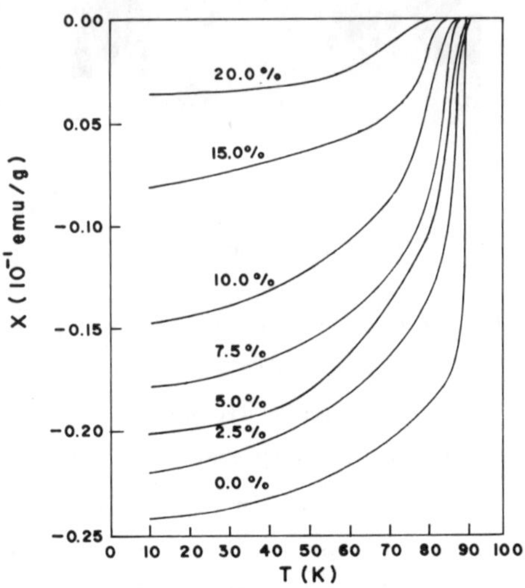

Figure 1 – Magnetic susceptibility for all samples studied.

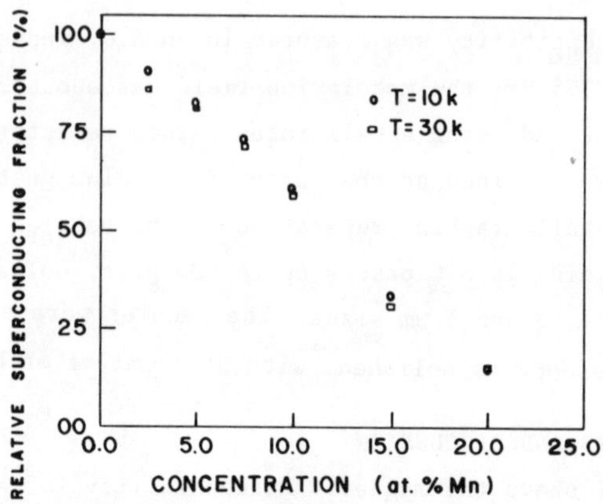

Figure 2 - Relative superconducting fraction versus Mn concentration at 10 and 30 K

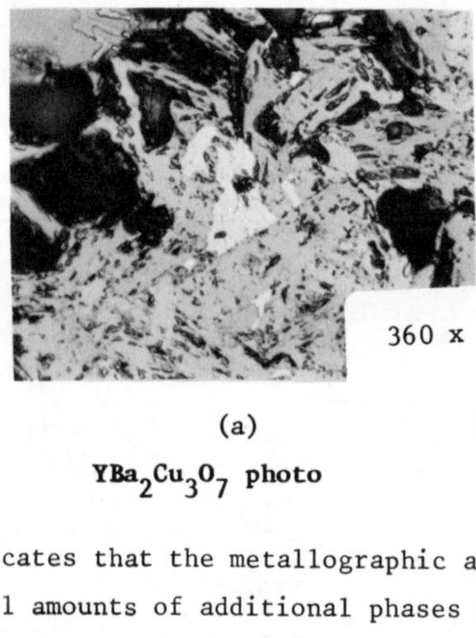

(a) **Figure 3** (b)

$YBa_2Cu_3O_7$ photo **x = 0.20 photo**

indicates that the metallographic analysis is more sensible to detect small amounts of additional phases than X-ray diffraction, since our data for this sample did not reveal any peak other than the ones belonging to the orthorhombic structure of $YBa_2Cu_3O_7$.

The pictures in Fig. 3 also show that the sample with x = 0.20 presents a higher level of porosity as compared with the "123" compound. This fact probably is indicative that the introduction of the Mn in the system Y-Ba-Cu-O increases the optimum sintering temperature. Recent experiments, to be published elsewhere /3/, qualitatively confirm this idea. A possible solubility limit of the Mn in the Y-Ba-Cu-O system is discussed by Silva et al /4/ in this conference.

In summary we show that the relative superconducting fraction decreases linearly with the Mn concentration in the samples with the form $YBa_2(Cu_{1-x}Mn_x)_3O_7$. Another interesting observation is that the porosity increases with the Mn content indicating that the introduction of the Mn in the system Y-Ba-Cu-O increases the optimum sintering temperature.

One of the authors (R.F.J.) acknowledges support from Pirelli, and a partial support from the FAPESP is acknowledged.

5. References

/1/ Jardim, R.F., Gama, S., de Lima, O.F. and Torriani, I., (this conference).

/2/ Cava, R.J., Batlogg, B., Van Dover, R.V., Murphy, D.W., Sunshine, S., Siegrist, T. Remeika, J.P., Rietman, E.A., Zahurak, S.M. and Espinosa, G.P., Phys. Rev. Lett. 58, 1676 (1987).

/3/ Jardim, R.F., Gama, S. and de Lima, O.F. (unpublished).

/4/ Silva, E.C., Vargas, H., Bento, A.C., Jardim, R.F., Gama, S., Pinheiro, E.A. and Galembeck, F. (this conference).

SINGLE CRYSTAL GROWTH OF La/Sr/Cu/O OXIDES BY THE FLUX GROWTH METHOD

E.Manghi, G.Polla, P.K.de Perazzo and G.Nievas

Departamento de Física, Comisión Nacional de Energía Atómica, Av. del Libertador 8250, 1429 Buenos Aires, Argentina.

La/Sr/Cu/O oxide single crystals were grown using a CuO flux method.
The starting material was a $La_{2-x}Sr_xCuO_4$ / CuO mixture at 20/80 mol % ratio.
Three essays were realized varying the starting nominal Sr concentration (x_n = 0.2, 0.4 and 0.6).
The influence of x_n, in nucleation density, chemical composition, morphology and size of crystals, were analyzed.
X-Ray diffraction, optical microscopy, SEM, and EDAX were used for the characterization of single crystal samples.

The interest to obtain the high T_c superconductor metal oxides as single crystal form has been emphasized by several authors in order to clarify the origin of layered superconductivity depicted by these compounds. Growth of single crystals in the La/Sr/Cu/O system has been succesfully accomplished in different laboratories[1][2][3][4].

To our knowledge, only few reports mention the actual Sr content x attained in crystal samples. Due to segregation effects during growth, x appear to be lower than the starting nominal composition x_n.

The aim of this work was to explore the possibility of controlling the final Sr content x in the crystals, for x_n running from 0.2 to 0.6.

This range of starting nominal compositions was chosen, hoping to obtain single crystals with similar superconducting behaviour to the one observed in sintered policrystalline specimens.

Experimental

Single crystals were grown by the flux method starting from $La_{2-x}Sr_xCuO_4$ powder samples of three different nominal compositions: x_n = 0.2, 0.4 and 0.6.

The following conditions were the same for all the runs:

- OCu was used as flux for crystallization with a flux-to-solute ratio of 4 : 1.
- Crucible : Platinum - Atmosphere : air
- Growth Temperature : 1180°C (2 hs.)
  Cooling rate: ~10°C/h from 1180 to 1100°C and then furnace cooled to room temperature.

The samples were characterized by Optical Microscopy (OM), Scanning Electron Microscopy (SEM), X-ray diffraction and Electron Dispersive Analysis of X-rays (EDAX).

GROWTH FEATURES (OM and SEM characterization)

From optical and SEM inspection we inferred that the nucleation takes place on the surface melt and then the crystal grow layer by layer, with a 2D mechanism.

When the melt was solidified, well developed square or rectangular plates were observed on surface, in the case x_n = 0.2. The flat exposed faces change to black colour under different light incidence. These crystals were identified by X-Ray diffraction and EDAX measurements as the tetragonal $La_{2-x}Sr_xCu\ O_4$ compound (Table I).

For x_n = 0.4 and 0.6 runs, the linear crystal size improves (L_{max}~4 mm) but the samples are thinner. No regular limits were observed for these series because the samples are partially covered by the solidified flux melt, masking the actual crystal contours (Fig.1 a,b,c).

The crystals being very thin and brittle, can be cracked during growth due to thermal stresses, specially in the last cooling process. Most of the cracks produced in the first growth stage (Fig.2a) were healed by the succesive spreading growth layers. Incomplete overgrowth of these layers was sometimes also observed (Fig.2b,c).

As no suitable preferential solvent for the flux was found for this system, the crystals could be extracted only by mechanical means. This was a major difficulty specially for isolating thin crystals (x_n= 0.4 or 0.6). Not only the crystals easily break but they also were not totally free of flux as seen in Fig.3a; it was difficult to isolate even those crystals emerging from the surface melt (Fig.3b). In our case, no perfect isolated crystal with appropriate dimensions was obtained for x_n= 0.4 or 0.6 precluding then, electrical or magnetic measurements on individual specimens of these series.

Segregation effects take place during growth for runs with larger Sr concentration x_n; we have observed in x_n = 0.4 and 0.6 series, needles or plate-like crystals (underlying the crystals nucleated on surface) which depicted a different morphology.

Besides, a few local policrystalline aggregates were observed on surface only for the x_n= 0.6 run (Fig.4c).

COMPOSITION ANALYSIS (X-Ray diffraction and EDAX)

Table I shows the results from EDAX (semiquantitative analysis, % in weight, neglecting oxygen) and X-Ray diffraction (a, c, c/a lattice parameters, precession diagrams). In order to facilitate the analysis, the data corresponding to others La/Sr/Cu/O compounds were also included.

- Crystals with different morphologies, as seen in Figs. 4a,b, have different compositions. The EDAX results were confirmed by the lattice parameters measurements: two different La/Sr/Cu/O compounds were obtained as single crystals under our experimental conditions. One of them, nucleating first on surface melt, was identified as tetragonal $La_{2-x}Sr_xCuO_4$ compound. The other not identified yet, appears to be related to the cubic perovskite cell in this system.
- In the case of (001) $La_{2-x}Sr_xCuO_4$ crystals it was possible to control in some extent, their actual x, Sr content, varying the starting nominal composition in the range x_n= 0.2 to 0.6 (for example, a larger x was observed by EDAX for the x_n= 0.6 run).

This fact was also confirmed by the corresponding increasing values of the c/a ratio, in agreement with other reported experiences (in the $La_{2-x}Sr_xCuO_4$ compound the Sr content, as well as T_c, are reflected in the lattice constants c/a ratio) [5][6].

Then, in accordance with our c/a ratio a superconducting behaviour can be expected in our samples with the following x and T_c values:

Nominal composition	x_n :	0.2 ; 0.4 ; 0.6
Expected actual Sr content	x :	0.075; 0.09; ~0.2
Expected Transition Temperature	T_c :	20 K ; 26 K ; 36 K

In the case of the x_n= 0.2 run, a T_c= 20 K was measured and confirmed[7], so, we believe that the two other T_c values could be also confirmed.

In the case of the second phase both measured composition (EDAX) and cell parameters do not match with those of any of the reported stable compounds of the La/Sr/Cu/O system [8][9][10] (See Table I). Then, the existance of a new phase can not be discarded a priori. Further research in this field is in progress.

We can conclude that for the La/Sr/Cu/O system, increasing nominal Sr content, under our experimental conditions, displaces the equilibrium reaction towards the nucleation of multiple phases: single crystals of $La_{2-x}Sr_xCuO_4$, crystals of a second phase and policrystalline aggregates of a third phase with higher content of Cu and Sr.

We are greatful to Dr. R. Baggio for useful discussions.

Table I

x_n	Observed phases	% in weight (measured) La	Sr	Cu	Measured Lattice parameters (Å) a	b	c	c/a
0.2	crystals on melt surface	75.8	4.80	19.4	3.802		13.154	3.460
0.4		75.6	4.40	20	3.774		13.185	3.494
0.6		72.1	6.50	21.4	3.764		13.207	3.509
0.4-0.6	crystals of the second phase	52	11	37	10.852		3.834	
0.6	local Polycristalline aggregates	4	11	85				
		% in weight (calculated)						
0.1	$La_{1.9}Sr_{0.1}CuO_4$	78.5	2.6	18.9				
0.2	$La_{1.8}Sr_{0.2}CuO_4$	75.5	5.3	19.2				
	$La_5SrCu_6O_{15}$	59.7	7.53	32.77	15.37	15.43	3.86	(11)
	$La_2SrCu_2O_6$	56.4	17.8	25.8	3.86		19.89	(12)

References

1. Tetsuo Inoue et al., Japan, J.Appl.Phys. 26 N.5 (1987) L732.
2. H.Hasegawa et al., Japan, J.Appl.Phys. 26 N.5 (1987) L673.
3. Y.Hidaka et al., Japan, J.Appl.Phys. 4 (1987) L377.
4. Y.Hidaka et al., submitted to J.Cryst.Growth.
5. Wang H.H. et al., Inorg.Chem. 26 (1987) 1190.
6. J.M.Tarascon et al., preprint.
7. G.Polla, E.Manghi, P.K.de Perazzo, L.Civale, H.Safar, 72 Reunión Asociación Física Argentina, Bariloche, Sept. 1987.
8. J.B.Torrance et al., Phys.Rev.Lett. 60 N.6 (1988) 542.
9. Er-Rakho-L., C.Michel, B.Raveau, submitted to Solid State Comm.
10. J.Hahn, T.O.Mason et al., submitted to CHEMTRONICS.
11. Y.Tokura, J.B.Torrance, J.Am.Chem.Soc. (1987) 7555.
12. N. Nguyen et al.,Mat. Res. Bull., 15 N°7 (1980) 981.

Figure Captions

Fig.1 As grown crystals still in the platinum crucible

a) $x_n = 0.2$ linear crystal size : $0.1 < L < 1$ mm

b) $x_n = 0.4$ linear crystal size : $0.5 < L < 4$ mm

c) $x_n = 0.6$ linear crystal size : $0.5 < L < 4$ mm

Note For the last two series the maximum crystal size is under estimated because the samples are partially covered by the flux melt.

Fig.2 $x_n = 0.2$

a) First growth stage: thin crystal showing cracks induced by thermal stresses.

b) As grown (001) crystal imbedded in the solidified flux melt, showing an elevated terrace corresponding to the first growth stage (initial nucleous).

c) Incomplete overgrowth of succesive growth layers.

Fig.3 a) Melt crystal interface of a mechanically isolated crystal. Conchoidal fracture is also observed in this sample.

b) Crystals emerging from the solidified melt.

Fig.4 a) (001) $La_{2-x}Sr_xCuO_4$ crystal on surface and the underlying flux melt including some needles or plate-like crystals of the second phase, appearing in x_n= 0.4 and 0.6 series.

b) Reverse of a (001) $La_{2-x}Sr_xCuO$ crystal showing the adjacent plate-like crystals of the second phase.

c) Cu and Sr rich local surface aggregates, adjacent to a (001) crystal seen only for the x_n= 0.6

Fig. 1

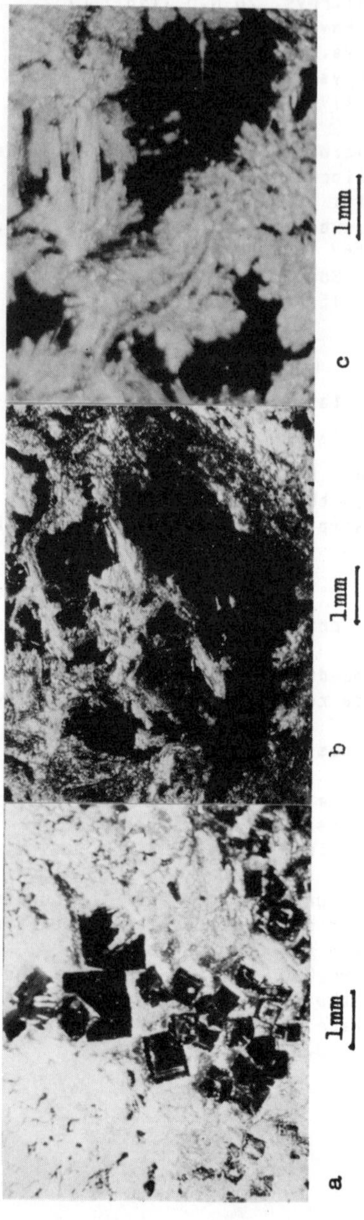

Fig. 2

Fig. 3

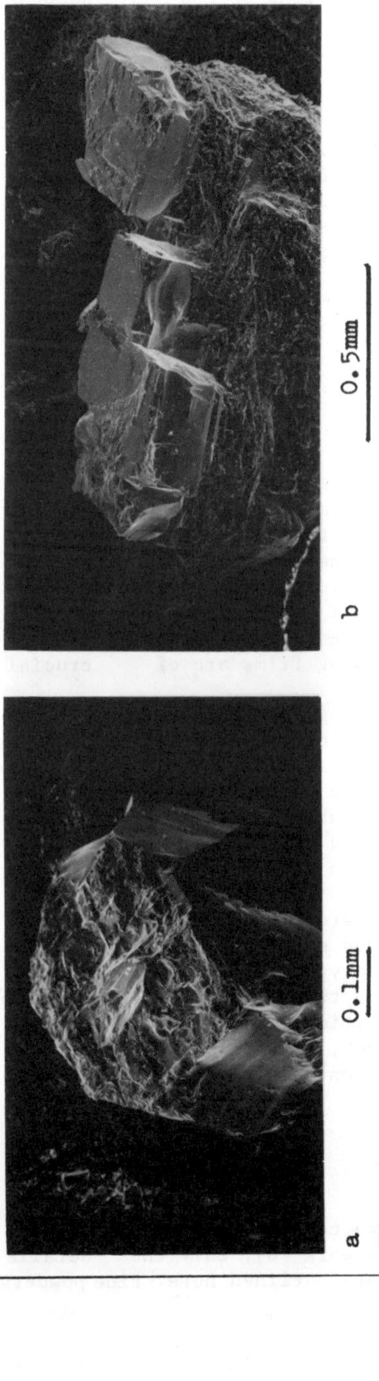

Fig. 4

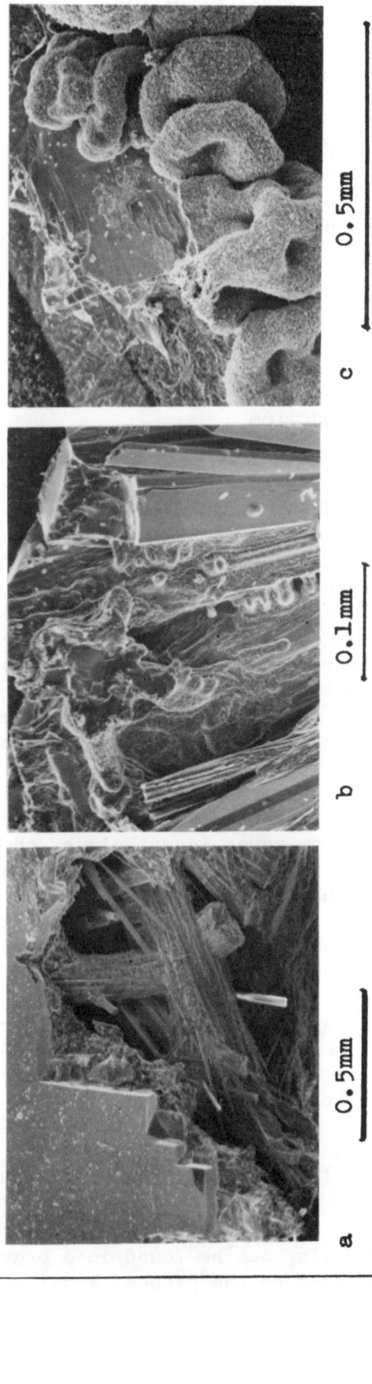

ANALYSIS OF THE COMPOSITION OF SURFACE LAYERS IN HIGH Tc SUPERCONDUCTING MATERIALS BY THE COMBINATION OF RBS AND NUCLEAR RESONANCE REACTION.

F.L. Freire Jr., C.V. Barros Leite, B.K. Patnaik and G.B. Baptista, Departamento de Física, Pontifícia Universidade Católica do Rio de Janeiro, 22453, Rio de Janeiro, R.J., Brasil.
D. Naugle, R.Q. Pandey and W. Kirq, Physics Department-Texas A&M University - College Station - Texas U.S.A.

INTRODUCTION

The recent discoveries of new high Tc superconducting compounds based on the La-Ba-Cu-O system [1] and the subsequent studies of related materials including the Ba-Y-Cu-O system with critical temperature around 90 K [2], have brought unprecedented interest, not only due to the scientific questions raised, but also on the potencial for technological applications of these materials. From both points of view the preparation and characterization of thin superconducting films are of crucial importance [3,4].

The Rutherford Backscattering Spectrometry (RBS) has been extensively used for the analysis of the composition of these films [5]. However, because of the low sensitivity of this technique to oxygen, the stoichiometric composition of the oxide is obtained indirectly via the contribution of the oxygen atoms to stopping power. Nevertheless, the method can be combined with other techniques sensitive to oxygen so as to determine the oxide stoichiometry unambiguously.

A direct method of oxygen profiling is by the $^{16}O(\alpha,\alpha)$ nuclear resonance [6] that can be used simultaneously with RBS. Besides being more sensitive to the oxygen content in the samples, it is a much finer probe as far as depth profiling is concerned. In this work, samples of nominal composition $YBa_2Cu_3O_\delta$ were analyzed by the $^{16}O(\alpha,\alpha)$ resonance at 3.045 MeV, coupled with the RBS, so as to obtain the oxygen atomic fraction as a function of depth near the surface region.

EXPERIMENTAL PROCEDURE

The experimental procedure concerning $^{16}O(\alpha,\alpha)$ resonant profiling can be found in a previous work [6]. Only the details relevant to the present experiment are outlined here. Fine powders of $YBa_2Cu_3O_\delta$ were pressed into the form of a pellet and sintered in oxygen atmosphere. The samples with an onset of Tc=95 K were irradiated with alpha particles in the energy range of 3.04 to 3.12 MeV at intervals of 2.6 keV. The targets were placed at normal incidence and the scattered particles were detected at 165°. A sample of silicon with a homogeneous dioxide surface layer of 3000 A° thickness was also irradiated and analyzed under the same conditions.

A number of incident particles were determined using a chopper, coated with a gold film which intercepts the beam 6% of the total time. The particles scattered by the chopper were detected by a second surface barrier detector placed at 160° with respect to the incident beam. The system was calibrated so as to establish the proportionality between the number of particles incident on the target to those scattered from the chopper.

RESULTS AND DISCUSSION

A typical energy spectrum of alpha particles scattered from the Y-Ba-Cu-O sample at incident energy greater than 3.045 MeV is shown in fig. 1. From this spectrum we obtain the yields corresponding to scattering from the elements Ba, Y and Cu at the surface. Their positions are indicated by arrows in the figure. These yields are directly proportional to the number of atoms of the element in the compound. The stoichiometric ratios obtained from these were found to be Y:Ba:Cu=1:1.8:3.0.

The pronounced peak observed in the spectrum of fig. 1 corresponds to the alpha particles resonantly scattered by ^{16}O nuclei. For a given incident energy, the total area under the peak is composed of the yields from the resonant and non-resonant elastic scattering. The yield from the latter is practically constant in the small range of variation of the incident energy. For energies of incidence less than the resonant energy E_r, only non-resonant elastic scattering occurs. The resonant yield, at higher energies of incidence, is obtained by subtracting the non-resoant contribution from the total area under the resonance peak. This yield corresponds to the oxygen content at a specific depth x, which is a function of the incident energy, over a layer of thickness that is determined by the width of the resonance. The resonant yield A_i at the incident energy E_i is given by [6]:

$$A_i = N_i \, n_o \, (x_i) \, \sigma \, (E_r) \, \Omega \, \frac{\Gamma}{2} \pi / n_{comp}(x_i) \, \varepsilon_{comp}(E_r) \qquad (1)$$

where N_i is the number of incident particles, $n_o(x_i)$ and $n_{comp}(x_i)$ are respectively the average atomic densities of oxygen and the compound at the depth x_i, Ω is the solid angle of the detector, Γ the width of the resonance and $\sigma(E_r)$ is the resonant scattering cross section. The stopping cross section of the compound at the resonant energy, $\varepsilon_{comp}(E_r)$, can be expressed in terms of those

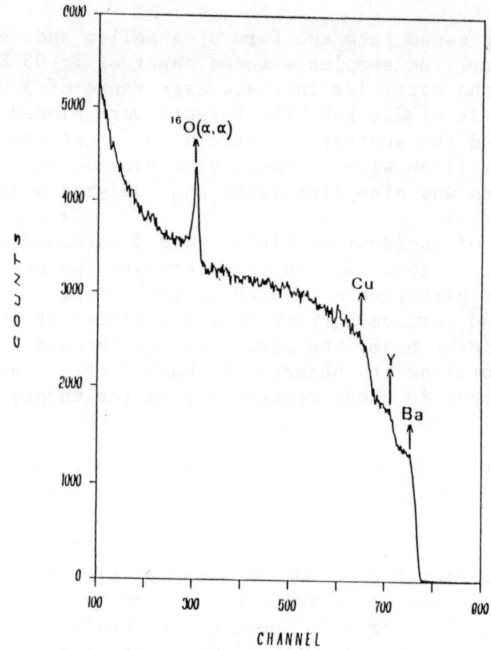

Figure 1 - Energy spectrum of alpha particles with 3.077Mev scattered at 165° from a Y-Ba-Cu-O superconducting sample. The surface position of the elements are shown by arrows.

of its constituents, by Bragg's rule:

$$\varepsilon(Y_1Ba_pCu_qO_\delta) = \varepsilon(Y) + p\ \varepsilon(Ba) + q\ \varepsilon(Cu) + \delta\ \varepsilon(O) \tag{2}$$

The ratio of the atomic densities $n_o(x_i)$ and $n_{comp}(x_i)$ in the equation (1) is the number of oxygen atoms, $\delta(x_i)$, in the compound at the depth x_i. Using the well known stoichiometry of the silicon oxide sample as standard, we obtain for the oxygen atomic abundance of the superconducting sample $Y_1Ba_pCu_qO_\delta$:

$$\delta(x_i) = R\,\frac{\varepsilon(Y) + p\ \varepsilon(Ba) + q\ \varepsilon(Cu)}{\varepsilon(Si)/2 + (1-R)\ \varepsilon(O)} \tag{3}$$

were R is the ratio of the normalized resonant yields (A_i/N_i) in the superconductor sample at the depth x_i to that of the SiO_2 sample of uniform composition. The elemental stopping powers ε are all evaluated at the resonant energy E_r using Ziegler's tabulation [7]. The stoichiometric ratios, p and q, of Y:Ba:Cu

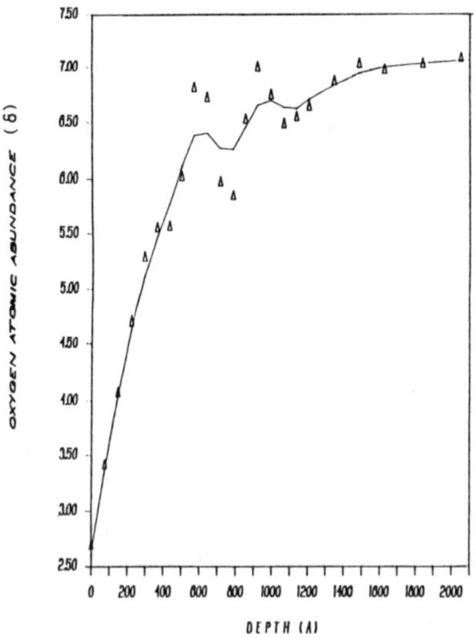

Figure 2 - Oxygen atomic abundance in the thin surface layer of the superconducting sample as a function of depth. The full line is a smoothed curve through the experimental points.

are obtained from the RBS data, viz.- the yields per channel corresponding to scattering from the respective elements at the surface.

The results obtained for the atomic abundance of oxygen as a function of depth in the superconductor sample are shown in fig. 2. The depth scale was obtained assuming a density of 6.68 g/cm^3 for the compound. The number of oxygen atoms per unit of the compound increases gradually from the surface to greater depths, attaining an approximately constant value of 7.1 for depths of the order of 1000 A°. Typical errors in the oxygen concentration are about 7%. These results are comparable to those obtained using other surface analytical techniques such as XPS or AES.

In summary, the analytical technique based on the nuclear resonant scattering $^{16}O(\alpha, \alpha)$ combined with RBS seems to be an appropriate tool to characterize superconducting thin surface layers. This is due to the high sensitivity for oxygen and the good depth resolution (100 A°) obtainable using the given resonance as a probe.

REFERENCES

[1] J.G. Bednorz and K.A. Muller, Z. Phys. B 64, 189 (1986)

[2] M.K. Wu, J.R. Ashbourn, C.J. Torng, P.H. Hor, R.L. Meng, L. Gao, Z.J. Huang, Y.Q. Wang and C.W. Chu, Phys. Rev. Lett. 58, 908 (1987)

[3] P. Chaudhari, R.H. Koch, R.B. Laibowitz, T.R. Mc Guire and R.J. Giambiano, Phys. Rev. Lett. 58, 2684 (1987)

[4] R.H. Koch, C.P. Umbach, G.J. Clark, P. Chaudhari and R.B. Laibowitz, Appl. Phys. Lett. 51, 200 (1987)

[5] See for example, M.Hong, S.H. Liou, J. Kwo and B.A. Davidson, Appl. Phys. Lett. 51, 694 (1987)

[6] B.K. Patnaik, C.V. Barros Leite, G.B. Baptista, E.A. Schweikert, D.L. Cocke, L. Quiñones and N. Magnussen, Nucl. Instr. Meth B (to be published)

[7] J.F. Ziegler, Helium Stopping Powers and Ranges in all Elemental Matter (Pergamon Press, 1977, New York).

AUTHOR INDEX